Short-Term Load Forecasting 2019

Short-Term Load Forecasting 2019

Editors

Antonio Gabaldón
María Carmen Ruiz-Abellón
Luis Alfredo Fernández-Jiménez

MDPI • Basel • Beijing • Wuhan • Barcelona • Belgrade • Manchester • Tokyo • Cluj • Tianjin

Editors
Antonio Gabaldón
Universidad Politecnica de Cartagena
Spain

María Carmen Ruiz-Abellón
Universidad Politecnica de Cartagena
Spain

Luis Alfredo Fernández-Jiménez
Universidad de La Rioja
Spain

Editorial Office
MDPI
St. Alban-Anlage 66
4052 Basel, Switzerland

This is a reprint of articles from the Special Issue published online in the open access journal *Energies* (ISSN 1996-1073) (available at: https://www.mdpi.com/journal/energies/special_issues/ STLF2019).

For citation purposes, cite each article independently as indicated on the article page online and as indicated below:

LastName, A.A.; LastName, B.B.; LastName, C.C. Article Title. *Journal Name* **Year**, *Article Number*, Page Range.

ISBN 978-3-03943-442-8 (Hbk)
ISBN 978-3-03943-443-5 (PDF)

Contents

About the Editors

Antonio Gabaldón Professor), Industrial Engineer, received his M.Sc. (1988) and Ph.D. (1991) from the Universidad Politécnica de Valencia (Spain). In 1989, he visited the Université de Montreal with a predoctoral grant. He has been a professor both at the University of Murcia (1990–1999) and the Spanish Air Force Academy of San Javier (1993–2008). Since 2003, he has been Professor in the Electrical Engineering Department at the Politécnica de Cartagena (UPCT). His lines of research are focused on the analysis of electrical distribution systems, electrical markets, demand response, energy efficiency, railway traction, and non-invasive monitoring techniques. He has participated as a researcher in several R&D projects funded by public calls, both national and international (European Commission, NATO Grants). He has been Main Investigator (IP) in four R&D projects funded by the Spanish Government (2007–2019), projects focused on demand response and energy efficiency. He is also a member of the research network REDYD-2050 (main topic: distributed energy resources in the energy horizon 2050). Moreover, he has participated in R&D contracts of special relevance with companies and utilities, serving as responsible researcher in some of them. Some of the results of his research activity are published in 30 research articles in international scientific journals, 20 of them in well-known indexed journals (IEEE, IET, Elsevier, Springer, MDPI, or Compel). These works have been referenced more than 880 times in 750 scientific articles (source: Scopus 2020). He was coordinator of doctorate programs with a "quality label" (Spanish Government). He is also a reviewer of Spanish National Research Agencies and scientific journals listed in the JCR-ISI. He has also held academic positions as Dean of the ETSII of Cartagena (1999–2008) and Vice-Rector for Academic Affairs and Doctorate (2008–2010).

María Carmen Ruiz-Abellón (Associate Professor) received her Ph.D. and M.Sc. degrees in Mathematics from the University of Murcia (Spain) in 2002 and 1998, respectively. Since October 1999, she has been with the Department of Applied Mathematics and Statistics, Universidad Politécnica de Cartagena, Spain. Her research interests mainly include forecasting methods, machine learning, time series analysis, clustering, statistical inference, and information theory. She has participated in 10 projects and 7 private R&D contracts, including the management of 5 of them. She has co-authored 25 research articles published in indexed journals. These works have been referenced 275 times in 243 scientific articles (source: Scopus 2020).

Luis Alfredo Fernández-Jiménez (Associate Professor), Industrial Engineer, received his M.Sc. (1992) from the University of Zaragoza (Spain) and Ph.D. (2007) from the University of La Rioja (Spain). Since 1992, he has been University Professor in the Electrical Engineering Department of the University of La Rioja. His lines of research are focused on planning, operation, and control of power systems. He has been Main Investigator (IP) in three R&D projects funded by the Spanish Government (2010–2019), projects focused on the development of forecasting models for applications in the electric power sector. He has authored or co-authored two university texts related to electrical engineering and 22 research articles published in indexed journals. These works have been referenced more than 680 times in 640 scientific articles (source: Scopus 2020). Since 2012, he has been the head of the Electrical Engineering Department of the University of La Rioja.

Preface to "Short-Term Load Forecasting 2019"

The future of power systems and markets is exciting but presents a number of risks for customers, utilities, network operators, and society as a whole. The integration of renewable generation sources by 2030–2050 [1] and the potentiation of "active customers" [2] will lead to quite different planning and operation tasks in this new scenario [3]. Traditional tools will no longer perform as they do at present. For instance, the ability of the future generation mix to meet load demands, at all times, becomes a more complex task with interesting opportunities for new actors both in the demand and supply sides of a power system. Uncertainties and randomness concerns related to electricity demand appear in the literature: how can we manage the availability of energy outputs from renewable generation resources and the flexibility of customers? New and complex forecasting methods [4] can provide a partial solution to this challenge. In our case, short-term load forecasting methods (STLF) are used to evaluate demand, and perhaps supply.

While STLF has usually been applied to non-responsive customers, this scenario is anticipated to change to "active customers". This concept refers to a new dynamic actor that can participate in electricity markets (energy, capacity, or balance), alone or through demand aggregators, and changes its demand due to economic and technical considerations. This participation requires an estimation of demand in the short term (much more complex) to avoid penalties for non-compliance at lower aggregation levels (from hundreds of kW to some MW), where STLF methodologies should be revisited and modified to improve their performance.

This book compiles thirteen papers published in the Special Issue titled "Short-Term Load Forecasting 2019", which represent a research advance inside a wide range of specific topics described below, all of them of great importance in the field of STLF. The usefulness and relevance of hybrid or combined models, together with multistep methodologies, is indisputable. There are many recent papers in this context. All of them try to overcome the drawbacks of other existing methods, while at the same time seeking to gain robustness and improve predictions. To some extent, all the articles of this Special Issue employ combined or/and multistep methods to provide predictions: in some cases, they employ them as an intermediate tool, and the novelty of the research is focused on other aspects; in other cases, the hybrid method proposed by the authors represents the main contribution. The latter group of studies includes the following papers: [5], where a novel model combining a data pre-processing technique, forecasting algorithms, and an advanced optimization algorithm is developed; [6], which proposes a very short-term bus load prediction model based on a phase space reconstruction and deep belief network; [7], which proposes a hybrid load-forecasting method that combines classical time series formulations with cubic splines to model electricity load; and [8], where the electricity demand time series is divided into two major components (deterministic and stochastic) and both components are estimated using different regression and time series methods with parametric and nonparametric estimation techniques. These last two papers remind us that we must not forget the usefulness of classical methods.

Despite the great number of papers on this topic, there is an issue that remains open: how to guide researchers to employ proper hybrid technology for different datasets [4]. Two review papers of this book represent an advance on this topic: [9], which discusses four categories of state-of-the-art STLF methodologies (similar pattern, variable selection, hierarchical forecasting, and weather station selection), where each of these methods proposes a specific solution for load prediction, and [10], where the authors highlight the necessity of developing additional and case-specific performance criteria for electricity load forecasting (better accuracy does not imply lower costs caused by forecasting errors).

Another aspect of interest related to the mix of methodologies can be found in the context of demand response programs in hybrid energy systems. In [11], a methodology is proposed that could help power systems or aggregators to make up for the lack of accuracy of the current forecasting methods when predicting renewable energy generation, whereas [12] utilizes both long and short data sequences to propose a model that supports the demand response program in hybrid energy systems, especially systems using renewable and fossil sources.

There are many features we must consider developing a good STLF model, such as climatic factors, seasonality, and calendar effects. The authors of [13] highlight the importance of distinguishing different types of special days (those on which working or social habits differ from the ordinary) to reduce the greatest forecasting errors and propose several ways to classify those special days.

Current forecasting methods have shown high efficiency and accuracy, mainly at the power system and great consumer levels. However, there is much to be done at the residential level due to the high volatility and uncertainty of the electric demand of a single household. This topic is dealt with by [14] and [15]: the former presents a scalable system for day-ahead household electrical energy consumption forecasting, based on a deep residual neural network, and extracts features from the historical load of the particular household and all households present in the dataset; the latter proposes a forecasting method based on convolution neural networks combined with a data-augmentation technique that can artificially enlarge the training data.

The issue caused by a lack of historical data or limited data is also addressed in [16] and [17]: in the first case, the authors propose a novel STLF model to predict energy consumption for buildings with limited data sets by using multivariate random forest to construct a transfer learning-based model; in the second case, the author introduces the problem of load "nowcasting" to the energy forecasting literature, where one predicts the recent past using limited available metering data from the supply side of the system.

References

1. IRENA. *Innovation Landscape Brief: Market Integration of Distributed Energy Resources*; International Renewable Energy Agency: Abu, Dhabi, 2019.

2. Directive (EU) 2019/944 of the European Parliament and of the Council of 5 June 2019 on Common Rules for the Internal Market for Electricity and Amending Directive 2012/27/EU. Available online: https://eur-lex.europa.eu/legal-content/EN/TXT/?qid=1570790363600&uri=CELEX:32019L0944 (accessed on 7 October 2020).

3. Michigan Public Service Commission, DTE Electric Company. Integrating Renewables into Lower Michigan Electric Grid. Available online: https://brattlefiles.blob.core.windows.net/files/15955_integrating_renewables_into _lower_michigans_electricity_grid.pdf (accessed on 7 October 2020).

4. Wei-Chiang Hong, Ming-Wei Li, Guo-Feng Fan, Short Term Load Forecasting by Artificial Intelligent Technologies, MDPI editors, ISBN 978-3-03897-582-3. Available online: https://www.mdpi.com/journal/energies/specialissues/Short Term Load Forecasting (accessed on 7 October 2020).

5. Zhang, Y.; Wang, J.; Lu, H. Research and Application of a Novel Combined Model Based on Multiobjective Optimization for Multistep-Ahead Electric Load Forecasting. *Energies* **2019**, *12*, 1931.

6. Shi, T.; Mei, F.; Lu, J.; Lu, J.; Pan, Y.; Zhou, C.; Wu, J.; Zheng, J. Phase Space Reconstruction Algorithm and Deep Learning-Based Very Short-Term Bus Load Forecasting. *Energies* **2019**, *12*, 4349.

7. Jornaz, A.; Samaranayake, V.A. A Multi-Step Approach to Modeling the 24-hour Daily Profiles of Electricity Load using Daily Splines. *Energies* **2019**, *12*, 4169.

8. Shah, I.; Iftikhar, H.; Ali, S.; Wang, D. Short-Term Electricity Demand Forecasting Using Components Estimation Technique. *Energies* **2019**, *12*, 2532.

9. Fallah, S.N.; Ganjkhani, M.; Shamshirband, S.; Chau, K.-W. Computational Intelligence on Short-Term Load Forecasting: A Methodological Overview. *Energies* **2019**, *12*, 393.

10. Koponen, P.; Ikäheimo, J.; Koskela, J.; Brester, C.; Niska, H. Assessing and Comparing Short Term Load Forecasting Performance. *Energies* **2019**, *13*, 2054.

11. Ruiz-Abellón, M.C.; Fernández-Jiménez, L.A.; Guillamón, A.; Falces, A.; García-Garre, A.; Gabaldón, A. Integration of Demand Response and Short-Term Forecasting for the Management of Prosumers' Demand and Generation. *Energies* **2019**, *13*, 11.

12. Pramono, S.H.; Rohmatillah, M.; Maulana, E.; Hasanah, R.N.; Hario, F. Deep Learning-Based Short-Term Load Forecasting for Supporting Demand Response Program in Hybrid Energy System. *Energies* **2019**, *12*, 3359.

13. López, M.; Sans, C.; Valero, S.; Senabre, C. Classification of Special Days in Short-Term Load Forecasting: The Spanish Case Study. Energies 2019, 12, 1253.

14. Kiprijanovska, I.; Stankoski, S.; Ilievski, I.; Jovanovski, S.; Gams, M.; Gjoreski, H. HousEEC: Day-Ahead Household Electrical Energy Consumption Forecasting Using Deep Learning. *Energies* **2020**, *13*, 2672.

15. Acharya, S.K.; Wi, Y.-M.; Lee, J. Short-Term Load Forecasting for a Single Household Based on Convolution Neural Networks Using Data Augmentation.
emphEnergies **2019**, *12*, 3560

16. Moon, J.; Kim, J.; Kang, P.; Hwang, E. Solving the Cold-Start Problem in Short-Term Load Forecasting Using Tree-Based Methods. Energies 2020, 13, 886.

17. Ziel, F. Load Nowcasting: Predicting Actuals with Limited Data. *Energies* **2020**, *13*, 2672.1443.

Antonio Gabaldón, María Carmen Ruiz-Abellón, Luis Alfredo Fernández-Jiménez
Editors

Article

Research and Application of a Novel Combined Model Based on Multiobjective Optimization for Multistep-Ahead Electric Load Forecasting

Yechi Zhang [1], Jianzhou Wang [1,*] and Haiyan Lu [2]

[1] School of Statistics, Dongbei University of Finance and Economics, Dalian 116025, China; derchi666@gmail.com
[2] School of Software, Faculty of Engineering and Information Technology, University of Technology, Sydney 2007, Australia; Haiyan.Lu@uts.edu.au
* Correspondence: wangjz@dufe.edu.cn; Tel.: +86-13009480823

Received: 10 April 2019; Accepted: 16 May 2019; Published: 20 May 2019

Abstract: Accurate forecasting of electric loads has a great impact on actual power generation, power distribution, and tariff pricing. Therefore, in recent years, scholars all over the world have been proposing more forecasting models aimed at improving forecasting performance; however, many of them are conventional forecasting models which do not take the limitations of individual predicting models or data preprocessing into account, leading to poor forecasting accuracy. In this study, to overcome these drawbacks, a novel model combining a data preprocessing technique, forecasting algorithms and an advanced optimization algorithm is developed. Thirty-minute electrical load data from power stations in New South Wales and Queensland, Australia, are used as the testing data to estimate our proposed model's effectiveness. From experimental results, our proposed combined model shows absolute superiority in both forecasting accuracy and forecasting stability compared with other conventional forecasting models.

Keywords: electric load forecasting; data preprocessing technique; multiobjective optimization algorithm; combined model

---

1. Introduction

It is known that the electric power industry plays a vital role in many aspects of people's lives [1]. Effective forecasting enables adjustments to be made of power generation according to market demand, and to the reduction of management and operational costs [2]. On this basis, accurate power load forecasting is necessary in daily operations of power systems [3]. However, due to various uncertainties and climate change, economic fluctuations, industrial structure, and national policy and other social environment complexity, it is difficult to meet expectations in terms of the accuracy of power load forecasting [4]. Inaccurate forecasting often results in considerable loss of power systems. For example, overestimated forecasts often result in wasted energy, while underestimated forecasts will result in economic loss [5]. With the development of society, the expansion of urbanization, and the continuous improvement of industry, the demand for electricity is continuously increasing, which poses a challenge to electric load prediction systems [6]. Accurate power load forecasting is indispensable to the whole society, which not only reflects the economic rationality of power dispatching, but can also be reflected in power construction planning and power supply reliability. Therefore, developing a novel and robust model to improve forecasting performance is essential for power load forecasting [7]. In the past few years, in order to achieve accurate short-term time series forecasting of power load, a lot of research has been carried out. There are mainly four types of related algorithms: (i) physical arithmetic, (ii) spatial correlation arithmetic, (iii) conventional statistical arithmetic, (iv) and artificial intelligence arithmetic [8].

2. Literature Review

"Physical algorithm" is a general term referring to models that primarily use physical data such as temperature, velocity, density, and terrain information based on a numerical weather prediction (NWP) model to predict wind speeds in subsequent periods [9]. The NWP model is a computer program designed to solve atmospheric equations. Based on the NWP wind resource assessment method, Cheng et al. [10] evaluated wind speed distribution by comparing three deterministic probabilities. From their experiment results, they found that NWP could not only achieve reliable probability assessment but also supply precise forecasting estimates. However, physical methods cannot handle time series for short-term horizons [11]. Moreover, when using an NWP model, much calculation time and many computing resources are required [12]. Spatial correlation models, which are applied to solve time series forecasting to make up for the shortcomings of physical algorithms, take the relationships of time series from different locations into consideration [13]. A classic case is a novel model proposed by Tascikaraoglu et al. [14] utilizing a spatiotemporal method and a wavelet transform, successfully improving the performance of forecasting compared to other benchmark models. However, spatial correlation arithmetic is always difficult to use in practice because of its requirements of strict measurements and a large amount of meticulous measuring in many spatially related sites [15].

Traditional prediction methods also include random time series models such as exponential smoothing, autoregressive (AR) methods, filtering methods, autoregressive moving average (ARMA) methods, and the well-known autoregressive integrated moving averages (ARIMA) and seasonal ARIMA models, mainly focusing on regression analysis [16,17]. The regression model is aimed at establishing a relationship between historical data, treated as dependent variables, and influencing factors, treated as independent variables [18]. For example, Lee and Ko [19] adopted an ARIMA-based model to forecast and simulate hourly electric load data of the Taipower system. Wang et al. [20] improved the accuracy of seasonal ARIMA applied to electricity demand forecasting by the use of residual modification models. They applied a seasonal ARIMA approach, an optimal Fourier model, and a combined model including seasonal ARIMA and the PSO optimal Fourier method. They used these three models to predict electric load time series data in northwestern China. After juxtaposing the results, they found that the combined model was the most accurate one. Brożyna et al. [21] used the TBATS model to overcome the seasonality in data, which may bring difficulties when doing time series forecasting by using models such as ARIMA.

Modern forecasting methods include artificial neural networks (ANNs), support vector machines (SVMs), fuzzy systems, expert system forecasting methods, chaotic time series methods, gray models, adaptive models, optimization algorithms, etc. [22]. These modern methods are getting more popular among researchers when dealing with time series forecasting [23]. These artificial intelligence models can achieve good forecasting performance because of their unique characteristics, such as memory, self-learning, and self-adaptability, since the neural networks are products of biological simulation that follow the behavior of the human brain [24]. Park [25] showed good performance of this type of model after first applying ANNs in power load forecasting in 1991. He concluded that ANNs were highly effective in electrical load forecasting. After that, many time series forecasting studies were performed using various artificial neural networks by a lot of researchers [26]. Lou and Dong [27] proved that electric load forecasting with RFNN showed much higher variability with hourly data in Macau. Okumus and Dinler [28] integrated ANNs and the adaptive neuro-fuzzy inference system to predict wind power, and forecasting results proved that their proposed hybrid model was better than the classical methods in forecasting accuracy. Hong [29] selected better parameters for SVR by using the CPSO algorithm, while Che and Wang [30] established a hybrid model that was a combination of ARIMA and SVM, called SVRARIMA. Liu et al. [31] built a model integrating EMD, extended extreme learning machine (ELM), Kalman filter, and PSO algorithm. Although the hybrid model seemed better than individual classical models, the limitations of each model due to the nature of the structure seemed inevitable [32]. In order to solve this problem, a combined forecasting model is proposed. The combined forecasting theory has been developed through the joint efforts of three generations of scientists. It was initiated by Bates and Granger [33] and developed by Diebold and Pauly [34], then further extended by Pesaran and Timmermann [35] as a combination of several individual models. Many kinds of ANNs have been combined into short-term forecasting models in order to fully utilize the advantages of individual models and at the same time overcome their shortcomings. There are some typical studies: Zhang et al. [36] successfully obtained promising results of wind speed forecasting by developing a combined model that consisted of CEEMDAN, five neural networks, CLSFPA, and no negative constraint theory (NNCT). In addition, Che et al. [37] developed a kernel-based SVR combination model in a study on electric load prediction.

It is obvious from the review of forecasting methods that there are shortcomings in both traditional and modern techniques. The shortcomings of these models are summarized as follows:

For physical algorithms, the main problem is that physical methods cannot deal with short-term horizons. Physical methods perform well when dealing with long-term forecasting problems [38]. Moreover, it costs a lot of computing time and resources when using NWP models because of their complex calculation process and high cost. Spatial correlation arithmetic requires detailed measurements from multiple spatially correlated sites, which increases the difficulty in searching for electric load data. Moreover, because of the strict measuring requirements and time delays, the model is always hard to implement [39].

For conventional statistical arithmetic, mainly known as the linear model, there are insurmountable shortcomings. First and foremost, these models cannot deal with nonlinear features of electric load time series [40]. Moreover, the regression method also fails to achieve the expected forecasting accuracy. Linear regression relies too much on historical data to cope with nonlinear forecasting problems; as time goes by, the forecasting effect of regression analysis models will become weaker and weaker [41]. In addition, when faced with complex objective data, it is hard to choose the appropriate influencing factors. The exponential smoothing model also has shortcomings, in that it cannot recognize the turning point of the data and does not perform well in long-term forecasting [42]. As for the autoregressive moving average model, it only gets results through historical and current data, ignoring potential influencing factors. In addition, strong random factors of the data may lead to instability of the model, which affects the accuracy of the forecasting performance [43]. All in all, none of these models meet the accuracy required by an electric load forecasting system.

For artificial intelligence arithmetic, although artificial intelligence neural network performance is superior to traditional forecasting techniques, ANNs are impeccable; the defects and shortcomings of their structure cannot be ignored. There are three major problems. First, it is hard to choose the parameters of ANN models, as a slight change in parameters may cause huge differences in the outcomes [44]. Second, ANNs are inclined to fall into local minima owing to their relatively slow self-learning convergence rate [45]. Lastly, the number of layers and neurons in a neural network structure has an effect on the forecasting result and computing time [46]. As to other models, SVM has a high requirement for storage space and expert systems strongly rely on knowledge databases, while gray forecasting models can produce decent results only under the condition of exponential growth trends [47]. To solve these problems, evolutionary algorithms are applied. When the optimization algorithms are combined with forecasting models, more reasonable parameters will be selected and more accurate results will be obtained.

To overcome the abovementioned drawbacks, in our proposed model, we use a data preprocessing method, no negative constraint theory (NNCT) [48], a multiobjective optimization algorithm, a linear forecasting method, autoregressive integrated moving average (ARIMA) [49], and three artificial intelligence forecasting algorithms, wavelet neural network (WNN) [50], extreme learning machine (ELM) [51], and back propagation neural network (BPNN) [52]. The proposed model improves forecasting performance by maximizing the benefits of both linear and nonlinear advantages by using each single model. It is worth mentioning that for the purpose of improving the forecasting effect of our model, a mechanism based on decomposition and reconstruction is employed to ensure that the main features of the original data are identified and extracted by removing high-frequency noise signals. Then, four individual models are applied to the electrical load forecasting. Lastly, a new weight decision technique based on the multiobjective grasshopper optimization algorithm and stay-one strategy was successfully used to integrate the four models. The experimental results show that our combined model has high forecasting accuracy and strong stability.

The main contributions and novelties of our proposed model are summarized as follows:

(1) *Applying the decomposition and reconstruction strategy, data preprocessing methods are adopted to extract main features of the original data by eliminating high-frequency signals, making predictions more accurate.* Decomposing the original power data and reconstructing it into a filtering sequence can eliminate the irregularity and uncertainty of the data and achieve better power load forecasting performance.

(2) *Applying the multiobjective optimization algorithm, the optimal weight coefficient of each single model can be optimized.* Our proposed combined model is not only robust, but also economical in power load forecasting. Moreover, it has higher precision and greater stability.

(3) *With the combination of the linear model (ARIMA) and nonlinear models (WNN, ELM, and BPNN), the developed model can reflect both the linearity and nonlinearity of electrical load data.* Our proposed model can use each individual model thoroughly and it spontaneously overcomes limitations such as low precision and instability to ensure the effectiveness of power load forecasting.

(4) *The new combined model beats other single models and will provide effective technical support for power system management.* The developed model was simulated and examined based on the electric load data of three different sites, which indicates its strong robustness and adaptability regardless of location and forecasting steps.

The rest of the paper is arranged as follows. In Section 2, we introduce the methodology we applied in the proposed model, including the data preprocessing technique, ARIMA, WNN, ELM, BPNN, the theory of combined models, and multiobjective grasshopper optimization. Section 3 describes the electric load time series we selected and three experiments aimed at verifying the effectiveness of our forecasting model. In Section 4, we provide an in-depth discussion of the proposed model, including a test of the performance of the proposed optimization algorithm, two tests of the effectiveness of the model, and a test showing the improvement of the model and a comparative experiment of the combination method.

3. Methods

In this section, we discuss the methods of the proposed combined model in detail, including the singular spectrum analysis (SSA) technique, the individual models used in the combined model, and the multiobjective grasshopper optimization algorithm (MOGOA). After that, a combined model that can significantly improve the definition of electric load forecasting is presented.

3.1. SSA Technique

SSA is a nonparametric spectral estimation method usually used for filtering in the preprocessing stage of time series forecasting. The advantage of SSA is that it always works well in both linear and nonlinear time series. Moreover, it performs well whether the time series is stationary or not. In short, the way SSA works is to identify the trend and noise parts of a time series, after which it reconstructs a new series.

3.2. Wavelet Neural Network

Wavelet neural network (WNN) is a modern artificial intelligence model. It is essentially a feed-forward neural network based on wavelet transform [53]. Its basic working principle is to use wavelet space as the feature space of pattern recognition to realize the feature extraction of signals by weighting the inner product of the wavelet base and the signal vector and combining the time-frequency localization of the wavelet transform and the self-learning function of the neural network. It has the advantage of being able to effectively learn the input/output characteristics of the system without the need for a priori information such as data structures and characteristics. In addition, compared with traditional neural networks, wavelet neural networks can often achieve better prediction accuracy, faster convergence, and better fault tolerance when forecasting in complex nonlinear, uncertain, and unknown systems. So, we applied WNN as an individual nonlinear model in our proposed model.

3.3. Extreme Learning Machine

Extreme learning machine (ELM) is a kind of machine learning algorithm based on feed-forward neuron network [54]. Its main feature is that the hidden layer node parameters can be given randomly or artificially and do not need to be adjusted. The learning process only needs to calculate the output weight. ELM has the advantages of high learning efficiency and strong generalization ability and is widely used in time series forecasting. As a result, we applied ELM as an individual nonlinear model in our proposed model.

3.4. Back Propagation Neural Network

The back propagation neural network (BPNN), composed of an input layer, a hidden layer, and an output layer, is a concept that was proposed by scientists led by Rumelhart and McClelland in 1986 [55]. It is a multilayer feed-forward neural network trained according to the error back propagation algorithm. Learning and working stages are the whole process of BPNN. It is the most widely used neural network. It has arbitrary complex pattern classification ability and excellent multidimensional function mapping ability, which solves the exclusive OR (XOR) and other problems that cannot be solved by simple perception. In essence, the BP algorithm uses the network error squared as the objective function and the gradient descent method to calculate the minimum value of the objective function. Moreover, because of its flexible structure and strong nonlinear mapping capability, BPNN is widely applied in the engineering field. So, we applied it as an individual nonlinear model in our proposed model.

3.5. Autoregressive Integrated Moving Average Model

The ARIMA model, also known as the autoregressive moving average model, is a model used for time series forecasting with relatively high prediction accuracy. The ARIMA model mainly consists of 3 forms, a moving average MA model, an autoregressive AR model, and a mixture of autoregressive moving average ARMA models. Before using this model, it is necessary to first analyze whether the time series is stable. If the sequence is a nonstationary time series, the first step is to differentiate the time series, and the difference must be smoothed before the model is established, otherwise it cannot be used.

The difference between the ARIMA model and the ARMA model is that the ARMA model is built for stationary time series and the ARIMA model is used for nonstationary time series. In other words, to establish an ARMA model for a nonstationary time series, you first need to transform into a stationary time series and then build an ARMA model. We applied ARIMA as an individual linear model in our proposed model.

3.6. Basic concepts of Multiobjective Optimization Problems

Conventional relational operators such as >, <, and =, which are always found in single-objective optimization problems, cannot be applied in multiobjective optimization. To address this problem, a new concept of dominates was proposed and then extended by Edgeworth in 1881 and Pareto in 1964. Details of Pareto dominance are as follows:

Definition 1 (*Pareto dominance*):

The definition of Pareto dominance is: vector $y = (y_1, y_{2,...}y_z)$ is dominated by vector $x = (x_1, x_{2,...}x_z)$ (i.e., $x > y$) when

$$\forall\, t \in [1, z], [f(x_t) \geq f(y_t)] \wedge [\exists t \in [1, z] : f(x_t) > f(y_t)] \tag{1}$$

where z represents the length of vectors.

3.7. Multiobjective Grasshopper Optimization Algorithm

MOGOA is the latest nature-inspired method, proposed by Mirjalili [56]. Essentially, MOGOA is a multiobjective version of GOA. GOA is a nature-inspired algorithm that simulates the swarming behavior of grasshoppers. The position of a grasshopper in the swarm representing a possible solution of a given single-objective optimization problem is the main principal of GOA. The details of MOGOA, and the main steps of building it, are as follows:

In order to replicate the real living conditions of grasshoppers in nature, MOGOA takes 3 factors—gravity force, social interaction, and wind advection—into the model. X_i means the position of the ith grasshopper and is represented by:

$$X_i = S_i + G_i + A_i \tag{2}$$

where S_i, G_i, and A_i mean social interaction, gravity force, and wind advection, respectively.

Social interaction is the most important factor, calculated by the following equation:

$$S_i = \sum_{\substack{j=1 \\ j \neq i}}^{N} s(d_{ij})\hat{d}_{ij} \tag{3}$$

$$d_{ij} = \left| \mathbf{x}_j - \mathbf{x}_i \right| \tag{4}$$

$$\hat{d}_{ij} = \left(\mathbf{x}_j - \mathbf{x}_i \right)/d_{ij} \tag{5}$$

$$s(r) = fe^{-r/l} - e^{-r} \tag{6}$$

where d_{ij} means the distance between the ith and jth grasshoppers, and $\hat{d}_{ij}$ is a unit vector of d_{ij}. Function s defines that the values of parameters f and l are changed, so the social forces can be changed too. The distance between grasshoppers is limited to the interval of [1,4], because, according to common sense, when 2 grasshoppers are far apart, they will not have a strong social influence on each other. Gravity force is defined as:

$$\mathbf{G}_i = -g\hat{e}_g \tag{7}$$

where g is the gravitational constant and $\hat{e}_g$ represents the unity vector toward the center of the earth. Wind advection is defined as:

$$\mathbf{A}_i = u\hat{e}_w \tag{8}$$

where u means constant drift and $\hat{e}_w$ represents the unity vector in the wind direction. After replacing Equation (2) with the above 3 equations, we can get:

$$X_i = \sum_{\substack{j=1 \\ j \neq i}}^{N} s(\left| \mathbf{x}_j - \mathbf{x}_i \right|)\frac{\mathbf{x}_j - \mathbf{x}_i}{d_{ij}} - g\hat{e}_g + u\hat{e}_w \tag{9}$$

Considering that the influence of gravity force on grasshoppers is too weak and assuming that the wind direction is always toward the best solution $\hat{T}_d$, some parameters are added to the mathematical model to enhance the ability to explore and exploit for the purpose of solving the optimization problem more effectively. After that, the mathematical model turns to:

$$X_i^d = c\left(\sum_{\substack{j=1 \\ j \neq i}}^{N} c\frac{ub_d - lb_d}{2} s\left(\left|x_j^d - x_i^d\right|\right) \frac{\mathbf{x}_j - \mathbf{x}_i}{d_{ij}} \right) + \hat{T}_d \tag{10}$$

where ub_d and lb_d are the upper and lower bound of the dth dimension, respectively, and $\hat{T}_d$ is the best solution's dth dimension value so far. For the purpose of reducing exploration and increasing exploitation proportional to c_{max} at the same time, the parameter c is updated with the following equation:

$$c = c_{max} - l\frac{c_{max} - c_{min}}{L} \tag{11}$$

Compared with finding solutions from a series of Pareto optimal solutions obtained by MOGOA, it is easier to find the best solution calculated so far in a single-objective search. Because the archive has all the Pareto optimal solutions, the position of the target is determined. Finding the target that can improve the solution's distribution becomes the biggest problem. The possibility of choosing the target from the archive is calculated by:

$$P_i = \frac{1}{N_i} \tag{12}$$

where Ni represents the neighborhood of the ith solution's total number. With this probability, there are 2 advantages to using the roulette method when selecting a target from a file: first, the roulette method can improve the distribution of less distributed areas of the search space, and second, when premature convergence occurs, a solution with a crowded neighborhood can be selected as a target to solve the problem.

When updating the content of the archive regularly in MOGOA, 2 criteria are implemented: (1) give up an external solution as long as this external solution is dominated by one archive solution; and (2) add an external solution to the archive when the external solution does not dominate all solutions inside the archive. Moreover, as long as an external solution dominates a solution inside the archive, the inside one should be replaced by the external one. All in all, MOGOA can not only find Pareto optimal solutions, but also store them in an archive.

The pseudocode of MOGOA is as follows:

Algorithm 1: *MOGOA*

Objective functions:

$$min\left\{\begin{array}{l} O_1 = \left|Bias(\hat{y})\right| \\ O_2 = std(y - \hat{y}) \end{array}\right.$$

Input:

$\hat{y}_B = (\hat{y}_B(1), \hat{y}_B(2), \cdots, \hat{y}_B(q))$- BPNN

$\hat{y}_E = (\hat{y}_E(1), \hat{y}_E(2), \cdots, \hat{y}_E(q))$- ELM

$\hat{y}_W = (\hat{y}_W(1), \hat{y}_W(2), \cdots, \hat{y}_W(q))$-WNN

$\hat{y}_A = (\hat{y}_A(1), \hat{y}_A(2), \cdots, \hat{y}_A(q))$-ARIMA

Output:

$\hat{y}_f = \left(\hat{y}_f(1), \hat{y}_f(2), \cdots, \hat{y}_f(l)\right)$- forecasting results

Parameters:

L—the maximum number of iterations

n—the number of grasshoppers

lb_i, ub_i—boundaries of the *i-th* variable

x_i—*i-th* grasshopper's position

l—current iteration number

d—dimension amount.

c_{max}—c's maximum value

c_{min}— c's minimum value

$\hat{T}_d$—best solution's d-th dimension value so far

d_{ij}—the distance between the *i-th* and the *j-th* grasshopper

s—social forces function

/*Set the parameters of MOGOA. */

/*Initialize x_i (i=1,2,...,n), c_{max}, c_{min}. */

/*Define the archive size. */

FOR EACH *i or j*: $1 \leq i\ (j) \leq n$ **DO**

    Calculate the corresponding fitness functions using ranking process

END FOR

    /*Determine the best grasshopper and suppose it as the elite. */

WHILE ($l < L$) **DO**

    FOR EACH *i*=1: n **DO**

        FOR EACH *j*=1: n **DO**

            /*Select a random grasshopper from the archive. */

            /*Select the elite using Roulette wheel from the archive. */

            /*Update c by applying equations. */

$$c = c_{max} - l\frac{c_{max} - c_{min}}{L}$$

            /* Normalize distances between grasshoppers. */

        END FOR

        /*Update the position of all grasshoppers. */

$$X_i^d = c\left(\sum_{\substack{j=1 \\ j \neq i}}^{n} c\frac{ub_d - lb_d}{2} s\left(\left|x_j^d - x_i^d\right|\right)\frac{x_j - x_i}{d_{ij}}\right) + \hat{T}_d$$

END FOR
/*Calculate the objective values of all grasshoppers. */
/*Find the non-dominated solutions. */
/*Update the archive in regard to the obtained non-dominated solutions. */
    IF the archive is full **DO**
      /* Delete some solutions to hold new solutions. */
      Applying Roulette wheel and $P_i = N_i / k$, $k > 1$
    END IF
    IF any new added solutions to the archive are outside boundaries **DO**
      /* Update the boundaries to cover the new solution(s). */
    END IF
    $l=l+1$
END WHILE
RETURN archive

3.8. SSA-MOGOA Combined Model

In our study, a new combined model applying a data preprocessing technique, a new parameter determination method, and several individual prediction algorithms, including both linear and nonlinear models, is successfully developed. The main steps are listed below. The flowcharts of the proposed model are depicted in Figure 1.

3.8.1. Stage 1: Data Preprocessing

SSA is a nonparametric spectral estimation method usually used for filtering in the preprocessing stage of time series forecasting [57]. The advantage of SSA is that it always works well in both linear and nonlinear time series. In addition, the processed data will be used in subsequent forecasting. The main steps of SSA are depicted in Figure 1.

3.8.2. Stage 2: Individual Models used for Forecasting

Three nonlinear models, BPNN, ELM, and WNN, and a linear model, ARIMA, are chosen as the individual models that together form the combined model. It is worth mentioning that all 4 models can achieve good prediction results in our electric load forecasting.

3.8.3. Stage 3: Optimization of Weight Parameters of Combined Model

Determining the parameter coefficients of each individual model is very important for construction of the combined model. In past combined models, a simple average coefficient allocation strategy was often used. In our research, we adopted a multiobjective optimization algorithm called MOGOA for the deciding parameters and made the combined model achieve good prediction results in electric load forecasting.

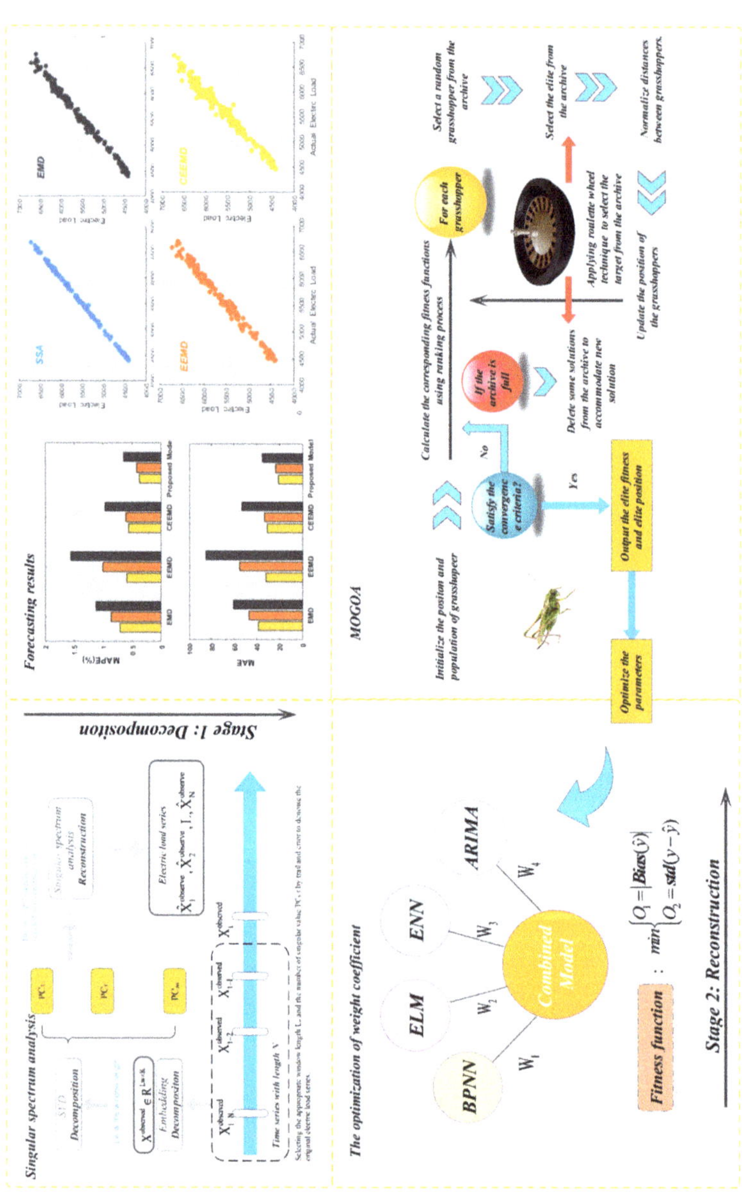

Figure 1. Structure of proposed singular spectrum analysis–multiobjective grasshopper optimization algorithm (SSA-MOGOA) combined model.

4. Experiments and Analysis

In this section, we introduce the electric load time series we selected and the performance metric and testing methods. We also present three experiments aimed at verifying the effectiveness of our forecasting model. The main steps and flowchart of the developed model are described in Figure 1, which includes a data preprocessing technique, application of several individual models, optimization of the combined model's weight coefficients, and forecasting results.

4.1. Datasets

In this paper, original electric load time series data were collected from two areas in Australia, New South Wales (NSW) and Queensland (QLD), on a half-hourly basis (48 data points per day). Two datasets were collected in New South Wales and Queensland, which were sampled from 13 to 31 July 2011, 19 days in all. The third dataset was sampled from 13 to 31 July 2010 in Queensland. Figure 2 presents a simple map of the study area, some descriptive statistical indicators of the datasets, and three general trends of testing samples. Specifically, in each dataset, the training set included 768 data points and the testing set consisted of 144 data points. There were 48 data points in a single day according to the data, therefore we selected the period T = 48 for the combined model. Statistical indicators including minimum, maximum, mean, and standard deviation are listed in Table 1. From one-step to three-step prediction, forecasting outcomes are all based on the historical data, which means this experimental outcome is not used as input data to forecast the subsequent values in this study, while in artificial intelligence models, based on plenty of experimental results, five historical data points are chosen as input so as to obtain the best forecasting performance in the multistep forecasting mechanism. The detailed data structure is presented in Figure 2.

Table 1. Statistical indicators of experimental samples for three sites.

Dataset.	Samples	Numbers	Statistical Indicator(kw)			
			Max	Min	Mean	Std.
QLD(2010)	All samples	912	7033.21	4316.89	5788.65	741.36
	Training	768	7033.21	4316.89	5803.04	746.40
	Testing	144	6476.49	4361.6	5711.87	708.99
QLD(2011)	All samples	912	7234.04	4399.42	5782.99	724.57
	Training	768	7234.04	4412.33	5834.35	729.96
	Testing	144	6718.05	4399.42	5509.06	627.75
NSW(2011)	All samples	912	12883.81	6821.4	9707.66	1337.86
	Training	768	12883.81	6821.4	9819.03	1346.71
	Testing	144	11314.46	6939.18	9113.68	1115.41

4.2. Performance Metrics

In our study, to evaluate the predictive power of the proposed model, we needed performance metrics in our time forecasting experiments. Because there is no general standard for evaluating a time forecasting model, we decided to apply three performance metrics: mean absolute error (MAE), root mean square error (RMSE), and mean absolute percent error (MAPE), as presented in Table 2 [58]. Next, we introduce these three performance metrics in detail.

From the definitions of MAE and RMSE, it is obvious that the advantage of these two performance metrics is that they can avoid canceling between positive and negative forecasting errors due to the use of absolute value symbols. They can evaluate the average dimension of the forecasted time series with actual data. MAPE, which is regarded as the most widely used performance metric in time series forecasting, is obtained by calculating the average of absolute error. The advantage of MAPE is that it can reflect the reliability and validity of the time series forecasting method. When observing the values of all three of these metrics, the smaller the value, the more accurate the prediction. Table 2 shows the formulas of the three error metrics. Here Ai means actual values of the time series and Fi means predicted values, and N means sample size.

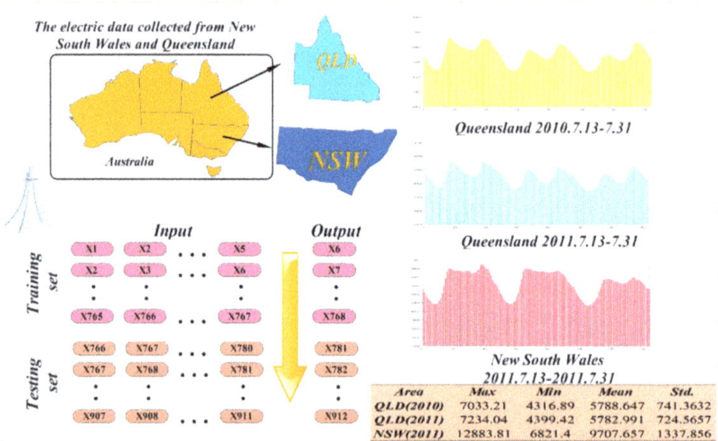

Area	Max	Min	Mean	Std.
QLD(2010)	7033.21	4316.89	5788.647	741.3632
QLD(2011)	7234.04	4399.42	5782.991	724.5657
NSW(2011)	12883.81	6821.4	9707.657	1337.856

Figure 2. Location of electric load and data structure.

Table 2. Three error metrics.

Metric	Definition	Equation
MAE	The mean absolute error of N forecasting results	$\text{MAE} = \frac{1}{N}\sum\limits_{i=1}^{N}|F_i - A_i|$
RMSE	The square root of the average of error squares	$\text{RMSE} = \sqrt{\frac{1}{N}\times\sum\limits_{i=1}^{N}(F_i - A_i)^2}$
MAPE	The average of N absolute percentage error	$\text{MAPE} = \frac{1}{N}\sum\limits_{i=1}^{N}\left|\frac{A_i - F_i}{A_i}\right|\times 100\%$

4.3. Testing Method

In this section, we introduce the Diebold–Mariano (DM) test and the forecasting effectiveness that were applied to statistically test the accuracy of our proposed model in time series forecasting.

4.3.1. Diebold–Mariano Test

Diebold and Mariano [59] developed a test to compare a model's prediction efficiency with that of other models. The main steps of the DM test are as follows:

Since the DM test is essentially a hypothesis test, the first things to introduce are the null hypothesis H_0 and alternative hypothesis H_1:

$$H_0 : E\big[F\big(e_t^1\big)\big] = E\big[F\big(e_t^2\big)\big] \tag{13}$$

$$H_1 : E\big[F\big(e_t^1\big)\big] \neq E\big[F\big(e_t^2\big)\big] \tag{14}$$

where e_t^1 and e_t^2 are subtracted from actual time series data and the different models' predicted time series values, also called forecasting errors, and F is the loss function of e_t^1 and e_t^2.

$$\bar{d} = \frac{1}{L}\sum_{t=1}^{L}\big[F\big(e_t^1\big) - F\big(e_t^2\big)\big] \tag{15}$$

$\bar{d}$ is obtained by calculating the average of the sum of differences between the two models' loss function, and L is the length of predicted values.

$$DM = \frac{\bar{d}}{\sqrt{2\pi \tilde{f}_d(0)/L}} \to N(0,1) \tag{16}$$

As shown in the above formula, the test statistic DM is convergent in the standard normal distribution N (0, 1). The null hypothesis will be rejected if $|DM|$ is bigger than $\left|Z_{\alpha/2}\right|$, where $z_{\alpha/2}$ stands for the critical z-value of the standard normal distribution and α denotes the significance level.

4.3.2. Forecasting Effectiveness

Forecasting effectiveness can be calculated by the accuracy of the mean squared deviation, which the DM test cannot do [60]. Forecasting effectiveness is also employed in our study. The principal ideas of forecasting effectiveness are as follows:

$m^k = \sum_{i=1}^{n} Q_i A_i^k$ is used to calculate the kth order forecasting effectiveness unit, where Ai means forecasting accuracy time i; Q_i object to $\sum_{i=1}^{n} Q_i = 1$, $Q_i > 0$, called discrete possibility distribution. Q_i will be defined as $Q_i = 1/n$, $i = 1, 2, \ldots, n$ when there is no prior information of Q_i. The kth order forecasting effectiveness is calculated by $H(m^1, m^2, \cdots, m^k)$, where H is a continuous function with k units. The first-order forecasting effectiveness will be defined as $H(m^1) = m^1$ when $H(x) = x$ is a continuous constant function. $H(m^1, m^2) = m^1 \left(1 - \sqrt{m^2 - (m^1)^2}\right)$ will be called as the second-order forecasting effectiveness when $H(x, y) = x\left(1 - \sqrt{y - x^2}\right)$ is a continuous function with two variables.

4.4. Experiments and Analysis

In this part, to examine our proposed model's performance in electric load time series forecasting, we did three experiments from corresponding power station sites.

4.4.1. Experiment I: Compare with Other Models Based on SSA

In order to determine the necessity of combining the models, we made this experiment comparing the electric load time series forecasting results of our new model with the four SSA-based models. The experimental results are shown in Table 3. Detailed descriptions are as follows:

- By observing the experimental results using the 2010 Queensland power data, the following results were found: First of all, the most obvious was that our proposed combined model had the best prediction performance whether the statistical indicator was MAE, RMSE, or MAPE; in other words, the smallest error metrics values. Second, if we look closely at the forecasting steps, we can find that the forecasting accuracy gets worse. In one-step forecasting, our proposed model's MAPE value is 0.37%, and it increases to 0.68% in three-step forecasting.
- For the 2011 Queensland power data, we found the following: First, our proposed model was still the most accurate one. It is worth mentioning that in one-step forecasting, the forecasting gap between our model and the SSA-ELM model was big. Specifically, the MAE values of our model and SSA-ELM were 20.79 and 23.90, respectively, while they were 21.26 and 22.35 with the 2011 Queensland power station data. The superiority of the proposed model can be more intuitively reflected in the Figure 3.
- Regarding the results using the 2011 New South Wales power data, compared to the first two experiments, which used electric load data from Queensland, the error metric values were significantly larger in the third experiment. This reflects the differences among different power plants. The great thing was that our proposed combined model still outperformed other SSA-based

models in one-step to three-step forecasting. This is powerful proof that our model is indeed superior. At the same time, we can also determine the necessity of combining models through this experiment by the fact that it really can improve forecast accuracy.

N.B. By comparing the forecasting results of our proposed combined model with other SSA-based models, there were many useful findings from Experiment I. Our model's overall performance in predicting accuracy demonstrates the need to combine models. Moreover, our proposed model greatly improves electric load forecasting accuracy with an average MAPE of 0.52% in all experiments.

Table 3. Comparison of proposed model with other SSA-based models.

Dataset	Model	MAE			RMSE			MAPE (%)		
		1-step	2-step	3-step	1-step	2-step	3-step	1-step	2-step	3-step
QLD(2010)	SSA-BP	24.78	32.79	62.15	30.87	39.70	80.51	0.44	0.58	1.10
	SSA-ELM	22.35	36.58	68.08	28.35	44.18	86.52	0.40	0.66	1.20
	SSA-WNN	26.23	52.69	95.45	36.09	68.78	126.16	0.47	0.93	1.71
	SSA-ARIMA	39.18	40.71	43.54	52.83	54.84	57.59	0.70	0.72	0.78
	Proposed Model	21.26	25.94	37.97	26.98	32.83	46.51	0.37	0.45	0.68
QLD(2011)	SSA-BP	26.21	35.21	71.09	34.48	45.07	86.81	0.50	0.65	1.31
	SSA-ELM	23.90	34.98	90.62	30.43	45.20	118.89	0.44	0.65	1.69
	SSA-WNN	35.30	80.94	169.98	45.58	100.84	216.67	0.68	1.53	3.16
	SSA-ARIMA	42.82	44.60	48.70	58.27	58.10	61.19	0.77	0.81	0.90
	Proposed Model	20.79	23.43	34.84	27.75	30.73	44.80	0.38	0.43	0.65
NSW(2011)	SSA-BP	47.49	77.03	130.75	62.67	97.79	159.54	0.53	0.86	1.49
	SSA-ELM	46.43	73.69	153.61	59.45	90.02	197.51	0.51	0.82	1.74
	SSA-WNN	58.74	125.89	258.21	75.26	163.72	324.61	0.66	1.43	2.94
	SSA-ARIMA	90.62	95.95	105.31	127.67	128.47	130.37	0.99	1.04	1.16
	Proposed Model	44.29	57.83	77.74	57.63	73.87	97.92	0.48	0.64	0.86

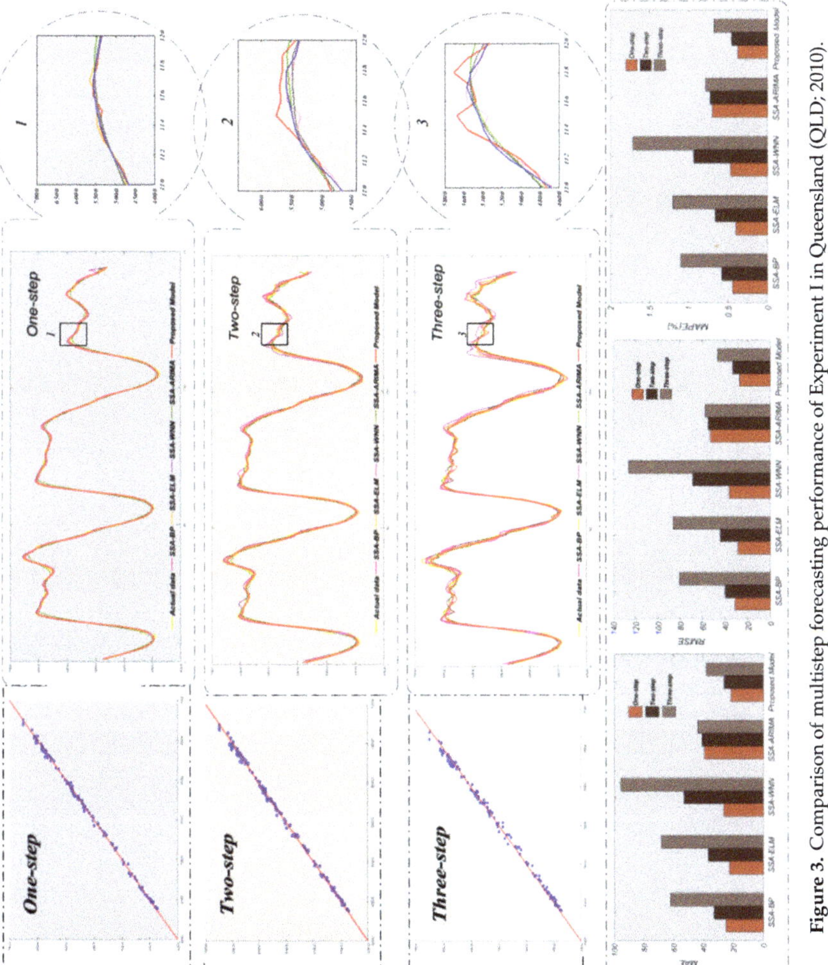

Figure 3. Comparison of multistep forecasting performance of Experiment I in Queensland (QLD; 2010).

4.4.2. Experiment II: Comparing Models using Other Data Preprocessing Methods

In order to verify whether singular spectrum analysis (SSA) is the best choice for data processing, we conducted an experiment comparing the electric load time series forecasting results of our proposed model with other data processing method–based models. The experimental results are shown in Table 4. Detailed descriptions are as follows:

- Observing the experimental results using the 2010 Queensland power data, the proposed combined model achieved the highest forecasting accuracy. In contrast, the CEEMD preprocessed model was the most effective among the other three data processing methods, with MAPE values of 0.54%, 0.64%, and 0.90% from one-step to three-step forecasting, respectively. For the proposed model, MAPE values were 0.37%, 0.45%, and 0.68% from one to three steps, respectively.

- For the experiment using the 2011 Queensland power data, according to the evaluation criteria, the proposed model outperformed the other models. The MAPE values of the models using EMD, EEMD, and CEEMD were, respectively, 0.33%, 0.21%, and 0.18% higher than those of the proposed model in one-step forecasting. Figure 4 shows a comparison of the one- to three-step forecasting performance of Experiment II. It can be concluded that the proposed combined model achieved the highest accuracy compared to the models using other data preprocessing methods in three-step forecasting.

- When using the 2011 New South Wales power data, similar to Experiment I, compared to the first two experiments using data from Queensland, the error metric values of the third experiment were significantly larger. This reflects the difference between different power plants. In addition, there were also some interesting conclusions. For example, in the first two sets of power plant data, the CEEMD model performed better than the EMD model, but in the third group, the EMD and CEEMD models performed almost the same. However, our model still the performed the best. We can also determine the necessity of applying singular spectrum analysis (SSA) in our model so that it performs better than the other three classic data processing methods.

N.B. In experiment II, by comparing the forecasting results of our proposed combined model with other models using different data processing methods, there are many useful findings. Our model's overall lead in predicting accuracy demonstrates that SSA is the best choice of data processing method.

Table 4. Comparison of forecasting performance of combined model and models using different data preprocessing methods.

Dataset	Model	MAE			RMSE			MAPE (%)		
		1-step	2-step	3-step	1-step	2-step	3-step	1-step	2-step	3-step
QLD(2010)	EMD	39.86	47.91	56.38	45.93	56.44	67.71	0.72	0.88	1.05
	EEMD	33.72	42.97	55.97	50.84	56.95	70.47	0.60	0.78	0.99
	CEEMD	30.04	35.90	51.47	36.74	44.14	64.19	0.54	0.64	0.90
	Proposed Model	21.26	25.94	37.97	26.98	32.83	46.51	0.37	0.45	0.68
QLD(2011)	EMD	38.37	46.63	60.59	50.13	57.51	72.93	0.71	0.86	1.12
	EEMD	31.56	55.11	84.83	38.57	73.23	107.65	0.59	1.00	1.56
	CEEMD	30.48	32.89	53.04	43.11	44.62	66.12	0.56	0.61	0.97
	Proposed Model	20.79	23.43	34.84	27.75	30.73	44.80	0.38	0.43	0.65
NSW(2011)	EMD	60.95	74.46	110.60	83.78	95.41	135.10	0.66	0.82	1.24
	EEMD	65.29	112.21	166.16	80.19	140.71	213.64	0.73	1.25	1.81
	CEEMD	62.29	82.48	100.39	82.51	100.77	125.26	0.67	0.92	1.12
	Proposed Model	44.29	57.83	77.74	57.63	73.87	97.92	0.48	0.64	0.86

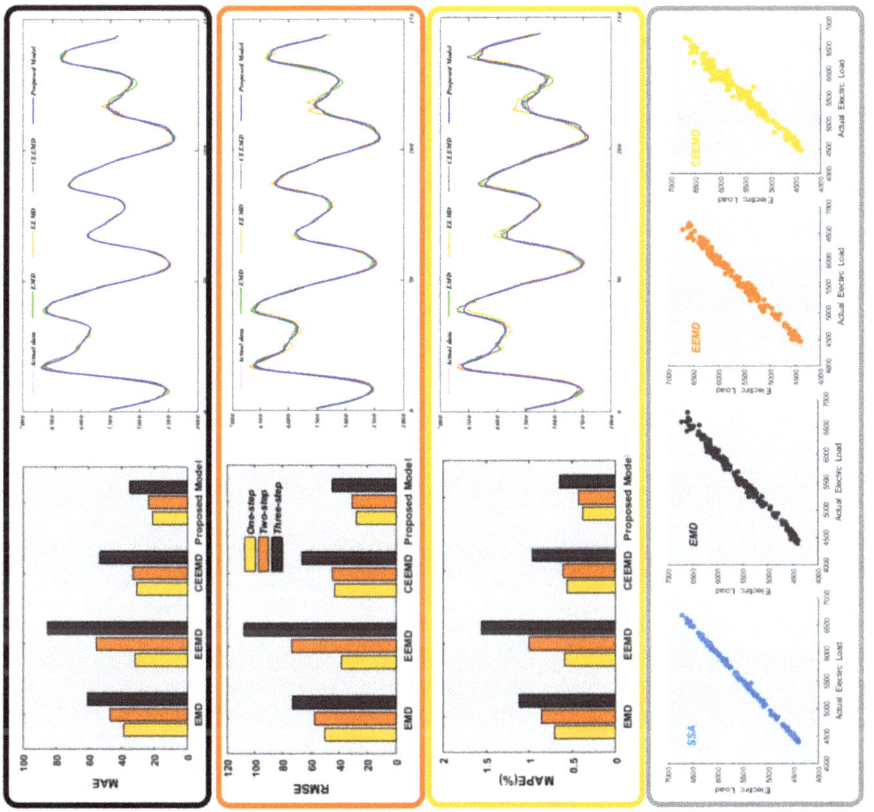

Figure 4. Comparison of multi-step forecasting performance of Experiment II for QLD (2011).

4.4.3. Experiment III: Comparing with Classic Models

In Experiment III we took the forecasting results of our proposed model and the artificial intelligence model to compare the BP, WNN, ELM, and ARIMA. In order to make the experiment more complete and persuasive, we also compared it with some classic conventional models. The experimental results are shown in Table 5. Figure 5 shows a comparison of the one- to three-step forecasting performance of Experiment III. Detailed descriptions are as follows:

- With the experiments using the 2010 Queensland power data, we found that, first, our proposed combined model had the best prediction performance whether the statistical indicator was MAE, RMSE, or MAPE. For instance, taking the one-step forecasting MAE values for comparison, the values were 46.12, 47.54, 51.24, 45.76, 46.24, 64.08, 38.971, and 21.26. The proposed model's MAE value was only about half of other methods' values. Second, in two- and three-step forecasting, the combined model was more effective than the other methods. The prediction performance of all other models was significantly worse than that of our model and there was still a big gap, which was sufficient to reflect the excellence of our model.
- With the 2011 Queensland power data, the results were as follows: First, our proposed model was still the most accurate. Second, although the data were from a different year, it is clear that forecasting results of the first two experiments are fairly similar, which reflects the stability of our method. The RMSE values of the proposed model were 27.75, 30.73, and 44.80 for one to three steps, respectively.
- For the 2011 New South Wales power data, the RMSE values of the proposed model were 57.63, 73.87, and 97.92 for one to three steps, respectively. The great thing is that our proposed combined model still outperformed the other data processing methods in one- to three-step forecasting. This is powerful proof that our model is indeed the best of all eight models. Although not as accurate as the predictions in the first two experiments, the degree of improvement in the prediction results did not change much at around 50%. This will be discussed in the next section.

Table 5. Comparison of forecasting performance of combined model and some classic individual models.

Dataset	Model	MAE			RMSE			MAPE (%)		
		1-step	2-step	3-step	1-step	2-step	3-step	1-step	2-step	3-step
QLD(2010)	BP	46.12	80.44	119.79	59.34	102.45	150.21	0.81	1.40	2.10
	BP-MODA	47.54	86.23	121.67	59.98	108.89	162.77	0.84	1.52	2.13
	WNN	51.24	99.72	145.12	64.72	136.56	189.77	0.90	1.75	2.58
	ENN	45.76	82.88	126.34	60.95	104.71	168.20	0.80	1.46	2.24
	ELM	46.24	85.19	130.82	59.28	110.41	171.58	0.81	1.50	2.34
	RBF	64.08	134.02	185.37	86.13	185.85	275.07	1.12	2.33	3.22
	ARIMA	38.97	73.96	88.00	49.44	91.95	105.91	0.70	1.32	1.58
	Proposed Model	21.26	25.94	37.97	26.98	32.83	46.51	0.37	0.45	0.68
QLD(2011)	BP	45.61	93.75	147.05	64.12	127.47	217.66	0.82	1.71	2.66
	BP-MODA	44.23	85.21	124.31	61.31	116.22	163.60	0.79	1.56	2.28
	WNN	62.72	138.62	200.49	80.68	181.26	268.88	1.16	2.55	3.75
	ENN	50.62	98.11	152.69	70.11	135.34	207.65	0.92	1.80	2.79
	ELM	48.77	96.45	158.82	66.91	133.68	216.32	0.89	1.76	2.91
	RBF	85.22	170.73	308.71	198.05	524.76	971.50	1.53	3.10	5.64
	ARIMA	37.50	68.01	87.55	46.94	87.81	107.32	0.71	1.28	1.64
	Proposed Model	20.79	23.43	34.84	27.75	30.73	44.80	0.38	0.43	0.65
NSW(2011)	BP	89.72	163.56	276.83	124.34	215.84	349.76	0.96	1.79	3.05
	BP-MODA	85.23	180.48	268.43	110.79	251.79	360.47	0.92	1.98	2.94
	WNN	92.97	243.25	400.39	118.54	323.58	538.95	1.02	2.71	4.52
	ENN	101.07	191.08	282.89	133.22	265.73	362.33	1.09	2.08	3.13
	ELM	98.76	205.09	317.34	130.92	274.27	410.74	1.06	2.24	3.53
	RBF	149.92	216.50	351.10	280.01	318.79	449.90	1.60	2.36	3.84
	ARIMA	78.00	130.15	159.76	95.68	161.82	203.99	0.88	1.46	1.80
	Proposed Model	44.29	57.83	77.74	57.63	73.87	97.92	0.48	0.64	0.86

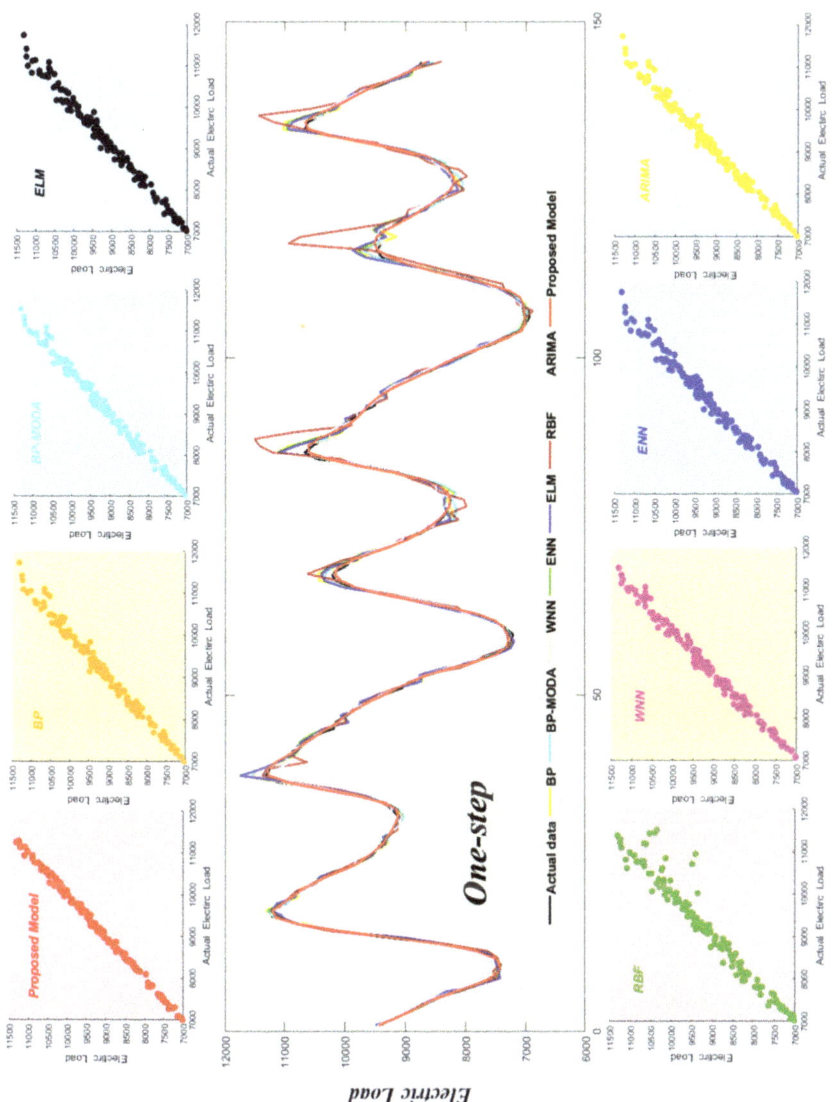

Figure 5. Comparison of one-step forecasting performance of Experiment III for New South Wales (NSW; 2011).

N.B. In experiment III, by observing the results including our proposed method, models that we applied in our model, and traditional models we did not use, we found that our model had an overall lead in predicting accuracy. This demonstrates that our proposed combined model will save a lot of energy for power systems and should be applied in actual electric load forecasting practice.

5. Discussion

This section provides an in-depth discussion of the proposed model, including a proposed optimization algorithm performance test, two proposed model effectiveness tests, and an experiment showing the improvements of the proposed model and a comparative experiment.

5.1. Multiobjective Grasshopper Algorithm Experiments

Four typical test functions (shown in Table 6) were applied to examine the excellence of the proposed algorithm. We chose multiobjective ant lion optimization (MOALO) and the multiobjective dragonfly algorithm (MODA) to compare with MOGOA to examine its optimization performance. To control variables, we set the maximum iterations and search agents as 100 and the size of archive as 150. We applied inverted generational distance (IGD), which is a metric showing the evaluation degree of multiobjective optimization algorithms. Table 7 shows the test results of IGD, for which we did 60 experiments for every test function [61]. Moreover, Figure 6 shows the obtained Pareto optimal solutions by these three algorithms.

Table 6. Test functions of algorithms.

ZDT1	ZDT2
Minimize: $f_1(x) = x_1$ Minimize: $f_2(x) = g(x) \times h(f_1(x), g(x))$ Where: $g(x) = 1 + \frac{9}{N-1}\sum_{i=2}^{N} x_i$ $h(f_1(x), g(x)) = 1 - \sqrt{\frac{f_1(x)}{g(x)}}$ $0 \leq x_1 \leq 1,\ 1 \leq i \leq 30$	Minimize: $f_1(x) = x_1$ Minimize: $f_2(x) = g(x) \times h(f_1(x), g(x))$ Where: $g(x) = 1 + \frac{9}{N-1}\sum_{i=2}^{N} x_i$ $h(f_1(x), g(x)) = 1 - \left(\frac{f_1(x)}{g(x)}\right)^2$ $0 \leq x_1 \leq 1,\ 1 \leq i \leq 30$
ZDT3	**ZDT1 with linear front**
Minimize: $f_1(x) = x_1$ Minimize: $f_2(x) = g(x) \times h(f_1(x), g(x))$ Where: $g(x) = 1 + \frac{9}{N-1}\sum_{i=2}^{N} x_i$ $h(f_1(x), g(x)) = 1 - \sqrt{\frac{f_1(x)}{g(x)}}$ $-\left(\frac{f_1(x)}{g(x)}\right)\sin(10\pi f_1(x))$ $0 \leq x_1 \leq 1,\ 1 \leq i \leq 30$	Minimize: $f_1(x) = x_1$ Minimize: $f_2(x) = g(x) \times h(f_1(x), g(x))$ Where: $g(x) = 1 + \frac{9}{N-1}\sum_{i=2}^{N} x_i$ $h(f_1(x), g(x)) = 1 - \frac{f_1(x)}{g(x)}$ $0 \leq x_1 \leq 1,\ 1 \leq i \leq 30$

From the experimental results we can see the following:

(a) MOGOA gets the best IGD values among the optimization algorithms in all four test functions, which proves that its optimizing ability is superior to that of MODA and MOALO.

(b) By observing the contrast of the number of the Pareto optimal solutions calculated by MOGOA, MODA, and MOALO shown in Figure 6, we find that MOGOA had the most Pareto optimal solutions among all three algorithms.

N.B. The optimization ability of MOGOA is proven to be good through the experiment and discussion above. Therefore, MOGOA can be widely applied to deal with multiobjective optimization problems.

Table 7. Results of multiobjective algorithms using inverted generational distance (IGD) on four test functions.

Algorithm	Ave.	Std.	Median	Min	Max
		ZDT1			
MOALO	0.006213	0.007038	0.005901	0.004272	0.024323
MODA	0.005826	0.005798	0.005082	0.002613	0.025404
MOGOA	0.004275	0.003089	0.004669	0.002573	0.024234
		ZDT2			
MOALO	0.009454	0.007343	0.008998	0.004738	0.022138
MODA	0.008173	0.005193	0.008532	0.003643	0.023234
MOGOA	**0.008015**	**0.003140**	**0.005395**	**0.002157**	**0.023118**
		ZDT3			
MOALO	0.027063	0.000867	0.026627	0.028135	0.026727
MODA	0.025089	0.000521	0.024982	**0.028182**	0.027322
MOGOA	**0.024270**	**0.000469**	**0.024246**	0.024186	**0.023801**
		ZDT1 with linear front			
MOALO	0.006821	0.005623	0.006532	0.005431	0.026626
MODA	0.006101	0.005541	0.005926	0.003863	0.024777
MOGOA	**0.005569**	**0.004986**	**0.003985**	**0.002211**	**0.024461**

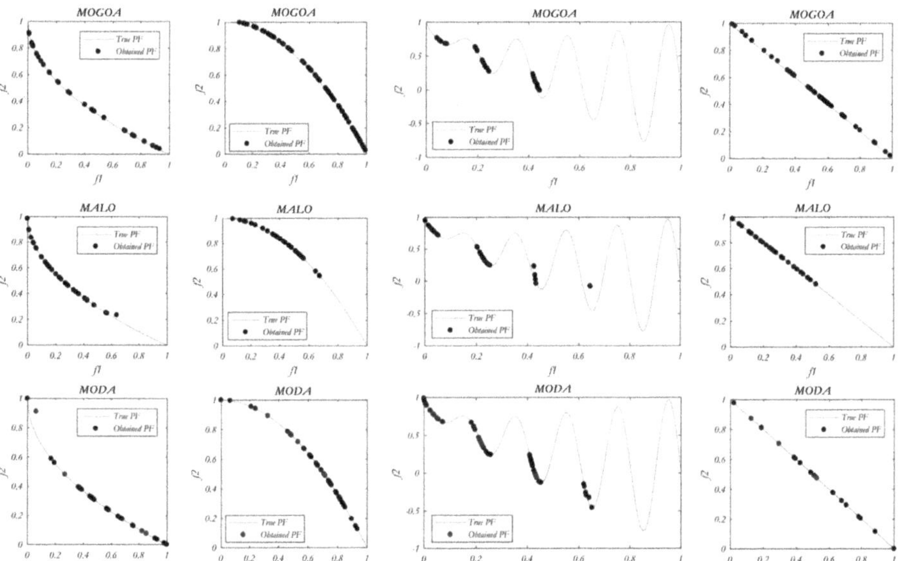

Figure 6. Pareto optimal solutions obtained by optimization algorithm for test functions.

5.2. Proposed Model's Effectiveness

The Diebold–Mariano test was used to verify the validity of the developed model, which means every model mentioned above was compared to the SSA-MOGOA combined model. The DM test is a kind of hypothetical test. The null hypothesis is that there is no significant difference in the models' forecasting performance. The opposite hypothesis is that there is a significant different in the models' forecasting performance. Table 8 shows average DM test values of all experiments for one- to three-step forecasting.

Table 8. DM test of different models.

Model	1-step	2-step	3-step
SSA-BP	2.7503 *	3.8971 *	6.1244 *
SSA-ELM	1.6379 **	3.9104 *	6.3244 *
SSA-WNN	4.0126 *	6.8486 *	7.3544 *
SSA-ARIMA	4.9261 *	5.0164 *	4.0033 *
EMD	5.4365 *	5.5545 *	5.7057 *
EEMD	4.3034 *	5.0669 *	5.21 *
CEEMD	3.7225 *	3.8063 *	4.3806 *
BP	4.7348 *	5.9805 *	6.3855 *
BP-MODA	5.2960 *	5.8782 *	6.2118 *
WNN	6.3481 *	6.2092 *	7.1581 *
ENN	5.3966 *	6.1685 *	6.3538 *
ELM	5.4820 *	5.7369 *	6.3290 *
RBF	3.3792 *	3.7372 *	3.5957 *
ARIMA	5.6641 *	7.0336 *	7.5187 *

* 1% significance level; ** 5% significance level.

Table 8 shows that except for the one-step SSA-ELM experiment, the DM value in all the other experiments is big enough to be rejected at the 1% significance level, while the null hypothesis of one-step SSA-ELM is rejected at 5%. Moreover, for the DM test of individual models, the value is 3.3792, which shows that the accuracy of the proposed model is fairly high.

To further evaluate our model, as introduced in Section 3.3, we also applied the forecasting effectiveness method in our testing experiments. Forecasting effectiveness can effectively reflect the accuracy of the forecasting performance of various models, making it easy to comparing their pros and cons. In Table 9, we record the detailed forecasting effectiveness values of all models in one- to three-step forecasting.

Table 9. Forecasting effectiveness of different models.

Model	1-step		2-step		3-step	
	1-order	2-order	1-order	2-order	1-order	2-order
Proposed Model	0.9959	0.9962	0.9949	0.9957	0.9927	0.9935
SSA-BP	0.9951	0.9950	0.9931	0.9935	0.9870	0.9869
SSA-ELM	0.9955	0.9956	0.9929	0.9935	0.9846	0.9831
SSA-WNN	0.9940	0.9932	0.9870	0.9847	0.9740	0.9684
SSA-ARIMA	0.9918	0.9923	0.9914	0.9919	0.9905	0.9910
EMD	0.9930	0.9929	0.9915	0.9914	0.9887	0.9888
EEMD	0.9936	0.9941	0.9899	0.9900	0.9855	0.9844
CEEMD	0.9941	0.9944	0.9928	0.9939	0.9900	0.9903
BP	0.9914	0.9918	0.9837	0.9829	0.9740	0.9734
BP-MODA	0.9915	0.9921	0.9831	0.9844	0.9755	0.9772
WNN	0.9897	0.9884	0.9766	0.9745	0.9638	0.9625
ENN	0.9906	0.9908	0.9822	0.9820	0.9728	0.9721
ELM	0.9908	0.9911	0.9817	0.9824	0.9707	0.9709
RBF	0.9858	0.9847	0.9740	0.9690	0.9589	0.9473
ARIMA	0.9924	0.9929	0.9865	0.9872	0.9833	0.9836

The forecasting effectiveness results in Table 9 show the following results: First, the most obvious is that our proposed combined model has the best prediction performance with the highest forecasting effectiveness values in all forecasting. Second, the prediction performance of other individual models is significantly worse than that of our model and there is still a big gap between them and our proposed model, which is sufficient to reflect the excellence of our model.

5.3. Proposed Combined Model's Improvements

In order to make the traditional MAPE criteria more clear in comparing the pros and cons of the models, in this paper we propose a new form of MAPE, defined as:

$$P_{MAPE} = \left| \frac{MAPE_1 - MAPE_2}{MAPE_1} \right| \tag{17}$$

This new MAPE criterion is used to compare the proposed model with the other models in the above experiments, including three data denoising algorithms, seven classic models, and four individual models with singular spectrum analysis. Table 10 shows the experimental results, and some interesting conclusions can be summarized as follows:

- Comparing the proposed model with other SSA-based models, it is obvious that the novel proposed model has lower MAPE values. For example, the average improvement of the proposed model's MAPE is 7.71%, 29.18%, and 51.88% compared with the SSA-ELM model, which is the least improved of the four models.
- Comparing the proposed model with the other three data preprocessing methods, its superiority is obvious. The lowest MAPE improvement is 22.09%, while the largest comes to 56.84%, which fully reflects the excellent prediction accuracy of our proposed model.
- Comparing the proposed model with the classic models, forecasting accuracy is greatly improved in every experiment. Compared with the ARIMA model, the proposed model improves by 45.75%, 62.68%, and 56.47% while the ARIMA model was the single model with the best prediction accuracy in the experiment.

Table 10. Percentage improvement of the proposed model.

Model	Site 1			Site 2			Site 3			Average		
	1-step	2-step	3-step	1-step	2-step	3-step	1-step	2-step	3-step	1-step	2-step	3-step
SSA-BP	22.64%	33.80%	50.34%	8.66%	25.58%	42.01%	14.66%	21.23%	38.48%	15.32%	26.87%	43.61%
SSA-ELM	12.61%	33.77%	61.43%	4.96%	22.46%	50.35%	5.55%	31.32%	43.85%	7.71%	29.18%	51.88%
SSA-WNN	43.95%	71.78%	79.41%	27.19%	55.24%	70.63%	19.65%	51.45%	60.49%	30.26%	59.49%	70.18%
SSA-ARIMA	50.37%	46.41%	27.33%	51.26%	38.70%	25.26%	46.33%	37.50%	13.76%	49.32%	40.87%	22.12%
EMD	45.58%	49.99%	41.74%	27.16%	22.09%	30.29%	48.34%	48.42%	35.46%	40.36%	40.17%	35.83%
EEMD	35.39%	56.84%	58.19%	34.38%	48.75%	52.29%	37.92%	41.76%	32.02%	35.90%	49.12%	47.50%
CEEMD	31.79%	28.77%	32.63%	28.47%	30.30%	23.02%	30.47%	29.66%	25.22%	30.24%	29.58%	26.96%
BP	53.29%	74.77%	75.57%	49.74%	64.28%	71.63%	53.78%	67.55%	67.77%	52.27%	68.87%	71.66%
BP-MODA	51.66%	72.40%	71.48%	47.87%	67.72%	70.57%	55.26%	70.17%	68.35%	51.60%	70.10%	70.13%
WNN	66.78%	83.08%	82.65%	52.83%	76.47%	80.86%	58.62%	74.15%	73.79%	59.41%	77.90%	79.10%
ENN	58.38%	76.02%	76.67%	55.70%	69.29%	72.41%	53.43%	68.92%	69.90%	55.84%	71.41%	72.99%
ELM	56.65%	75.48%	77.67%	54.65%	71.56%	75.49%	53.92%	69.74%	71.14%	55.07%	72.26%	74.77%
RBF	74.90%	86.07%	88.46%	69.96%	72.97%	77.51%	66.48%	80.54%	79.01%	70.45%	79.86%	81.66%
ARIMA	45.65%	66.20%	60.38%	45.25%	56.24%	51.93%	46.34%	65.59%	57.11%	45.75%	62.68%	56.47%

5.4. Combined Strategy

We selected and applied a simple averaging strategy to calculate the prediction results of all individual models to compare with the results of MOGOA optimization to test the effectiveness of the proposed combination strategy. The results of the comparison between the two methods are shown in Table 11.

From Table 11, we can easily find from the prediction results that the proposed combined model using MOGOA always outperformed the model applying a simple average strategy, no matter which sites and forecasting steps were used in all three error metrics. For instance, in the three-step forecasting of NSW (2011), the MAPE of the proposed model is 0.8648% while the corresponding MAPE is 1.8336%, which shows the excellence of the model's combined strategy.

Table 11. Comparison between proposed model and simple average strategy.

Dateset	Multi-Step	Model	MAE	RMSE	MAPE (%)
QLD(2010)	1-step	Simple average strategy	28.13	37.04	0.50
		Proposed model	21.26	26.98	0.37
	2-step	Simple average strategy	40.69	51.88	0.72
		Proposed model	25.94	32.83	0.45
	3-step	Simple average strategy	67.30	87.69	1.20
		Proposed model	37.97	46.51	0.68
QLD(2011)	1-step	Simple average strategy	32.06	42.19	0.60
		Proposed model	20.79	27.75	0.38
	2-step	Simple average strategy	48.93	62.31	0.91
		Proposed model	23.43	30.73	0.43
	3-step	Simple average strategy	95.10	120.89	1.76
		Proposed model	34.84	44.80	0.65
NSW(2011)	1-step	Simple average strategy	60.82	81.26	0.67
		Proposed model	44.29	57.63	0.48
	2-step	Simple average strategy	93.14	120.00	1.04
		Proposed model	57.83	73.87	0.64
	3-step	Simple average strategy	161.97	203.01	1.83
		Proposed model	77.74	97.92	0.86

6. Conclusions

As an indispensable part of the economic operation of power systems, electric load prediction has developed a lot in the past few years. Many studies have been developed and have contributed to improving forecasting accuracy. Establishing a model with perfect forecasting performance and strong stability can provide huge economic and social benefits. At the same time, it can help managers to develop blueprints for future power system construction to ensure the reliability and efficiency of the power supply. As a result, developing a new, robust model with high forecasting accuracy means a lot to the whole world. However, classic and individual models do not always produce satisfactory results. A combined model using data preprocessing technology, a combination of four individual models optimized by an intelligence algorithm called the multiobjective grasshopper optimization algorithm, and the multistep forecasting strategy was used for electric load forecasting in our study. Specifically, the technique of singular spectrum analysis, based on decomposition and reconstruction, was employed to get basic features of the time series by removing high-frequency signals. Moreover, the weight coefficients of individual models in the combined model were optimized by the latest advanced optimization algorithms to obtain both high precision and strong stability. With regard to the individual models in the combined model, the ARIMA model was selected to reflect the linearity of the sequence and artificial intelligence models were selected to reflect the nonlinearity. Furthermore, the combined model was employed in multistep forecasting to validate its forecasting performance. The experimental results show that the new combined model performed significantly better than the other benchmark models on the basis of multiple comparisons and analysis. Additionally, by comparing the outcomes of DM and forecasting effectiveness tests, we found that our model performed best among all the models applied in the experiments. The proposed combined model, with its brilliant prediction performance, can yield tremendous economic benefits and lead to a dramatic reduction in the consumption of environmental resources. Apart from that, it is certain that wide application of this model will contribute to the management of power systems, rational electric dispatching, and electric power scheduling. In conclusion, our proposed combined model can improve the performance of electric load time series forecasting and provide a new feasible choice for smart power distribution planning.

Author Contributions: Conceptualization, Y.Z. and J.W.; Methodology, J.W.; Software, Y.Z.; Validation, J.W., H.L.; Formal Analysis, Y.Z. and J.W.; Investigation, H.L.; Resources, Y.Z.; Data Curation, H.L.; Writing-Original Draft Preparation, Y.Z.; Writing-Review & Editing, Y.Z. and J.W.; Visualization, Y.Z. and J.W.; Supervision, J.W.; Project Administration, Y.Z.; Funding Acquisition, J.W. and H.L.

Funding: This work was funded by the National Natural Science Foundation of China (grant number 71671029).

Conflicts of Interest: The authors declare that there is no conflict of interest regarding the publication of this paper.

References

1. Song, J.; Wang, J.; Lu, H. A novel combined model based on advanced optimization algorithm for short-term wind speed forecasting. *Appl. Energy* **2018**, *215*, 643–658. [CrossRef]
2. Koprinska, I.; Rana, M.; Agelidis, V.G. Correlation and instance based feature selection for electricity load forecasting. *Knowl.-Based Syst.* **2015**, *82*, 29–40. [CrossRef]
3. Kaytez, F.; Taplamacioglu, M.C.; Cam, E.; Hardalac, F. Forecasting electricity consumption: A comparison of regression analysis, neural networks and least squares support vector machines. *Int. J. Electr. Power* **2015**, *67*, 431–438. [CrossRef]
4. Ou, T.C.; Hong, C.M. Dynamic operation and control of microgrid hybrid power systems. *Energy* **2014**, *66*, 314–323. [CrossRef]
5. Raviv, E.; Bouwman, K.E.; Dijk, D.V. Forecasting day-ahead electricity prices: Utilizing hourly prices. *Energy Econ.* **2015**, *50*, 227–239. [CrossRef]
6. Wang, J.; Du, P.; Niu, T.; Yang, W. A novel hybrid system based on a new proposed algorithm—Multi-Objective Whale Optimization Algorithm for wind speed forecasting. *Appl. Energy* **2017**. [CrossRef]
7. Takeda, H.; Tamura, Y.; Sato, S. Using the ensemble Kalman filter for electricity load forecasting and analysis. *Energy* **2016**, *104*, 184–198. [CrossRef]
8. Lei, M.; Shiyan, L.; Chuanwen, J.; Hongling, L.; Yan, Z. A review on the forecasting of wind speed and generated power. *Renew. Sustain. Energy Rev.* **2009**, *13*, 915–920. [CrossRef]
9. Zhao, J.; Guo, Z.H.; Su, Z.Y.; Zhao, Z.Y.; Xiao, X.; Liu, F. An improved multi-step forecasting model based on WRF ensembles and creative fuzzy systems for wind speed. *Appl. Energy* **2016**, *162*, 808–826. [CrossRef]
10. Cheng, W.Y.Y.; Liu, Y.; Bourgeois, A.J.; Wu, Y.; Haupt, S.E. Short-term wind forecast of a data assimilation/weather forecasting system with wind turbine anemometer measurement assimilation. *Renew. Energy* **2017**, *107*, 340–351. [CrossRef]
11. Landberg, L. Short-term prediction of local wind conditions. *J. Wind Eng. Ind. Aerodyn.* **2001**, *89*, 235–245. [CrossRef]
12. Negnevitsky, M.; Johnson, P.; Santoso, S. Short term wind power forecasting using hybrid intelligent systems. In Proceedings of the 2007 IEEE Power Engineering Society General Meeting, Tampa, FL, USA, 24–28 June 2007; pp. 1–4. [CrossRef]
13. Zhang, C.; Zhou, J.; Li, C.; Fu, W.; Peng, T. A compound structure of ELM based on feature selection and parameter optimization using hybrid backtracking search algorithm for wind speed forecasting. *Energy Convers. Manag.* **2017**, *143*, 360–376. [CrossRef]
14. Tascikaraoglu, A.; Sanandaji, B.M.; Poolla, K.; Varaiya, P. Exploiting sparsity of interconnections in spatio-temporal wind speed forecasting using Wavelet Transform. *Appl. Energy* **2016**, *165*, 735–747. [CrossRef]
15. Jung, J.; Broadwater, R.P. Current status and future advances for wind speed and power forecasting. *Renew. Sustain. Energy Rev.* **2014**, *31*, 762–777. [CrossRef]
16. Babu, C.N.; Reddy, B.E. A moving-average filter based hybrid ARIMA-ANN model for forecasting time series data. *Appl. Soft Comput.* **2014**, *23*, 27–38. [CrossRef]
17. Niu, X.; Wang, J. A combined model based on data preprocessing strategy and multi-objective optimization algorithm for short-term wind speed forecasting. *Appl. Energy* **2019**, *241*, 519–539. [CrossRef]
18. Soman, S.S.; Zareipour, H.; Malik, O.; Mandal, P. A review of wind power and wind speed forecasting methods with different time horizons. *N. Am. Power Symp.* **2010**, 1–8. [CrossRef]
19. Lee, C.M.; Ko, C.N. Short-term load forecasting using lifting scheme and ARIMA models. *Expert Syst. Appl.* **2011**, *38*, 5902–5911. [CrossRef]

20. Wang, Y.Y.; Wang, J.Z.; Zhao, G.; Dong, Y. Application of residual modification approach in seasonal ARIMA for electricity demand forecasting: A case study of China. *Energy Policy* **2012**, *48*, 284–294. [CrossRef]

21. Brożyna, J.; Mentel, G.; Szetela, B.; Strielkowski, W. Multi-Seasonality in the TBATS Model Using Demand for Electric Energy as a Case Study. *Econ. Comput. Econ. Cybern. Stud. Res.* **2018**, *52*, 229–246. [CrossRef]

22. Ahmad, A.S.; Hassan, M.Y.; Abdullah, M.P.; Rahman, H.A.; Hussion, F.; Abdullah, H.; Saidur, R. A review on applications of ANN and SVM for building electrical energy consumption forecasting. *Renew. Sust. Energy Rev.* **2014**, *33*, 102–109. [CrossRef]

23. Zhao, X.; Wang, C.; Su, J.; Wang, J. Research and application based on the swarm intelligence algorithm and artificial intelligence for wind farm decision system. *Renew. Energy* **2019**, *134*, 681–697. [CrossRef]

24. Sadaei, H.J.; de Lima e Silva, P.C.; Guimarães, F.G.; Lee, M.H. Short-term load forecasting by using a combined method of convolutional neural networks and fuzzy time series. *Energy* **2019**. [CrossRef]

25. Park, D.C.; Sharkawi, M.A.; Marks, R.J. Electric load forecasting using a neural network. *IEEE Trans. Power Syst.* **1991**, *6*, 442–449. [CrossRef]

26. Wang, J.; Yang, W.; Du, P.; Li, Y. Research and application of a hybrid forecasting framework based on multi-objective optimization for electrical power system. *Energy* **2018**, *148*, 59–78. [CrossRef]

27. Lou, C.W.; Dong, M.C. A novel random fuzzy neural networks for tackling uncertainties of electric load forecasting. *Int. J. Electr. Power Energy Syst.* **2015**, *73*, 34–44. [CrossRef]

28. Okumus, I.; Dinler, A. Current status of wind energy forecasting and a hybrid method for hourly predictions. *Energy Convers. Manag.* **2016**, *123*, 362–371. [CrossRef]

29. Hong, C.M.; Ou, T.C.; Lu, K.H. Development of intelligent MPPT (maximum power point tracking) control for a grid-connected hybrid power generation system. *Energy* **2013**, *50*, 270–279. [CrossRef]

30. Che, J.; Wang, J. Short-term electricity prices forecasting based on support vector regression and Auto-regressive integrated moving average modeling. *Energy Convers. Manag.* **2010**, *51*, 1911–1917. [CrossRef]

31. Liu, H.; Tian, H.Q.; Liang, X.F.; Li, Y.F. New wind speed forecasting approaches using fast ensemble empirical model decomposition, genetic algorithm, mind evolutionary algorithm and artificial neural networks. *Renew. Energy* **2015**, *83*, 1066–1075. [CrossRef]

32. Wu, C.; Wang, J.; Chen, X.; Du, P.; Yang, W. A Novel Hybrid System Based on Multi-objective Optimization for Wind Speed Forecasting. *Renew. Energy* **2019**. [CrossRef]

33. Bates, J.M.; Granger, C.W.J. The combination of forecasts. *Oper. Res. Q* **1969**, *20*, 451–468. [CrossRef]

34. Diebold, F.X. *Element of Forecasting*, 4th ed.; Thomson South-Western: Cincinnati, OH, USA, 2007; pp. 257–287.

35. Pesaran, M.H.; Pick, A.; Timmermann, A. Variable selection, estimation and inference for multi-period forecasting problems. *J. Econom.* **2011**, *164*, 173–187. [CrossRef]

36. Zhang, W.; Qu, Z.; Zhang, K.; Mao, W.; Ma, Y.; Fan, X. A combined model based on CEEMDAN and modified flower pollination algorithm for wind speed forecasting. *Energy Convers Manag.* **2017**, *136*, 439–451. [CrossRef]

37. Yang, Y.; Che, J.; Li, Y.; Zhao, Y.; Zhu, S. An incremental electric load forecasting model based on support vector regression. *Energy* **2016**, *113*, 796–808. [CrossRef]

38. Liu, H.; Tian, H.Q.; Li, Y.F. Comparison of two new ARIMA-ANN and ARIMA-Kalman hybrid methods for wind speed prediction. *Appl. Energy* **2012**, *98*, 415–424. [CrossRef]

39. Cardenas-Barrera, J.L.; Meng, J.; Castillo-Guerra, E.; Chang, L. A neural network approach to multi-step-ahead, short-term wind speed forecasting. In Proceedings of the IEEE 2013 12th International Conference on Machine Learning and Applications, Miami, FL, USA, 4–7 December 2013; Volume 2, pp. 243–248.

40. Wang, J.; Gao, Y.; Chen, X. A novel hybrid interval prediction approach based on modified lower upper bound estimation in combination with multi-objective salp swarm algorithm for short-term load forecasting. *Energies* **2018**, *11*, 1561. [CrossRef]

41. Barbounis, T.G.; Theocharis, J.B. A locally recurrent fuzzy neural network with application to the wind speed prediction using spatial correlation. *Neurocomputing* **2007**, *70*, 1525–1542. [CrossRef]

42. Yang, D.; Sharma, V.; Ye, Z.; Lim, L.I.; Zhao, L.; Aryaputera, A.W. Forecasting of global horizontal irradiance by exponential smoothing, using decompositions. *Energy* **2015**, *81*, 111–119. [CrossRef]

43. Li, R.; Jin, Y. A wind speed interval prediction system based on multi-objective optimization for machine learning method. *Appl. Energy* **2018**, *228*, 2207–2220. [CrossRef]

44. Patterson, D.W. *Artificial Neural Networks: Theory and Applications*; Prentice Hall PTR: Upper Saddle River, NJ, USA, 1998.

45. Vel_azquez, S.; Carta, J.A.; Matías, J.M. Influence of the input layer signals of ANNs on wind power estimation for a target site: A case study. *Renew. Sustain. Energy Rev.* **2011**, *15*, 1556–1566. [CrossRef]

46. Koo, J.; Han, G.D.; Choi, H.J.; Shim, J.H.; Lund, H.; Kaiser, M.J. Wind-speed prediction and analysis based on geological and distance variables using an artificial neural network: A case study in South Korea. *Energy* **2015**, *93*, 1296–1302. [CrossRef]

47. Wang, J.; Hu, J. A robust combination approach for short-term wind speed forecasting and analysisecombination of the ARIMA (autoregressive integrated moving average), ELM (extreme learning machine), SVM (support vector machine) and LSSVM (least square SVM) forecasts using a GPR (gaussian process regression) model. *Energy* **2015**, *93*, 41–56.

48. Xiao, L.; Wang, J.; Hou, R.; Wu, J. A combined model based on data pre-analysis and weight coefficients optimization for electrical load forecasting. *Energy* **2015**, *82*, 524–549. [CrossRef]

49. Zhang, S.; Wang, J.; Guo, Z. Research on combined model based on multi-objective optimization and application in time series forecast. *Soft Comput.* **2018**. [CrossRef]

50. Wang, J.-J.; Zhang, W.-Y.; Liu, X.; Wang, C.-Y. Modifying Wind Speed Data Observed from Manual Observation System to Automatic Observation System Using Wavelet Neural Network. *Phys. Procedia* **2012**, *25*, 1980–1987. [CrossRef]

51. Li, S.; Goel, L.; Wang, P. An ensemble approach for short-term load forecasting by extreme learning machine. *Appl. Energy* **2016**, *170*, 22–29. [CrossRef]

52. Rumelhart, D.E.; Hinton, G.E.; Williams, R.J. Learning representations by back-propagating errors. *Nature* **1986**, *323*, 533–536. [CrossRef]

53. Meng, A.; Ge, J.; Yin, H.; Chen, S. Wind speed forecasting based on wavelet packet decomposition and artificial neural networks trained by crisscross optimization algorithm. *Energy Convers. Manag.* **2016**, *114*, 75–88. [CrossRef]

54. Heng, J.; Wang, C.; Zhao, X.; Xiao, L. Research and Application Based on Adaptive Boosting Strategy and Modified CGFPA Algorithm: A Case Study for Wind Speed Forecasting. *Sustainability* **2016**, *8*, 235. [CrossRef]

55. McClelland, J.L.; Rumelhart, D.E. An Interactive Activation Model of Context Effects in Letter Perception: Part I. An Account of Basic Findings. *Read. Cogn. Sci.* **1988**, 580–596. [CrossRef]

56. Mirjalili, S.Z.; Mirjalili, S.; Saremi, S.; Faris, H.; Aljarah, I. Grasshopper optimization algorithm for multi-objective optimization problems. *Appl. Intell.* **2017**, *48*, 805–820. [CrossRef]

57. Niu, M.; Sun, S.; Wu, J.; Yu, L.; Wang, J. An innovative integrated model using the singular spectrum analysis and nonlinear multi-layer perceptron network optimized by hybrid intelligent algorithm for short-term load forecasting. *Appl. Math. Model.* **2016**, *40*, 4079–4093. [CrossRef]

58. Xu, Y.; Yang, W.; Wang, J. Air quality early-warning system for cities in China. *Atmos. Environ.* **2017**, *148*, 239–257. [CrossRef]

59. Diebold, F.X.; Mariano, R. Comparing predictive accuracy. *J. Bus. Econ. Stat.* **1995**, *13*, 253–265.

60. Yang, Z.; Wang, J. A combination forecasting approach applied in multistep wind speed forecasting based on a data processing strategy and an optimized artificial intelligence algorithm. *Appl. Energy* **2018**, *230*, 1108–1125. [CrossRef]

61. Yang, X.S. *A New Metaheuristic Bat-Inspired Algorithm. Nature Inspired Cooperative Strategies for Optimization (NICSO)*; Studies in Computational Intelligence; Springer: Berlin/Heidelberg, Germany, 2010; Volume 284, pp. 65–74.

Article

Phase Space Reconstruction Algorithm and Deep Learning-Based Very Short-Term Bus Load Forecasting

Tian Shi [1], Fei Mei [2], Jixiang Lu [3], Jinjun Lu [3], Yi Pan [1], Cheng Zhou [1], Jianzhang Wu [1] and Jianyong Zheng [1,*]

[1] School of Electrical Engineering, Southeast University, Nanjing 210096, China; stx@seu.edu.cn (T.S.); 230159517@seu.edu.cn (Y.P.); 220172740@seu.edu.cn (C.Z.); 220182734@seu.edu.cn (J.W.)
[2] College of Energy and Electrical Engineering, Hohai University, Nanjing 211100, China; meifei@hhu.edu.cn
[3] State Key Laboratory of Smart Grid Protection and Control, NARI Group Corporation, Nanjing 211000, China; lujixiang@sgepri.sgcc.com.cn (J.L.); lujinjun@sgepri.sgcc.com.cn (J.L.)
* Correspondence: jy_zheng@seu.edu.cn

Received: 20 October 2019; Accepted: 13 November 2019; Published: 15 November 2019

Abstract: With the refinement and intelligence of power system optimal dispatching, the widespread adoption of advanced grid applications that consider the safety and economy of power systems, and the massive access of distributed energy resources, the requirement for bus load prediction accuracy is continuously increasing. Aiming at the volatility brought about by the large-scale access of new energy sources, the adaptability to different forecasting horizons and the time series characteristics of the load, this paper proposes a phase space reconstruction (PSR) and deep belief network (DBN)-based very short-term bus load prediction model. Cross-validation is also employed to optimize the structure of the DBN. The proposed PSR-DBN very short-term bus load forecasting model is verified by applying the real measured load data of a substation. The results prove that, when compared to other alternative models, the PSR-DBN model has higher prediction accuracy and better adaptability for different forecasting horizons in the case of high distributed power penetration and large fluctuation of bus load.

Keywords: Load forecasting; VSTLF; bus load forecasting; DBN; PSR; deep learning

---

1. Introduction

Electricity cannot be stored in large quantities, and the investment recovery cycle of large-scale energy storage equipment is long. Therefore, in order to ensure the safe operation of power systems and power quality on the user side, the operators must have knowledge of future power loads [1]. Power system load forecasting is an important method to understand the trend of future electric load. In addition, power load forecasting is of great significance for the planning of power systems and scheduling of generation and transmission maintenance. Power system load forecasting is generally divided into long-term forecasting, medium-term forecasting, short-term forecasting, and very short-term forecasting [2]. Among them, short-term load forecasting (STLF) and very short-term load forecasting (VSTLF) are of great significance for economic dispatch, optimal power flow, and electricity market trading. The higher the accuracy of load forecasting is, the more beneficial it is to improve the utilization rate of power generation equipment and the effectiveness of economic dispatch, and reduce the operation cost of smart grid.

In the past decades, experts and scholars have made systematic and effective research on traditional deterministic and probabilistic STLF and VSTLF. Deterministic forecasting methods can be divided into two main categories [3]: The first category uses statistical forecasting models, such as linear regression [4], curve extrapolation [5], Autoregressive Integrated Moving Average (ARIMA) model [6,7], and other time series methods; the second category uses artificial intelligent forecasting models, such as Bayesian estimation [8], Random Forests [9], Support Vector Regression (SVR) [10,11], Artificial Neural Network (ANN) [12,13], Deep Belief Network (DBN) [14,15], and Long Short Term Memory (LSTM) Network [16,17]. These methods have achieved high forecasting accuracy and good robustness in day-ahead and hour-ahead load forecasting. However, most of the studies are focused on system-level load forecasting and there are relatively few on bus load forecasting. Generally, bus loads can refer to loads supplied by transmission and distribution systems transformers [18]. Bus load forecasting can be conducive to the optimal scheduling of decentralized generation, network congestion studies, and others [19].

With the refinement and intelligence of power grid optimal dispatching and the wide application of advanced smart grid applications, which take into account the security and economy of power system, the demand for bus load forecasting accuracy is increasing. Since bus load base is much smaller than that of a system, the uncertainty of bus load and the multi-dimensional nonlinearity [20] are more obvious. The traditional method of distributing the predicted value of system through bus load ratio often fails to achieve satisfactory results [21]. In this regard, the literature [18] modifies the ANN model for the aggregated load of the interconnected system and proposes two novel hybrid forecasting models, which can capture and successfully treat the special characteristics of bus load patterns. In Ref. [19], a bus load forecasting model based on clustering and ANN is proposed; the day-ahead load forecasting and hour-ahead load forecasting are carried out, achieving high prediction accuracy.

With a large number of distributed power access, the uncertainty and nonlinear characteristics of system [22] and bus load are further enhanced. As the collection time interval of distributed photovoltaic power, plant data is generally several minutes. The short-term and day-ahead load forecasting in hours of bus cannot make full use of historical information and have low prediction accuracy. In order to ensure the reliable operation of power system real-time security analysis and economic dispatch, more detailed VSTLF is needed. The authors of [23] proposed a chaotic-radial basis function (RBF) photovoltaic power generation prediction model and verified its prediction accuracy under different weather conditions. However, the author only validated the prediction accuracy of the model in the case of single-step prediction and did not involve the forecasting horizon problem of the model [24]. Ref. [20] proposes a novel load forecasting model based on phase space reconstruction (PSR) algorithm and bi-square kernel (BSK) regression, and achieves high prediction accuracy on different data sets. However, after phase space reconstruction, different BSK models are used to independently predict the various dimensional data, which neglect the time series characteristics of the load.

In view of the shortcomings of the above forecasting model, considering the adaptability to different forecasting horizons, the time series characteristics of load, and the volatility brought by large-scale access of new energy sources, this paper proposes a novel very short-term bus load forecasting model based on phase space reconstruction and deep belief network (PSR-DBN). Because the amount of historical data in VSTLF is relatively large and closely related to future load trend, the impact of weather, electricity price, and other factors on VSTLF is not considered in this paper. Firstly, the proposed PSR-DBN model performs phase space reconstruction on bus load history data, and projects the historical data to the motion track of a moving point in the phase space. Then the model takes advantage of the excellent nonlinear fitting ability of the deep belief network to fit the moving point trajectory and provide a prediction of the trajectory. Finally, the predicted value of the load is obtained. At the same time, the structure of the DBN is optimized by cross-validation. In order to test the validity and superiority of the proposed PSR-DBN very short-term bus load forecasting model, this paper applies the measured load data of a substation in China to verify the forecasting effectiveness of the model under different forecasting horizons (5 min–1 h). In addition, other six alternative forecasting models are employed to further compare with the proposed PSR-DBN model.

The major contributions of this paper are as follows:

- A novel hybrid VSTLF model based on phase space reconstruction ensemble deep belief network is proposed, which can maintain high prediction accuracy in the case of high distributed power penetration and large fluctuation of bus load.
- The Levenberg-Marquardt backpropagation (LMBP) algorithm is used to fine-tune the DBN, which can make DBN convergence faster and more accurate, compared with a BP algorithm.
- A practical method based on cross-validation is proposed to tune the structure of DBN for better forecasting performance.
- The PSR algorithm is adopted to make a regular pattern that could not be obtained in one-dimensional time series appear in a high-dimensional phase space, which improves the adaptability of a forecasting model to different forecasting horizons, especially long estimation.

The rest of this paper is organized as follows. In Section 2, the relevant theory of the PSR-DBN model is introduced. Section 3 provides the principal steps of the PSR-DBN model and covers the tuning method of network hyperparameter based on cross-validation. Section 4 presents the evaluation criteria of forecasting accuracy, case study settings, forecasting results, and a comparison. Section 5 gives the conclusion of the paper and an outlook for future research.

2. Methodology

2.1. Phase Space Reconstruction (PSR)

Phase space reconstruction (PSR) is an efficient method for analyzing nonlinear time series. The basic idea of phase space reconstruction is to regard the time series as a component generated by a nonlinear dynamic system. The variation law of the component can reconstruct the equivalent high-dimensional phase space of the dynamic system, and the time series can be projected into a moving point trajectory in the high-dimensional phase space. If there is a one-dimension time series $x = \{x_1, x_2 \cdots x_N\}$, the embedding dimension is m, and the delay time is t, then the set of time series reconstructed by phase space can be expressed as:

$$\begin{bmatrix} \mathbf{X}_1 \\ \mathbf{X}_2 \\ \vdots \\ \mathbf{X}_M \end{bmatrix} = \begin{bmatrix} x_1 & x_{1+t} & \cdots & x_{1+(m-1)t} \\ x_2 & x_{2+t} & \cdots & x_{2+(m-1)t} \\ \vdots & \vdots & \cdots & \vdots \\ x_M & x_{M+t} & \cdots & x_N \end{bmatrix} \tag{1}$$

where $M = N - (m-1)t$.

The key to PSR is to determine the optimal embedding dimension m_{opt} and optimal delay t_{opt}. In this paper, the C-C method [25] is employed to determine the optimal embedding dimension m_{opt} and delay t_{opt} at the same time.

Based on Equation (1), the associated integral is defined as:

$$C(m, N, r_k, t) = \frac{2}{M(M-1)} \sum_{1 \leq i < j \leq M} \theta\left(r_k - \|\mathbf{X}_i - \mathbf{X}_j\|\right) \tag{2}$$

where $\theta(x) = \begin{cases} 0 & x < 0 \\ 1 & x \geq 0 \end{cases}$.

According to BDS (Brock-Dechert-Scheinkman) statistical conclusions [26,27], when $N > 3000$, the range of values of m and r_k can be obtained, $m \in \{2, 3, 4, 5\}$, $r_k = k \times 0.5\sigma$, where σ is the standard deviation of the time series and $k \in \{1, 2, 3, 4\}$.

Based on matrix partitioning average strategy, the test statistics S is defined as:

$$S(m, N, r_k, t) = \frac{1}{t} \sum_{i=1}^{t} C_i\left(m, \frac{N}{t}, r_k, t\right) - C_i^m\left(m, \frac{N}{t}, r_k, t\right) \tag{3}$$

For $N \to \infty$, Equation (3) can be deformed to:

$$S(m, r_k, t) = \frac{1}{t}\sum_{i=1}^{t} C_i(m, r_k, t) - C_i^m(m, r_k, t) \tag{4}$$

For the fixed m and t, $S(m, r_k, t)$ will equal for 0 for all r, if the data are iid and $N \to \infty$. However, the real data set is not infinite, and there may be a correlation between data. Thus, the optimal delay time may be either the zero crossing of $S(m, r_k, t)$ or the times at which $S(m, r_k, t)$ shows the least variation with r [25].

To represent the variation of $S(m, r_k, t)$ with r, the test statistics ΔS is defined as:

$$\Delta S(m, t) = \max\left[S\left(m, r_{k_1}, t\right)\right] - \min\left[S\left(m, r_{k_2}, t\right)\right] \tag{5}$$

where $k_1 \in \{1, 2, 3, 4\}$, $k_2 \in \{1, 2, 3, 4\}$.

The means of S and ΔS are defined as $\overline{S}$ and $\overline{\Delta S}$, and the equations are shown as:

$$\begin{cases} \overline{S}(t) = \frac{1}{4\times 4}\sum\limits_{m=2}^{5}\sum\limits_{k=1}^{4} S(m, r_k, t) \\ \overline{\Delta S}(t) = \frac{1}{4}\sum\limits_{m=2}^{5} \Delta S(m, t) \end{cases} \tag{6}$$

For all values of t, $\overline{S}(t)$ and $\overline{\Delta S}(t)$ can find corresponding values. Wherein, the t value corresponding to the first zero point of $\overline{S}(t)$ or the first minimum point of $\overline{\Delta S}(t)$ is rounded to be the optimal delay t_{opt}.

The test statistic S_{cor} is defined as:

$$S_{cor}(t) = \overline{\Delta S}(t) + \left|\overline{S}(t)\right| \tag{7}$$

where the t value corresponding to the global minimum point of $S_{cor}(t)$ is the optimal embedded window t_ω.

When the optimal delay t_{opt} is determined by Equation (6) and the optimal embedded window t_ω is determined by Equation (7), the optimal embedding dimension m_{opt} can be determined by rounding the value of Equation (8).

$$t_\omega = (m_{opt} - 1)t_{opt} \tag{8}$$

2.2. Deep Belief Network (DBN)

The Deep Belief Network (DBN) is a deep learning model proposed by Geoffrey Hinton [28] in 2006 and is a stack of multiple Restricted Boltzmann Machines (RBM). Compared with the Artificial Neural Network (ANN), DBN employs pre-training technology combined with Backpropagation (BP) algorithm to solve network parameters. Therefore, it is not easy for DBN to fall into a local optimal solution and has higher convergence accuracy. Furthermore, when the number of layers and the number of neurons in each layer are large, DBN also has a fast convergence speed, which makes it more suitable for the fitting problem of complex nonlinear time series [29].

2.2.1. Restricted Boltzmann Machine (RBM)

The RBM consists of a visible layer V and a hidden layer H. As shown in Figure 1, the visible layer consists of n_v neurons and the hidden layer consists of m_h neurons, each of which takes a value of 0 or 1 and obeys the Bernoulli distribution, i.e., $v_i \in \{0, 1\}(i = 1, 2\cdots n)$, $h_j \in \{0, 1\}(j = 1, 2\cdots m)$. There is no connection between the neurons in each layer, and the neurons between the layers are connected by weights ω.

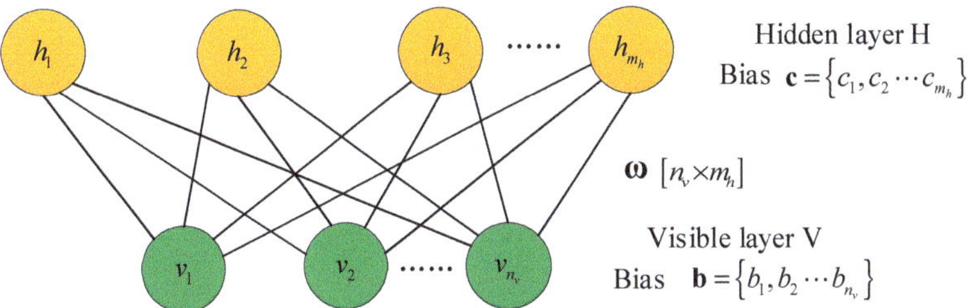

Figure 1. Structure of Restricted Boltzmann Machine (RBM).

RBM is a kind of probabilistic unsupervised learning. Its network parameters are composed of visible layer bias **b**, weight matrix **ω**, and hidden layer bias **c**, and the optimal value of network parameters is determined by the minimum energy function. The energy function is defined as:

$$E(\mathbf{v}, \mathbf{h} | \theta) = -\sum_{i=1}^{n} b_i v_i - \sum_{j=1}^{m} c_j h_j - \sum_{i=1}^{n} \sum_{j=1}^{m} v_i \omega_{ij} h_j \tag{9}$$

where ω_{ij} is the connection weight of the i-th visible layer neuron and the j-th hidden layer neuron, and $\theta = \{\mathbf{b}, \boldsymbol{\omega}, \mathbf{c}\}$.

Based on Equation (9), the joint probability distribution of the visible neuron state and the hidden neuron state is shown as:

$$P(\mathbf{v}, \mathbf{h} | \theta) = \frac{e^{-E(\mathbf{v}, \mathbf{h} | \theta)}}{Z} \tag{10}$$

where normalization factor $Z = \sum_{\mathbf{v}, \mathbf{h}} e^{-E(\mathbf{v}, \mathbf{h} | \theta)}$, which represents the sum of the energy function negative exponents under all possible values of visible layer neuron state variable **v** and hidden layer neuron state variable **h**.

The probability distribution $P(\mathbf{v})$ of **v** can be derived from Equation (10) as:

$$P(\mathbf{v} | \theta) = \sum_{\mathbf{h}} \frac{e^{-E(\mathbf{v}, \mathbf{h} | \theta)}}{Z} \tag{11}$$

Thus, the objective function of RBM training can be expressed as a likelihood function of the probability distribution of visible layer state variable **v** on the training set, and the likelihood function can be derived from the Equation (11) as:

$$L(\theta) = \sum_{\mathbf{v} \in T} \log P(\mathbf{v} | \theta) \tag{12}$$

where T represents the set of sample inputs on the training set, and when the objective function takes the maximum value, the energy function is the minimum.

According to the network structure in Figure 1, the activation probability of v_i in a given hidden layer neuron state **h** and the activation probability of h_j in a given visible layer neuron state **v** can be derived as:

$$P(v_i = 1 | \mathbf{h}) = \sigma\left(b_i + \sum_{j=1}^{m} \omega_{ij} h_j \right) \tag{13}$$

$$P(h_j = 1 | \mathbf{v}) = \sigma\left(c_i + \sum_{i=1}^{n} \omega_{ij} v_i \right) \tag{14}$$

where σ represents the sigmoid function, $\sigma(x) = \frac{1}{1+e^{-x}}$.

Since the gradient cannot be directly obtained when using stochastic gradient ascent algorithm to seek the maximum of Equation (12), the training of RBM usually applies Contrastive Divergence (CD) algorithm to approximate the gradient of likelihood function [30]. The specific steps of RBM training are as follows:

Step 1: Substitute the input of the training set as $\mathbf{v}_1$ in Equation (14) to obtain $P(\mathbf{h}_1 = 1|\mathbf{v}_1)$, then employ random sampling to acquire the reconstructed value of $\mathbf{h}_1$.

Step 2: Substitute $\mathbf{h}_1$ in Equation (13) to obtain $P(\mathbf{v}_2 = 1|\mathbf{h}_1)$ and then employ random sampling to acquire the reconstructed value of $\mathbf{v}_2$.

Step 3: Substitute $\mathbf{v}_2$ in Equation (14) to obtain $P(\mathbf{h}_2 = 1|\mathbf{v}_2)$.

Step 4: Update network parameters. The iteration algorithm of network parameters is as follows:

$$\begin{cases} \boldsymbol{\omega}^{(k+1)} = \boldsymbol{\omega}^{(k)} + \varepsilon\left(\mathbf{h}_1\mathbf{v}_1^{\mathrm{T}} - P(\mathbf{h}_2 = 1|\mathbf{v}_2)\mathbf{v}_2^{\mathrm{T}}\right) \\ \mathbf{b}^{(k+1)} = \mathbf{b}^{(k)} + \varepsilon(\mathbf{v}_1 - \mathbf{v}_2) \\ \mathbf{c}^{(k+1)} = \mathbf{c}^{(k)} + \varepsilon(\mathbf{h}_1 - P(\mathbf{h}_2 = 1|\mathbf{v}_2)) \end{cases} \tag{15}$$

where ε is the learning rate, which takes the value of 0.8 in this paper, and the superscript k represents the k-th iteration.

2.2.2. DBN based on Levenberg-Marquardt backpropagation (LMBP) Algorithm

Traditional DBN is formed by stacking multiple RBMs, in which the hidden layer of the previous RBM is used as the visible layer of the next RBM. CD algorithm is used to determine the network parameters layer by layer during pre-training, which is unsupervised learning. Then the pre-trained network parameters are assigned to the neural network as the initial training value of network parameters. The network parameters are fine-tuned by using the sample labels in the training set combined with BP algorithm, which is supervised learning. The structure of a traditional DBN is shown in Figure 2.

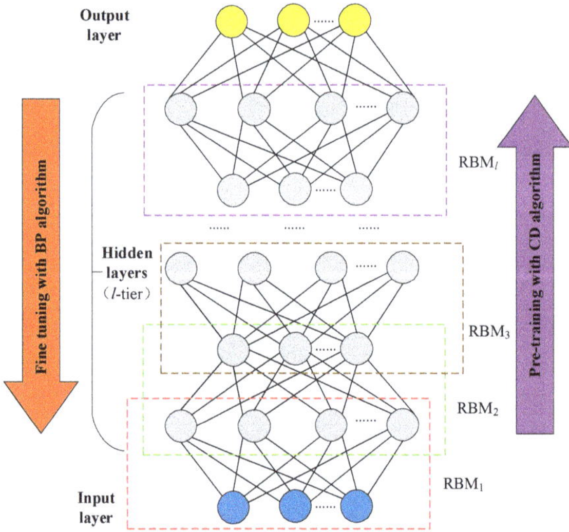

Figure 2. Structure of the traditional Deep Belief Network (DBN).

In this paper, the LM (Levenberg-Marquardt) BP algorithm [31] is used to replace the traditional BP algorithm to fine-tune the DBN. Compared with the traditional BP algorithm, the LMBP algorithm has faster convergence speed and higher convergence reliability, and is more suitable for training neural networks with many hidden layers and neurons.

Different from the traditional BP algorithm, the LMBP algorithm is based on the Gauss-Newton method in the least square solution. The square of error $\mathbf{v}$ is taken as the objective function and the second-order Taylor expansion of objective function is derived. After approximating the gradient of the objective function (ignoring the high-order term), the correction of weight $\boldsymbol{\omega}$ is:

$$\Delta\boldsymbol{\omega} = -\left[\mathbf{J}^T(\boldsymbol{\omega})\mathbf{J}(\boldsymbol{\omega}) + \mu\mathbf{I}\right]^{-1}\mathbf{J}^T(\boldsymbol{\omega})\mathbf{v}(\boldsymbol{\omega}) \tag{16}$$

where μ is the correction coefficient, which is set to prevent $\mathbf{J}^T(\boldsymbol{\omega})\mathbf{J}(\boldsymbol{\omega})$ from being irreversible; $\mathbf{I}$ is the identity matrix; $\mathbf{J}(\boldsymbol{\omega})$ is the Jacobian matrix of $\mathbf{v}(\boldsymbol{\omega})$, which can be written as:

$$\mathbf{J}(\boldsymbol{\omega}) = \begin{bmatrix} \frac{\partial v_1}{\partial \omega_1} & \frac{\partial v_1}{\partial \omega_2} & \cdots & \frac{\partial v_1}{\partial \omega_a} \\ \frac{\partial v_2}{\partial \omega_1} & \frac{\partial v_2}{\partial \omega_2} & \cdots & \frac{\partial v_2}{\partial \omega_a} \\ \vdots & \vdots & \ddots & \vdots \\ \frac{\partial v_a}{\partial \omega_1} & \frac{\partial v_a}{\partial \omega_2} & \cdots & \frac{\partial v_a}{\partial \omega_a} \end{bmatrix} \tag{17}$$

where $\frac{\partial v_i}{\partial \omega_j}$ represents the partial derivative of v_i to ω_j.

Similar to BP algorithm, the modifier expression of weight $\boldsymbol{\omega}^{(k+1)}$ in the k-th iteration is:

$$\boldsymbol{\omega}^{(k+1)} = \boldsymbol{\omega}^{(k)} + \Delta\boldsymbol{\omega}^{(k)} \tag{18}$$

μ needs to be adjusted in each iteration to obtain a better convergence effect. When μ is small, the algorithm is standard Gauss-Newton method, which has higher convergence accuracy. However, if the difference between the objective function and the approximate quadratic function is too large in the iteration, the convergence effect will be poor. When μ is large, the algorithm becomes traditional BP algorithm. When the Gauss-Newton method has a poor convergence performance, the gradient descent BP algorithm can be used as an auxiliary solution.

3. PSR-DBN Forecasting Model

3.1. The Procedure of the PSR-DBN Model

For a list of bus load historical data time series $p = \{p_1, p_2 \cdots p_N\}$, the prediction process of the PSR-DBN forecasting model is as follows:

Step 1 normalization: The load time series is normalized to prepare for the training of deep belief network, and the maximum and minimum values of data are saved for subsequent denormalization of the load predicted value to restore real value.

Step 2 PSR: The C-C method is adopted to process the load time series to find the optimal embedding dimension m and the optimal delay t of time series. Then the load time series is reconstructed according to the obtained embedding dimension m and delay t. The reconstructed load time series is as follows:

$$\begin{bmatrix} \mathbf{P}_1 \\ \mathbf{P}_2 \\ \vdots \\ \mathbf{P}_M \end{bmatrix} = \begin{bmatrix} p_1 & p_{1+t} & \cdots & p_{1+(m-1)t} \\ p_2 & p_{2+t} & \cdots & p_{2+(m-1)t} \\ \vdots & \vdots & \cdots & \vdots \\ p_M & p_{M+t} & \cdots & p_N \end{bmatrix} \tag{19}$$

where $M = N - (m-1)t$.

Step 3 DBN: The DBN is constructed by using the phase space matrix in step 2 as the training set and employing cross-validation to optimize the hyperparameters of the network. The details of hyperparameters tuning are introduced in Section 3.2. Finally, the trained DBN is adopted to predict the load value of the future moment.

Step 4 denormalization: The load prediction value returned by the deep belief network in step 3 is denormalized by applying the maximum and minimum values saved in step 1, then the actual load forecasting value is obtained.

The flow chart corresponding to the above steps is shown in Figure 3:

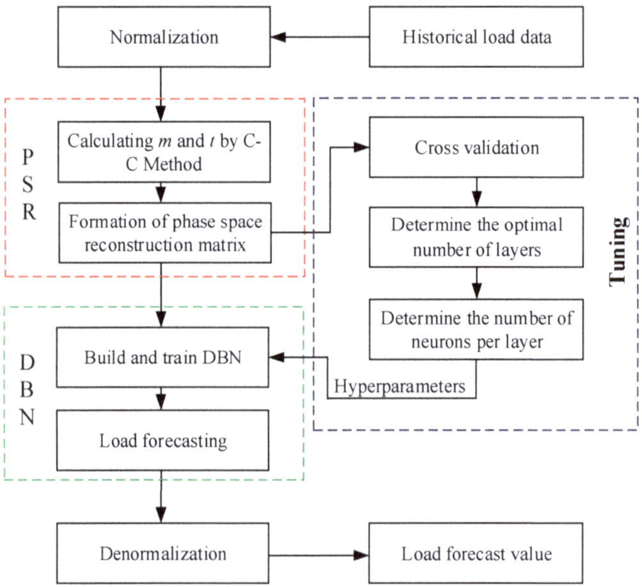

Figure 3. Flowchart of the PSR-DBN forecasting model.

3.2. Determination of DBN Network Structure

Similar to the neural network, DBN also has many hyperparameters to be set. The rationality of hyperparameters adjustment determines the prediction accuracy of the prediction model. Unreasonable hyperparameters will lead to a significant increase in prediction error. The determination of the network structure is an important part of DBN hyperparameters adjustment.

For the input layer of the DBN, since the original load data is input to DBN after passing through PSR, the number of the input layer neurons of DBN does not need to be tuned and can be directly set to embedding dimensions m, that is, one line of elements in Equation (19) is input each time. For the DBN output layer, when the DBN input is a row of elements in Equation (19), it is equivalent to inputting the position vector of the moving point in phase space at a certain moment. Then it needs to output the predicted value of the moving point position vector at the next moment.

However, if the input of the model is $\mathbf{p}_i(1 \leq i \leq M)$ in the phase space matrix of Equation (19), only $p_{i+1+(m-1)t}$ in the position vector $\mathbf{p}_{i+1}$ at the next moment is unknown, so the output layer only needs to output the predicted value $\hat{p}_{i+1+(m-1)t}$ of the load at time $i + 1 + (m-1)t$. If $i+1$ is greater than M, then the phase space matrix of Equation (19) needs to be extended downwards, $\mathbf{p}_{i+1}$ is added as a new row, and $\mathbf{p}_{i+1}$ is taken as the input of the model to obtain the predicted value of $p_{i+2+(m-1)t}$; Then the matrix is augmented and the predicted value of $p_{i+3+(m-1)t}$ is obtained, and so on until the end of the forecasting. The expression of $\mathbf{p}_{i+1}$ in the augmented matrix is:

$$\mathbf{p}_{i+1} = \begin{bmatrix} p_{i+1} & p_{i+1+t} & \cdots & p_{i+1+(m-1)t} \end{bmatrix} \tag{20}$$

where the value of $p_{i+1+(m-1)t}$ is determined by the forecasting horizons. If it is one-step forecasting, $p_{i+1+(m-1)t}$ takes the real value of load measured at time $i + 1 + (m-1)t$. If it is multi-step forecasting and the number of forecasting steps is k, $p_{i+1+(m-1)t} = \hat{p}_{i+1+(m-1)t}$ is taken until the k-th step forecasting. and then the predicted value in the augmented matrix is replaced by the real value of the measured load.

For the hidden layer of DBN, the number of hidden layers and neurons in each hidden layer has a significant influence on the prediction result, and it is found that the effect of optimizing the number of hidden layers is usually more obvious. Too few hidden layers will cause the under-fitting to affect the forecasting performance, and too many hidden layers will lead to over-fitting and will make forecasting performance worse. Therefore, in order to improve the prediction accuracy of the PSR-DBN model, cross-validation is used to optimize the number of hidden layers and neurons in each hidden layer. The flow is shown in the blue dotted line in Figure 3. The specific steps are as follows:

1. Cross-validation method: Considering that the load data is a time series, it is not appropriate to use a K-fold cross-validation method to disrupt the order. Therefore, the last part of the training set is eliminated and used as a verification set by hold-out cross-validation. The rest of the data is kept as a training set.

2. Determine the optimal number of layers: The enumeration is used to determine the optimal number of hidden layers. Many researchers find that a shallow network requires exponential width (number of neurons in each layer) to implement a function that a deep network of polynomial width could implement [32]. That is, compared with the number of layers, the number of neurons in each layer has less influence on prediction, so it is fixed during the enumeration. The number of neurons in each layer is set to be $2m$, and the number of hidden layers is increased layer by layer until a significant over-fitting occurs. Then the number of hidden layers with the smallest forecasting error is selected.

3. Determine the number of neurons in each layer: Since the forecasting performance of DBN varies with the initial value, the effect of changing the number of neurons one by one on forecasting performance is easily submerged in the fluctuation of forecasting performance caused by different initial values. Therefore, this paper uses a fixed step size to search for the superior number of neurons roughly. After determining the number of hidden layers according to step 2, the combination of the number of neurons with minimum prediction error is searched in steps of m in each layer. Because too many neurons will make the training of network slow and bring the risk of over-fitting, the selected search range of this paper is m to $5m$, and a good combination of the number of neurons is determined by testing.

In summary, the structure, input, and output of the proposed DBN are shown in Figure 4, in which the number of the hidden layers and the number of neurons in each hidden layer are obtained by cross-validation.

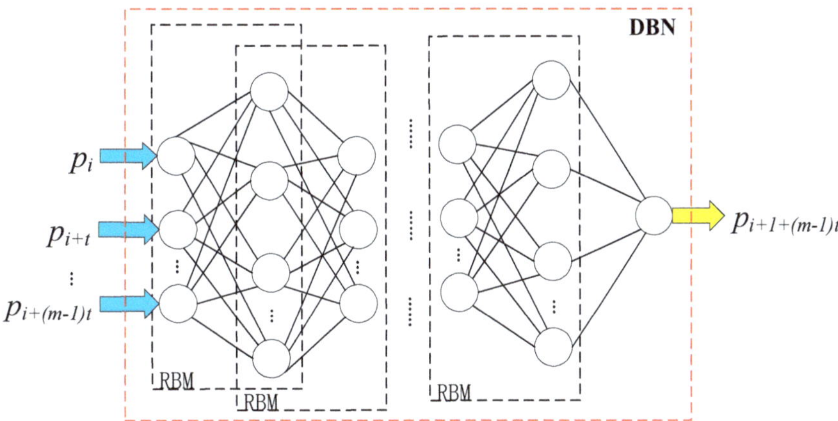

Figure 4. The input, output, and structure of the network.

4. Case Study

4.1. Bus Load Data

In order to test the validity of the model, the load data of a 220-kV substation bus in a city of China from 1 May to 18 May 2017 are used in this paper. The bus is connected with a distributed photovoltaic power station with an installed capacity of about 50 MW, and the sampling time interval of load data is 5 min. During the period, the substation had no overhaul or fault shutdown. The reliability of historical data is high, and 3σ criteria is used to detect that there is no abnormal data.

In this paper, the data of 1–14 May are selected as the training set of the PSR-DBN forecasting model, and cross-validation is used to adjust the model hyperparameters. The data of 15–18 May are selected as the forecasting test set. If only one-step forecasting of the load in the next 5 min is used as given in Ref. [22], the prediction horizons are too short to meet the real-time safety analysis and economic dispatching requirements of the power system. However, short-term load forecasting on an hourly scale combined with the interpolation has poor forecasting accuracy. Therefore, the forecasting horizons of very short-term load forecasting in this paper are from 5 min to 1 h, and the proposed model is validated in the MATLAB (R2018a, MathWorks Inc., Massachusetts, USA) environment.

4.2. Forecasting Evaluation Index

In order to more intuitively and accurately evaluate the forecasting performance of the model and the accuracy of prediction, this paper adopts Mean Absolute Percentage Error (MAPE), Mean Absolute Scaled Error (MASE), Symmetric Mean Absolute Error (sMAPE), Geometric Mean Absolute Error (GMAE), and Root Mean Square Error (RMSE) [33] as evaluation indicators.

$$\textbf{MAPE} = \frac{1}{n}\sum_{i=1}^{n}\left|\frac{p_i - \hat{p}_i}{p_i}\right| \times 100\% \tag{21}$$

$$\textbf{MASE} = \frac{1}{n}\sum_{i=1}^{n}\left(\frac{|p_i - \hat{p}_i|}{\frac{1}{n-1}\sum_{j=2}^{n}|p_j - p_{j-1}|}\right) \tag{22}$$

$$\textbf{sMASE} = \frac{100\%}{n}\sum_{i=1}^{n}\frac{|p_i - \hat{p}_i|}{0.5\left(|p_i| + |\hat{p}_i|\right)} \tag{23}$$

$$\textbf{GMAE} = \sqrt[n]{\prod_{i=1}^{n}|p_i - \hat{p}_i|} \tag{24}$$

$$\textbf{RMSE} = \sqrt{\frac{1}{n}\sum_{i=1}^{n}(p_i - \hat{p}_i)^2} \tag{25}$$

where n represents the number of predicted samples, p_i represents the real value of the load at time i, and $\hat{p}_i$ represents the predicted value of the load at time i.

The smaller the values of each metric, the higher the prediction accuracy of the model. However, these indicators are relative values and need to be compared under the same data scale to be meaningful.

4.3. PSR Reconstruction Results

Based on the theoretical analysis of PSR in Section 2.1, the C-C method is employed to reconstruct the phase space of the bus load data from 1 May to 14 May. The corresponding statistics of $\overline{\Delta S}(t)$ and $S_{cor}(t)$ are shown in Figure 5. It can be seen that the first minimum point of $\overline{\Delta S}(t)$ is $t = 18$, while $S_{cor}(t)$ cannot get the optimal embedding window t_ω without an obvious minimum point. However, from the

BDS statistics, when $N > 3000$, $m \in \{2, 3, 4, 5\}$, so the maximum value of m can only be 5. According to Equation (8), the final optimal embedding dimension $m_{opt} = 5$ and the optimal delay $t_{opt} = 18$.

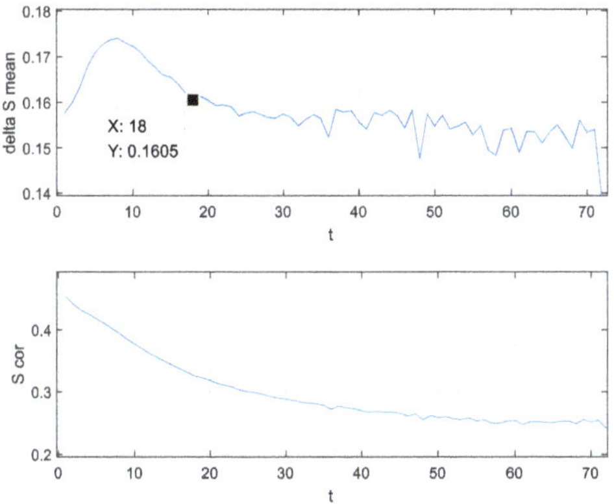

Figure 5. Curves of $\overline{\Delta S}(t)$ and $S_{cor}(t)$.

4.4. DBN Hyperparameter Setting

In Section 4.3, the optimal embedding dimension obtained by PSR is $m_{opt} = 5$. The structure of DBN is determined by the method described in Section 3.2. The number of neurons in the input layer and output layer of DBN is 5 and 1, respectively.

To determine the hidden layer hyperparameter, 13–14 May of the training set is eliminated and used as a verification set; the rest are reserved as the training set. The number of hidden layer neurons of DBN is fixed to 10, the number of layers is increased one by one, and the MAPE of the prediction result on the verification set is used as the criterion. The forecasting horizon is 1 h, and the MAPE of different hidden layers models is shown in Table 1.

Table 1. Mean Absolute Percentage Error (MAPE) of models with various number of hidden layers.

Number of Hidden Layers	2	3	4	5	6	7	8
MAPE	1.0387	1.0129	0.9358	1.0443	1.0413	1.0857	1.1336

In Table 1, when the number of hidden layers equals 8, the MAPE is significantly increased, and it can be inferred that over-fitting occurs at this time, so the continuation of increasing the number of layers is stopped. When the number of hidden layers is 4, MAPE is the smallest and is 0.9358. Therefore, the optimal number of hidden layers equals 4.

Based on the four hidden layers, the optimal number of neurons was roughly searched by a fixed step size, and the step size was set to 5 neurons. The MAPE of prediction result on the verification set is used as the criterion, and the forecasting horizon is 1 h. The sample space of the search is $5^4 = 625$. When the MAPE is the smallest, the number of neurons in each hidden layer is [25, 15, 20, 15]. At this time, the MAPE of predicted value on the verification set is 0.8447, which is obviously better than the MAPE = 0.9358 of 4 hidden layers and the number of neurons in each layer is 10. Therefore, the number of neurons in each hidden layer is [25, 15, 20, 15].

In summary, the hyperparameter setting of DBN in this paper is shown in Table 2.

Table 2. Hyperparameter setting of DBN.

Hyperparameter	Value
DBN Network structure	[5, 25, 15, 20, 15]
learning rate	0.8
Maximum epochs of RBM	100
NN Network structure	[5, 25, 15, 20, 15, 1]
LMBP(μ)	0.001
Maximum epochs of NN	150

4.5. Forecasting Result

4.5.1. One Hour Ahead Load Forecasting

In order to verify the forecasting performance of the proposed method, the ARIMA model, NN model, PSR-NN model, LSTM model, PSR-DBN model (without tuning), and DBN model are adopted to predict the load of 15–18 May after training with 1–14 May as historical data. The forecasting horizon is 1 h, and the corresponding evaluation indicators of each model are calculated. The ARIMA model employs the Akaike Information Criterion (AIC) to determine the optimal autoregressive model (AR model) order p and moving average model (MA model) order q. The hyperparameter of the NN model, LSTM model, and DBN model are also optimized by cross-validation. The PSR-DBN model (without tuning) has four hidden layers and 10 neurons per layer. The curves of the load predicted value and the load measured value corresponding to different models on 15–18 May are shown in Figure 6.

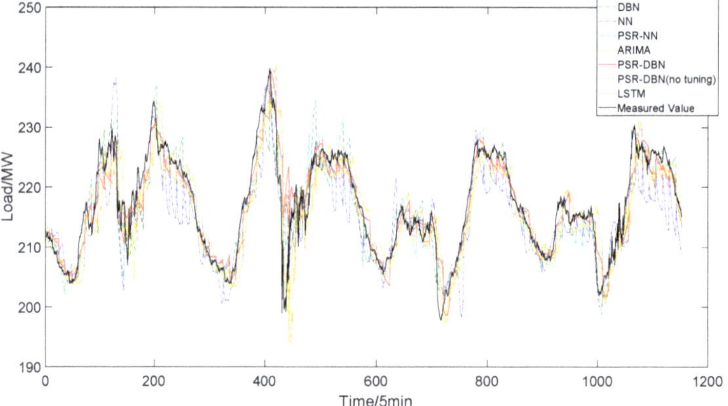

Figure 6. Load forecasting curve one hour ahead of 15–18 May.

The solid black line in Figure 6 is the measured value. It can be inferred that the active output of the photovoltaic power station has large fluctuations, and the bus load curve is severely distorted by the general saddle type, showing irregular fluctuations. If short-term load forecasting is used at this time, it will cause a large error and waste a lot of information. The remaining curves are the forecasting values of the ARIMA model, NN model, PSR-NN model, LSTM model, PSR-DBN model (no tuning), and DBN model. It can be clearly seen that the predicted curves of the ARIMA model with the golden dashed line and the NN model with the blue dashed line deviate significantly from the black measured values, while the forecasting performance of other models cannot be directly judged by curves. In order to more intuitively see the prediction accuracy of each model, the forecasting evaluation indicators and training time of each model are calculated as shown in Table 3.

Table 3. Forecasting performance of each model.

Model	MAPE (%)	RMSE	MASE	sMASE	GMAE	Training Time (s)
PSR-DBN	0.9892	3.2316	2.5403	0.9856	1.2027	12.3397
PSR-NN	1.0125	3.2310	2.6098	1.0126	1.2259	18.9233
DBN	1.1322	3.4684	2.9353	1.1358	1.3815	7.2980
ARIMA	1.5929	4.8657	4.0986	1.5922	Inf	20.0625
NN	1.6222	4.6757	4.2132	1.6333	Inf	15.1189
LSTM	1.0736	3.3266	2.7877	1.0799	Inf	17.2867
PSR-DBN (no tuning)	1.0380	3.2792	2.6737	1.0358	1.3279	10.2336

In Table 3, it is seen that the training time of all models meet the requirements of VSTLF. The PSR-DBN model proposed in this paper has the smallest indicators, except for the second smallest RMSE in all seven models. The RMSE of the PSR-DBN model proposed in this paper is only 0.0006 more than the minimum RMSE from the PSR-NN model. The 'Inf' of GAME indicates that the product of local error exceeds the upper limit of double type. It can be seen that the prediction accuracy of the DBN pre-trained by the CD method is better than that of the ordinary NN. The forecasting evaluation indicators of the PSR-DBN model and PSR-NN model, which adds PSR link to reconstruct original data, are also significantly less compared with the DBN model and the NN model. Compared with the PSR-DBN model without tuning, the tuned PSR-DBN model also has less evaluation indicators. Therefore, it can be inferred that the proposed method has higher prediction accuracy and better forecasting performance in the one-hour very short-term prediction of bus load with high distributed energy permeability and large fluctuation.

Figure 7 is a bar graph of the relative error of the predicted values of each model on 17 May. It can be seen from Figure 7 that the prediction errors of the ARIMA model and NN model are larger than those of the other five models, and the time with large prediction errors is concentrated in the noon period. At this time, the power output of the photovoltaic power station is large and vulnerable to clouds and other weather factors, resulting in great fluctuations of output and difficulties in prediction. Although the prediction accuracy of the noon period is not improved after adding the PSR algorithm, the reconstruction of the data reduces the influence of the fluctuation of the historical load data at noontime on the load forecasting of other time periods, thus effectively reducing the relative error of other time periods and improving the prediction accuracy of the model.

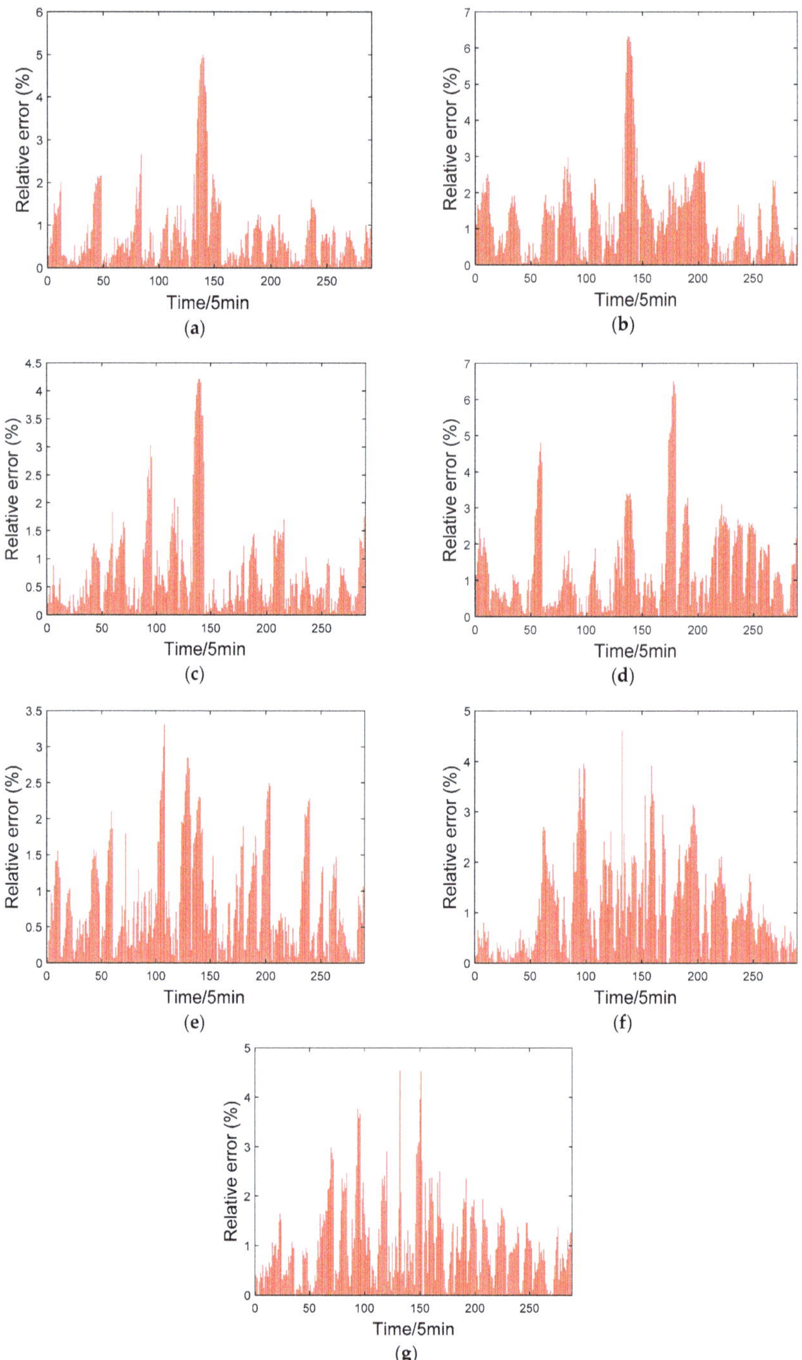

Figure 7. Relative error of load forecasting on 17 May. (**a**) PSR-DBN; (**b**) ARIMA; (**c**)PSR-NN; (**d**) NN; (**e**) DBN; (**f**) LSTM; (**g**) PSR-DBN (no tuning).

4.5.2. Prediction of Different Forecasting Horizons (5 min to 1 h ahead)

In order to verify the adaptability of the proposed PSR-DBN model to different forecasting horizons, the DBN and NN models are still employed in this paper, and the load of 15–18 May is forecasted by using historical data of 1–14 May. The forecasting horizons are 5 min to 1 h, and the corresponding MAPE is calculated, respectively. The MAPE curves of different models vary with forecasting horizons (shown in Figure 8).

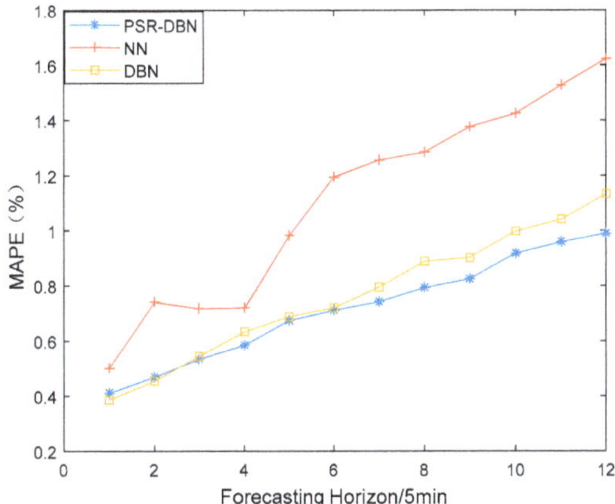

Figure 8. The curve of MAPE varies with forecasting horizons.

As can be seen in Figure 8, MAPE generally increases with increase of the forecasting horizon. The model proposed in this paper has higher prediction accuracy in most of the forecasting horizon, while the forecasting performance of the NN model is not ideal. When the forecasting horizon is 5 min to half an hour, the DBN model and PSR-DBN model do not have much difference in prediction accuracy. However, when the forecasting horizon is further increased from half an hour to one hour, the advantage of reconstructing data by PSR begins to appear. At this time, the MAPE of the PSR-DBN model is obviously less than that of the pure DBN model. Therefore, the model proposed in this paper can also have a small prediction error within the forecasting horizon of 5 min to 1 h and better adaptability in different forecasting horizons.

As can be seen in Figure 8, MAPE generally increases with increase of the forecasting horizon. The model proposed in this paper has higher prediction accuracy in most of the forecasting horizon, while the forecasting performance of the NN model is not ideal. When the forecasting horizon is 5 min to half an hour, the DBN model and PSR-DBN model do not have much difference in prediction accuracy. However, when the forecasting horizon is further increased from half an hour to one hour, the advantage of reconstructing data by PSR begins to appear. At this time, the MAPE of the PSR-DBN model is obviously less than that of the pure DBN model. Therefore, the model proposed in this paper can also have a small prediction error within the forecasting horizon of 5 min to 1 h and better adaptability in different forecasting horizons.

5. Conclusions

In this paper, aiming at the adaptability of forecasting horizons, the time series characteristics of the load, and the fluctuation caused by large amounts of distributed power access in bus load forecasting, a very short-term bus load forecasting model based on phase space reconstruction and deep belief network is proposed. The time series is projected by phase space reconstruction as the trajectory of a moving point in phase space, then the excellent non-linear fitting ability of DBN network is applied to fit the trajectory, so as to realize bus load forecasting. This paper also employs a practical method based on cross-validation to optimize the DBN network structure, and the real bus load data are employed to verify that:

- The PSR-DBN forecasting model proposed in this paper can still maintain relatively high prediction accuracy under the condition of high distributed power penetration and large fluctuation of bus load. The prediction accuracy of the proposed model is greatly improved, when compared to the ARIMA model of traditional time series models and the general neural network model.
- The proposed practical tuning method, which is based on cross-validation, can effectively improve prediction accuracy of the model compared with the random structure selection strategy.
- Under different forecasting horizons (5 min to 1 h), the PSR-DBN model proposed in this paper can still have a small prediction error. Compared with the model only using DBN, the phase-space reconstruction technique improves the adaptability of the model to long forecasting horizons. Therefore, the PSR-DBN model in this paper can maintain a small prediction error even in long forecasting horizons.

In this paper, the hyperparameters, such as the network structure of DBN, are only optimized by a roughly tuning method, and it is difficult to find the optimal value of hyperparameters. In practice, the corresponding load regular pattern will change greatly with the change of bus operation mode. The temperature elements also have an impact on load forecasting. Therefore, there are several factors of the bus load very short-term prediction model proposed in this paper that need to be considered and improved upon in the future.

Author Contributions: Data curation, F.M.; Formal analysis, C.Z.; Funding acquisition, J.L. (Jixiang Lu) and J.L. (Jinjun Lu); Investigation, F.M.; Project administration, J.Z.; Software, T.S.; Supervision, J.Z.; Validation, Y.P.; Writing—original draft, T.S.; Writing—review & editing, F.M. and J.W.

Funding: This research was funded by the "State Key Laboratory of Smart Grid Protection and Control, SKL of SGPC (No. 20195021212)" and "National Key Research and Development Project (No. 2018YFB0905000)".

Conflicts of Interest: The authors declare no conflict of interest.

References

1. Lee, K.Y.; Cha, Y.T.; Park, J.H.; Kurzyn, M.S.; Park, D.C.; Mohammed, O.A. Short-term load forecasting using an artificial neural network. *IEEE Trans. Power Syst.* **1992**, *7*, 124–132. [CrossRef]
2. Kang, C.; Xia, Q.; Zhang, B. Review of power system load forecasting and its development. *Autom. Electr. Power Syst.* **2004**, *28*, 1–11.
3. Hong, T.; Fan, S. Probabilistic electric load forecasting: A tutorial review. *Int. J. Forecast.* **2016**, *32*, 914–938. [CrossRef]
4. Moghram, I.; Rahman, S. Analysis and evaluation of 5 short-term load forecasting techniques. *IEEE Trans. Power Syst.* **1989**, *4*, 1484–1491. [CrossRef]
5. Ding, Q.; Lu, J.; Qian, Y.; Zhang, J.; Liao, H. A practical method for ultra-short term load forecasting. *Autom. Electr. Power Syst.* **2004**, *28*, 83–85.
6. Fard, A.K.; Akbari-Zadeh, M.-R. A hybrid method based on wavelet, ANN and ARIMA model for short-term load forecasting. *J. Exp. Theor. Artif. Intell.* **2014**, *26*, 167–182. [CrossRef]
7. Lee, C.-M.; Ko, C.-N. Short-term load forecasting using lifting scheme and ARIMA models. *Expert Syst. Appl.* **2011**, *38*, 5902–5911. [CrossRef]
8. Douglas, A.P.; Breipohl, A.M.; Lee, F.N.; Adapa, R. The impacts of temperature forecast uncertainty on Bayesian load forecasting. *IEEE Trans. Power Syst.* **1998**, *13*, 1507–1513. [CrossRef]
9. Lahouar, A.; Slama, J.H. Day-ahead load forecast using random forest and expert input selection. *Energy Convers. Manag.* **2015**, *103*, 1040–1051. [CrossRef]
10. Jiang, H.; Zhang, Y.; Muljadi, E.; Zhang, J.J.; Gao, D.W. A short-term and high-resolution distribution system load forecasting approach using support vector regression with hybrid parameters optimization. *IEEE Trans. Smart Grid* **2018**, *9*, 3341–3350. [CrossRef]
11. Wang, X.; Wang, Y. A hybrid model of EMD and PSO-SVR for short-term load forecasting in residential quarters. *Math. Probl. Eng.* **2016**, *2016*, 9895639. [CrossRef]

12. Bento, P.M.R.; Pombo, J.A.N.; Calado, M.R.A.; Mariano, S.J.P.S. Optimization of neural network with wavelet transform and improved data selection using bat algorithm for short-term load forecasting. *Neurocomputing* **2019**, *358*, 53–71. [CrossRef]

13. Feng, Y.; Xu, X.; Meng, Y. Short-term load forecasting with tensor partial least squares-neural network. *Energies* **2019**, *12*, 990. [CrossRef]

14. Zhang, X.; Wang, R.; Zhang, T.; Liu, Y.; Zha, Y. Short-term load forecasting using a novel deep learning framework. *Energies* **2018**, *11*, 1554. [CrossRef]

15. Pan, Y.; Mei, F.; Miao, H.; Zheng, J.; Zhu, K.; Sha, H. An approach for HVCB mechanical fault diagnosis based on a deep belief network and a transfer learning strategy. *J. Electr. Eng. Technol.* **2019**, *14*, 407–419. [CrossRef]

16. Yu, R.; Gao, J.; Yu, M.; Lu, W.; Xu, T.; Zhao, M.; Zhang, J.; Zhang, R.; Zhang, Z. LSTM-EFG for wind power forecasting based on sequential correlation features. *Future Gener. Comput. Syst.* **2019**, *93*, 33–42. [CrossRef]

17. Yang, L.; Yang, H. Analysis of different neural networks and a new architecture for short-term load forecasting. *Energies* **2019**, *12*, 1433. [CrossRef]

18. Panapakidis, I.P. Application of hybrid computational intelligence models in short-term bus load forecasting. *Expert Syst. Appl.* **2016**, *54*, 105–120. [CrossRef]

19. Panapakidis, I.P. Clustering based day-ahead and hour-ahead bus load forecasting models. *Int. J. Electr. Power Energy Syst.* **2016**, *80*, 171–178. [CrossRef]

20. Fan, G.-F.; Peng, L.-L.; Hong, W.-C. Short term load forecasting based on phase space reconstruction algorithm and bi-square kernel regression model. *Appl. Energy* **2018**, *224*, 13–33. [CrossRef]

21. Yang, L.; Zhang, W.; Zhou, Y.; Liao, F.; Xu, C.; Cheng, Y.; Yao, J. Application of indirect forecasting method in bus load forecasting. *Power Syst. Technol.* **2011**, *35*, 177–182.

22. He, Y.; Qin, Y.; Lei, X.; Feng, N. A study on short-term power load probability density forecasting considering wind power effects. *Int. J. Electr. Power Energy Syst.* **2019**, *113*, 502–514. [CrossRef]

23. Wang, Y.; Fu, Y.; Sun, L.; Xue, H. Ultra-short term prediction model of photovoltaic output power based on chaos-RBF neural network. *Power Syst. Technol.* **2018**, *42*, 1110–1116.

24. Wang, H.Z.; Wang, G.B.; Li, G.Q.; Peng, J.C.; Liu, Y.T. Deep belief network based deterministic and probabilistic wind speed forecasting approach. *Appl. Energy* **2016**, *182*, 80–93. [CrossRef]

25. Kim, H.S.; Eykholt, R.; Salas, J.D. Nonlinear dynamics, delay times, and embedding windows. *Phys. D Nonlinear Phenom.* **1999**, *127*, 48–60. [CrossRef]

26. Kim, H.S.; Kang, D.S.; Kim, J.H. The BDS statistic and residual test. *Stoch. Environ. Res. Risk Assess.* **2003**, *17*, 104–115. [CrossRef]

27. Durlauf, S.N. Nonlinear dynamics, chaos, and instability—Statistical-theory and economic evidence. *J. Econ. Lit.* **1993**, *31*, 232–234.

28. Hinton, G.E.; Osindero, S.; Teh, Y.-W. A fast learning algorithm for deep belief nets. *Neural Comput.* **2006**, *18*, 1527–1554. [CrossRef]

29. Yang, Z.; Liu, J.; Liu, Y.; Wen, L.; Wang, Z.; Ning, S. Transformer load forecasting based on adaptive deep belief network. *Proc. Chin. Soc. Electr. Eng.* **2019**, *39*, 4049–4060.

30. Bengio; Y, Learning deep architectures for AI. *Found. Trends Mach. Learn.* **2009**, *2*, 1–127. [CrossRef]

31. Mei, F.; Wu, Q.; Shi, T.; Lu, J.; Pan, Y.; Zheng, J. An ultrashort-term net load forecasting model based on phase space reconstruction and deep neural network. *Appl. Sci.* **2019**, *9*, 1487. [CrossRef]

32. Zhong, G.; Ling, X.; Wang, L.-N. From shallow feature learning to deep learning: Benefits from the width and depth of deep architectures. *Wiley Interdiscip. Rev. Data Min. Knowl. Discov.* **2019**, *9*, e1255. [CrossRef]

33. Hyndman, R.J. Another look at forecast accuracy metrics for intermittent demand. *Foresight Int. J. Appl. Forecast.* **2006**, *4*, 43–46.

 *energies*

Article

A Multi-Step Approach to Modeling the 24-hour Daily Profiles of Electricity Load using Daily Splines

Abdelmonaem Jornaz [1],* and V. A. Samaranayake [2]

[1]   Department of Mathematics and Statistics, Northwest Missouri State University, Maryville, MO 64468, USA
[2]   Department of Mathematics and Statistics, Missouri University of Science and Technology, Rolla, MO 65409, USA; vsam@mst.edu
*   Correspondence: ajornaz@nwmissouri.edu

Received: 8 October 2019; Accepted: 30 October 2019; Published: 1 November 2019

Abstract: Forecasting of real-time electricity load has been an important research topic over many years. Electricity load is driven by many factors, including economic conditions and weather. Furthermore, the demand for electricity varies with time, with different hours of the day and different days of the week having an effect on the load. This paper proposes a hybrid load-forecasting method that combines classical time series formulations with cubic splines to model electricity load. It is shown that this approach produces a model capable of making short-term forecasts with reasonable accuracy. In contrast to forecasting models that utilize a multitude of regressor variables observed at multiple time points within a day, only the hourly temperature is used in the proposed model and predictive power gains are achieved through the modeling of the 24-hour load profiles across weekends and weekdays while also taking into consideration seasonal variations of such profiles. Long-term trends are accounted for by using population and economic variables. The proposed approach can be used as a stand-alone predictive platform or be used as a scaffolding to build a more complex model involving additional inputs. The data cover the period from 1 January 1993 through 31 December 2013 from the Atlantic City Electric zone.

Keywords: forecasting; time series; cubic splines; real-time electricity load; seasonal patterns

---

1. Introduction

There is a long history of research on the modeling of hourly real-time electricity load. They range from standard regression and time series approaches to methods that use machine learning algorithms, such as artificial neural networks (ANNs), which require training by experts familiar with the algorithms being utilized. In contrast to naive regression approaches or the use of more sophisticated machine learning algorithms, a hybrid method that amalgamates regression splines with time series methods is proposed. One advantage of the proposed method is that it is implementable by using standard off-the-shelf software that does not require specialized training to be an effective user. Another is that it utilizes temperature as the only weather-related variable. Moreover, the proposed time-varying spline approach allows one to model the profile of daily electricity load for weak days as well as weekends for winter, summer, and shoulder months, providing valuable information about the daily electricity use patterns and how they evolve across days and seasons. In addition, the proposed method can be used as a platform for building more sophisticated models with additional variables. Finally, the model is readily interpretable as opposed to a forecasting model that utilizes a "black box" type algorithm.

The literature on load forecasting is extensive, and therefore, a complete discussion of the literature is not possible in this paper. However, a sample of the approaches to load forecasting is presented herein to demonstrated the variety of available methods. For early classical work, the reader is referred to Bunn and Farmer [1], which summarized approaches developed for short-term forecasting of electricity load. An important reference that classifies different methods of load forecasting is

Alfares and Nazeeruddin [2]. They categorized the various approaches into nine classes, which are: (1) multiple regression, (2) exponential smoothing, (3) iterative reweighted least-squares, (4) adaptive load forecasting, (5) stochastic time series, (6) Autoregressive Moving Average models with exogenous inputs (ARMAX models) and those with optimal model selection using the genetic algorithm, (7) fuzzy logic, (8) neural networks, and (9) expert systems. Alfares and Nazeeruddin also commented that while the pure time series approach is widely used, hybrid approaches, which combine several techniques, have become more common.

As mentioned by Alfares and Nazeeruddin, there are many instances of the use of hybrid methods. For example, El-Keib et al. [3] presented a hybrid approach where exponential smoothing was augmented with power spectrum analysis and adaptive autoregressive modeling. On the other hand, Dash et al. [4] utilized an expert system modeled fuzzy neural network and a hybrid neural network to forecast electricity load. Other publications that employed hybrid approaches are: Kim et al. [5] Chow et al. [6], and Choueiki et al. [7]. A more recent two-stage approach to forecasting the hourly electricity load for 24 hours ahead was developed by Gajowniczek and Zabkowski [8]. In this approach, the peak load values were determined by using the generic function quantile in the first stage, followed by building of three classification models, corresponding to the 99th, 95th, and 90th percentile of the distribution, to identify the load level. They used two machine learning algorithms, namely support vector machine (SVM) and artificial neural networks (ANN), to forecast the 24-hour electricity load. Another recently introduced approach, based on an extreme learning technique, was proposed by Das et al. [9]. This method considered relative difference in percentage of load at different intervals in its modeling approach. More recently, Annamareddi et al. [10] proposed a hybrid model based on a wavelet transform technique and double exponential smoothing to forecast the electricity load. Another hybrid method for predicting the electricity load using Support Vector Regression (SVR) and the Krill-Herd (KH) algorithm was proposed by Baziar and Kavousi-Fard [11]. The first step used training data, and the KH algorithm was used to optimize the SVR parameters. Consequently, in the second step, the optimized SVR was used to forecast the electricity load.

A research paper that influenced our approach to electricity load modeling is the publication by Nowicka-Zagrajek and Weron [12], which proposed a two-step procedure based on removing the trend and seasonal effects first and then fitting an autoregressive moving average (ARMA) model to the de-seasonalized data to obtain day-ahead predictions for the real-time load. In addition to removing trend and seasonal effects, our approach uses spline regression to model daily load profiles. In contrast, Liu et al. [13] utilized a semi-parametric model for nonlinear time series data, with the model consisting of two components. One of the components is nonparametric, while the other is a parametric Autoregressive Integrated Moving Average (ARIMA) component. Another approach that accounts for periodicity is a generalization of the logistic Smooth Transition Autoregressive (STAR) model for short term forecasting, developed by Amaral et al. [14]. This approach combines periodic models with a smooth transition between the regimes. Another publication that deals with cyclical behavior is that by Dordonnat et al. [15], which presented a periodic state space model, with different equations and different parameters for each hour, for forecasting of the hourly electricity load. The multi-equation linear model with autoregressive order two AR(2) approach developed by Chapagain and Kittipiyakul [16] uses 48 separate equations to forecast every half hour electricity load for one day ahead. Two different techniques, namely the ordinary least square (OLS) and a Bayesian approach, were used to estimate the model parameters for each type of day separately weekdays, weekends, and holidays.

The daily electricity use profile over a 24-hour period has prompted researchers to use functional approaches to modeling electricity load. Kosiorowski [17] compared methods of load forecasting that utilizes such approaches and concluded that the moving functional median is the appropriate approach for functional time series that contain outliers and nonstationary functional time series. In comparison, the other three approaches, functional autoregressive, fully functional regression, and the method proposed by Hyndman and Shang [18], work for a stationarity functional time series, and the prediction accuracy of the Hyndman and Shang method was the best overall. Our proposed methodology treats the 24-hour load profile as a function, which changes according to the type of day and season; however, we model these changing profiles using the well-understood cubic spline approach.

More recently, Papadopoulos and Karakatsanis [19] compared four different approaches, namely the seasonal autoregressive moving average (SARIMA), seasonal autoregressive moving average with exogenous variable (SARIMAX), random forests (RF), and gradient boosting regression trees (GBRT). Among the methods compared, GBRT showed the most accurate results based on mean absolute percentage error (MAPE) and root mean square error (RMSE). Alkhathami [20] also discussed the merits of various forecasting methodologies for load forecasting. He mentioned that the complex methods give more accurate results. Yang et al. [21] developed a hybrid method to forecast the half-hour electricity load, and they applied autocorrelation function (ACF) to select the important input features of least square support vector machines (LSSVM), followed by the grey wolf optimization algorithm (GWO) and cross validation (CV) to optimize the parameters of LSSVM. This method was more accurate compared to nine other approaches over three electricity load data sets from Australia.

In addition to the literature discussed in the previous paragraphs, there are a plethora of publications on the topics of load forecasting. Nevertheless, for the sake of brevity, we presented only a limited sample to illustrate the diversity of approaches taken by researchers in this area.

It is worth noting that many of the RTOs (Regional Transmission Organizations) and ISOs (Independent System Operators), as well as utility companies, have tended to use multiple regression models with a multitude of weather-related inputs for short-term prediction in spite of recent research that tend to include more sophisticated approaches. Possible reasons for this are discussed later in this section.

The Pennsylvania-New Jersey-Maryland (PFM) RTO uses multiple regression models with many regressor, such as 26 calendar variables, which are the days of the week (6), month of the year (11), holidays (15), and daylight-saving time impacts (see PJM Manual 19 [22]). Different variables for heating, cooling and shoulder seasons are included in the PJM model. Moreover, several formulae are used to calculate some of the weather, economic, and end-use variables. Another variable labeled load adjustment has also been used in PJM model.

It is evident that while many sophisticated models have been proposed, at least some practitioners, such as the modelers at PJM, seem to prefer models based on classical statistical approaches. One reason for this may be that the less sophisticated multiple regression models work reasonably well and are both interpretable as well as easy to modify and re-train compared to those that are based on ANNs (Artificial neural networks) or RFs (Random Forests). Keeping this perspective in mind, an approach for short-term forecasting of electricity load using classical techniques that are relatively easy to implement, is proposed. One of the goals is to avoid using "black box" approaches that result in non-interpretable formulations, but instead to utilize methodologies that result in easy to understand models. The proposed method, while somewhat more complex than straightforward regression approaches, is nevertheless based on regression and basic time series modeling that can be executed using widely available software. In addition, it uses a minimum amount of weather variables and drives the forecasting power by capturing the effect of such variables implicitly embedded in the lagged values of the load series as well as by exploiting the cyclical patterns inherent in the data. While relatively simple when compared to the more sophisticated models described earlier, the proposed approach nevertheless provides flexibility to model non-linear and non-stationary components that exhibit seasonal variability. In addition, it provides a platform on which more complex models, involving regressors such as additional weather variables, can be built.

The rest of the paper follows the following format. In Section 2, the main factors that affect electricity load are discussed and their impact on the load is illustrated graphically using empirical data. Section 3 describes the sources of the electricity load data employed in the analysis as well as weather and macro-economic data utilized in the proposed model. The proposed modeling approach is detailed in Section 4 and concluding remarks are made in Section 5.

2. The Factors Affecting Electricity Load

There are many factors that affect the electricity load, some in the short term and others in the long term. Fahad and Arbab [23] described the impact of various factors on the short-term load and grouped those factors into four categories, namely time, weather, economy, and random disturbances. Several economic and macroeconomic factors influence the electricity load over the long-term and researchers have utilized these to obtain long-term forecasts. Some examples of such variables are gross domestic product (GDP), gross domestic product per capita (GDP per capita), and population size. At the initial stages of the proposed modeling approach, long-term trend is removed using a regression model that accounts for several macro-economic variables and population size.

The time related changes in electricity load are not only due to long-term trend. Daily variations in human activity due to working, leisure, and sleeping periods (see Figure 1) can introduce a cyclical pattern with a 24-hour period. Other time related factors including day of the week, holidays, and seasonal changes in consumer behavior can affect this 24-hour cycle as seen in Figure 1.

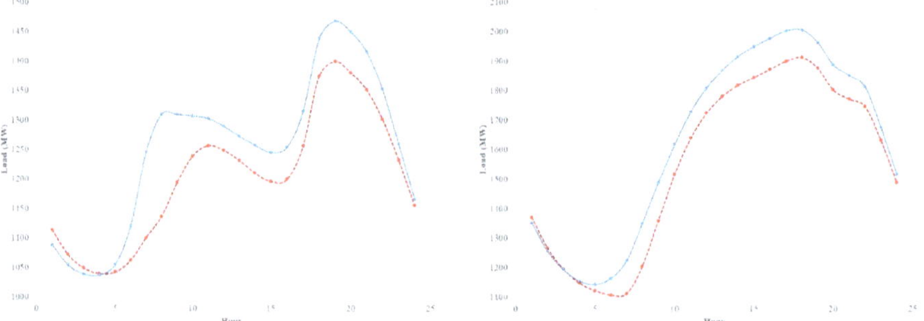

Figure 1. The average of a 24-hour of load curve of weekdays (blue solid) and weekends (red dashed) 1993–2012 (**left**: January; **right**: July).

The weather variables, such as temperature, humidity, precipitation, and wind speed, have played a significant role in electricity load forecasting, such as in the models used by the PJM TRO. Out of these factors, the temperature plays a major role (Figure 2). The effect of seasons on the electricity load, as seen in Figure 3, can be mainly attributed to seasonal fluctuation of the temperature, even though seasonal changes in human behavior can also play a role. The proposed approach to modeling electricity load strives to capture these effects due to seasonality, week-day and weekend differences, as well as the intra-day fluctuation of temperature on the 24-hour load curve. Details of how this is accomplished are given in Section 4.

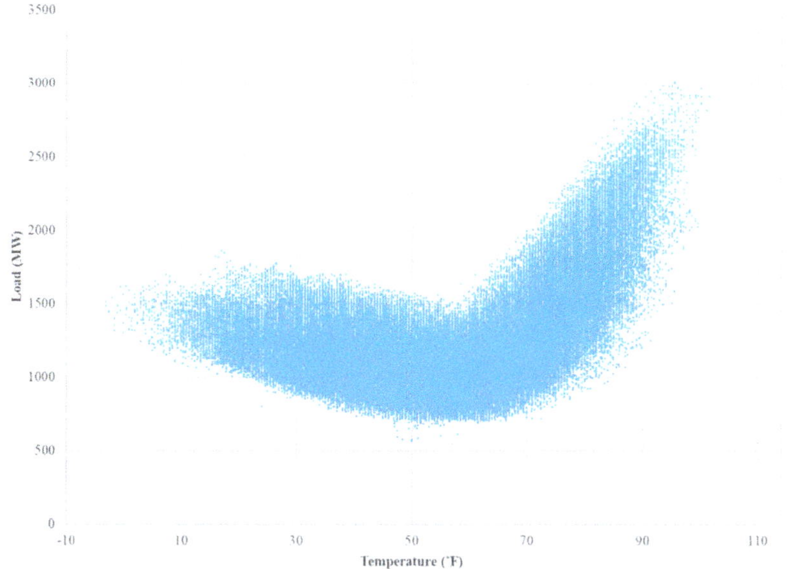

Figure 2. The relationship between the hourly load and hourly temperature—South New Jersey 1993–2012.

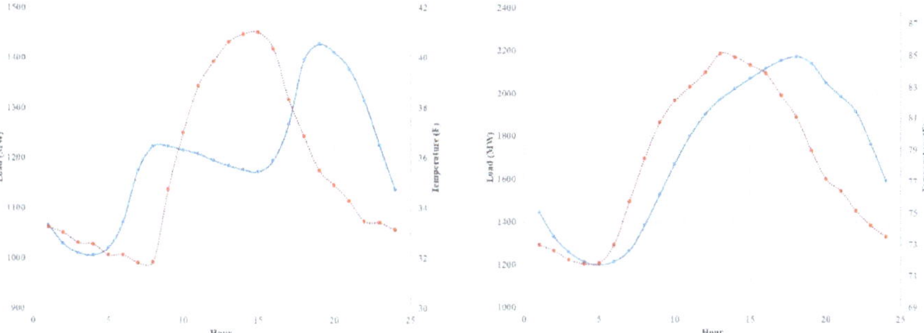

Figure 3. The average of 24-hour load curves (blue solid) and the temperature (red doted) (**left**: January 2013; **right**: July 2013).

3. Data Sources

The historical load dataset used in this study was obtained from the Pennsylvania-New Jersey-Maryland RTO website (PJM) [24]. The data cover a sub region of PJM (see Figure 4), namely the Atlantic City Electric company (AE), which serve approximately 556,000 customers in eight counties (Atlantic, Burlington, Camden, Cape May, Cumberland, Gloucester, Ocean, and Salem), in southern New Jersey. This dataset includes hourly observations measured in megawatts (MW) over 20 years from 1 January 1993 through 31 December 2012 (see Figure 5), which were used for modeling purposes (i.e., as training data), and data from 1 January 2013 through 31 December 2013, shown in Figure 6, which were used as test data for computing forecasting error. The weather data were obtained from the National Oceanic Atmospheric Administration (NOAA) based on four weather stations in different locations of the study area, southern New Jersey. These stations are located in Atlantic City, Millville, Mount Holly, and Wildwood.

The economic data were obtained from Federal Reserve Bank of St. Louis. The specific economic variables used in this study are: industrial production index in the US (IPI) which is an economic indicator that measures the amount of the output from manufacturing, mining, electric and gas industries; government employment in New Jersey (NJGOVTN), which is defined as the total body of employees in all government agencies apart from the military; home vacancy rate in New Jersey (NJHVAC), which is defined as the percentage of all available units in a rental property that are vacant or unoccupied at a particular time.

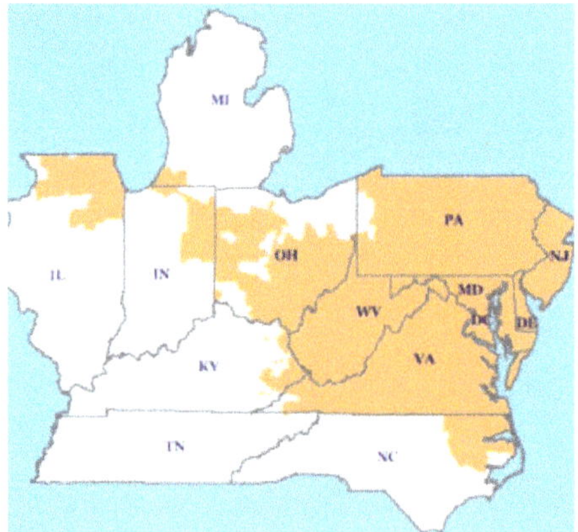

Figure 4. Map of Pennsylvania-New Jersey-Maryland (PJM) [25].

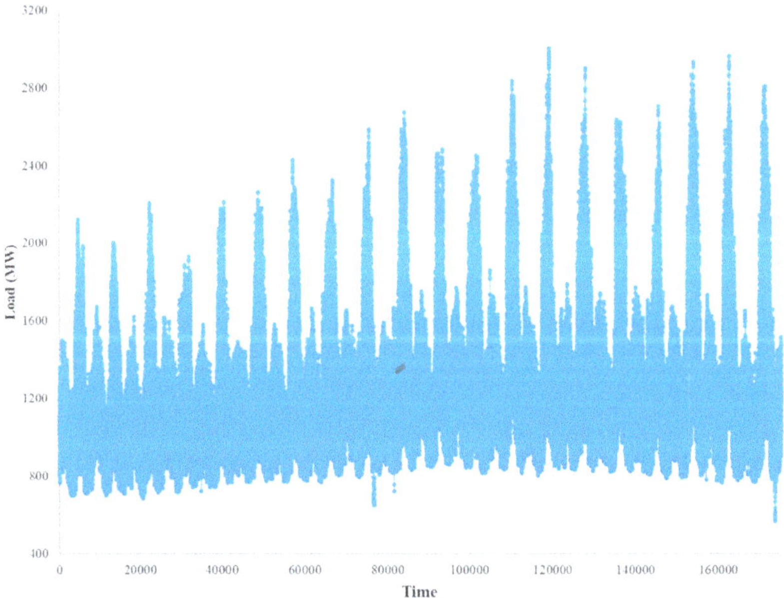

Figure 5. The hourly observed load over 20 years.

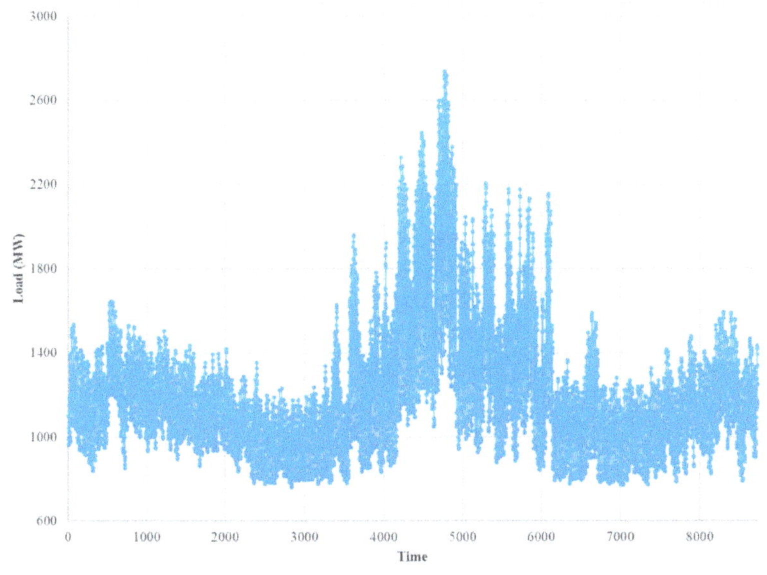

Figure 6. The hourly load of Atlantic City Electric Company (AE) in 2013.

4. The Proposed Approach to Modeling Electricity Load

The approach used in the following assumes a traditional structural time series model with trend, seasonal, and cyclical components, but utilizes a variation of cubic splines to estimate the 24-hour load profile for weekdays and weekends in a given season with hourly temperature playing an explanatory role. Specifically, model parameters are estimated for the mean load curve separately for weekdays and weekend days within each season after performing a de-trending operation. This approach can be considered suitable for short term prediction because of the need to have good estimates of hourly temperature.

The model assumes that $Y_{t,i}$, the real-time load at hour i on day t, is a composite of structural components consisting of a long-term trend τ_t, a seasonal component S_t, a weekly cycle w_t, a set of functions $f_{s,d}(x)$ representing the hourly load profile at time x for season s and day of the week d (taking one value for week-days and a different value for weekends), and an irregular stochastic component $u_{t,i}$. Thus, $Y_{t,i}$ can be expressed as:

$$Y_{t,i} = \tau_t + S_t + w_t + f_{s,d}(i) + u_{t,i},$$

where $t = 1, 2, \ldots, N$ and $i = 1, 2, \ldots, 24$. Note that N denotes the number of days in the training data set and i denotes the hour of the day.

The long-term trend was modeled using classical regression with select economic variables as regressors. The weakly seasonal component was modeled using a vector autoregressive moving average with exogenous terms (ARMAX) formulation with the average weekly temperature and its square as exogenous variables. The 24-hour load profiles were modeled by using a separate set of cubic splines for each season and weekdays/weekend combinations. Different spline models were used for each season because the 24-hour load profile within a season has almost the same pattern but differs across seasons. The weekdays were modeled separately within each season because they have quite different load profiles as well when compared to weekend patterns. The assumption of only one functional form for the load profile of weekdays can be relaxed by adding unique functions for each day of the week or for Monday, Friday, and the rest of the weekdays. Similarly, one can assume separate functional forms for Saturday and Sundays. Since such an approach can reduce the accuracy of estimates due to reduced sample sizes, the number of different functions was kept to a minimum.

Details of the modeling process are described below, beginning with the detrending process followed by the estimation of the seasonal components and concluding with the spline modeling of the 24-hour load profile.

4.1. Predicting Long-Term Trend

The first step included modeling the hourly average electricity load per year, $\tau_l^* = \frac{1}{24N_l} \sum_{t=1}^{N_l} \sum_{i=1}^{24} Y_{t,i}$, using classical regression analysis. Note that in the above expression for the average load, l denotes the year with $l = 1, 2, \ldots, 20$, and N_l denotes the total number of days in that year. A stepwise selection method was used to determine the independent variables to be included in the model. Out of more than 20 economic variables plus population size and the average monthly temperature, the following variables were selected: government employment in New Jersey (NJGOVTN), industrial production index in the US (IPI), home vacancy rate in New Jersey (NJHVAC), and the average temperature of September (Temp_Sep). Table 1 provides the results of the multiple linear regression analysis.

Table 1. The results for the regression model for annual load.

Analysis of Variance					
Source	**DF**	**Sum of Squares**	**Mean Square**	**F-Ratio**	**Prob > F**
Model	4	117546	29386	310.41	<0.0001
Error	15	1420.061	94.67q		
Corrected Total	19	118966			
Root MSE	9.73	**R-Square**	0.988	**Adjusted R-Square**	0.985
AIC	95.255	**Dependent Mean**	1237.84	**Coefficient of Variation**	0.78604

Variable	**DF**	**Parameter Estimate**	**Standard Error**	**t-Statistic**	**Prob > |t|**	**Variance Inflation**
Intercept	1	−422.011	91.167	−4.63	0.0003	0
NJGOVTN	1	1.61	0.109	14.71	<0.0001	2.429
NJHVAC	1	−36.776	4.975	−7.39	<0.0001	1.041
IPI	1	2.769	0.343	8.08	<0.0001	2.399
Sep_M	1	7.247	1.276	5.68	<0.0001	1.079

Note: The above results are for the training data set only.

The estimated regression model for the annual data is:

$$\hat{\tau}_l^* = -422.01 - 36.78 \, NJHVAC + 1.61 \, NJGOVTN + 2.77 \, IPI + 7.25 \, Temp\_Sep.$$

The selected independent variables explain 98.5% of the variation in the average annual load and the root mean square error (RMSE) is 9.7, which is relatively small. Moreover, no serious multicollinearity among the independent variables was detected. The residual analysis is shown in the Figure A1 in the Appendix A, and while two outliers (with high Cook's distance) are shown, no major concern is raised. In addition, the model does a good job of predicting the annual load, as seen in Figure 7.

Figure 7 displays the average annual real-time load per hour for the 20 years of training data and one year of test data and the average annual load predicted, using the estimated regression model. The figure shows the predicted trend using actual macroeconomic data for the test year, but the macroeconomic data for the test year can be predicted very accurately using an ARMA model, and the results do not change by much. The display shows very good in-sample agreement between the observed and predicted load and a reasonable agreement between the two for the test year. One word of caution is that we observed that the electricity load decreased in the last three years, but one of the two most important variables in the model, the government employment in New Jersey (NJGOVTN), decreased only slightly, while the industrial production index in the US (IPI) increased. Those two variables explained 92% of the variation in the electricity load. Thus, some delinking of these variables with electricity load may be occurring and developing an annual model with more recent data may be pragmatic.

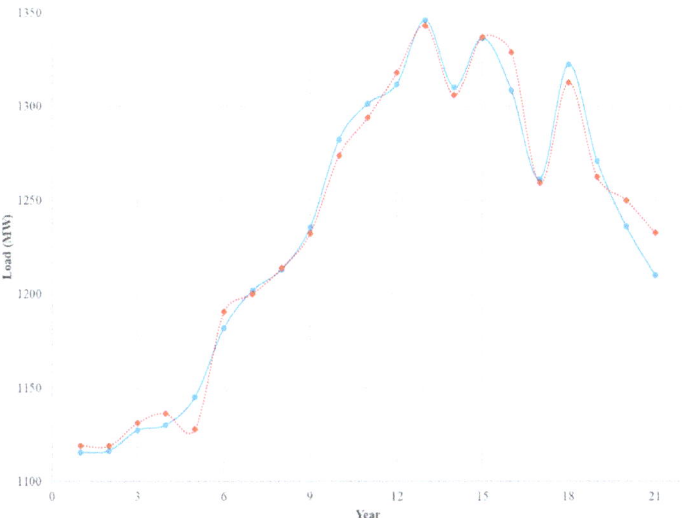

Figure 7. The annual average of hourly load (blue solid) and the predicted load (red doted) 1993–2013.

4.2. Estimating Seasonal Variation in Data

The trend estimates for each year were transposed onto a weekly series, and a 52-week moving average was applied to this series to smooth the predictions from a step function to a smooth one. The smoothed trend, $\widetilde{\tau}_w^*$, for week w, was subtracted from the average load $\overline{Y}_{t,\bullet} = (24)^{-1} \sum_{i=1}^{24} Y_{t,i}$ for each day within the corresponding week, with the process repeated for all weeks, yielding the detrended daily averages $\overrightarrow{Y}_{t,\bullet}$. The resulting data can be represented in a vector series S_w^*, where each vector contains the seven detrended daily averages $\overrightarrow{Y}_{t,\bullet}$ corresponding to that week. Note that $w = 1, 2, \ldots$, W, where W is the total number of weeks in the training data set. The de-trended weekly time series, $\hat{S}_w^*$, was then used to fit a subset ARMAX model (see Baillie, R. T [26] for details) given below:

$$S_w^* = 1023 \quad + \; 1.13S_{(w-1)}^* - 0.23S_{(w-2)}^* + 0.76S_{(w-52)}^* - 0.67S*_{(w-53)} + 0.75Z_{(w-1)}$$

$$- \; 0.06Z_{(w-50)} + 0.71Z_{(w-52)} - 0.44Z_{(w-53)} - 0.11Z_{(w-54)} - 50.36T + 0.53T^2,$$

where $L_{(w-lag)}$ denotes the autoregressive lag term, $Z_{(w-lag)}$ denotes the moving average lag terms, and T denotes the weekly average temperature. The residuals of the fit do not show any major autocorrelations and the test for white noise (bottom right hand corner of Figure A2 in the Appendix A) shows no evidence that the residuals are anything other than white noise. The check for normality of the residuals, given in Figure A3 in the Appendix A, shows some deviation from normality, but this is not much of a concern because the model performed an adequate job of extracting the seasonal component as indicated by white noise residual.

The forecasted weekly vector values for the test data set (year 2013) were then obtained using the estimated ARMAX model. These estimated vectors $\hat{S}_w^*$ contain forecasted daily values for each week. These daily values were averaged to get a forecast weekly average. The trend model was then employed to forecast a yearly trend for the test data (year 2013). Note that to predict a trend, we needed macroeconomic data for the test year and in practice, these have to be predicted. We found that applying an ARMA model to the past macroeconomic data would yield accurate forecasts for the test year. This yielded a constant forecast across all the weeks of the test year. These were them smoothed using a 52-week moving average that utilized previous year's data for the smoothing. The smoothed weekly trend data were then added to the forecast weekly average obtained from averaging the daily values from the ARMAX vector forecasts. The resulting weekly averages were then compared with the observed weekly average load for the test year (Figure 8). These out-of-sample checks show that the seasonal (weekly) model provides a satisfactory estimation of the seasonal component.

Note that the smoothing of the yearly trend data allows for a smooth transition from one year to the next. It also reduces any bias due to poor estimation of the trend value for the test year, when computing trend values for the early part of the test year. The trend estimates can be updated later in the test year by reforecasting the trend value by using updated macroeconomic variables that may become available after the first quarter of that year and later after the second and third quarter.

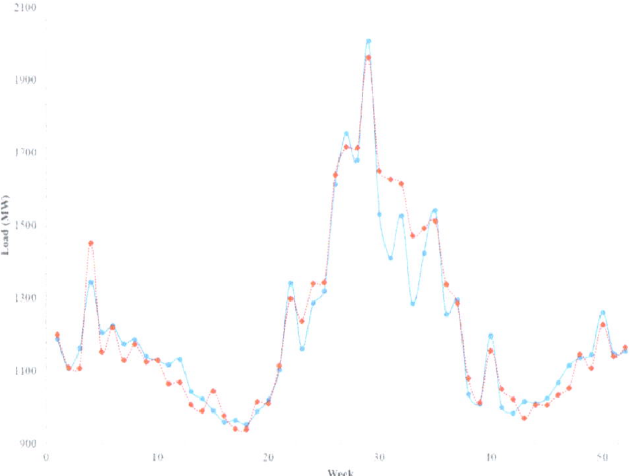

Figure 8. The weekly average of the hourly load (blue solid) and the predicted load (red dashed) in 2013.

4.3. Modeling the Hourly Load

At this point, the weekly smoothed trend $\widetilde{\tau}_w^*$ and the estimated seasonal component $\hat{S}_w^*$ were removed from the hourly data $Y_{t,i}$ for both training and test data years, and a new de-trended and de-seasonalized time series, $Y_{t,i}^*$, was obtained. The times series $Y_{t,i}^*$ was modeled using the training data set, by fitting cubic splines to model the 24-hour daily profile. Different spline estimates were obtained for each season, weekday, and weekend combination. Temperature and its interaction with time were also fitted as regressors.

Two scenarios were applied here. The first one modeled each season and each day type (weekday or weekend) separately. We denoted the resulting model as Model 1. The second scenario modeled each season separately and ignored the day type but added a dummy variable to identify the type of day. This approach provided us with Model 2.

4.3.1. The First Scenario

The general spline Model 1 is:

$$
\begin{aligned}
Y^*_{t,i} = b_0 \ & + \ b_1 i + b_2 i^2 + b_3 i^3 + b_4 (i - \kappa_1)^2 + b_5 (i - \kappa_2)^2 + b_6 (i - \kappa_3)^2 + b_7 (i - \kappa_1)^3 + b_8 (i - \kappa_2)^3 \\
& + \ b_9 (i - \kappa_3)^3 + b_{10} T_{t,i} + b_{11} T_{t,i} * i + b_{12} T_{t,i} * i^2 + b_{13} T_{t,i} * i^3 + b_{14} T_{t,i} * (i - \kappa_1)^2 + b_{15} T_{t,i} * (i - \kappa_2)^2 \\
& + \ b_{16} T_{t,i} * (i - \kappa_3)^2 + b_{17} T_{t,i} * (i - \kappa_1)^3 + b_{18} T_{t,i} * (i - \kappa_2)^3 + b_{19} T_{t,i} * (i - \kappa_3)^3 ,
\end{aligned}
$$

where i is the hour, $\kappa'_j s$ are the knots that change according to season and day type, and $T_{t,i}$ is temperature at hour i on day t. The knot positions were chosen by inspection for each season and are given, together with the parameter estimates, in the Tables A2–A5 in the Appendix A. Note that non-significant terms were dropped from the model and what is given above is the reduced model.

4.3.2. The Second Scenario

The general spline Model 2 is:

$$
\begin{aligned}
Y^*_{t,i} = b_0 \ & + \ b_1 i + b_2 i^2 + b_3 i^3 + b_4 (i - \kappa_1)^2 + b_5 (i - \kappa_2)^2 + b_6 (i - \kappa_3)^2 + b_7 (i - \kappa_1)^3 + b_8 (i - \kappa_2)^3 \\
& + \ b_9 (i - \kappa_3)^3 + b_{10} T_{t,i} + b_{11} T_{t,i} * i + b_{12} T_{t,i} * i^2 + b_{13} T_{t,i} * i^3 + b_{14} T_{t,i} * (i - \kappa_1)^2 + b_{15} T_{t,i} * (i - \kappa_2)^2 \\
& + \ b_{16} T_{t,i} * (i - \kappa_3)^2 + b_{17} T_{t,i} * (i - \kappa_1)^3 + b_{18} T_{t,i} * (i - \kappa_2)^3 + b_{19} T_{t,i} * (i - \kappa_3)^3 + b_{20} w_t ,
\end{aligned}
$$

where i is the hour, κ is a knot that changes according to the season and the knot positions and parameter estimates are given in the Tables A6 and A7 in the Appendix A. Note that $T_{t,i}$ is the temperature at time i on day t, and w_t is a dummy variable denoting weekend day. Note that non-significant terms were dropped from the model and the results reported are for the reduced model.

As mentioned previously, the Table A2 through Table A7, given in the Appendix A, present the knot positions and the results of the regression model for each season and each type of day. The tables for weekdays and weekends for a given season are paired together for easy comparison.

Model obtained for each season is different from the others, reflecting changes in the daily load profiles across seasons. The comparison between Model 1 and Model 2, based on Akaike Information Criteria (*AIC*) and Root Mean Square Error (*RMSE*) is presented in Table 2. The results show very little difference between the two models. In addition, the Figures 10 and 11 show the comparison between the two models based on the Coefficient of Variation (*CV*) for each month and each hour, respectively.

Table 2. The comparison between the two models.

Season	Day Type	RMSE		AIC	
		Model 1	**Model 2**	**Model 1**	**Model 2**
Winter	Weekdays	72.961	72.997	260089.07	260119.13
	Weekends	74.395	74.743	100751.11	100862.34
Spring	Weekdays	73.524	73.484	271284.41	271249.53
	Weekends	84.087	84.273	111701.59	111757.18
Summer	Weekdays	121.68	121.754	303082.91	303118.57
	Weekends	138.232	138.465	124225.17	124267.61
Fall	Weekdays	94.403	92.402	282449.53	282449.54
	Weekends	100.777	101.231	115155.18	115268.54

Figure 9 shows very close agreement between the two models when compared using the *CV* by month. There is a slight drop in the *CV* for Model 1 suggesting a slight gain in accuracy when the weekdays and weekends are modeled separately. Figure 10 given below shows the *CV* for the two models by each hour of the day. Again, Model 1 shows a slight advantage with the *CV* for Model 2 showing higher values for hours before 10 am. This may be because the load profile for weekends shows a two-hour shift in the morning load profile and the inclusion of a dummy variable is not sufficient to account for this difference in the shape of the load profile.

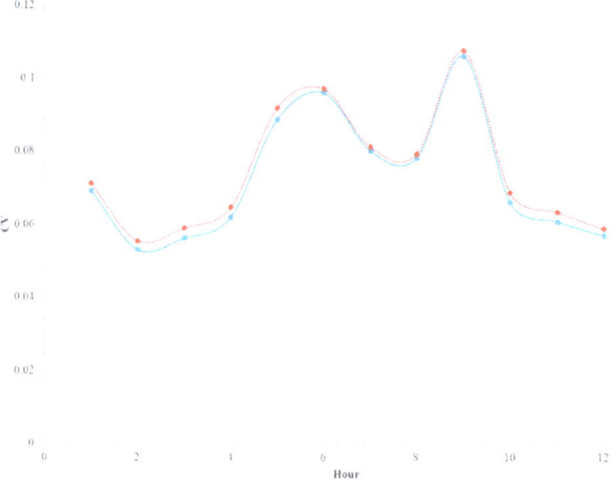

Figure 9. The comparison between the *CV* of the predicted monthly average load of Model 1 (blue solid) and Model 2 (red doted).

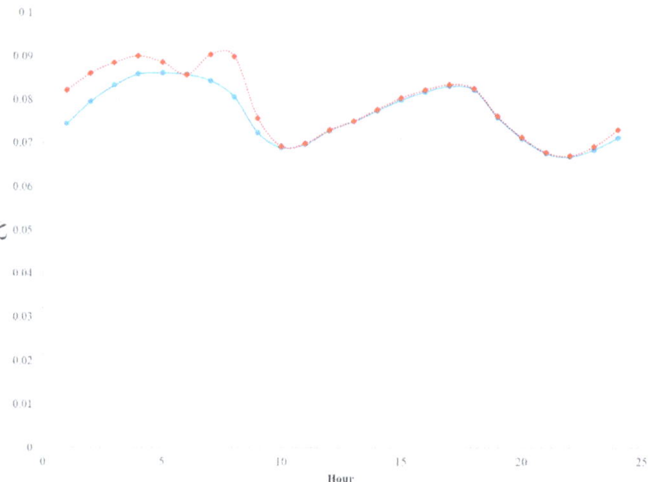

Figure 10. The comparison between the *CV* of the predicted hourly average load of Model 1 (blue solid) and Model 2 (red doted).

The Figures 11–14 provide a comparison between the two models for four different weeks of the test year. The weeks were chosen from the middle of each season. The forecasts based on each model were very close to one another, which suggests that adding a dummy variable for the day type instead of building the extra models for the type of day provides satisfactory forecast overall; however, for all seasons except summer, Model 2 yielded forecasts that fall below the observed load during the weekends (last two days in the graph), especially in the morning period. However, except for the weekends, both models underestimated the afternoon peak in winter and spring. For the summer season (Figure 13), the afternoon peak was overestimated by both models on Fridays.

Table 3 provides an additional contrast between the two models based on *CV*. It is immediately apparent that any difference between the two models is quite marginal and may not have any practical consequences. The cells highlighted in light blue indicate places where the *CV* for a given model is lower than that for the competing model. If any conclusion can be made based examining these results, it may be that Model 1 is slightly better than Model 2 across most hours in winter and fall, and Model 1 appears to perform slightly better before noon during spring. One reason for this may be the somewhat poor performance of model two during the weekend mornings. If pressed to select one model over the other, the natural choice would be Model 1, but a strong argument cannot be made that it is much more superior to Model 2.

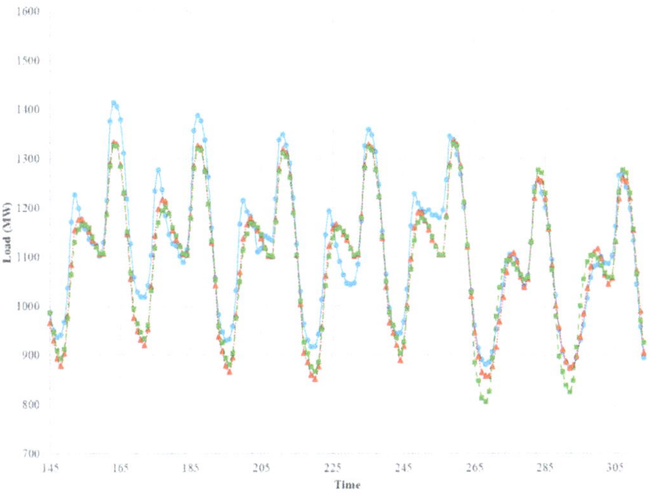

Figure 11. The comparison between the observed hourly load (blue solid), the predicted hourly load of Model 1 (red doted), and Model 2 (green dashed). 7 January 2013–13 January 2013.

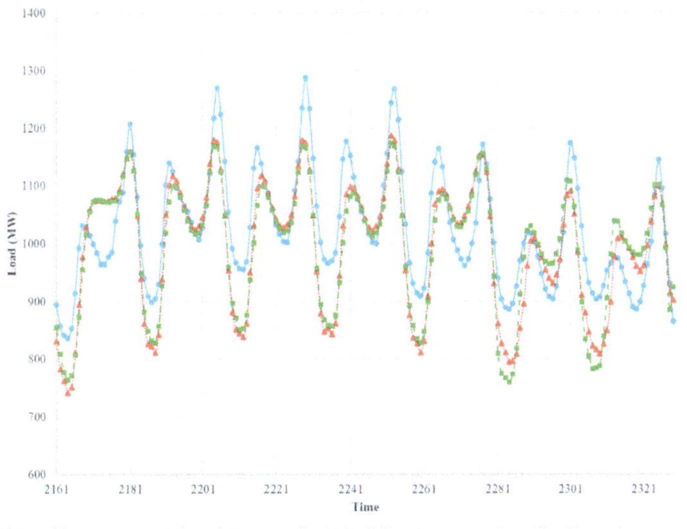

Figure 12. The comparison between the observed hourly load (blue solid), the predicted hourly load of Model 1 (red doted), and Model 2 (green dashed). 1 April 2013–7 April 2013.

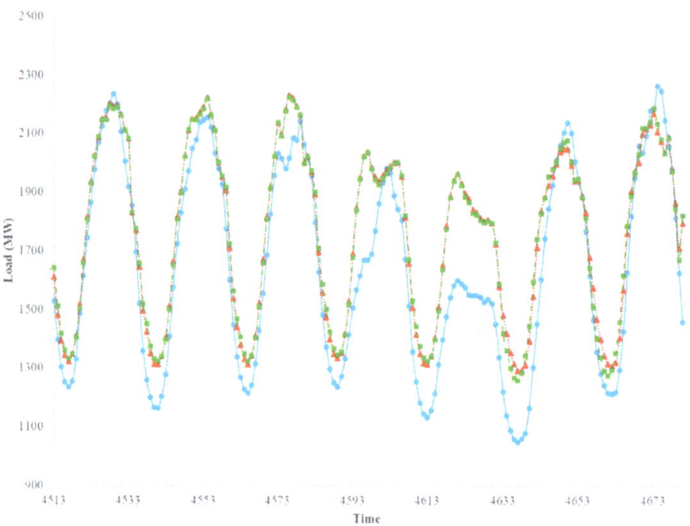

Figure 13. The comparison between the observed hourly load (blue solid), the predicted hourly load of Model 1 (red doted), and Model 2 (green dashed). 8 July 2013–14 July 2013.

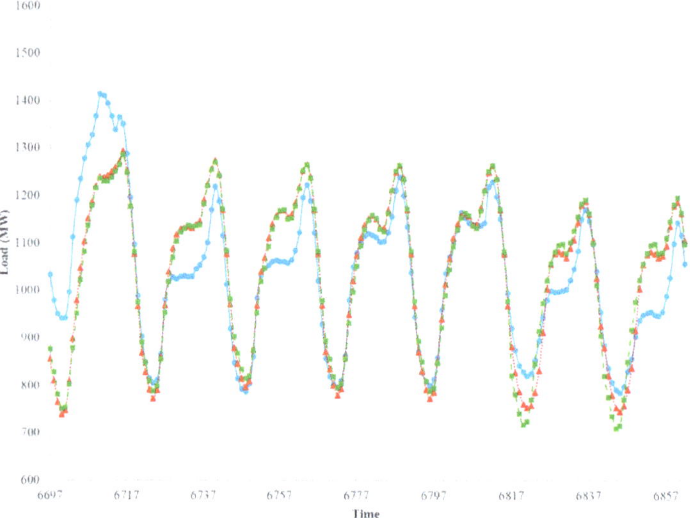

Figure 14. The comparison between the observed hourly load (blue solid), the predicted hourly load of Model 1 (red doted), and Model 2 (green dashed). 7 October 2013–13 October 2013.

Table 3. The comparison between the CV of the two models for each season by hour.

Hour	Winter		Spring		Summer		Fall	
	Model 1	Model 2	Model 1	Model 2	Model 1	Model 2	Model 1	Model 2
1	0.064	0.068	0.077	0.082	0.087	0.090	0.083	0.092
2	0.061	0.063	0.076	0.090	0.093	0.094	0.077	0.086
3	0.063	0.065	0.075	0.080	0.093	0.093	0.078	0.086
4	0.065	0.067	0.079	0.080	0.093	0.091	0.081	0.088
5	0.063	0.063	0.077	0.077	0.092	0.089	0.081	0.085
6	0.059	0.059	0.077	0.079	0.087	0.088	0.081	0.081
7	0.066	0.075	0.087	0.093	0.091	0.092	0.088	0.093
8	0.062	0.076	0.076	0.090	0.096	0.090	0.088	0.095
9	0.056	0.064	0.072	0.080	0.104	0.098	0.088	0.088
10	0.063	0.066	0.078	0.081	0.104	0.106	0.092	0.091
11	0.072	0.071	0.083	0.084	0.100	0.101	0.096	0.097
12	0.076	0.074	0.090	0.089	0.102	0.102	0.105	0.107
13	0.076	0.076	0.099	0.096	0.107	0.104	0.119	0.120
14	0.073	0.075	0.105	0.104	0.101	0.100	0.132	0.132
15	0.071	0.072	0.107	0.107	0.096	0.095	0.139	0.138
16	0.069	0.069	0.110	0.111	0.093	0.093	0.144	0.145
17	0.068	0.070	0.114	0.115	0.090	0.091	0.150	0.150
18	0.072	0.073	0.118	0.117	0.087	0.086	0.156	0.155
19	0.060	0.062	0.116	0.115	0.085	0.085	0.134	0.135
20	0.059	0.060	0.106	0.105	0.083	0.083	0.116	0.117
21	0.061	0.062	0.099	0.097	0.079	0.078	0.107	0.109
22	0.062	0.062	0.090	0.085	0.080	0.078	0.096	0.098
23	0.060	0.060	0.080	0.074	0.092	0.088	0.086	0.088
24	0.060	0.060	0.078	0.075	0.085	0.084	0.081	0.086

Note: The numbers highlighted in blue indicates the lower of the two CV values for each hour of each season.

5. Conclusions

A multi-step approach to modeling the hourly electricity load using a structural time series model that utilizes only standard statistical modeling techniques was introduced. While the proposed methods require multiple steps to model the load data, every step can be implemented using commonly available statistical software packages, and therefore, they are within the reach of empirical modelers who do not have training in the use of sophisticated machine learning algorithms or have the time required to master complex analytical techniques. The results of modeling observed real-time load data from the PJM market show that the proposed method performs reasonably well in modeling the training data and short-term forecasting out-of-sample data. In addition, the proposed methodology utilized only macroeconomic and temperature data, and the use of additional input variables has the potential to further improve the performance of the models considered in this study. One shortcoming of the proposed study is the need to know macroeconomic data and the population figures to predict long-term trend, but in the context of forecasting in the short-term, this may not be a great drawback, because near-term forecasts of these can be quite reliable or one can use the most recent data without sacrificing much accuracy. In spite of the above shortcoming, it is seen that the cubic spline model worked very well in capturing the 24-hour load curve, and therefore, the proposed methodology can provide a framework for modeling other phenomena that exhibit a daily cycle, especially if long-term trend forecasting is not needed.

Author Contributions: The contributions made by the two authors are as follows: conceptualization, A.J. and V.A.S.; methodology, A.J. and V.A.S.; software, A.J.; validation, A.J.; formal analysis, A.J.; investigation, A.J.; resources, V.A.S.; data curation, A.J.; writing—original draft preparation, A.J.; writing—V.A.S.; visualization, A.J.; supervision, V.A.S.; project administration, V.A.S.

Funding: This research received no external funding.

Conflicts of Interest: The authors declare no conflict of interest.

Appendix A

In this appendix, additional figures and tables relevant to material presented in this paper are given.

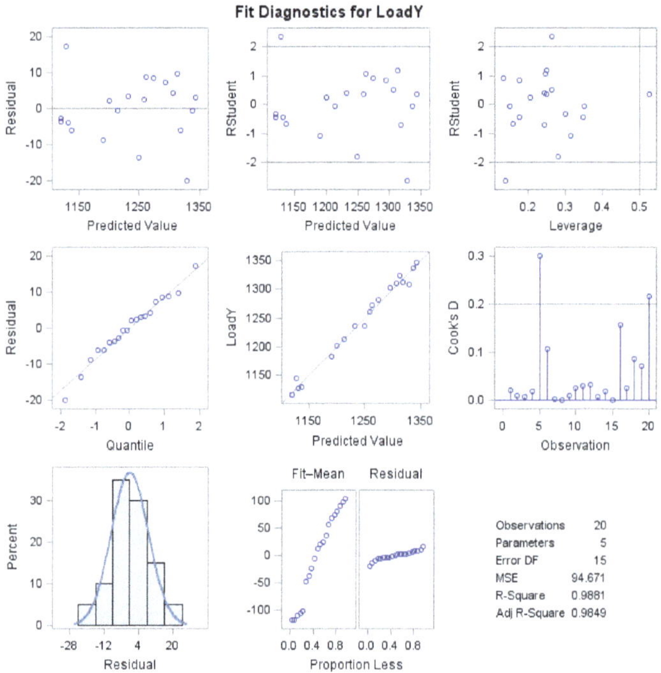

Figure A1. Residual Analysis of the Annual Regression Model.

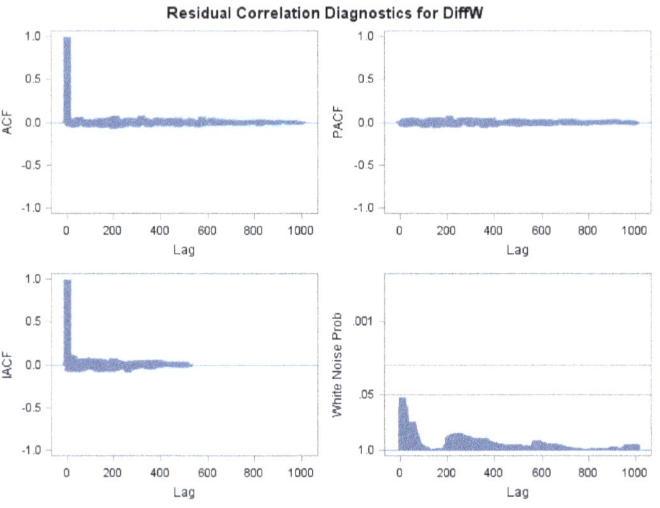

Figure A2. Analysis of Residual for the Weekly ARMAX Model.

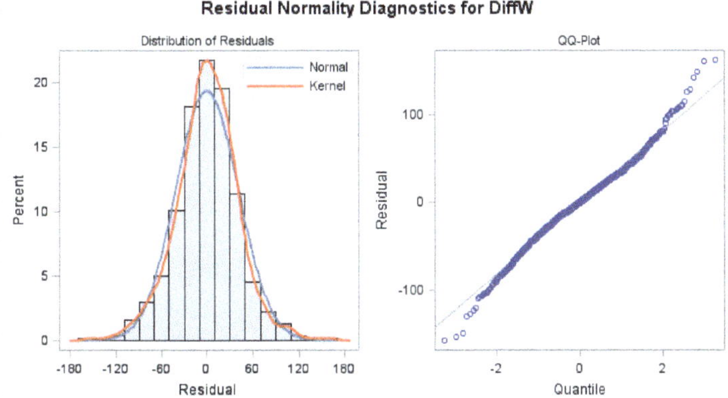

Figure A3. Normality Diagnostics of the Residuals of the Weekly ARMAX Model.

Table A1. The Results for the Weekly ARMAX Model.

Maximum Likelihood Estimation					
Parameter	**Estimate**	**Standard Error**	**t Value**	**Approx. Pr > |t|**	**Lag**
MU	1023.0	28.446	35.96	<0.0001	0
MA1,1	0.751	0.06	12.55	<0.0001	1
MA1,2	−0.056	0.019	−2.99	0.0028	50
MA1,3	0.712	0.037	19.22	<0.0001	52
MA1,4	−0.440	0.048	−9.22	<0.0001	53
MA1,5	−0.105	0.025	−4.24	<0.0001	54
AR1,1	1.128	0.057	19.67	<0.0001	1
AR1,2	−0.227	0.033	−6.87	<0.0001	2
AR1,3	0.762	0.027	28.06	<0.0001	52
AR1,4	−0.673	0.031	−22.11	<0.0001	53
T	−50.357	0.958	−52.56	<0.0001	0
T^2	0.533	0.01	57.50	<0.0001	0
Constant Estimate	10.074	**Std Error Estimate**	41.473		
AIC	10765.4	**SBC**	10824.8		

Table A2. The Results for the Regression Model for the Winter Season (Model 1).

Winter—Weekdays			Winter—Weekends		
Variable	**Parameter Estimate**	**Pr > |t|**	**Variable**	**Parameter Estimate**	**Pr > |t|**
Intercept	−43.94700	0.0003	Intercept	33.16159	0.0438
i	43.00126	<0.0001	i	−68.74540	<0.0001
i^2	−28.47236	<0.0001	i^2	10.86316	0.0456
i^3	3.90525	<0.0001	i^3	−0.34335	0.5244
$(i-6)^2$	−72.28854	<0.0001	$(i-5)^2$	4.53027	0.1994
$(i-14)^2$	−42.29815	<0.0001	$(i-12)^2$	21.97808	<0.0001
$(i-17)^2$	−103.8987	<0.0001	$(i-17)^2$	−42.63870	<0.0001
$(i-6)^3$	−1.95948	<0.0001	$(i-5)^3$	−0.98781	0.0461
$(i-14)^3$	8.92152	<0.0001	$(i-12)^3$	2.63976	<0.0001
$(i-17)^3$	−9.58272	<0.0001	$(i-17)^3$	−0.96957	0.0194
temp	−2.89175	<0.0001	temp	−2.59769	<0.0001
T^*i	−0.71365	<0.0001	T^*i	−0.15635	0.0027
$T^* i^2$	0.17868	<0.0001	$T^*(i-5)^2$	0.06344	<0.0001
$T^* i^3$	−0.01462	<0.0001	$T^*(i-12)^2$	−0.56360	<0.0001
$T^*(i-6)^2$	0.26047	<0.0001	$T^*(i-17)^2$	−0.67866	0.0005
$T^*(i-14)^2$	0.23657	<0.0001	$T^*(i-12)^3$	0.06180	<0.0001
$T^*(i-17)^3$	0.02458	<0.0001	$T^*(i-17)^3$	−0.03642	0.0005
RMSE	72.961		**RMSE**	74.388	
Adj-R2	0.767		**Adj-R2**	0.707	
AIC	260089.07		**AIC**	100750.97	

Table A3. The Results for the Regression Model for the Spring Season (Model 1).

Spring—Weekdays			Spring—Weekends		
Variable	**Parameter Estimate**	**Pr > |t|**	**Variable**	**Parameter Estimate**	**Pr > |t|**
Intercept	93.65097	<0.0001	Intercept	−7.13119	0.6187
i	−150.22285	<0.0001	i	−48.84602	<0.0001
i^2	68.15250	<0.0001	i^2	13.06001	<0.0001
i^3	−8.97118	<0.0001	i^3	−0.74388	<0.0001
$(i-4)^2$	115.05545	<0.0001	$(i-9)^2$	−13.54758	0.0001
$(i-8)^2$	61.51265	<0.0001	$(i-18)^2$	124.02630	0.0021
$(i-19)^2$	−124.08785	<0.0001	$(i-20)^2$	231.54827	0.0011
$(i-4)^3$	−3.47963	0.0009	$(i-9)^3$	2.62643	<0.0001
$(i-8)^3$	13.74708	<0.0001	$(i-18)^3$	−66.42324	<0.0001
$(i-19)^3$	9.48140	<0.0001	$(i-20)^3$	65.77067	<0.0001
temp	−2.50879	<0.0001	Temp	−1.45289	<0.0001
$T^* i^2$	−0.41258	<0.0001	$T^* i^2$	−0.22542	<0.0001
$T^* i^3$	0.07054	0.0001	$T^* i^3$	0.02324	<0.0001
$T^*(i-4)^2$	−1.05263	<0.0001	$T^*(i-9)^2$	−0.43314	<0.0001
$T^*(i-8)^2$	−1.02330	<0.0001	$T^*(i-18)^2$	−2.77329	<0.0001
$T^*(i-19)^2$	1.75668	<0.0001	$T^*(i-20)^2$	−5.14219	<0.0001
$T^*(i-4)^3$	0.06608	<0.0001	$T^*(i-9)^3$	−0.03102	<0.0001
$T^*(i-8)^3$	−0.14706	<0.0001	$T^*(i-18)^3$	1.26328	<0.0001
$T^*(i-19)^3$	−0.19409	<0.0001	$T^*(i-20)^3$	−1.22318	<0.0001
RMSE	73.524		**RMSE**	84.087	
Adj-R2	0.75		**Adj-R2**	0.625	
AIC	271284.41		**AIC**	111701.59	

Table A4. The Results for the Regression Model for the Summer Season (Model 1).

Summer—Weekdays			Summer—Weekends		
Variable	**Parameter Estimate**	**Pr > |t|**	**Variable**	**Parameter Estimate**	**Pr > |t|**
Intercept	−1031.8216	<0.0001	Intercept	−1121.8108	<0.0001
i	20.83889	0.6311	i	92.24404	0.0765
i^2	23.34519	0.0028	i^2	−4.50357	0.5201
i^3	−2.15146	<0.0001	i^3	0.23011	0.4594
$(i-9)^2$	−58.35710	<0.0001	$(i-7)^2$	−50.56548	0.0002
$(i-14)^2$	−52.55395	0.0212	$(i-12)^2$	79.89184	<0.0001
$(i-19)^2$	−55.43427	0.1113	$(i-20)^2$	−45.09066	0.0476
$(i-9)^3$	13.63412	<0.0001	$(i-7)^3$	0.75179	0.4427
$(i-14)^3$	−10.04996	<0.0001	$(i-12)^3$	−1.41147	0.3334
$(i-19)^3$	−3.42800	<0.0001	$(i-20)^3$	−5.65133	<0.0001
temp	12.30673	<0.0001	temp	13.92763	<0.0001
$T{*}i$	−1.67881	0.0056	$T{*}i$	−2.61695	<0.0001
$T{*}\,i^2$	−0.23403	0.0326	$T{*}\,i^2$	0.15085	0.0283
$T{*}\,i^3$	0.03886	<0.0001	$T{*}(i-7)^2$	1.43653	<0.0001
$T{*}(i-14)^2$	1.07409	0.0002	$T{*}(i-20)^2$	0.72256	0.0058
$T{*}(i-19)^2$	1.20714	0.0053	$T{*}(i-7)^3$	−0.14516	<0.0001
$T{*}(i-9)^3$	−0.17607	<0.0001	$T{*}(i-12)^3$	0.15275	<0.0001
$T{*}(i-14)^3$	0.09527	<0.0001			
RMSE	121.682		**RMSE**	138.235	
Adj-R2	0.866		**Adj-R2**	0.815	
AIC	303082.91		**AIC**	124226.62	

Table A5. The Results for the Regression Model for the Fall Season (Model 1).

Fall—Weekdays			Fall—Weekends		
Variable	**Parameter Estimate**	**Pr > |t|**	**Variable**	**Parameter Estimate**	**Pr > |t|**
Intercept	−124.73508	<0.0001	Intercept	−21.40477	0.3987
i	49.54582	0.0001	h	−27.84008	0.0282
i^2	−20.69335	<0.0001	h^2	7.44891	0.0016
i^3	2.88465	<0.0001	h^3	−0.32170	0.0803
$(i-6)^2$	−62.12202	<0.0001	$(i-8)^2$	−14.30060	0.3538
$(i-15)^2$	106.77851	<0.0001	$(i-10)^2$	5.40145	0.5837
$(i-20)^2$	190.28792	<0.0001	$(i-18)^2$	−147.68332	<0.0001
$(i-6)^3$	−1.41533	<0.0001	$(i-8)^3$	−1.67366	0.6195
$(i-15)^3$	−17.40696	<0.0001	$(i-10)^3$	5.28608	0.1368
$(i-20)^3$	4.95084	<0.0001	$(i-18)^3$	5.18245	<0.0001
temp	−0.55227	0.0901	temp	−1.22595	0.0026
$T{*}\,i$	−0.95896	<0.0001	$T{*}\,i$	−0.87935	<0.0001
$T{*}\,i^2$	0.05788	0.0049	$T{*}\,i^2$	0.00681	<0.0001
$T{*}(i-6)^2$	0.31932	<0.0001	$T{*}(i-8)^2$	0.68040	0.0036
$T{*}(i-15)^2$	−1.29167	<0.0001	$T{*}(i-18)^2$	1.72536	<0.0001
$T{*}(i-20)^2$	−3.25219	<0.0001	$T{*}(i-8)^3$	−0.14824	0.0002
$T{*}(i-6)^3$	−0.02245	<0.0001	$T{*}(i-10)^3$	0.11613	0.0105
$T{*}(i-15)^3$	0.23814	<0.0001	$T{*}(i-18)^3$	−0.10069	<0.0001
RMSE	92.402		**RMSE**	100.78	
Adj-R2	0.745		**Adj-R2**	0.653	
AIC	282449.54		**AIC**	115156.88	

Table A6. The Results for the Regression Model for the Winter and Spring Seasons (Model 2).

Winter			Spring		
Variable	**Parameter Estimate**	**Pr > |t|**	**Variable**	**Parameter Estimate**	**Pr > |t|**
Intercept	−16.55248	0.1207	Intercept	−28.71512	0.3281
i	23.25116	0.0014	i	43.44180	0.1746
i^2	−20.66445	<0.0001	i^2	−18.69310	0.0568
i^3	2.96367	<0.0001	i^3	2.56499	0.0042
$(i-6)^2$	−52.98180	<0.0001	$(i-5)^2$	17.48659	<0.0001
$(i-15)^2$	−16.66555	<0.0001	$(i-8)^2$	42.40157	<0.0001
$(i-17)^2$	−112.22096	<0.0001	$(i-19)^2$	−123.9761	<0.0001
$(i-6)^3$	−1.83272	<0.0001	$(i-5)^3$	−13.11806	<0.0001
$(i-15)^3$	13.03173	<0.0001	$(i-8)^3$	11.98398	<0.0001
$(i-17)^3$	−12.65773	<0.0001	$(i-19)^3$	9.16725	<0.0001
temp	−2.76643	<0.0001	temp	−0.64997	0.2414
T^*i	−0.64320	<0.0001	T^*i	−2.15721	0.0003
T^*i^2	0.13092	<0.0001	T^*i^2	0.50905	0.0049
T^*i^3	−0.00834	<0.0001	T^*i^3	−0.04807	0.0027
$T^*(i-6)^2$	0.12037	0.0070	$T^*(i-8)^2$	−0.96311	<0.0001
$T^*(i-15)^2$	0.13705	0.0009	$T^*(i-19)^2$	1.72975	<0.0001
$T^*(i-17)^3$	0.01452	<0.0001	$T^*(i-5)^3$	0.18007	<0.0001
Weekend	−49.01851	<0.0001	$T^*(i-8)^3$	−0.14305	<0.0001
			$T^*(i-19)^3$	−0.18822	<0.0001
			Weekend	−59.90003	<0.0001
RMSE	75.781		**RMSE**	79.687	
Adj-R2	0.741		**Adj-R2**	0.705	
AIC	363556.63		**AIC**	386694.58	

Table A7. The Results for the Regression Model for the Summer and Fall Seasons (Model 2).

Summer			Fall		
Variable	**Parameter Estimate**	**Pr > |t|**	**Variable**	**Parameter Estimate**	**Pr > |t|**
Intercept	−1131.9170	<0.0001	Intercept	−84.22522	<0.0001
i	115.72585	<0.0001	i	40.45021	0.0004
i^2	1.12711	0.2755	i^2	−17.21285	<0.0001
i^3	−0.77157	<0.0001	i^3	2.41275	<0.0001
$(i-9)^2$	−83.52495	<0.0001	$(i-6)^2$	−51.38721	<0.0001
$(i-14)^2$	−46.86818	0.0578	$(i-15)^2$	120.69047	<0.0001
$(i-19)^2$	−45.76871	0.1566	$(i-20)^2$	205.64318	<0.0001
$(i-9)^3$	13.02938	<0.0001	$(i-6)^3$	−1.34485	<0.0001
$(i-14)^3$	−11.66866	<0.0001	$(i-15)^3$	−18.15278	<0.0001
$(i-19)^3$	−3.20285	<0.0001	$(i-20)^3$	4.97629	<0.0001
temp	13.81380	<0.0001	temp	−0.70591	0.0145
T^*i	−2.73699	<0.0001	T^*i	−0.99218	<0.0001
$T^* i^3$	0.02435	<0.0001	$T^* i^2$	0.06113	0.0008
$T^* (i-9)^2$	0.30996	0.0564	$T^* (i-6)^2$	0.30423	<0.0001
$T^*(i-14)^2$	1.08035	0.0008	$T^*(i-15)^2$	−1.37326	<0.0001
$T^*(i-19)^2$	1.11159	0.0059	$T^*(i-20)^2$	−3.36051	<0.0001
$T^*(i-9)^3$	−0.17592	<0.0001	$T^*(i-6)^3$	−0.02131	<0.0001
$T^*(i-14)^3$	0.11706	<0.0001	$T^*(i-15)^3$	0.24565	<0.0001
Weekend	−55.87929	<0.0001	Weekend	−71.82423	<0.0001
RMSE	128.175		**RMSE**	97.147	
Adj-R2	0.85		**Adj-R2**	0.717	
AIC	428670.67		**AIC**	399798.26	

References

1. Bunn, D.; Farmer, E. *Comparative Models for Electrical Load Forecasting*; John Wiley & Sons: New York, NY, USA, 1985.
2. Alfares, H.; Nazeeruddin, M. Electric load forecasting: Literature survey and classification of methods. *Int. J. Syst. Sci.* **2002**, *33*, 23–34. [CrossRef]
3. El-Keib, A.A.; Ma, X.; Ma, H. Advancement of statistical based modeling techniques for short-term load forecasting. *Electr. Power Syst. Res.* **1995**, *35*, 51–58. [CrossRef]
4. Dash, P.K.; Liew, A.C.; Rahman, S. A comparison fuzzy neural network for the generation of daily average and peak load profiles. *Int. J. Syst. Sci.* **1995**, *26*, 2091–2106. [CrossRef]
5. Kim, K.H.; Park, J.K.; Hwang, K.J.; Kim, S.H. Implementation of hybrid short-term load forecasting system using artificial neural networks and fuzzy expert system. *IEEE Trans. Power Syst.* **1995**, *10*, 1534–1539.
6. Cho, H.; Goude, Y.; Brossat, X.; Yao, Q. Modeling and forecasting daily electricity load curves: A hybrid approach. *J. Am. Stat. Assoc.* **2013**, *108*, 7–21. [CrossRef]
7. Choueiki, M.H.; Mount-Campell, C.A.; Ahalt, S. Implementing a weighted least squares procedure in training a neural network to solve the short-term load forecasting problem. *IEEE Trans. Power Syst.* **1997**, *12*, 1689–1694. [CrossRef]
8. Gajowniczek, K.; Zabkowski, T. Two-stage electricity demand modeling using machine learning algorithms. *Energies* **2017**, *10*, 1–25.

9. Das, S.P.; Laharika, V.; Achray, N.S. Improved short-term electricity load forecasting using extreme learning machine. In Proceedings of the International Conference on Big Data Analytics and Computational Intelligence (ICBDAC), Chirala, India, 23–25 March 2017.

10. Annamareddi, S.; Gopinathan, S.; Dora, B. A simple hybrid model for short-term load forecasting. *J. Eng.* **2013**, *5*, 23–34. [CrossRef]

11. Baziar, A.; Kavousi-Fard, A. Short term load forecasting using a hybrid model based on support vector regression. *Int. J. Sci. Technol. Res.* **2015**, *4*, 189–195.

12. Nowicka-Zagrajek, J.; Weron, R. Modeling electricity loads in California: ARMA models with hyperbolic noise. *Signal Process.* **2002**, *82*, 1903–1915. [CrossRef]

13. Liu, J.M.; Chen, R.; Liu, L.; Harris, J. A semi-parametric time series approach in modeling hourly electricity loads. *J. Forecast.* **2006**, *25*, 537–559. [CrossRef]

14. Amaral, L.F.; Souza, R.C.; Stevenson, M. A smooth transition periodic autoregressive (ATPAR) model for short-term load forecasting. *Int. J. Forecast.* **2008**, *24*, 603–615. [CrossRef]

15. Dordonnat, V.; Koopman, S.J.; Ooms, M.; Dessertaine, A.; Collet, J. An hourly periodic state space model for modeling French national electricity load. *Int. J. Forecast.* **2008**, *24*, 566–587. [CrossRef]

16. Chapagain, K.; Kittipiyakul, S. Short-term electricity load forecasting model and Bayesian estimation for Thailand data. *MATEC Web Conf.* **2016**, *55*, 06003. Available online: https://www.matec-conferences.org/articles/matecconf/abs/2016/18/matecconf_acpee2016_06003/matecconf_acpee2016_06003.html (accessed on 1 November 2019). [CrossRef]

17. Kosiorowski, D. Functional regression in short-term prediction of economic time series. *Stat. Transit.* **2014**, *15*, 611–626.

18. Hyndman, R.J.; Shang, H.L. Functional time series forecasting. *J. Korean Stat. Soc.* **2009**, *38*, 199–211. [CrossRef]

19. Papadopoulos, S.; Karakatsanis, I. Short term electricity load forecasting using time series and ensemble learning methods. In Proceedings of the IEEE Power and Energy Conference at Illinois (PECI), Champaign, IL, USA, 20–21 February 2015; ISBN 978-1-4799-7949-3.

20. Alkhathami, M. Introduction to electric load forecasting methods. *J. Adv. Electr. Comput. Eng.* **2015**, *2*, 1–12.

21. Yang, A.; Li, W.; Yang, X. Short-term electricity load forecasting based on feature selection and least squares support vector machines. *Knowl. Based Syst.* **2019**, *163*, 159–173. [CrossRef]

22. PJM. *PJM Manual 19. Load Forecasting and Analysis*; PJM: Audubon, PA, USA, 2018.

23. Fahad, M.U.; Arabab, N. Factor affecting short term load forecasting. *J. Clean Energy Technol.* **2014**, *2*, 305–309. [CrossRef]

24. PJM. Historical Load Data. Available online: http://www.pjm.com/markets-and-operations/ops-analysis/historical-load-data.aspx (accessed on 6 May 2014).

25. Kunkel, C. Reforms Across PJM's 13-State Market are Costly and Regressive. Institute for Energy Economics and Financial Analysis. Available online: http://ieefa.org/pjms-reform/ (accessed on 10 December 2018).

26. Baillie, R.T. Predictions from ARMAX models. *J. Econom.* **1980**, *12*, 365–374. [CrossRef]

Article

Short-Term Electricity Demand Forecasting Using Components Estimation Technique

Ismail Shah [1,2,*,†], **Hasnain Iftikhar** [1,†], **Sajid Ali** [1,†] **and Depeng Wang** [3,*,†]

1   Department of Statistics, Quaid-i-Azam University, Islamabad 45320, Pakistan
2   Department of Statistical Sciences, University of Padua, 35121 Padova, Italy
3   College of Life Science, Linyi University, Linyi 276000, China
*   Correspondence: ishah@qau.edu.pk (I.S.); wangdepeng@lyu.edu.cn (D.W.)
†   These authors contributed equally to this work.

Received: 27 May 2019; Accepted: 28 June 2019; Published: 1 July 2019

Abstract: Currently, in most countries, the electricity sector is liberalized, and electricity is traded in deregulated electricity markets. In these markets, electricity demand is determined the day before the physical delivery through (semi-)hourly concurrent auctions. Hence, accurate forecasts are essential for efficient and effective management of power systems. The electricity demand and prices, however, exhibit specific features, including non-constant mean and variance, calendar effects, multiple periodicities, high volatility, jumps, and so on, which complicate the forecasting problem. In this work, we compare different modeling techniques able to capture the specific dynamics of the demand time series. To this end, the electricity demand time series is divided into two major components: deterministic and stochastic. Both components are estimated using different regression and time series methods with parametric and nonparametric estimation techniques. Specifically, we use linear regression-based models (local polynomial regression models based on different types of kernel functions; tri-cubic, Gaussian, and Epanechnikov), spline function-based models (smoothing splines, regression splines), and traditional time series models (autoregressive moving average, nonparametric autoregressive, and vector autoregressive). Within the deterministic part, special attention is paid to the estimation of the yearly cycle as it was previously ignored by many authors. This work considers electricity demand data from the Nordic electricity market for the period covering 1 January 2013–31 December 2016. To assess the one-day-ahead out-of-sample forecasting accuracy, Mean Absolute Percentage Error (MAPE), Mean Absolute Error (MAE), and Root Mean Squared Error (RMSE) are calculated. The results suggest that the proposed component-wise estimation method is extremely effective at forecasting electricity demand. Further, vector autoregressive modeling combined with spline function-based regression gives superior performance compared with the rest.

Keywords: Nordic electricity market; electricity demand; component estimation method; univariate and multivariate time series analysis; modeling and forecasting

---

1. Introduction

Liberalization of the energy sector, changes in climate policies, and the upgrade of renewable energy resources have completely changed the structure of the previous strictly-controlled energy sector. Today, most energy markets have been liberalized and privatized with the purpose of gaining consistent and inexpensive facilities for power trades. Within the energy sector, the liberalization of the electricity market has also introduced new challenges. In particular, electricity demand and price forecasting have become extremely important issues for producers, energy suppliers, system operators, and other market participants. In many electricity markets, electricity demand is fixed a day before the physical delivery by concurrent (semi-)hourly auctions. Further, electricity cannot be stored in an efficient manner, and the end-user demand must be satisfied instantaneously; thus, accurate forecast for electricity demand is crucial for effective power system management [1,2].

The electricity demand forecast can be broadly divided into three time horizons: (a) short-term, (b) medium-term, and (c) long-term load forecasting. Long-Term Load Forecast (LTLF) includes horizons from a few months to several years ahead. LTLF is generally used for planning and investment profitability analysis, determining upcoming sites, or acquiring fuel sources for production plants [3]. Medium-Term Load Forecast (MTLF) normally considers horizons from a few days to months ahead and is usually preferred for risk management, balance sheet calculations, and derivatives pricing [4]. Finally, Short-Term Load Forecast (STLF) generally includes horizons from a few minutes up to a few days ahead. In practice, the most attention in electricity load forecasting has been paid to STLF since it is an essential tool for the daily market operations [5].

However, electricity demand forecasting is a difficult task due to the features demand time series exhibit. These features include non-constant mean and variance, calendar effects, multiple periodicities, high volatility, jumps, etc. For example, the yearly, weekly, and daily periodicities can be seen from Figure 1. The weekly phase is comprised of comparatively lower variation in the data. The load curves are comparatively different on different days of the week, and the demand varies throughout the day. The demand is high on weekdays as compared to weekends. Moreover, electricity demand is also affected by calendar effects (bank/bridging holidays) and by seasons. In general, the demand is considerably lower during bank holidays and bridging holidays (a day among two non-working days). From the figure, high volatility in electricity demand can also be observed in almost all load periods. In addition, different environmental, geographical, and meteorological factors have a direct effect on electricity demand. Further, as electricity is a secondary source of energy, which is retrieved by converting prime energy sources like fossil fuels, natural gas, solar, wind power, etc. [6], the cost related to each source is different. Thus, a consistent electricity supply mechanism for different levels of demand with short periods of high and rather longer periods of moderate demand is necessary.

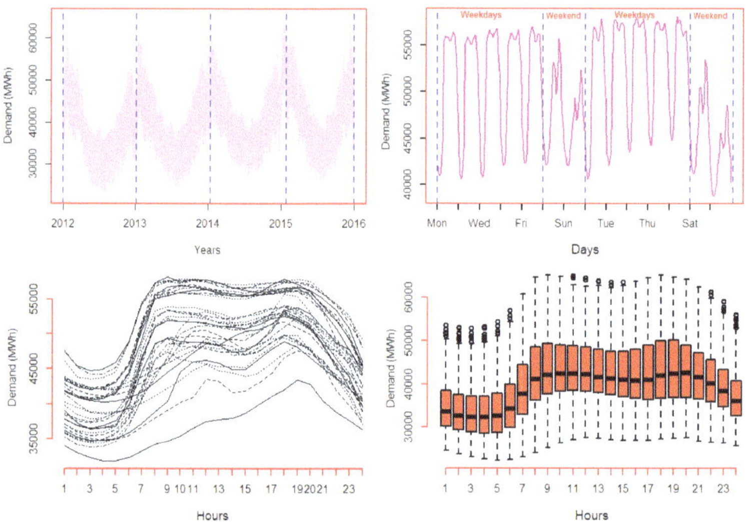

Figure 1. Yearly seasonality for the period 01-01-2012–31-12-2015 (**top left**), weekly periodicity for the period 01-01-2013–14-01-2013 (**top right**), box plot of hourly electricity load for the period 01-01-2013–31-12-2016 (**bottom right**), daily load curves for the period 01-01-2013–31-01-2013, weekdays (solid lines), Saturdays (dashed lines), Sundays (dotted lines), and bank holidays at the bottom (solid) representing 1 January (**bottom left**).

To account for the different features of the demand series, in the last two decades, researchers suggested different methods and models to forecast electricity demand [7–11]. For example, the work in [12] proposed a semi-parametric component-based model consisting of a non-parametric (smoothing spline) and a parametric (autoregressive moving average model) component. Exponential smoothing techniques are also widely used in forecasting electricity demand [13,14]. Multiple equations time series models, e.g., the Double Seasonal Autoregressive Moving Average (DSARIMA) model, the Double Holt–Winters (D-HW) model, and Multiple Equations Time Series (MET) approaches are also used for short-term load forecasting [15,16]. Regression methods are easy to implement and have been widely used for electricity demand forecasting in the past. For example, the work in [17] used parametric regression models to forecast electricity demand for the Turkish electricity market. Some authors included exogenous variables in the time series models to improve the forecasting performance [18–20]. Several researchers compared the classical time series models and computational intelligence models [21–23]. For example, the work in [24] compared the Seasonal Autoregressive Moving Average (SARIMA) and Adaptive Network-based Fuzzy Inference System (ANFIS) models. For short-term load forecasting, the work in [25] introduced a new hybrid model that combines SARIMA and the Back Propagation Neural Network (BPNN) model. Some authors suggested the use of functional data analysis to predict electricity demand [26–28]. The main idea behind this approach is to consider the daily demand profile as a single functional object; thus, functional approaches can be applied to electricity load series. Other approaches used for demand forecasting can be seen in [29–33]. Apart from the forecasting models, Distributed Energy Resources (DERs) that are directly connected to a local distribution system and can be used for electricity producing or as controllable loads are also discussed in the literature [34,35]. DERs include solar panels, combined heat and power plants, electricity storage, small natural gas-fueled generators, and electric water heaters.

The main objective of this work is to compare different modeling techniques for electricity demand forecasting. The main attention is paid to the yearly cycle, which in many cases is ignored. The authors suggest to estimate jointly the effect of the long-term trend and yearly cycle using one component [36,37]. In practice, however, the yearly component shows regular cycles, while the long-term component highlights the trend structure of the data. Thus, these two components must be modeled separately [26]. Further, in our case, some pilot analyses suggested that modeling these two components separately can significantly improve the forecasting results. Thus, the main contribution of this paper is the thorough investigation of the impact of yearly component estimation on one-day-ahead out-of-sample electricity demand forecasting. Within the framework of the components estimation method, we compare models in terms of forecasting ability considering both univariate and multivariate, as well as parametric and non-parametric models. Moreover, for the considered models, the significance analysis of the difference in predication accuracy is also conducted.

The rest of the article is organized as follows: Section 2 contains a description of the proposed modeling framework and of the considered models. Section 3 provides an application of the proposed modeling framework. Section 4 contains a summary and conclusions.

2. Component-Wise Estimation: General Modeling Framework

The main objective of this study is to forecast one-day-ahead electricity demand using different forecasting models and methods. To this end, let $\log(D_{t,j})$ be the series of the log demand for the tth day and the jth hour. Following [28,33], the dynamics of the log demand, $\log(D_{t,j})$, can be modeled as:

$$\log\left(D_{t,j}\right) = F_{t,j} + R_{t,j} \tag{1}$$

That is, the $\log(D_{t,j})$ is divided into two major components: a deterministic component $F_{t,j}$ and a stochastic component $R_{t,j}$. The deterministic component, $F_{t,j}$, is comprised of the long-run trend, annual, seasonal, and weekly cycles, and calendar effects and is modeled as:

$$F_{t,j} = l_{t,j} + a_{t,j} + s_{t,j} + w_{t,j} + b_{t,j} \tag{2}$$

where $l_{t,j}$ represents the long-run (trend component), $a_{t,j}$ represents the annual cycles, $s_{t,j}$ represents the seasonal cycles, $w_{t,j}$ is the weekly cycles, and $b_{t,j}$ represents the bank holidays. On the other hand, $R_{t,j}$ is a (residual) stochastic component that describes the short-run dependence of demand series. Concerning the estimation of the deterministic component, apart from the yearly component $a_{t,j}$, the remaining components are estimated using parametric regression. For the estimation of $a_{t,j}$, six different methods including the sinusoidal function-based regression techniques, three local polynomial regression models, and two regression spline function-based models are used. All the components in Equation (2) are estimated using the back fitting algorithm. In the case of stochastic component $R_{t,j}$, four different methods, namely the Autoregressive Model (AR), the Non-Parametric Autoregressive model (NPAR), the Autoregressive Moving Average Model (ARMA), and the Vector-Autoregressive model (VAR) are used. Combining the models for deterministic and stochastic components estimations leads us to comparing twenty four $(6^F \times 4^R =)$ 24 different combinations. Note that in the case of univariate models, each load period is modeled separately to account for the intra-daily periodicity [38].

2.1. Modeling the Deterministic Component

This section will explain the estimation of the deterministic component. The long-run (trend) component $l_{t,j}$, which is a function of time t, is estimated using Ordinary Least Squares (OLS). Dummy variables are used for seasonal periodicities, weekly periodicities, and for bank holidays, i.e., $s_t = \sum_{i=1}^{4} \alpha_i I_{i,t}$, with $I_{i,t} = 1$ if t refers to the ith season of the year and zero otherwise, $w_t = \sum_{i=1}^{7} \beta_i I_{i,t}$, with $I_{i,t} = 1$ if t refers to the ith day of the week and zero otherwise, and $b_t = \sum_{i=1}^{2} \gamma I_{i,t}$, with $I_{i,t} = 1$ if t refers to a bank holiday or zero otherwise. The coefficients α's, β's, and γ's are estimated by OLS. On the other hand, the annual component $a_{t,j}$, which is a function of the series $(1, 2, 3, \ldots, 365, 1, 2, 3, \ldots, 365, \ldots)$, is estimated by six different methods that include Sinusoidal function-based Regression (SR), local polynomial regression models with three different kernels, namely: tri-cubic ((L1), Gaussian (L2), and Epanechnikov (L3), Regression Splines (RS), and Smoothing Splines (SS).

2.1.1. Sinusoidal Function Regression Model

Sinusoidal function Regression (SR) is widely used in the literature to capture the periodicity of a periodic component [39–44]. In this method, we consider that the annual cycle can be estimated using q sine and cosine functions given as:

$$a_t = \sum_{i=1}^{q} \left(\alpha_{1,i} \sin wa_t + \alpha_{2,i} \cos wa_t \right), \tag{3}$$

where $w = \frac{2\pi}{365.25}$. The unknown parameters $\alpha_{1,i}$ and $\alpha_{2,i}$ $(i = 1, \ldots, q)$ are estimated by OLS.

2.1.2. Local Polynomial Regression

Local polynomial regression is a flexible non-parametric technique that approximates a_t at a point a_0 by a low-order polynomial (say, q), fit using only points in some neighborhood of a_0.

$$\hat{a}_t = \sum_{j=1}^{q} \hat{\alpha}_j (a_t - a_0)^j. \tag{4}$$

Parameters $\hat{\alpha}_j$ are estimated by Weighted Least Squares (WLS) by minimizing:

$$\sum_{t=1}^{N}(a_t - \hat{a}_t)^2 K_{\delta(a)}(a_t - a_0), \tag{5}$$

where $K_{\delta(a)}(a_t - a_0)$ is a weighting kernel function, which depends on the smoothing parameter δ, also known as the bandwidth. It controls the size of the neighborhood around a_0 [45] and, thus, of the locality of the approximation. In this work, the value of the bandwidth is selected by using the cross-validation technique. Three different weighting kernel functions, namely the tri-cubic kernel (L1), the Epanechnikov (L2), and Gaussian kernels (L3) are used. It is worth mentioning that these types of local kernel-based regression techniques have been extensively used in the literature [31,39,40,44,46].

2.1.3. Regression Spline Models

Spline Regression (RS) is a popular non-parametric regression technique, which approximates a_t by means of piecewise polynomials of order p, estimated in the subintervals delimited by a sequence of m points called knots. Any spline function $Z(a)$ of order p can be described as a linear combination of functions $Z_j(a)$ called basis functions and is expressed in the following way:

$$Z(a) = \sum_{j=1}^{m+p+1} \alpha_j Z_j(a).$$

The unknown parameters α_j are estimated by the OLS. The most important choice is the number of knots and their location because they define the smoothness of the approximation. Again, we chose it by the cross-validation approach. In the literature, many authors considered this approach for long-run component prediction [11,12,26]. The annual cycle component for regression splines is estimated as,

$$\hat{a}_t = \hat{Z}(a)$$

2.1.4. Smoothing Splines

To overcome the requirement for fixing the number of knots, spline functions can alternatively be estimated by using the penalized least squares condition to minimize the sum of squares. Hence, the expression to minimize becomes:

$$\sum_{j=1}^{N}(a_t - Z(a))^2 + \lambda \int (Z''(a))^2 dt \tag{6}$$

where $(Z''(a))$ is the second derivative of $Z(a)$. The first term accounts for the degree of fitting, while the second one penalizes the roughness of the function through the smoothing parameter λ. The selection of smoothing parameter is an important task, which in this work is done using the cross-validation approach. Smoothing Splines (SS) have been previously used by some authors to estimate the long-run dynamics of the series, e.g., [11,47,48].

To see the performance of all six models defined above for the estimation of the annual component $a_{t,j}$, the observed log demand and the estimated annual components are depicted in Figure 2. From the figure, we can see that all six models for $a_{t,j}$ were capable of capturing the annual seasonality, as the annual cycles can be seen clearly from the figure.

Finally, it is worth mentioning that one-day-ahead forecast for the deterministic component is straightforward as the elements of $F_{t,j}$ are deterministic functions of time or calendar conditions, which are known at any time. Once all these components are estimated, the residual (stochastic) component $R_{t,j}$ is obtained as:

$$R_{t,j} = \log(D_{t,j}) - (\hat{l}_{t,j} + \hat{a}_{t,j} + \hat{s}_{t,j} + \hat{w}_{t,j} + \hat{b}_{t,j})$$

(7)

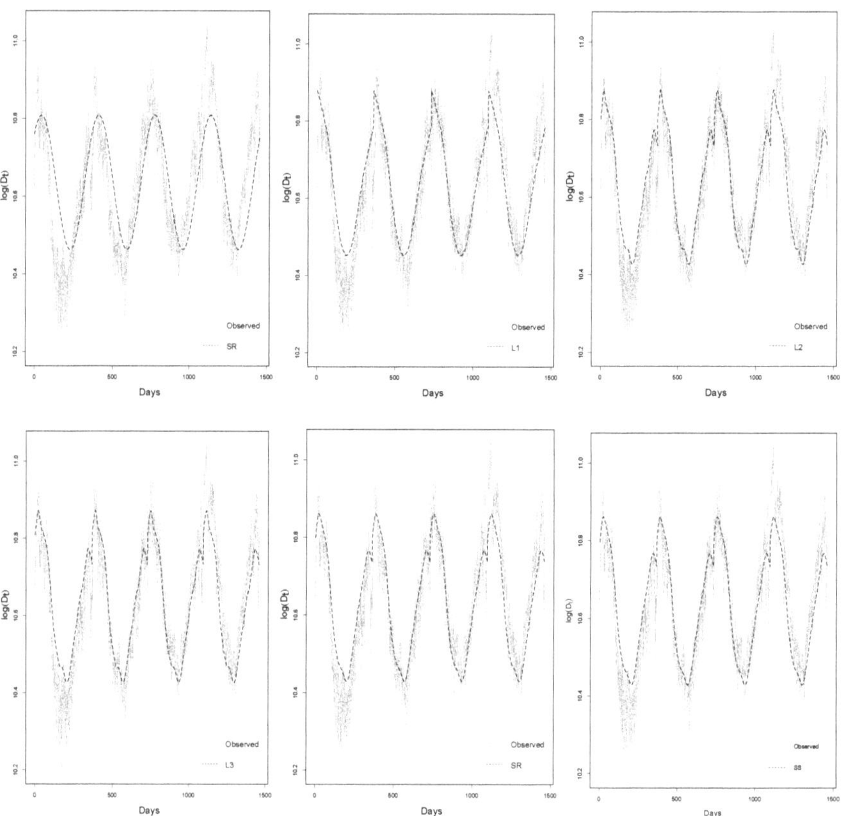

Figure 2. Observed $log(D_{t,21})$ with superimposed estimated $a_{t,j}$ using: (**first row**) Sinusoidal Regression (SR) (**left**), Local regression (L1) (**middle**), L2 (**right**), and (**second row**) L3 (**left**), Regression Splines (RS) (**middle**), and Smoothing Splines (SS) (**right**).

2.2. Modeling the Stochastic Component

Once the stochastic (residual) component is obtained, different types of parametric and non-parametric time series models can be considered. In our case, from the univariate class, we consider parametric AutoRegressive (AR), Non-Parametric AutoRegressive (NPAR), and Autoregressive Moving Average (ARMA). On the other hand, the Vector AutoRegressive (VAR) model is used to compare the performance of the multivariate model with the univariate models.

2.2.1. Autoregressive Model

A linear parametric Autoregressive (AR) model defines the short-run dynamics of $R_{t,j}$ taking into account a linear combination of the past r observations of $R_{t,j}$ and is given by:

$$R_{t,j} = c + \beta_1 R_{t-1,j} + \beta_2 R_{t-2,j} + ... + \beta_r R_{t-r,j} + \epsilon_t$$

(8)

where c is the intercept, β_i ($i = 1, 2, \ldots, r$) are the parameters of the AR(r) model, and ϵ_t is a white noise process. In our case, the parameters are estimated using the maximum likelihood estimation method. After some pilot analysis on different load periods, we concluded that the lags 1, 2, and 7 were significant in most cases and hence were used to estimate the model.

2.2.2. Non-Parametric Autoregressive Model

The additive non-parametric counterpart of AR is an additive model (NPAR), where the relation between $R_{t,j}$, and its lagged values do not have a particular parametric form, allowing, potentially, for any type of non-linearity and given by:

$$R_{t,j} = g_1(R_{t-1,j}) + g_2(R_{t-2,j}) + \ldots + g_r(R_{t-r,j}) + \epsilon_{t,j} \tag{9}$$

where g_i are smoothing functions describing the relation between each past values and $R_{t,j}$. In our case, functions g_i are represented by cubic regression splines. As in the parametric case, we used the lags 1, 2, and 7 to estimate NPAR. To avoid the so-called "curse of dimensionality", we used the back fitting algorithm to estimate the model [49].

2.2.3. Autoregressive Moving Average Model

The Autoregressive Moving Average (ARMA) model not only includes the lagged values of the series, but also considers the past error terms in the model. In our case, the stochastic component $R_{t,j}$ is modeled as a linear combination of the past r observations, as well as the lagged error terms. Mathematically,

$$R_{t,j} = c + \beta_1 R_{t-1,j} + \beta_2 R_{t-2,j} + \ldots + \beta_r R_{t-r,j} + \epsilon_{t,j} + \phi_1 \epsilon_{t-1,j} + \phi_2 \epsilon_{t-2,j} + \ldots + \phi \epsilon_{t-s,j} \tag{10}$$

where c is the intercept, β_i ($i = 1, 2, \ldots, r$) and ϕ_j ($j = 1, 2, \ldots, s$) are parameters of the AR and MA components, respectively, and $\epsilon_t \sim \mathcal{N}(0, \sigma_\epsilon^2)$. In this case, some pilot analyses suggest that the lags 1, 2, and 7 are significant for the AR part, while only the lag 1 for the MA part, thus a constrained ARMA(7,1) where $\beta_3 = \cdots = \beta_6 = 0$ is fitted to $R_{t,j}$ using the maximum likelihood estimation method.

2.2.4. Vector Autoregressive Model

In the Vector Autoregressive (VAR) model, both the response and the predictors are vectors, and hence, they contain information on the whole daily load profile. This allows one to account for possible interdependence among demand levels at different load periods. In our context, the daily stochastic component R_t is modeled as a linear combination of the past r observations of R_t, i.e.,

$$\mathbf{R_t} = \mathbf{G_1 R_{t-1}} + \mathbf{G_2 R_{t-2}} + \cdots + \mathbf{G_r R_{t-r}} + \epsilon_t \tag{11}$$

where $\mathbf{R_t} = \{R_{t,1}, \ldots, R_{t,24}\}$, $\mathbf{G_j}$ ($j = 1, 2, \cdots, r$) are coefficient matrices and $\epsilon_t = (\epsilon_{t,1}, \ldots, \epsilon_{t,24})$ is a vector of the disturbance term, such that $\epsilon_t \sim \mathcal{N}(0, \Sigma_\epsilon)$. Estimation of the parameters is done using the maximum likelihood estimation method.

Finally, once estimation of both, deterministic and stochastic, components is done, the final day-ahead electricity demand forecast is obtained as:

$$\begin{aligned} \hat{D}_{t+1,j} &= \exp\left(\hat{l}_{t+1,j} + \hat{a}_{t+1,j} + \hat{s}_{t+1,j} + \hat{w}_{t+1,j} + \hat{b}_{t+1,j} + \hat{R}_{t+1,j}\right) \\ &= \exp\left(\hat{F}_{t+1,j} + \hat{R}_{t+1,j}\right) \end{aligned} \tag{12}$$

For the stochastic component $R_{t,j}$ and the final model error $\epsilon_{t,j}$, examples of the Autocorrelation Function (ACF) and Partial Autocorrelation Function (PACF) are plotted in Figure 3 and 4. Note that in the case of $\epsilon_{t,j}$, both ACF and PACF refer to the models when VAR is used as a stochastic model.

The reason for plotting the residual obtained after applying the VAR model to $R_{t,j}$ is the superior forecasting performance of the multivariate model (see Table 1). Overall, the residuals $\epsilon_{t,j}$ of each model have been whitened. In some cases, residuals still show some significant correlation, but with an absolute value so small that it is useless for prediction.

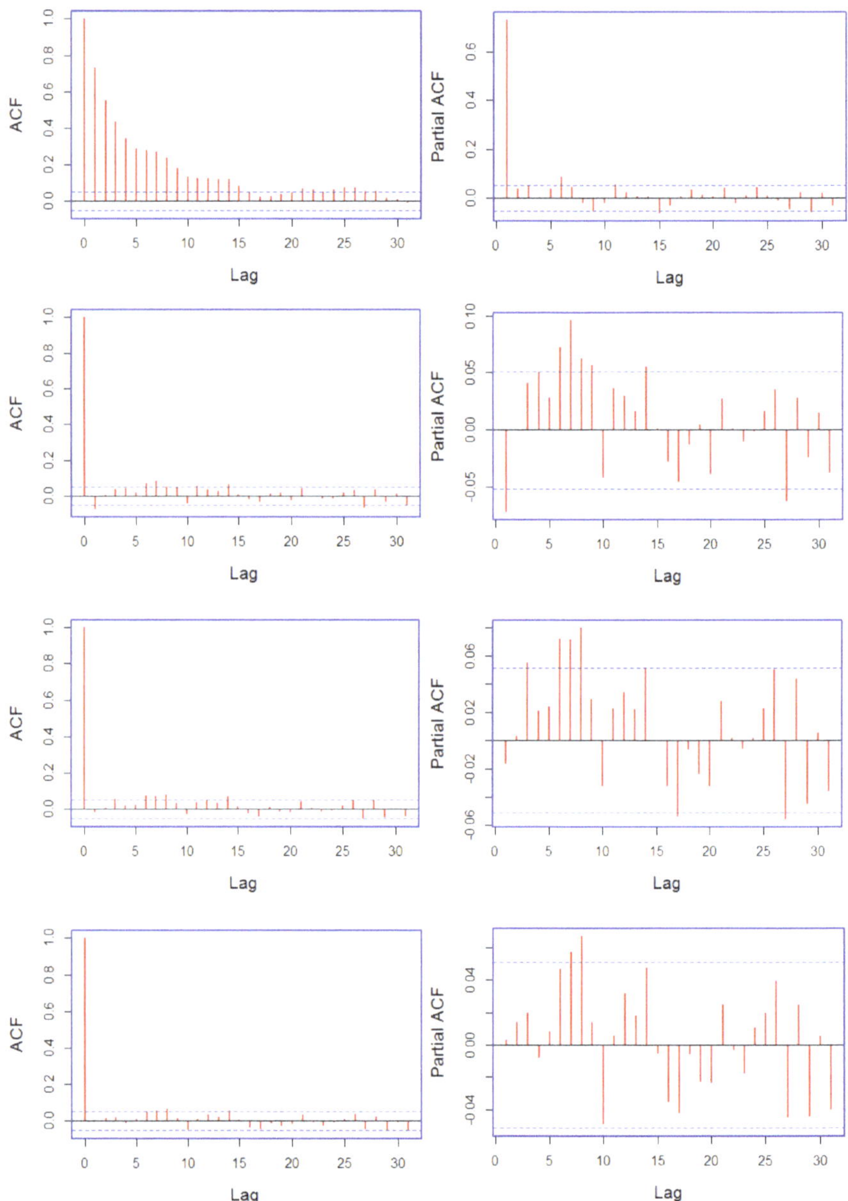

Figure 3. ACF and Partial Autocorrelation Function (PACF) plots for $R_{t,21}$ (**first row**), ACF and PACF plots for $\epsilon_{t,21}$ obtained with L1-VAR (**second row**), L2-VAR (**third row**), and L3-VAR (**fourth row**).

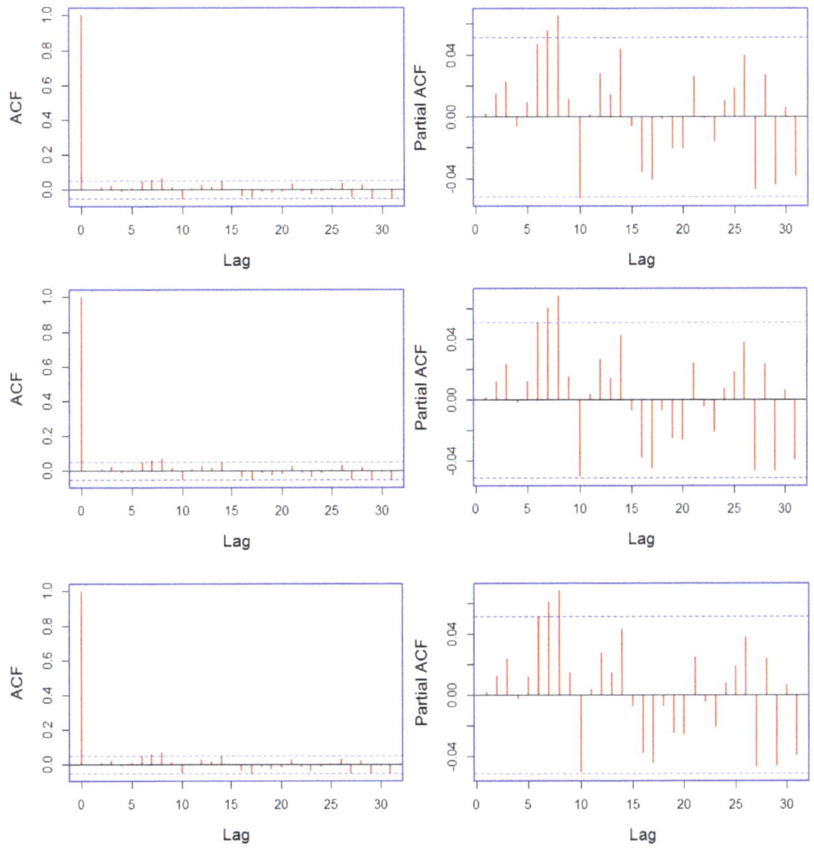

Figure 4. ACF and PACF plots for $\epsilon_{t,21}$ obtained with SR-VAR (**first row**), RS-VAR (**second row**), and SS-VAR (**third row**).

Table 1. Descriptive statistics for one-day-ahead out-of-sample forecasting: The column represents the estimation of the yearly component through Sinusoidal Regression (SR), Local regression (L1), Local regression (L2), Local regression (L3), Regression Spline (RS), and Smoothing Spline (SS). The row represents the estimation of the stochastic component thorough Autoregressive (AR), Non-Parametric Autoregressive (NPAR), Autoregressive Moving Average (ARMA), and Vector Autoregressive (VAR).

ERRORS	MODELS	SR	L1	L2	L3	RS	SS
	AR	2.503	2.466	2.413	2.412	2.411	2.412
MAPE	**NPAR**	2.510	2.434	2.413	2.411	2.399	2.395
	ARMA	2.514	2.435	2.413	2.418	2.405	2.396
	VAR	2.143	2.109	1.997	1.995	1.994	1.995
	AR	1081.184	1069.125	1044.341	1044.686	1044.177	1044.611
MAE	**NPAR**	1086.336	1056.392	1045.753	1046.899	1041.275	1039.804
	ARMA	1084.869	1055.017	1048.107	1045.881	1042.705	1038.553
	VAR	922.405	907.187	856.497	856.135	856.082	856.088
	AR	1486.580	1493.652	1450.551	1454.510	1450.676	1453.358
RMSE	**NPAR**	1476.394	1450.813	1436.108	1439.677	1434.670	1433.172
	ARMA	1468.908	1443.367	1431.686	1431.296	1427.437	1422.794
	VAR	1219.608	1211.225	1146.302	1146.002	1145.979	1146.014

3. Out-of-Sample Forecasting

This work considers the electricity demand data for the Nord Pool electricity market. The data cover the period from 1 January 2013–31 December 2016 (35,064 hourly demand levels for 1461 days). A few missing observations in the load series were replaced by averages of the neighboring observations. The whole dataset was divided into two parts: 1 January 2013-31 December 2015 (26,280 data points, covering 1095 days) for model estimation and 1 January 2016–31 December 2016 (8784 data points, covering 366 days) for one-day-ahead out-of-sample forecasting.

In the first step, the deterministic component was estimated separately for each load period as described in Section 2.1. An example of estimated deterministic components, as well as of $R_{t,j}$ is plotted in Figure 5. In the figure, along with the log demand at the top left, the long trend, yearly, seasonal, and weekly components are plotted on top right, middle left, middle right, and bottom left, respectively. Note that the elements of the deterministic components capture different dynamics of the log demand. An example of the series $R_{t,21}$ is plotted at the bottom right in Figure 5. In the second step, the previously-defined models for stochastic component were applied to the residual series $R_{t,j}$. In both steps, models were estimated and one-day-ahead forecasts were obtained for 366 days using the rolling window technique. Final demand forecasts were obtained using Equation (12).

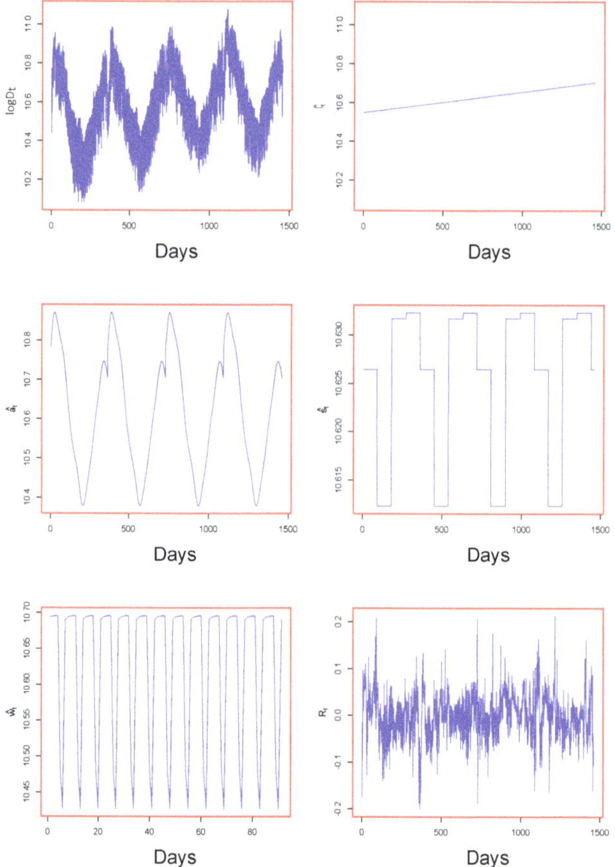

Figure 5. $log(D_{t,21})$ (**top left**), $\hat{l}_{t,21}$ (**top right**), $\hat{a}_{t,21}$ (**middle left**), $\hat{s}_{t,21}$ (**middle right**), $\hat{w}_{t,21}$ (**bottom left**), and $R_{t,21}$ (**bottom right**).

To evaluate the forecasting performance of the final models obtained from different combinations of deterministic and stochastic components, three accuracy measures, namely Mean Absolute Percentage Error (MAPE), Mean Absolute Error (MAE), and Root Mean Squared Error (RMSE) were computed as:

$$\text{MAPE} = \text{mean}\left(\frac{|D_{t,j} - \hat{D}_{t,j}|}{D_{t,j}}\right) \times 100$$

$$\text{MAE} = \text{mean}\left(|D_{t,j} - \hat{D}_{t,j}|\right)$$

$$\text{RMSE} = \sqrt{\text{mean}(D_{t,j} - \hat{D}_{t,j})^2} ,$$

where $D_{t,j}$ and $\hat{D}_{t,j}$ are the observed and the forecasted demand for the tth day ($t = 1, 2, \ldots, 366$) and the jth ($j = 1, 2, \ldots, 24$) load period.

As within the deterministic component, this work used six different estimation methods for $a_{t,j}$, whereas the estimation of other elements was the same; six different combinations were obtained. On the other hand, four different models were used to model the stochastic component. Hence, the estimation of both, deterministic and stochastic, components led us to compare twenty four different models. For these twenty four models, one-day-ahead out-of-sample forecast results are listed in Table 1. From the table, it is evident that the multivariate VAR model combined with any estimation technique used for $a_{t,j}$ led to a better forecast compared to the univariate models. The best forecasting model was obtained by combining VAR and RS, which produced 1.994, 856.082, and 1145.979 for MAPE, MAE, and RMSE, respectively. VAR combined with SS or with L3 produced the second best results. Within the univariate models, NPAR combined with the spline-based regression models performed better than the other two parametric counterparts. Finally, any stochastic model combined with SR or with L1 led to the worst forecast in their respective classes (univariate and multivariate). Considering only MAPE, a graphical representation of the results for the twenty four combination is given in Figure 6. From the figure, we can easily see that multivariate models performed better than the univariate models. To assess the significance of the difference among accuracy measures listed in Table 1 for different combinations, we performed the Diebold and Mariano (DM) [50] test of equal forecast accuracy. The DM test is a widely-used statistical test for comparing forecasts obtained from different models. To understand it, consider two forecasts, $\hat{y}_{1t}$ and $\hat{y}_{2t}$, that are available for the time series y_t for $t = 1, \ldots, T$. The associated forecast errors are $\epsilon_{1t} = y_t - \hat{y}_{1t}$ and $\epsilon_{2t} = y_t - \hat{y}_{2t}$. Let the loss associated with forecast error $\{\epsilon_{it}\}_{i=1}^2$ by $L(\epsilon_{it})$. For example, time t absolute loss would be $L(\epsilon_{it}) = |\epsilon_{it}|$. The loss differential between Forecasts 1 and 2 for time t is then $\eta_t = L(\epsilon_{1t}) - L(\epsilon_{2t})$. The null hypothesis of equal forecast accuracy for two forecast is $E[\eta_t] = 0$. The DM test requires that the loss differential be covariance stationary, i.e.,

$$\begin{aligned} E[\eta_t] &= \mu, \quad \forall\, t \\ \text{cov}(\eta_t - \eta_{t-\tau}) &= \gamma(\tau), \quad \forall\, t \\ \text{var}(\eta_t) &= \sigma_\eta, \quad 0 < \sigma_\eta < \infty \end{aligned}$$

Under these assumptions, the DM test of equal forecast accuracy is:

$$\text{DM} = \frac{\bar{\eta}}{\hat{\sigma}_\eta} \xrightarrow{d} N(0,1)$$

where $\bar{\eta} = \frac{1}{T}\sum_{t=1}^{T} \eta_t$ is the sample mean loss differential and $\hat{\sigma}_\eta$ is a consistent standard error estimate of η_t.

The results for the DM test are listed in Table 2 and Table 3. The elements of these tables are p-values of the Diebold and Mariano test where the null hypothesis assumes no difference in the accuracy of predictors in the column and row against the alternative hypothesis that the predictor in the column is more accurate than the predictor in the row. From Table 2, it is clear that the multivariate VAR models outperform their univariate counterparts. When looking at the results of VAR using different methods of estimation for $a_{t,j}$ in Table 3, it can be seen that, except for SR-VAR and L1-VAR, the remaining four combinations had the same predictive ability. In the case of SR-VAR and L1-VAR, the remaining four combinations performed statistically better.

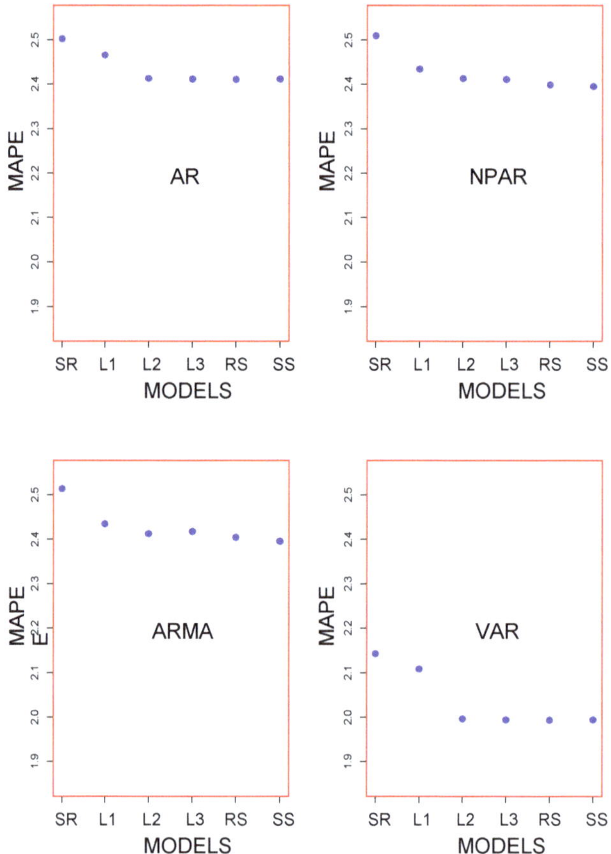

Figure 6. One-day-ahead out-of-sample MAPE for electricity demand using SR, L1, L2, L3, RS, SS, AR, NPAR, ARMA, and VAR.

Table 2. *p*-values for the Diebold and Mariano test. H_0: the forecasting accuracy for the model in the row and the model in the column is the same; H_1: the forecasting accuracy of the model in the column is greater than that of the model in the row.

MODELS	RS-AR	RS-NPAR	RS-ARMA	RS-VAR
RS-AR	-	0.33	0.40	< 0.01
RS-NPAR	0.67	-	0.75	< 0.01
RS-ARMA	0.60	0.25	-	< 0.01
RS-VAR	0.99	0.99	0.99	-

Table 3. *p*-values for the Diebold and Mariano test. H_0: the forecasting accuracy for the model in the row and the model in the column is the same; H_1: the forecasting accuracy of the model in the column is greater than that of the model in the row.

Models	SR-VAR	L1-VAR	L2-VAR	L3-VAR	RS-VAR	SS-VAR
SR-VAR	-	.28	< 0.01	< 0.01	< 0.01	< 0.01
L1-VAR	0.72	-	< 0.01	< 0.01	0.01	< 0.01
L2-VAR	0.99	0.99	-	0.93	0.85	0.83
L3-VAR	0.99	0.99	0.07	-	0.47	0.45
RS-VAR	0.99	0.99	0.15	0.53	-	0.48
SS-VAR	0.99	0.99	0.17	0.55	0.52	-

The day-specific MAPE, MAE, and RMSE are tabulated in Table 4. From this table, we can see that day-specific MAPE was relatively higher on Monday and Sunday and smaller on other weekdays. As the VAR model performed better previously, the day-specific MAPE values for this model were considerably lower compared to univariate models, except on Wednesday, Thursday, and Friday. For these three days, both the univariate and multivariate models produced lower errors. The same findings can be seen by looking at day-specific MAE and day-specific RMSE. The day-specific MAPE values are also depicted in Figure 7. The figure clearly indicates that the MAPE value was lower in the middle of the week and was higher on Monday and Sunday.

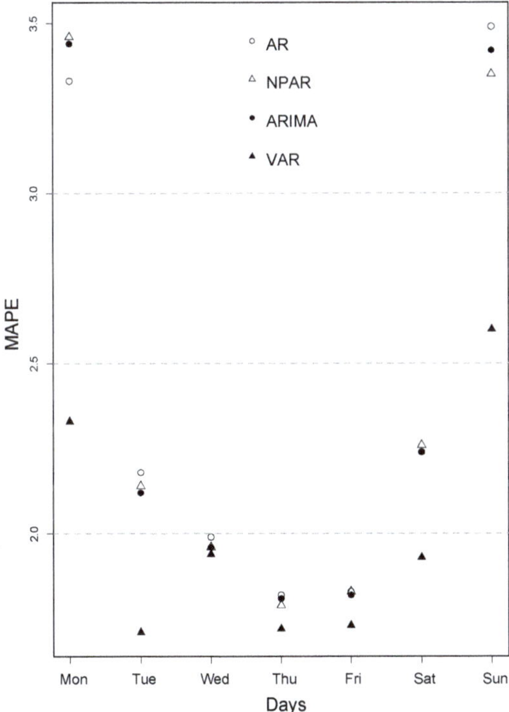

Figure 7. Day-specific MAPEs for all stochastic component models: AR, NPAR, ARIMA, and VAR.

Table 4. Electricity demand: hourly day-specific MAPE, MAE, and RMSE.

ERRORS	MODELS	Monday	Tuesday	Wednesday	Thursday	Friday	Saturday	Sunday
	AR	3.33	2.18	1.99	1.82	1.83	2.24	3.49
MAPE	NPAR	3.46	2.14	1.96	1.79	1.83	2.26	3.35
	ARMA	3.44	2.12	1.96	1.81	1.82	2.24	3.42
	VAR	2.33	1.71	1.94	1.72	1.73	1.93	2.60
	AR	1728.10	1002.69	755.78	633.24	634.54	905.95	1649.36
MAE	NPAR	1811.31	973.38	740.55	618.19	638.20	921.64	1585.79
	ARMA	1798.48	969.16	743.45	626.00	631.97	912.87	1615.01
	VAR	1225.76	799.34	739.52	592.69	601.85	791.00	1250.86
	AR	2194.77	1288.65	980.01	774.78	798.23	1176.36	2142.07
RMSE	NPAR	2252.48	1253.27	950.24	764.92	795.33	1187.31	2038.49
	ARMA	2232.41	1248.31	952.02	775.92	789.30	1181.01	2028.24
	VAR	1601.44	999.26	963.13	733.99	747.15	981.53	1619.33

To conclude this section, the hourly RMSE and forecasted demand for the best four combinations including one for each stochastic model is plotted in Figure 8. From the figure (left), note that hourly RMSE are considerably lower at the low load periods, while they are high at peak load periods. Further, note the best forecasting performance of the SR-VAR model compared to the competing stochastic models. For these models, the observed and the forecasted demand are also plotted in Figure 8 (right). The forecasted demand was following the actual demand very well, especially when VAR was used as a stochastic model. Thus, we can conclude that the multivariate model VAR outperformed the univariate counterparts.

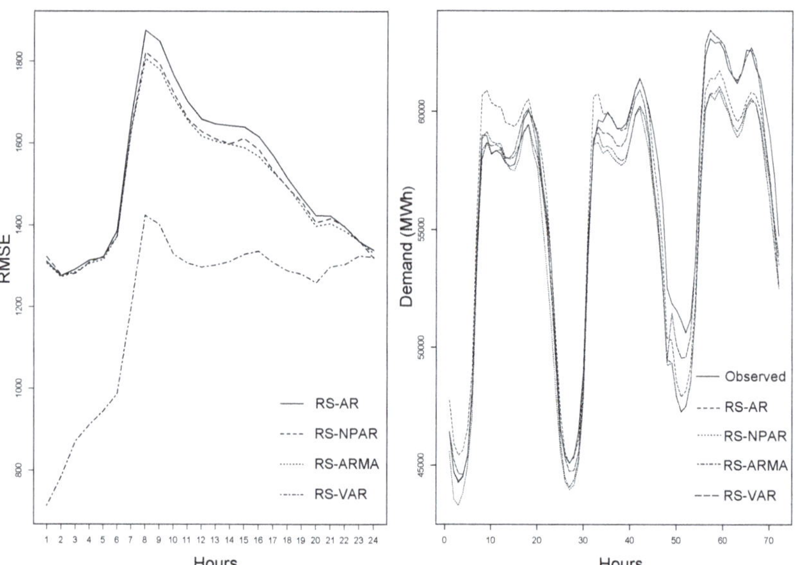

Figure 8. (**left**) Hourly RMSE for: RS-AR (solid), RS-NPAR (dashed), RS-ARMA (dotted), and RS-VAR (dotted-dashed). (**Right**) Observed demand (solid) and forecasted demand for: RS-AR (dashed), RS-NPAR (dotted), RS-ARMA (dotted-dashed), and RS-VAR (long dash).

4. Conclusions

The main aim of this work was to model and forecast electricity demand using the component estimation method. For this purpose, the log demand was divided into two components: deterministic and stochastic. The elements of the deterministic component consisted of a long trend, multiple periodicities due to annual, seasonal, and weekly regular cycles, and bank holidays. Special attention was paid to the estimation of the yearly seasonality as it was previously ignored by many authors. The estimation of yearly components was based on six different estimation methods, whereas other elements of the deterministic component were estimated using ordinary least squares. In particular, for the estimation of annual periodicity, this work used the sinusoidal function-based model (SR), the local polynomial regression models with three different kernels: tri-cubic (L1), Gaussian (L2), and Epanechnikov (L3), Regression Splines (RS), and Smoothing Splines (SS). For the stochastic component, we used four univariate and multivariate models, namely the Autoregressive Model (AR), the Non-Parametric Autoregressive Model (NPAR), the Autoregressive Moving Average model (ARMA), and the Vector Autoregressive model (VAR). The estimation of both, deterministic and stochastic, components led us to compare twenty four different combinations of these models. To see the predictive performance of different models, demand data from the Nord Pool electricity market were used, and one-day-ahead out-of-sample forecasts were obtained for a complete year. The forecasting accuracy of the models was assessed through the MAPE, MAE, and RMSE. To assess the significance of the differences in the predictive performance of the models, the Diebold and Mariano test was performed. Results suggested that the component-wise estimation method was extremely effective for modeling and forecasting electricity demand. The best results were produced by combining RS and the VAR model, which led to the lowest error values. Further, all the combinations of the multivariate model VAR completely outperformed the univariate counterparts, suggesting the superiority of multivariate models. Within the combination of VAR, however, the results were not statistically different for all models.

Author Contributions: Conceptualization and Methodology I.S. ; Software, S.A.; Validation, I.S., S.A.; Formal Analysis, H.I. and I.S.; Investigation, H.I.; Resources, D.W.; Data Curation, S.A.; Writing—Original Draft Preparation, I.S.; Writing—Review & Editing, I.S. S.A., and D.W.; Visualization, H.I. and S.A.; Supervision, I.S.; Project Administration, I.S.; Funding Acquisition, S.A. and D.W.

Funding: The work of Ismail Shah is partially funded by Quaid-i-Azam University, Islamabad, Pakistan through university research fund.

Acknowledgments: The authors would like to thank Francesco Lisi, department of statistical sciences, University of Padua, for his valuable suggestions and comments. We are also grateful to the anonymous referees for their constructive comments, which greatly improved the manuscript.

Conflicts of Interest: The authors declare no conflicts of interest.

References

1. Serrallés, R.J. Electric energy restructuring in the European Union: Integration, subsidiarity and the challenge of harmonization. *Energy Policy* **2006**, *34*, 2542–2551. [CrossRef]
2. Bunn, D.W. Forecasting loads and prices in competitive power markets. *Proc. IEEE* **2000**, *88*, 163–169. [CrossRef]
3. Herbst, A.; Toro, F.; Reitze, F.; Jochem, E. Introduction to energy systems modelling. *Swiss J. Econ. Stat.* **2012**, *148*, 111–135. [CrossRef]
4. Gonzalez-Romera, E.; Jaramillo-Moran, M.A.; Carmona-Fernandez, D. Monthly electric energy demand forecasting based on trend extraction. *IEEE Trans. Power Syst.* **2006**, *21*, 1946–1953. [CrossRef]
5. Kyriakides, E.; Polycarpou, M. Short term electric load forecasting: A tutorial. In *Trends in Neural Computation*; Springer: Berlin/Heidelberg, Germany, 2007; pp. 391–418.
6. Pollak, J.; Schubert, S.; Slominski, P. Die Energiepolitik der EU; Facultas WUV/UTB; 2010. Available online: http://www.utb-shop.de/die-energiepolitik-der-eu-2646.html (accessed on 30 April 2019).
7. Bosco, B.P.; Parisio, L.P.; Pelagatti, M.M. Deregulated wholesale electricity prices in Italy: An empirical analysis. *Int. Adv. Econ. Res.* **2007**, *13*, 415–432. [CrossRef]
8. Janczura, J.; Weron, R. An empirical comparison of alternate regime-switching models for electricity spot prices. *Energy Econ.* **2010**, *32*, 1059–1073. [CrossRef]

9. Trueck, S.; Weron, R.; Wolff, R. Outlier treatment and robust approaches for modeling electricity spot prices. In Proceedings of the 56th Session of the ISI, Lisbon, Portugal, 22–29 August 2007. Available online: http://mpra.ub.uni-muenchen.de/4711/ (accessed on 28 March 2019).

10. Sigauke, C.; Chikobvu, D. Prediction of daily peak electricity demand in South Africa using volatility forecasting models. *Energy Econ.* **2011**, *33*, 882–888. [CrossRef]

11. Lisi, F.; Nan, F. Component estimation for electricity prices: Procedures and comparisons. *Energy Econ.* **2014**, *44*, 143–159. [CrossRef]

12. Liu, J.M.; Chen, R.; Liu, L.M.; Harris, J.L. A semi-parametric time series approach in modeling hourly electricity loads. *J. Forecast.* **2006**, *25*, 537–559. [CrossRef]

13. Taylor, J.W.; De Menezes, L.M.; McSharry, P.E. A comparison of univariate methods for forecasting electricity demand up to a day ahead. *Int. J. Forecast.* **2006**, *22*, 1–16. [CrossRef]

14. Taylor, J.W.; McSharry, P.E. Short-term load forecasting methods: An evaluation based on european data. *IEEE Trans. Power Syst.* **2007**, *22*, 2213–2219. [CrossRef]

15. Clements, A.E.; Hurn, A.; Li, Z. Forecasting day-ahead electricity load using a multiple equation time series approach. *Eur. J. Oper. Res.* **2016**, *251*, 522–530. [CrossRef]

16. Ismail, M.A.; Zahran, A.R.; El-Metaal, E.M.A. Forecasting Hourly Electricity Demand in Egypt Using Double Seasonal Autoregressive Integrated Moving Average Model. In Proceedings of the First International Conference on Big Data, Small Data, Linked Data and Open Data, Barcelona, Spain, 19–24 April 2015; pp. 42–45.

17. Yukseltan, E.; Yucekaya, A.; Bilge, A.H. Forecasting electricity demand for Turkey: Modeling periodic variations and demand segregation. *Appl. Energy* **2017**, *193*, 287–296. [CrossRef]

18. Hinman, J.; Hickey, E. Modeling and forecasting short-term electricity load using regression analysis. *Inst. Regul. Policy Stud.* **2009**, 1–51, Available online https://irps.illinoisstate.edu/downloads/research/documents/LoadForecastingHinman-HickeyFall2009.pdf (accessed on 31 May 2018) .

19. Feng, Y.; Ryan, S.M. Day-ahead hourly electricity load modeling by functional regression. *Appl. Energy* **2016**, *170*, 455–465. [CrossRef]

20. Weron, R.; Misiorek, A. Forecasting spot electricity prices: A comparison of parametric and semiparametric time series models. *Int. J. Forecast.* **2008**, *24*, 744–763. [CrossRef]

21. Kandananond, K. Forecasting electricity demand in Thailand with an artificial neural network approach. *Energies* **2011**, *4*, 1246–1257. [CrossRef]

22. Meng, M.; Niu, D.; Sun, W. Forecasting monthly electric energy consumption using feature extraction. *Energies* **2011**, *4*, 1495–1507. [CrossRef]

23. Ryu, S.; Noh, J.; Kim, H. Deep neural network based demand side short term load forecasting. *Energies* **2016**, *10*, 3. [CrossRef]

24. Debusschere, V.; Bacha, S. One week hourly electricity load forecasting using Neuro-Fuzzy and Seasonal ARIMA models. *IFAC Proc. Vol.* **2012**, *45*, 97–102.

25. Yang, Y.; Wu, J.; Chen, Y.; Li, C. A new strategy for short-term load forecasting. *Abstr. Appl. Anal.* **2013**, *2013*, 208964. [CrossRef]

26. Lisi, F.; Shah, I. Forecasting Next-Day Electricity Demand and Prices Based on Functional Models; Working Paper; 2019; pp. 1–30. Available online https://www.researchgate.net/profile/Ismail_Shah2/publications (accessed on 30 March 2019)

27. Shang, H.L. Functional time series approach for forecasting very short-term electricity demand. *J. Appl. Stat.* **2013**, *40*, 152–168. [CrossRef]

28. Shah, I.; Lisi, F. Day-ahead electricity demand forecasting with non-parametric functional models. In Proceedings of the 12th International Conference on the European Energy Market, Lisbon, Portugal, 19–22 May 2015; pp. 1–5.

29. Bisaglia, L.; Bordignon, S.; Marzovilli, M. *Modelling and Forecasting Hourly Spot Electricity Prices: Some Preliminary Results*; Working Paper Series; University of Padua: Padova, Italy, 2010.

30. Gianfreda, A.; Grossi, L. Forecasting Italian electricity zonal prices with exogenous variables. *Energy Econ.* **2012**, *34*, 2228–2239. [CrossRef]

31. Bordignon, S.; Bunn, D.W.; Lisi, F.; Nan, F. Combining day-ahead forecasts for British electricity prices. *Energy Econ.* **2013**, *35*, 88–103. [CrossRef]

32. Laouafi, A.; Mordjaoui, M.; Laouafi, F.; Boukelia, T.E. Daily peak electricity demand forecasting based on an adaptive hybrid two-stage methodology. *Int. J. Electr. Power Energy Syst.* **2016**, *77*, 136–144. [CrossRef]

33. Lisi, F.; Pelagatti, M.M. Component estimation for electricity market data: Deterministic or stochastic? *Energy Econ.* **2018**, *74*, 13–37. [CrossRef]

34. Dominguez-Garcia, A.D.; Cady, S.T.; Hadjicostis, C.N. Decentralized optimal dispatch of distributed energy resources. In Proceedings of the 2012 IEEE 51st IEEE Conference on Decision and Control (CDC), Maui, HI, USA, 10–13 December 2012; pp. 3688–3693.

35. Basak, P.; Chowdhury, S.; nee Dey, S.H.; Chowdhury, S. A literature review on integration of distributed energy resources in the perspective of control, protection and stability of microgrid. *Renew. Sustain. Energy Rev.* **2012**, *16*, 5545–5556. [CrossRef]

36. Nowotarski, J.; Weron, R. On the importance of the long-term seasonal component in day-ahead electricity price forecasting. *Energy Econ.* **2016**, *57*, 228–235. [CrossRef]

37. Marcjasz, G.; Uniejewski, B.; Weron, R. On the importance of the long-term seasonal component in day-ahead electricity price forecasting with NARX neural networks. *Int. J. Forecast.* **2018**, in press. [CrossRef]

38. Ramanathan, R.; Engle, R.; Granger, C.W.; Vahid-Araghi, F.; Brace, C. Short-run forecasts of electricity loads and peaks. *Int. J. Forecast.* **1997**, *13*, 161–174. [CrossRef]

39. Pilipovic, D. *Energy Risk: Valuing and Managing Energy Derivatives*; McGraw-Hill: New York, NY, USA, 1998; Volume 300.

40. Lucia, J.J.; Schwartz, E.S. Electricity prices and power derivatives: Evidence from the nordic power exchange. *Rev. Deriv. Res.* **2002**, *5*, 5–50. [CrossRef]

41. Weron, R.; Bierbrauer, M.; Trück, S. Modeling electricity prices: Jump diffusion and regime switching. *Phys. A Stat. Mech. Its Appl.* **2004**, *336*, 39–48. [CrossRef]

42. Kosater, P.; Mosler, K. Can Markov regime-switching models improve power-price forecasts? Evidence from German daily power prices. *Appl. Energy* **2006**, *83*, 943–958. [CrossRef]

43. De Jong, C. The nature of power spikes: A regime-switch approach. *Stud. Nonlinear Dyn. Econom.* **2006**, *10*. [CrossRef]

44. Escribano, A.; Ignacio Peña, J.; Villaplana, P. Modelling electricity prices: International evidence. *Oxf. Bull. Econ. Stat.* **2011**, *73*, 622–650. [CrossRef]

45. Avery, M. Literature review for local polynomial regression. Unpublished manuscript, 2013.

46. Veraart, A.E.; Veraart, L.A. Modelling electricity day-ahead prices by multivariate Lévy semistationary processes. In *Quantitative Energy Finance*; Springer: New York, NY, USA, 2014; pp. 157–188.

47. Shah, I. Modeling and Forecasting Electricity Market Variables. Ph.D Thesis, University of Padova, Padua, Italy, 2016.

48. Dordonnat, V.; Koopman, S.J.; Ooms, M. Intra-daily smoothing splines for time-varying regression models of hourly electricity load. *J. Energy Mark.* **2010**, *3*, 17. [CrossRef]

49. Hastie, T.; Tibshirani, R. Generalized additive models: Some applications. *J. Am. Stat. Assoc.* **1987**, *82*, 371–386. [CrossRef]

50. Diebold, F.; Mariano, R. Comparing predictive accuracy. *J. Bus. Econ. Stat.* **1995**, *13*, 253–263.

Review

Computational Intelligence on Short-Term Load Forecasting: A Methodological Overview

Seyedeh Narjes Fallah [1], Mehdi Ganjkhani [2], Shahaboddin Shamshirband [3,4,*] and Kwok-wing Chau [5]

[1] Independent Researcher, Sari 4816783787, Iran; s.narjes.f@gmail.com
[2] Department of Electrical Engineering, Sharif University of Technology, Tehran P.O. Box 11365-11155, Iran; ganjkhani.mehdi@alum.sharif.edu
[3] Department for Management of Science and Technology Development, Ton Duc Thang University, Ho Chi Minh City, Vietnam
[4] Faculty of Information Technology, Ton Duc Thang University, Ho Chi Minh City, Vietnam
[5] Department of Civil and Environmental Engineering, Hong Kong Polytechnic University, Hong Kong, China; cekwchau@polyu.edu.hk
* Correspondence: shahaboddin.shamshirband@tdtu.edu.vn

Received: 17 December 2018; Accepted: 25 January 2019; Published: 27 January 2019

Abstract: Electricity demand forecasting has been a real challenge for power system scheduling in different levels of energy sectors. Various computational intelligence techniques and methodologies have been employed in the electricity market for short-term load forecasting, although scant evidence is available about the feasibility of these methods considering the type of data and other potential factors. This work introduces several scientific, technical rationales behind short-term load forecasting methodologies based on works of previous researchers in the energy field. Fundamental benefits and drawbacks of these methods are discussed to represent the efficiency of each approach in various circumstances. Finally, a hybrid strategy is proposed.

Keywords: short-term load forecasting; demand-side management; pattern similarity; hierarchical short-term load forecasting; feature selection; weather station selection

---

1. Introduction

Short-Term load forecasting (STLF) is an integral part of the energy planning sector. Designing a time-ahead power market requires demand-scheduling for various energy divisions, namely, generation, transmission, and distribution. STLF helps power system operators with various decision-making in the power system, including supply planning, generation reserve, system security, dispatching scheduling, demand-side management, financial planning, and so forth. While STLF is particularly essential for the time-ahead power market operation, inaccurate demand forecasting will cost the utility a tremendous financial burden [1].

Traditionally, engineering approaches were employed to predict the future demand manually with the help of charts and tables. These traditional methods mainly considered weather impacts as well as calendar effects. Nowadays, these features are still determined for developing load models with novel methods [2].

With the advent of statistical software packages and artificial intelligence techniques, several outstanding pieces of research are devoted to statistical [3] and computational intelligence (CI) approaches [4] to model the future load. Some examples of statistical regression-based STLF approaches in the literature, including auto-regressive moving average (ARMA) [5,6], auto-regressive integrated moving average (ARIMA) [7], and seasonal ARIMA (SARMIA) [8]. Artificial neural network

(ANN) [4], support vector machine (SVM) [9], fuzzy logic [10], etc., are considered prevailing CI-based forecasting techniques.

CI-based load models, regardless of underlying computational algorithms, can be further categorized into several methodological outlines. Correspondingly, it must be acknowledged that different forecasting techniques cannot be interpreted as different methodological approaches. A method is defined as a structured procedural solution designed for specific cases of forecasting practices; while a technique refers to a certain model that can be categorized with all other similar models in one technical category such as regression or neural network techniques. For example, Fan & Hyndman [11] and Mandal et al. [12] both applied ANN architecture to develop a 24-hour ahead load model whereas different methodological approaches were considered in each of these papers. In [11], a stepwise method, which locates the lowest error in the model, is applied for selecting the optimal subset of variables including the historical load and meteorological variables. However, in [12], only daily load profiles similar to day-ahead load recognized by a similarity index (similar day type and similar weather) are fed into the engine. The solution is not always narrowed down to the technique that the forecasters use. Instead, the strategy to implement those techniques is important as well.

Generally, both methods and techniques are important when it comes to accurate estimation. However, limited literature is available for STLF methodologies. Most surveys in the literature are devoted to the investigation of different STLF techniques [13–16]. For example, Mogharm et al. [14] investigated STLF techniques by classifying them into two categories of statistical approaches and CI-based techniques. Hippert et al. [13] reviewed ANN-based STLF. Although these surveys addressed most applicable STLF techniques, this still might be unclear for young researchers to understand the merits behind developing any specific load model.

This paper explains the main framework of state-of-the-art methodologies applied for the CI-based STLF via examples of several case studies. A comprehensive overview of technical and computational difficulties for STLF is presented as well as the proposed strategies by various researchers to unravel them. These strategies are categorized into four main groups based on their identical topologies. The robustness of each method to deal with different type of load data is identified.

The rest of the paper is organized as follows. Section 2 presents a general overview of four principal methodologies, followed by four subsections wherein details of each method are fully described. Section 3 discusses the main advantages and disadvantages of STLF methods. Moreover, in Section 3, advantages of hybrid methods are highlighted, and a hybrid method is proposed. Finally, the concluding remarks are drawn in the last section.

2. STLF Methodologies

Load forecasting can be conducted by various methodologies. The selection of a forecasting method relies on several factors including the relevance and availability of historical data, the forecast horizon, the level of accuracy for weather data, desired prediction accuracy, and so forth. Accordingly, selecting the proper load forecasting approach primarily depends on the time horizon of the prediction.

Different time horizons are adopted for load forecasting based on their specific applications in power system planning. For instance, the distribution and transmission planning are involved with STLF, while for longer durations, i.e., more than a year up to a few decades, the load prediction provides a decision platform for financial or power supply planning. Likewise, the required level of accuracy in these time horizons are not equal, as the decisions in the long-term are preliminary and may need significant changes in subsequent planning stages due to very uncertain input information while a short-term forecast provides information to the day-ahead market, which requires a high-level of accuracy. Moreover, different kinds of predictor variables are considered for each horizon of the forecast. For example, a long-term load forecasting takes into account population variations, economic development, industrial construction, and technology development whereas a short-term forecast mainly considers calendar variables, weather data, and customers' behaviors [17].

Generally, both time categories of load forecasts are important for power system operation and development especially by the integration of distributed generators into the grid, which adds further intermittency and vulnerability in power provision. This study exclusively investigates the STLF approaches by reviewing original papers in this field. Even though some of the artificial intelligence (AI) techniques that are used for STLF might be also applicable for long-term load forecasting, on a methodological foundation, they are not comparable due to the aforementioned reasons.

Hong and Fan [2] identified four general categories of STLF methodologies, which can be applied to several techniques to solve the STLF problem. The four categories, i.e., similar day, variable selection, hierarchical forecasting, and weather station selection are specified based on different realizations of forecast problem. For example, similar-day method determines the load data as a sequence of various similar daily load profiles, while variable selection method presumes that the load data behaves like a series of variables either correlated or independent from each other. Hierarchical method, on the other hand, considers the data as an aggregated load, which is highly varying by changes in the load at lower levels of the hierarchy. Finally, weather station selection is a method, which determines the best-fitted weather data into the load model.

Hong and Fan outlined these general methodological approaches in a review [2], describing two or three examples of each, while an extensive literature of STLF has not been assessed accordingly. More investigation of the adopted STLF methodologies in the literature reveals that there are some novel approaches that could further be subcategorized into the four root categories. For example, the classic similar-day method, which distinguished the similarity between daily load profiles by assigning the day type index, was later developed as similar-pattern method, in which similar load profiles are extracted by using either a minimization algorithm or clustering techniques. Other novel approaches, such as pattern-sequence and sequence learning, are also recognized to be in the category of similar-pattern method, if their algorithms try to find or learn similar sequences of patterns within the dataset.

Moreover, the majority of STLF researchers chose the variable selection method, while different algorithms were employed for selecting prominent features of candidate variables. This research distinguished five state-of-the-art feature selection approaches for STLF. These approaches are specifically important to create the optimal subseries of data and leading to more accurate results.

Another category of STLF methodologies is assigned to hierarchical short-term load forecasting (HSTLF), which has been limitedly addressed in the literature. HSTLF methodology addresses forecasting at several levels of aggregate data. Although at each aggregate level, other methodologies, i.e., similar pattern and variable selection are used for individual predictions, the novelty of HSTLF is related to the applied combination method. Thus, four approaches are identified for HSTLF while each one proposes a different strategy. The classical top-down and bottom-up approaches are two common algorithms for hierarchical forecasting, with the latter aggregating the data and the former aggregating the forecast. Yet, recent approaches for hierarchical forecasting, i.e., weighted combination and ensemble model try to capture the model at each aggregate level individually and find the correlation between the individual models at different levels of the hierarchies. Despite limited literature, HSTLF lately received more attention by distribution and transmission operators for power system control and planning. It takes into account recent advances in communication infrastructures for remote measurement and automated metering, which enables operators with high granular data at user ends. Thus, the most recent challenges of HSTLF methodology is highlighted in this study to help young researchers find the competing research direction in this field.

By drawing attention towards HSTLF, a question might come to minds that by aggregating the data, what happens to other exogenous variables such as meteorological variables, as they cannot be aggregated. This challenge was raised for the first time by Hong and Pinson [18] and a competition was launched to address this question. Results of this competition are further discussed in this paper to draw a conclusion.

Figure 1 shows the tree diagram of these four forecasting methods. As can be seen, each method can be carried out via multiple strategies. For example, there are various approaches to predict a hierarchical structure including bottom-up, top-down, ensemble, and weighted combination. A full description of these four recognized categories of STLF methodologies is presented in the following subsections with examples of several case studies.

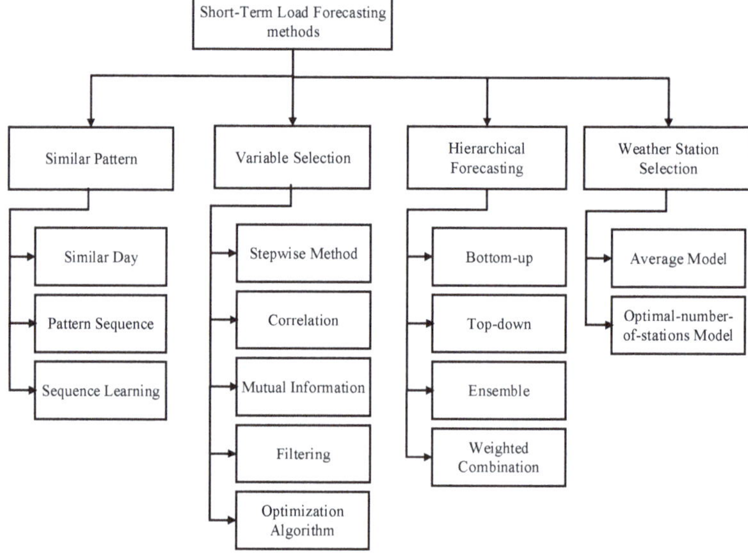

Figure 1. Tree diagram of the STLF methods.

2.1. Similar-Pattern Method

Similarity-based methods are generalized forms of minimum distance approaches applied in machine learning and pattern recognition. These methods have also been used for STLF by finding similar demand patterns within the data set and predicting the future load using interpolation or weighting [19]. There are different strategies for finding similar load profiles; in the simplest case, it can be achieved by assigning a similarity index to the type-of-the-day in the calendar or meteorological factors. Similar patterns will then be achieved by searching between those days with similar indexes. Searching space is generally within a close neighborhood, although sometimes annual lagged data is also determined. For example, Dudek et al. [20] developed a similarity-based forecasting model by using the similarity between seasonal patterns of a load time series based on the calendar-lagged load data. The search space in [20] was limited to the nearest neighbor of the forecast day as well as the nearest neighbor of the same calendar day in the previous year. In fact, assigning the day-of-the-year index besides the weekday index is essential to avoid seasonal variations. A typical search space for similar-day method is illustrated in Figure 2.

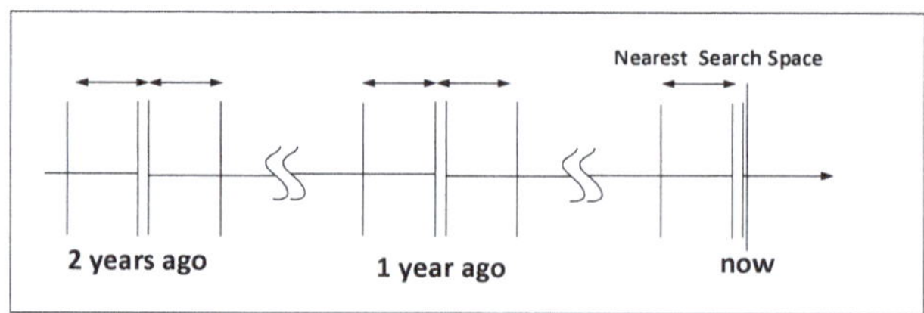

Figure 2. Limitation of search space for similar days.

Figure 3 illustrates the methodology applied by Dudek et al. [20]. In the first step, days similar to the forecast day with similar weekday and day-of-the-year indices are extracted from the load time series (first series). Thereupon, a sequence of days following these similar days (second series) is created. In the second step, days with similar patterns within the first series (similar-day series) are chosen by a selection strategy, and those followed by these newly selected days within the second series (sequence series). The outcome of the third step is a regression model of load data extracted from the sequence series. Eventually, the load of the next day in the original time series is forecasted by decoding the final model.

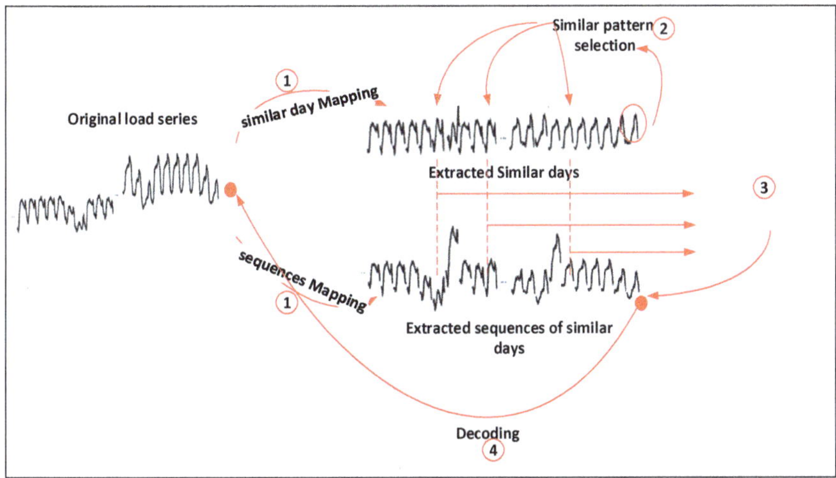

Figure 3. Similar day-based prediction algorithm developed by Dedek et al. [20].

Besides the calendar index as the similarity indicator, other characteristics such as weather similarities can be considered as well. For instance, Ying Chen et al. [21] proposed a similar-day selection method based on the weather similarity of the forecast day. In their proposed method, which was designed to forecast the load in a short-term period (two working days excluding the weekend) by hourly resolution, the search for the similar days was limited to days with the same weekday and weather indices to the forecast day. Days with similar weather condition were selected based on a minimization process, while the meteorological condition was defined by wind chill, temperature, humidity, wind speed, and cloud cover variables. In addition, the same index was assigned for some of the weekdays with similar load pattern. It has also been shown that relying only on similar days' data without establishing the initial status of tomorrow's demand leads to an inaccurate forecast result. Thus, the 24-hour today's load has been fed as an input to the forecasting engine. Figure 4 illustrates the schematic diagram of similar-day method developed in [21].

As already mentioned, the selection of similar load profiles between days with similar indexes (weekday, the day-of-the-year and weather indexes) can be made by a distance minimization technique. Some works in the literature applied Euclidean norm to measure the match level between similar days [12,21,22]. As listed in Table 1, Chen et al. [21] used the Euclidean norm to evaluate the weather similarity between the forecast day and previous days. Senjyu et al. [22] also applied a weighted Euclidian to investigate the similarity of load patterns using load deviations between forecast day and historical days, weather deviation, and the slope of load deviations. The assigned weights (w) in Equation (2) is determined based on a regression model using the trend of load and temperature.

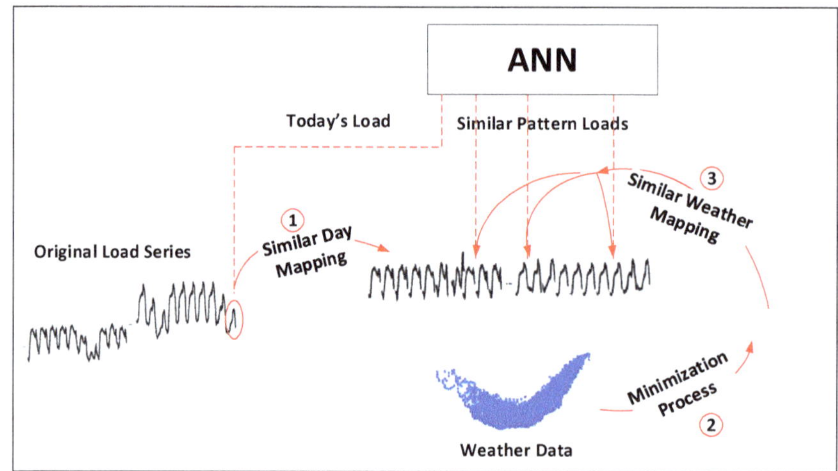

Figure 4. Schematic diagram of similar-day method developed by Chen et al. [21].

Table 1. Distance minimization technique for similarity measurement.

Paper	Method	Formulae	Parameters
[21]	Euclidean distance minimization	$\min\limits_{i} \sum_{t=1}^{24} \left(w(t)^f - w(t)^i \right),\ i \in \theta$ (1)	θ: historical days f: forecast day i: historical day in θ w: weather factor under consideration
[12,22]	Weighted Euclidean distance minimization	$\min\limits_{i} \sqrt{w_1(\Delta L_t)^2 + w_2(\Delta L_s)^2 + w_3(\Delta T_t)^2}$ (2)	ΔL_t: deviation of load of forecast day and historical day ΔL_s: deviation of slope between load on forecast day and load of historical day ΔT_t: deviation of temperature between forecast day and historical day w_n: Weight factor

Dynamic time warping (DTW) is another method to measure the similarity for those time series with similar values not exactly at the same time point. Using DTW method might end up finding several similar patterns of load profiles within the dataset. Teeraratkul et al. [23] indicated that by using DTW method, the number of groups for similar profiles reduced by 50%.

More recently, clustering algorithms are used to find similar sequence of load patterns within the dataset [24,25]. These clustering techniques are used to group data into a specific number of categories of daily load patterns, which were termed pattern-sequence-based STLF method. Under this method, a label indexes the load for each day in the dataset. Consequently, a sequence of labels is created in the dataset. Alvarez et al. [26] applied K-means clustering technique to create different clusters of load patterns and extracted a sequence of labels from the dataset as a pattern to search and predict the next day's load. A schematic diagram of pattern-sequence-based forecasting method is depicted in Figure 5. According to Figure 5, all weekdays in a dataset are labeled by using a clustering method. To predict the next day's load, a window of a sequence of labels before the forecast day is selected. By using this window, similar sequence of labels is searched within the dataset. Eventually, the load of the target day can be predicted by averaging the next day's load of the discovered sequences.

The prevalence of smart meters in a smart grid facilitated market planners with fine-grained data in hourly and sub-hourly resolution. Load profiles at the customer-end provide sophisticated information about the type of customers and their consumption behaviors. Quilumba et al. [27] used a clustering technique to group smart meter customers according to their similar energy pattern consumption. Temperature information was interpolated between neighbor values to become as granular as the smart data.

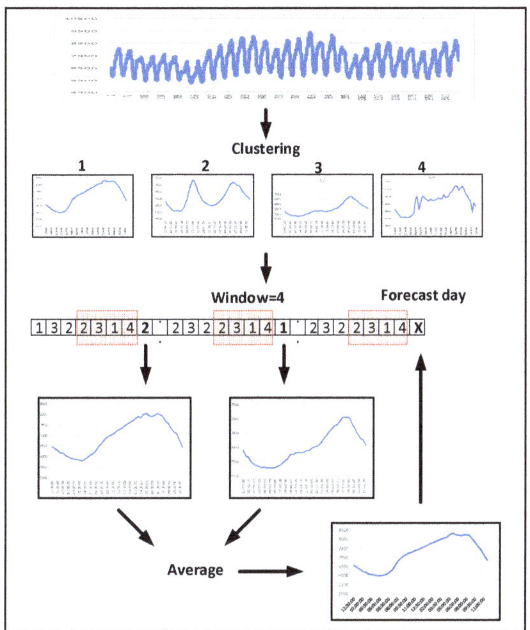

Figure 5. Schematic diagram of pattern sequence-based forecasting method [26].

Clustering methods can distinguish similar sequences within a dataset as discussed earlier; however, they cannot differentiate among the main features of these patterns. More recently, adding memory to the structure of learning engines such as recurrent neural network and deep learning, outweighed this drawback.

Liu et al. [28] considered sequence learning approach for developing a load model by using recurrent neural network structure (RNN). Kong et al. [29] recommended that long short-term memory of RNN was a powerful engine to learn the look back sequences due to its memory cells, remember important features, and forget gates to reset the cells for redundant features. Shi et al. [30] applied deep RNN to map the sequence of input data into the corresponding output sequence. Zheng et al. [31] proposed a hybrid method, by applying a clustering technique to capture similar days within a dataset, and then used the sequence-to-sequence structure of the long short-term memory structure to adjust the length of the input and output sequences. A sequence-to-sequence structure was primarily designed to map sequences with different length [32,33]. Marino et al. [33] suggested that the main advantage of the sequence-to-sequence structure was related to its ability to predict an arbitrary number of future time steps having an arbitrary length of an input sequence. Satish et al. [34] investigated the optimum learning sequence for the training stage. Results indicated that the number of patterns in a sequence affected the accuracy of the model.

Table 2 lists some highly cited publications in which similar-pattern method was applied for load prediction. These publications are categorized based on three common techniques, namely, "similar-day", "pattern-sequence", and "sequence learning".

In general, pattern similarity method is an efficient approach to capture repeated patterns of the load series in the short term. The overall pattern of a system is rarely changing in the short term; however, in longer periods, some significant deviations might lessen the similarity of future load to past load.

Table 2. Published articles employing similar-pattern method.

Method	Publications	Technique
	[12,20–23,35,36]	similar day
Similar-Pattern Method	[24–27,37–41]	pattern-sequence
	[28–34]	sequence learning

2.2. Variable Selection Method

Variable selection is the process of selecting the most influential variables or features (predictor variables) within the dataset while they can adequately capture the relationship between the available data and the output. Despite time series forecasting relies only on past data, variable selection method determines external variables besides historical load in order to embed into the model [42]. Some of these external variables, which are termed explanatory variables to explain the reason of load fluctuations, are calendar variables (time of the day, day of the week, month of the year, and day of the year etc.), meteorological variables (temperature, humidity, cloud cover, wind chill, solar radiation etc.) and so forth [43].

Several studies also considered the lagged load data into their model [44,45]. The lagged variables determine the recency effect by incorporating alteration of demand level throughout load time series into the model. For example, Ceperic et al. [44] proposed a feature selection algorithm to select the optimum number of lagged loads in order to embed the sequential correlation of load variables into the model. Another example is the work of Fan and Hyndman [11], which considered the following variables as candidate predictors: the lagged load demand for each of the preceding 12 hours, lagged values for the same hours of the two previous days, maximum and minimum load values in the past 24 hours, and the average load in the preceding week. Consequently, a selection algorithm was applied to choose between potential variables and create a subset of optimal predictor variables.

Besides the lagged demand, some studies embedded lagged temperatures as input variables. The electricity demand is remarkably impacted by the recent temperature as well as the current temperature. That is why in the forecasting model developed by Fan and Hyndman [11], besides the lagged demand, the current and 12-hour lagged temperature for the preceding day and the former two days were involved in the model. However, the main concern about weather variables was its level of validation, which depends partly on the weather station selection. It is discussed more in Section 2.4.

By nominating multiple input variables and considering a large amount of available data for every variable, the predictor engine might not be able to converge to an accurate predictive model. Therefore, an effective subset of the data with the optimal number of predictor variables will help the forecast accuracy [46]. An efficient predictor variable is highly explanatory and independent of other variables. The aim is to select the optimal subset of predictor variables with fewer numbers, which suitably describes the characteristics of the output variable. Optimal input subset favors model accuracy as well as cost efficiency and model interpretability [47].

In the literature, researchers employed different methods and techniques to select explanatory variables optimally.

One of the methods used for variable selection is the stepwise refinement which is a step by step approach for input selection. In this method, the primary model is a full model consisting of all measured variables. Hence, based on the predictive capability of individual variables, redundant terms from the model are omitted. The retained variables consequently lead to the best model. One example is the work of Fan and Hyndman [11], who carried out a step by step variable selection method to extract the best-suited model. The nominated inputs were the calendar variables, actual demand and lagged demand (from the National Electricity Market of Australia-NEM), and forecasted temperature data from more than one site in the target area. Assuming the selection of temperature differentials, in the first step, the temperature differentials form the same period of the last six days were dropped one at a time, and the one leading to the lowest error was selected. Consequently, in the next step, the temperature variable was frozen to only the selected day from the previous step, and temperatures of the last six hours were considered for the trial. This procedure was continued until the final group of variables was selected.

Nedellec et al. [48] followed the same strategy of stepwise refinement for variable selection as well, but in a three-step procedure while the variables in each stage were selected based on the scale of forecast. In a long-term module, monthly load and temperature time series for every region and weather station were selected to extract long-term trend and low-frequency effects. The residual of the first stage with no seasonality and weather effects were considered for a medium-term estimation. Variables such as a type-of-the-day, type-of-the-year, de-trended electrical load, real temperature, and lagged temperature were predictor variables in this medium-term model. In a short-term stage, more localized factors, which remained from previous stages, were captured by selecting variables such as year, month, day, hour, time-of-the-year, and day type as well as real and smoothed weather variables. This stepwise algorithm is illustrated in Figure 6 for better understanding. As can be seen, the final forecasted load is an additive model of three components.

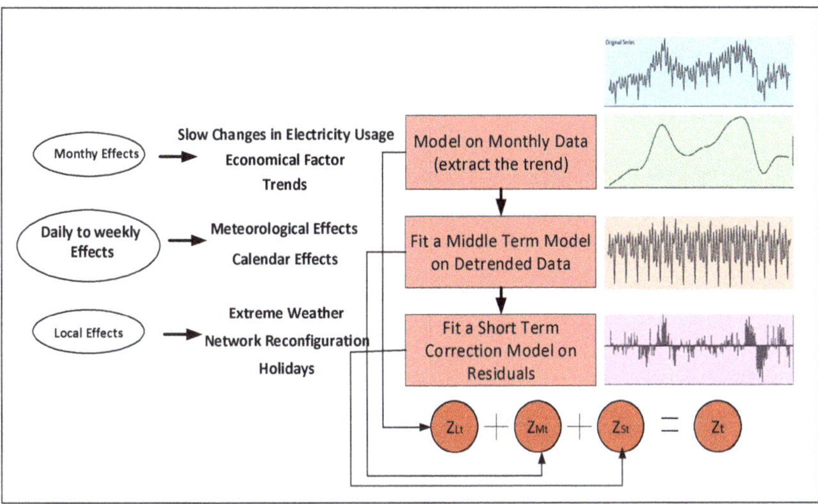

Figure 6. Stepwise algorithm for STLF [48].

Xiao et al. [49] also developed an ensemble load model by applying a group of STLF techniques to capture the trend of the load series. Consequently, the highly nonlinear characteristics of the residual subseries were modeled by using various data handling techniques.

Moreover, there are other approaches to identify the maximum relevance between different variables. Correlation-based methods use a heuristic algorithm to find a subset of variables, which are highly correlated with the output but are not correlated with each other [50]. Chen et al. [9] used correlation method to measure the dependency of the peak demand to temperature. Kouhi et al. [51] developed a correlation-based feature selection method to reduce chaotic structure of load time series and selected highly relevant variables within this reconstructed space. Amjady et al. [3] used a correlation approach to create a subseries of load data to develop a hybrid forecast model.

Mutual information (MI) is an information theoretic-based approach to measure the interdependency between two random variables. If MI is zero, two variables are independent and contain no mutual information about each other. Higher MI values indicate higher relevance with more information about the target feature [52]. Wang et al. [53] used MI method to obtain initial weights of the developed ANN-based load forecast model. Elattar et al. [54] reconstructed a load time series by embedding dimension and time delay computed by MI approach. Young-Min Wi et al. [55] adopted MI method to evaluate mutual information between dominant weather features and loads at different seasons.

Moreover, filtering methods can be applied to data to find the correlation among variables independent of any learning machine. Filter-based feature selection algorithms use general characteristics of the training data, i.e., statistical dependencies to select highly ranked features by applying a threshold for the number of features [56]. Reis et al. [57] applied wavelet filter to reconstruct a subseries of data after selecting input variables by using autocorrelation function. Amjady et al. [58] proposed a hybrid load prediction algorithm, in which a filter-based technique was selected for a minimum subset of inputs. Zhongyi Hu et al. [59] proposed a hybrid filter method for feature selection procedure.

More recently, developing bio-inspired optimization tools as well as evolutionary optimization algorithms led to improvement of CI-based feature selection techniques for STLF. Some examples of developed optimization algorithms for feature selection in the literature include ant colony [60], particle swarm [61,62], differential evolution [63], hybrid genetic and a colony [64] and so forth.

Some of the highly cited publications for STLF, which are categorized based on the applied feature selection techniques, are listed in Table 3.

Table 3. List of publications employing different feature selection techniques.

Publication	Technique
[11,48,49]	Stepwise
[57,58]	Filter
[3,9,47,51,65,66]	Correlation
[47,53–55,67]	Mutual Information
[60–64,68,69]	Optimization Algorithms

Selecting proper variables is sometimes time-dependent, while variables have significant impacts on load behavior of several hours and subtle effects on loads of other hours during a 24-hour period. Thus, a suitable architecture for a forecasting engine can provide a simpler model to decrease the number of redundant data [70]. A general idea is that instead of creating one subseries of data, different subsets of variables can be created for each category of time, while data in each category is affected by the same variables. For example, Khotanzad et al. [71] proposed two different parallel architectures for load forecasting. The first design, as illustrated in Figure 7, was a three-module structure to model hourly, daily, and weekly trends. In their developed architecture for prediction of the hourly load of the next day, each of three modules would be trained by 24 ANN engines. Each of them represented an hour of a day. The second architecture divides 24 hours into four categories, i.e., 1–9, 10–14 and 19–22, 15–18, and 23–24 while different input variables are determined for each group of hourly loads, as depicted in Figure 7.

Some other papers in the literature also applied the so-called parallel architecture for 24-hour-ahead load forecasting [44,72]. The reasons for using this design are smaller number of training data for each module with omitted parameters for each hour of the day, and a simpler model for each hour of the day, compared to a general model for all 24 hours.

In overall, developing an explanatory model via variable selection method is appropriate when forecasters have fundamental knowledge about the system. To forecast the variable of interest, one needs to identify different exogenous variables. Generally, there are no rules implied for the selection of input variables. The forecaster's experiences in analyzing the type of data from a specific market as well as a preliminary testing might help to select a proper group of variables. Thus, professional judgment is undoubtedly part of the process.

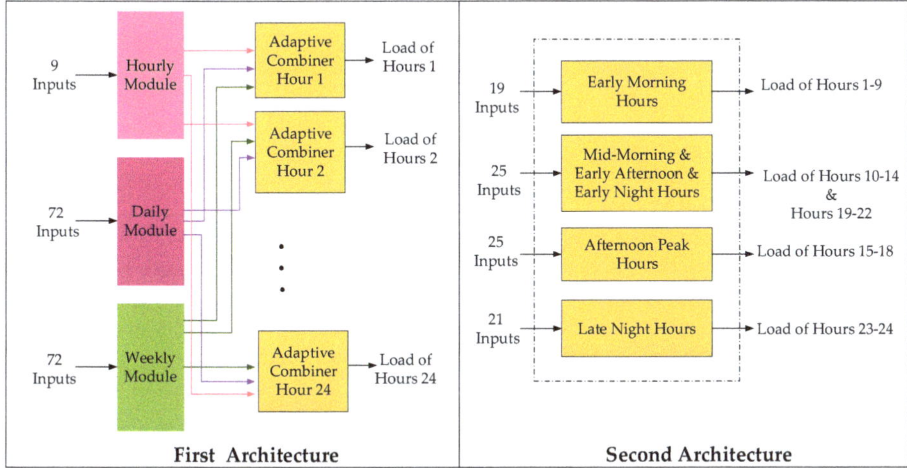

Figure 7. Parallel architecture for 24-hour ahead forecasting proposed in [71].

2.3. Hierarchical Forecasting

Previous methods presume load data as single time series, while these time series can be inherently disaggregated by different attributes of interest [42]. Load time series naturally are organized based on different hierarchies such as geographic, temporal, circuit connection, and revenue. Figure 8 depicts a typical hierarchical structure of a time series divided into aggregate and disaggregate levels.

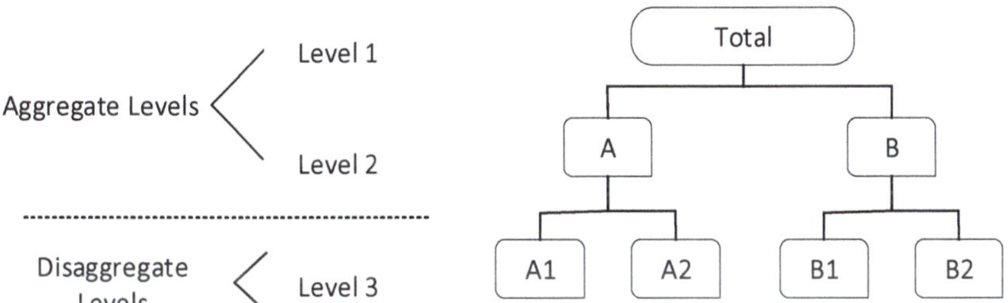

Figure 8. Schematic diagram of hierarchical structure of a load time series.

An example of hierarchical load structure can be found in a study conducted by Zhang et al. [73]. The load data was recorded consumption of three hundred smart meter customers of a subsection in Australian utility within three years. The customers were clustered into 30 nodes according to their postcodes. These 30 nodes were grouped into three nodes. Besides, these three nodes were summed up at the final level to an aggregated time series. In the distribution level, however, the hierarchical levels were specified as load of substations, feeders, transformers and, customers [74].

Recently, there has been a prevailing attention to HSTLF due to market considerations for decision-making in different levels of the power system including independent system operator, distribution operator, and customer-end. Utilities require load forecasting at low voltage levels to effectively perform distribution operation such as circuit switching and load control. An accurate load forecasting at low level could even increase the prediction accuracy at independent system operator level [75]. In fact, the independent system operator in the upper level in a power system covered a large geographical area, with extensive load diversities throughout the area. Hence, a single model was not able to guarantee the prediction accuracy.

The state-of-the-art HSTLF methods to address hierarchical load structure are sub-grouped into bottom-up and top-down approaches [27,76]. The bottom-up approach aggregates forecasts from low level to aggregated level, while the top-down method aggregates historical load prior to forecasting. The former approach does not miss out any information due to the aggregation, although high volatility of bottom level is challenging for prediction [77]. The top-down method, on the other hand, is simpler for less noisiness due to the aggregation. However, some features of the individual series are lost [42]. For instance, Quilumba et al. [27] used the bottom-up approach for forecasting load of the customers disaggregated by similar consumption patterns.

Some of the advantages and disadvantages of bottom-up and top-down approaches were highlighted by Hyndman et al. [78] who referenced early works in the literature. Generally, the bottom-up approach was robust when the data in bottom level was reliable without missing information. Otherwise, the forecast at a low level was error-prone and the top-down approach resulted in a more accurate forecast. Overall, the superiority of a method over another was not uniform.

HSTLF can also be conducted at all levels of hierarchies individually, which is termed "base forecast". However, the challenge here is that the prediction at aggregated level might not be consistent with the summed base forecasts [79].

Zhang et al. [73] proposed a solution to optimally adjust base forecast at each node in order to be consistent across the aggregation structure. This goal was accomplished by minimizing the redundancy between the forecast at the aggregated level and the sum of the base forecasts, by using quadratic programming in a post-processing scheme. The method was tested on two electricity networks; one bulk system of a large area with several dispatch zones at the bottom level, and the other was a distribution network covering a small area with hundreds of individual customers. Results indicate that for more than 85% of nodes in the bulk network, the proposed method was more accurate. For distribution network with more volatile load, the improvement was more obvious, especially at upper aggregated level where the error was significantly decreased. Nose-Filho et al. [80] also developed a load model for a sub-distribution system in New Zealand by finding participation factors between local forecasts and global forecast.

Another example is the study by Fan et al. [81], who proposed a strategy to forecast load of sub-regions within a large geographical area independently by finding the optimal region partition in the combination procedure. It was reported in [81] that the weather condition was a dominant factor for load variations and therefore, in a large geographical region, the extreme weather condition throughout the area caused high load diversity. Another factor that rendered regional load profiles vastly different was identified in [81] as non-coincident load peaks.

Sun et al. [74] proposed a strategy to predict loads of different nodes in a power distribution system by a top-down approach. Firstly, loads of parent nodes were forecasted. Subsequently, by finding the similarity between the parent node (aggregated level) and its child nodes (correspondent disaggregated levels), two classes of regular and irregular nodes were identified. Thus, for regular nodes, the load is a fraction of the origin load computed by a distribution factor. For those irregular loads, which did not follow leading characteristics of the parent node, individual models were forecasted. The similarity between nodes was identified by using distance minimization method for both weather parameter and historical load.

More recently, with the dominance of smart meters, fine-grained data at sub-levels revealed more information at the aggregate level. Wang et al. [82] used granular smart meter data to construct a forecast model at an aggregated level. In their proposed model, data was clustered into different groups of loads with similar patterns, and the aggregated forecast was obtained by adding the forecast of individual clusters. However, instead of the bottom-up strategy, a weight was assigned to each model while varying the number of clusters. The final forecast was an optimally weighted combination of these individual forecasts. Their proposed method was implemented on a data set consisting of 5237 residential consumers' information with half-hourly resolution for 75-week duration. It was shown that results of the direct aggregated load were more accurate than the clustering strategy although their proposed methodology outweighed the conventional bottom-up method. Besides this data set, the method was tested on 155 substations' load data for a 103-week duration. In contrast to the first data set, the outcomes of the forecast on the second dataset indicated that the bottom-up model was more accurate than other individual clustering models. It was concluded that this contrast was due to regularity in substation load in comparison to residential load profiles.

Table 4 illustrates two combination methods, which were applied to sum up base forecasts for maintaining its coherence with the aggregated forecast. Both of these methods minimized the error between the summed up base forecasts and aggregated forecast, either by linear [82] or quadratic [73] programming. Other combination methods were discussed in [83] with further theoretical explanations. This is suggested that new HSTLF methods might be expressed by selecting an appropriate combination algorithm.

Table 4. Combination methods for base forecasts.

Combination Method	Formulae	Parameters
Linear Programming	$w = \arg\min_{w} \sum_{t=1}^{T} \frac{1}{T} \frac{\lvert \hat{Y} - \tilde{Y} \rvert}{\hat{Y}}$ $\tilde{Y} = \sum_{n=1}^{N} w_n \tilde{Y}_n$	$\hat{Y}$: base forecast $\tilde{Y}$: adjusted forecast w: weight factor
Quadratic Programming	$\min_{\tilde{Y}} \frac{1}{2} (\hat{Y} - \tilde{Y})^T \Sigma^{-1} (\hat{Y} - \tilde{Y})$ $\tilde{Y} = \left[\tilde{a}^T, \tilde{b}^T \right]^T$ $\tilde{a} = p\,\tilde{b}$	$\hat{Y}$: base forecast $\tilde{Y}$: adjusted forecast $\tilde{a}$: load of the aggregated level $\tilde{b}$: Load of the disaggregated level p: participation factor

Different levels of a hierarchical structure interacted with each other in a complicated fashion, whereas a change in one series at one level could sequentially change the series at the same level as well as other levels of a hierarchy. Sun et al. [74] considered the change that switching operation might cause on the load trend by adjusting the forecast whenever a switching was detected. Abnormal changes in the demand were identified by measuring the mean and standard deviation of the load by using statistical process control. The load participation factor was then computed based on the new data. Comparably, deviations in the meteorological conditions in a large geographical area caused base forecasts to vary, leading to changes in the aggregated load accordingly. However, meteorological information might not be available at every sub-level. There were usually several meteorological services available at a geographical area for providing weather forecast information. Hong et al. [18] recommended that, in a hierarchical structure with various nodes to be forecasted, the best-related weather information could not be selected manually for each node. Weather station selection method was one of the main objectives in the Global Energy Forecasting Competition 2012 (GEFC) [84]. More about this is discussed in the next section.

2.4. Weather Station Selection

In a large electricity market covering an expanded area, a single forecasting model cannot capture the load pattern. HSTLF method, which is discussed in the previous section, ensures a more satisfactory forecast across different levels of hierarchy. However, in HSTLF method that disaggregates the load based on geographical divisions or zonal hierarchies, meteorological hierarchies that are definitely a dominant factor in load diversity cannot be easily captured. The challenge is to assign the most related weather station information to each zone or area in the hierarchy.

Fan et al. [81] proposed a combination method to select the best adapted individual weather forecast between multiple forecasts provided by different meteorological services. Several papers in the literature [85,86] used the average data from multiple services for its simple and effective result compared to other weighted averaging methods.

In Hong & Pinson's planned competition (GEFC competition) [18], weather station selection was one of the addressed issues. Data provided in the competition was the hourly load history of 20 zones in the U.S. along with weather data gathered from 11 weather stations, without specifying locations of weather stations.

Among the winning teams, Charlton et al. [87] built 11 energy models for each zone based on the weather data of 11 weather stations provided in the competition. The best-fitted weather station for each zone was not a single station, rather, it was a linear combination of up to five best-fitting weather stations for each group. Lloyd [88] also developed a forecast model based on data from all weather stations and used a Bayesian model averaging to integrate these models into one final average model. Moreover, in the proposed model by Nedellec et al. [48], one station was selected for each zone, considering that other combination strategies led to unsatisfactory outcomes. Taieb et al. [89] selected the best-fitted station for each zone by testing the temperature data from previous week for each zone. The demand was modeled by using average temperature data of three best weather sites. Hong et al. [18], on the other hand, proposed a method for weather station selection that, instead of assigning the same number of weather station to all nodes at the same level of hierarchy (as it was the common strategy in the GEFC competition), different numbers of weather stations were selected for individual load zones. Yet, the result was not always superior to other alternatives.

3. Method Evaluation and Future Work

A comprehensive explanation of STLF methodologies is provided in the previous sections. Generally, the logic behind every specific method helps the forecaster to choose the best-fitted method based on their application. For example, similar-pattern method mainly relies on historical values, whereas variable selection method incorporates information about explanatory variables. Therefore, the forecaster might consider similar-pattern method in cases where the system might not be comprehensive enough, or if it is explanatory, it is extremely difficult to extract the main features that govern the demand behavior. In this situation, there are always some variations in the load that cannot be captured by explanatory variables. In similar-pattern strategy, on the other hand, the focus is on what is going to happen rather than why it happens. Still, when there is a correlation between exogenous variables and load data, explanatory model, i.e., variable selection method is an appropriate approach.

Some of the main advantages and disadvantages of these four methods are listed in Table 5. For example, in variable selection method, despite efforts to find independent variables in the dataset by using feature selection algorithms, the selected variables might still be partly correlated with each other. This matter is expressed as one of the drawbacks in Table 5. Similar-pattern method, on the other hand, presumes that the past values of a variable are important in predicting the future, although the algorithms can only look back for a few steps for a limited sequence of data.

Despite the unique characteristics of these four categories of STLF methodologies, they were not independent of each other and there might be some overlap between them. For example, in similar-pattern method, the similarity of exogenous variables such as temperature or humidity were used to find similar patterns [21]. Consequently, selection of highly correlated exogenous variables is essential for detecting similar load patterns within a dataset.

Sometimes the selection of exogenous variables in variable selection method was conducted by using similarity method. For example, Fujimoto et al. [90] applied the minimum distance technique to find the relationship between exogenous variables and residential demands of multiple houses.

Another example was HSTLF method, as already discussed in Section 2, wherein either variable selection method or similar-pattern method was applied to forecast the load at each level of aggregation. Similarly, for weather station selection, a forecaster addressed a subset of exogenous variables, i.e., meteorological variables.

Table 5. Advantages and disadvantages of STLF methodologies.

Method	Advantages	Disadvantages
Similar-Pattern Method	• Adapt to exceptional circumstances and random events • No previous knowledge about the system is needed	• The horizon of forecast is limited up to a couple of days ahead • Limited search space • Not explanatory
Variable Selection Method	• Embedding exterior variables into the model • Increase prediction accuracy by reducing overfitting and addressing the curse of dimensionality • Simpler prediction (smaller number of predictors or smaller size of input space) • Improves the understanding of the prediction model	• The load data might not be comprehensive (difficult to measure the relationship that govern the load's behavior) • Difficulty in identifying exogenous variables • Huge number of variables • Redundancy • It is unrealistic to find a set of input variables with zero correlation to each other
Hierarchical Method	• Help the power system operators to perform the load control and circuit switching at different levels of the hierarchies • Enhancing the model accuracy by using the information at the lower levels	• Lack of coherency across the aggregated structure • Loss of information due to aggregation in the top levels • The high irregular data at the bottom levels of the hierarchies
Weather Station Selection	• Find the best-fitted weather data for each level of the hierarchy	• Uncertainty about optimal number of stations for each hierarchy

Hyndman et al. [78] discussed that taking advantage of the prominent features of different methods and combining them in a hybrid scheme was what we needed to do now. Some examples of this combination were available in the literature. For example, Quilumba et al. [27] applied similar-pattern method in one step to group smart meter load profiles into an optimal number of groups and then feature selection method in the next step to forecast the aggregated load at each group of data.

In the proposed load model by Wang et al. [82], a three-stage combined model was applied. The hierarchical structure of the load series was extracted by applying hierarchical clustering technique based on similar consumption behavior of customers. Different load models were developed at each subgroup of data by using variable selection method. Eventually, the final model was undertaken by adding a weight factor to individual models to be coherent across the aggregate level.

Another example of the hybrid methodology could be found in the work of Zheng et al. [31], in which feature selection method was used to help find similar days' clusters. Each cluster was shaped based on feature values of the data, whereas a weighted parameter was assigned to each feature.

In this paper, a hybrid method is represented based on some of the main features of methods reviewed in the previous sections. The schematic diagram of the method is illustrated in Figure 9. As can be seen, this method is proposed to find base forecasts at each level of the hierarchical structure by applying similar-pattern method, and then by using a strategy to keep the coherency between the loads at different levels. The strategy is performed in seven steps as shown in Figure 9. In the first step, the patterns similar to today's load profile are extracted from each load series at the disaggregate level. Considering that n number of similar patterns are obtained for each subseries, and by assuming that there is N number of subseries at the disaggregate level, n^N number of aggregated profiles is created. Between these aggregated profiles, the one with the minimum distance from today's profile at the aggregated level is selected. Subsequently, in the next step, the combined profile will be matched to the real aggregated profile by finding the weighting factor. Eventually, to forecast the next day's load at the aggregated level, load profiles of sequential days (days after similar-pattern days), which are selected in the optimal combination, will be summed up by using the weighting factor of step 5. This method finds similar patterns in the disaggregated level, but measures the similarity distance again at the aggregated level.

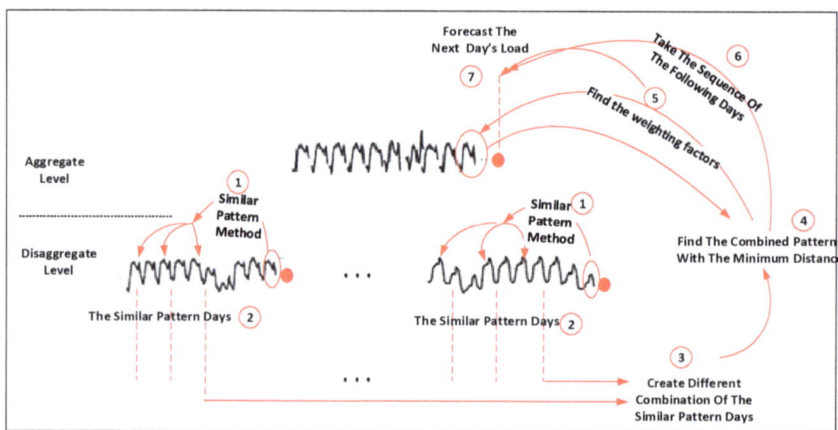

Figure 9. Schematic diagram of the proposed hybrid method.

The novelty of the proposed method over the aforementioned hybrid method is that it neither aggregates the data from bottom levels nor aggregates low-level forecasts. Following the hybrid method developed in [80], either forecast results of similar-pattern method is aggregated or a weighted averaged result of similar patterns are aggregated. However, the proposed method creates multiple subsets of data at the disaggregate level; consequently, the optimal subset is selected by comparing the combination results to the upper-level data. In this way, distinguishing the degree of similarity is not limited to one subset of data with averaging results. However, it is still not clear that this method might be more accurate than the conventional hybrid methods [82].

The proposed method assumes that finding the optimal subset of data might result in a more accurate forecast than averaging similar patterns in each low-level subseries. In fact, the idea of selecting an optimal subset of data at every disaggregate level for prediction of the next level's load is interesting; although, the technical difficulties for implementation need to be investigated in future works.

4. Conclusions

This paper discusses four categories of state-of-the-art STLF methodologies, i.e., similar-pattern, variable selection, hierarchical forecasting, and weather station selection while each of these methods proposes a specific solution for load prediction. Similar-pattern method, which is rooted from the minimum distance approach, presumes that the load trend is unlikely to vary during a short period. Hence, by searching within close vicinity of today's load, some similar patterns can be distinguished. In fact, forecasting the future load is based on the subsequent behavior of the discovered similar patterns in the load series.

Variable selection method, on the other hand, tries to find prominent and independent features in a dataset with the lowest correlation with each other and the highest correlation with the output. Constructing a subseries of these features helps to improve the forecast accuracy.

Hierarchical forecasting methods address the aggregated loads in different levels of the hierarchical structure. Predicting loads in various zonal level help power system operators to effectively perform the switching operation and load control. In addition, improving the forecast at sub-levels enhances the prediction accuracy at upper levels.

Besides geographical and zonal hierarchies, the weather hierarchy is another vital factor in STLF, which cannot be captured easily for each geographical zone. Various weather services in a large geographical area provide different weather forecast information. Selecting the best-suited weather information is substantially important for STLF, considering the influence of weather variables on the load trend.

Eventually, by highlighting the main advantages and disadvantages of each approach, it is concluded that the load model can benefit from the robustness of individual methods in a hybrid scheme. Finally, the general outline of a hybrid strategy is proposed for future evaluation.

Author Contributions: S.N.F. came up with the primary idea. The initial idea further developed with the collaboration of M.G., S.S., and K.-w.C. The contribution is as follows: methodology, S.N.F.; investigation, S.N.F., M.G., S.S.; supervision, S.S. and K.-w.C.; original draft preparation: S.N.F.; writing and editing: S.N.F., M.G., S.S. and K.-w.C.; visualization, M.G., S.S.

Funding: This research received no external funding.

Conflicts of Interest: The authors declare no conflict of interest.

Nomenclature

ANN	Artificial Neural Network
ARIMA	Auto-Regressive Integrated Moving Average
ARMA	Auto-Regressive Moving Average
CI	Computational Intelligence
DTW	Dynamic Time Warping
GEFC	Global Energy Forecasting Competition
HSTLF	Hierarchical Short-Term Load Forecasting
STLF	Short-Term Load Forecasting
MI	Mutual Information
RNN	Recurrent Neural Network
SARIMA	Seasonal Auto-Regressive Integrated Moving Average
SVM	Support Vector Machine

References

1. Shahidehpour, M.; Yamin, H.; Li, Z. *Market Operations in Electric Power Systems: Forecasting, Scheduling, and Risk Management*; John Wiley & Sons: Hoboken, NJ, USA, 2003.
2. Hong, T.; Fan, S. Probabilistic electric load forecasting: A tutorial review. *Int. J. Forecast.* **2016**, *32*, 914–938. [CrossRef]
3. Amjady, N. Short-term hourly load forecasting using time-series modeling with peak load estimation capability. *IEEE Trans. Power Syst.* **2001**, *16*, 498–505. [CrossRef]
4. Khotanzad, A.; Afkhami-Rohani, R.; Lu, T.-L.; Abaye, A.; Davis, M.; Maratukulam, D.J. ANNSTLF-a neural-network-based electric load forecasting system. *IEEE Trans. Neural Netw.* **1997**, *8*, 835–846. [CrossRef]
5. Huang, S.-J.; Shih, K.-R. Short-term load forecasting via ARMA model identification including non-Gaussian process considerations. *IEEE Trans. Power Syst.* **2003**, *18*, 673–679. [CrossRef]
6. Pappas, S.S.; Ekonomou, L.; Karamousantas, D.C.; Chatzarakis, G.; Katsikas, S.; Liatsis, P. Electricity demand loads modeling using AutoRegressive Moving Average (ARMA) models. *Energy* **2008**, *33*, 1353–1360. [CrossRef]
7. Lee, Y.-S.; Tong, L.-I. Forecasting time series using a methodology based on autoregressive integrated moving average and genetic programming. *Knowl.-Based Syst.* **2011**, *24*, 66–72. [CrossRef]
8. Chakhchoukh, Y.; Panciatici, P.; Mili, L. Electric load forecasting based on statistical robust methods. *IEEE Trans. Power Syst.* **2011**, *26*, 982–991. [CrossRef]
9. Chen, B.-J.; Chang, M.-W. Load forecasting using support vector machines: A study on EUNITE competition 2001. *IEEE Trans. Power Syst.* **2004**, *19*, 1821–1830. [CrossRef]
10. Khosravi, A.; Nahavandi, S.; Creighton, D.; Srinivasan, D. Interval type-2 fuzzy logic systems for load forecasting: A comparative study. *IEEE Trans. Power Syst.* **2012**, *27*, 1274–1282. [CrossRef]
11. Fan, S.; Hyndman, R.J. Short-term load forecasting based on a semi-parametric additive model. *IEEE Trans. Power Syst.* **2012**, *27*, 134–141. [CrossRef]
12. Mandal, P.; Senjyu, T.; Urasaki, N.; Funabashi, T. A neural network based several-hour-ahead electric load forecasting using similar days approach. *Int. J. Electr. Power Energy Syst.* **2006**, *28*, 367–373. [CrossRef]
13. Moghram, I.; Rahman, S. Analysis and evaluation of five short-term load forecasting techniques. *IEEE Trans. Power Syst.* **1989**, *4*, 1484–1491. [CrossRef]

14. Raza, M.Q.; Khosravi, A. A review on artificial intelligence based load demand forecasting techniques for smart grid and buildings. *Renew. Sustain. Energy Rev.* **2015**, *50*, 1352–1372. [CrossRef]
15. Fallah, S.; Deo, R.; Shojafar, M.; Conti, M.; Shamshirband, S. Computational intelligence approaches for energy load forecasting in smart energy management grids: state of the art, future challenges, and research directions. *Energies* **2018**, *11*, 596. [CrossRef]
16. Hippert, H.S.; Pedreira, C.E.; Souza, R.C. Neural networks for short-term load forecasting: A review and evaluation. *IEEE Trans. Power Syst.* **2001**, *16*, 44–55. [CrossRef]
17. Feinberg, E.A.; Genethliou, D. Load forecasting. In *Applied Mathematics for Restructured Electric Power Systems*; Springer: New York, NY, USA, 2005; pp. 269–285.
18. Hong, T.; Wang, P.; White, L. Weather station selection for electric load forecasting. *Int. J. Forecast.* **2015**, *31*, 286–295. [CrossRef]
19. Mu, Q.; Wu, Y.; Pan, X.; Huang, L.; Li, X. Short-term load forecasting using improved similar days method. In Proceedings of the 2010 Asia-Pacific Power and Energy Engineering Conference, Chengdu, China, 28–31 March 2010; pp. 1–4.
20. Dudek, G. Pattern similarity-based methods for short-term load forecasting–Part 1: Principles. *Appl. Soft Comput.* **2015**, *37*, 277–287. [CrossRef]
21. Chen, Y.; Luh, P.B.; Guan, C.; Zhao, Y.; Michel, L.D.; Coolbeth, M.A.; Friedland, P.B.; Rourke, S.J. Short-term load forecasting: similar day-based wavelet neural networks. *IEEE Trans. Power Syst.* **2010**, *25*, 322–330. [CrossRef]
22. Senjyu, T.; Takara, H.; Uezato, K.; Funabashi, T. One-hour-ahead load forecasting using neural network. *IEEE Trans. Power Syst.* **2002**, *17*, 113–118. [CrossRef]
23. Teeraratkul, T.; O'Neill, D.; Lall, S. Shape-based approach to household electric load curve clustering and prediction. *IEEE Trans. Smart Grid* **2018**, *9*, 5196–5206. [CrossRef]
24. Iglesias, F.; Kastner, W. Analysis of similarity measures in times series clustering for the discovery of building energy patterns. *Energies* **2013**, *6*, 579–597. [CrossRef]
25. Seem, J.E. Pattern recognition algorithm for determining days of the week with similar energy consumption profiles. *Energy Build.* **2005**, *37*, 127–139. [CrossRef]
26. Alvarez, F.M.; Troncoso, A.; Riquelme, J.C.; Ruiz, J.S.A. Energy time series forecasting based on pattern sequence similarity. *IEEE Trans. Knowl. Data Eng.* **2011**, *23*, 1230–1243. [CrossRef]
27. Quilumba, F.L.; Lee, W.-J.; Huang, H.; Wang, D.Y.; Szabados, R.L. Using Smart Meter Data to Improve the Accuracy of Intraday Load Forecasting Considering Customer Behavior Similarities. *IEEE Trans. Smart Grid* **2015**, *6*, 911–918. [CrossRef]
28. Liu, C.; Jin, Z.; Gu, J.; Qiu, C. Short-term load forecasting using a long short-term memory network. In Proceedings of the 2017 IEEE PES Innovative Smart Grid Technologies Conference Europe (ISGT-Europe), Torino, Italy, 26–29 September 2017; pp. 1–6.
29. Kong, W.; Dong, Z.Y.; Hill, D.J.; Luo, F.; Xu, Y. Short-term residential load forecasting based on resident behaviour learning. *IEEE Trans. Power Syst.* **2018**, *33*, 1087–1088. [CrossRef]
30. Shi, H.; Xu, M.; Li, R. Deep learning for household load forecasting–a novel pooling deep RNN. *IEEE Trans. Smart Grid* **2018**, *9*, 5271–5280. [CrossRef]
31. Zheng, H.; Yuan, J.; Chen, L. Short-term load forecasting using EMD-LSTM neural networks with a Xgboost algorithm for feature importance evaluation. *Energies* **2017**, *10*, 1168. [CrossRef]
32. Sutskever, I.; Vinyals, O.; Le, Q.V. Sequence to sequence learning with neural networks. In Proceedings of the Advances in Neural Information Processing Systems; 2014; pp. 3104–3112.
33. Marino, D.L.; Amarasinghe, K.; Manic, M. Building energy load forecasting using deep neural networks. In Proceedings of the IECON 2016-42nd Annual Conference of the IEEE Industrial Electronics Society, Florence, Italy, 23–26 October 2016; pp. 7046–7051.
34. Satish, B.; Swarup, K.; Srinivas, S.; Rao, A.H. Effect of temperature on short term load forecasting using an integrated ANN. *Electr. Power Syst. Res.* **2004**, *72*, 95–101. [CrossRef]
35. Barman, M.; Choudhury, N.D.; Sutradhar, S. A regional hybrid GOA-SVM model based on similar day approach for short-term load forecasting in Assam, India. *Energy* **2018**, *145*, 710–720. [CrossRef]
36. Ghofrani, M.; Ghayekhloo, M.; Arabali, A.; Ghayekhloo, A. A hybrid short-term load forecasting with a new input selection framework. *Energy* **2015**, *81*, 777–786. [CrossRef]

37. Jin, C.H.; Pok, G.; Lee, Y.; Park, H.-W.; Kim, K.D.; Yun, U.; Ryu, K.H. A SOM clustering pattern sequence-based next symbol prediction method for day-ahead direct electricity load and price forecasting. *Energy Convers. Manag.* **2015**, *90*, 84–92. [CrossRef]

38. Panapakidis, I.P. Clustering based day-ahead and hour-ahead bus load forecasting models. *Int. J. Electr. Power Energy Syst.* **2016**, *80*, 171–178. [CrossRef]

39. Goia, A.; May, C.; Fusai, G. Functional clustering and linear regression for peak load forecasting. *Int. J. Forecast.* **2010**, *26*, 700–711. [CrossRef]

40. Mori, H.; Itagaki, T. A precondition technique with reconstruction of data similarity based classification for short-term load forecasting. In Proceedings of the IEEE Power Engineering Society General Meeting, Denver, CO, USA, 6–10 June 2004; pp. 280–285.

41. Verdú, S.V.; Garcia, M.O.; Senabre, C.; Marín, A.G.; Franco, F.G. Classification, filtering, and identification of electrical customer load patterns through the use of self-organizing maps. *IEEE Trans. Power Syst.* **2006**, *21*, 1672–1682. [CrossRef]

42. Hyndman, R.J.; Athanasopoulos, G. Forecasting: Principles and Practice. 2018. Available online: https://otexts.org/fpp2/ (accessed on 28 November 2018).

43. Lusis, P.; Khalilpour, K.R.; Andrew, L.; Liebman, A. Short-term residential load forecasting: Impact of calendar effects and forecast granularity. *Appl. Energy* **2017**, *205*, 654–669. [CrossRef]

44. Ceperic, E.; Ceperic, V.; Baric, A. A strategy for short-term load forecasting by support vector regression machines. *IEEE Trans. Power Syst.* **2013**, *28*, 4356–4364. [CrossRef]

45. Espinoza, M.; Joye, C.; Belmans, R.; De Moor, B. Short-term load forecasting, profile identification, and customer segmentation: a methodology based on periodic time series. *IEEE Trans. Power Syst.* **2005**, *20*, 1622–1630. [CrossRef]

46. May, R.; Dandy, G.; Maier, H. Review of input variable selection methods for artificial neural networks. *Artificial Neural Networks-Methodological Advances and Biomedical Applications*. 2011. Available online: https://www.intechopen.com/ (accessed on 28 November 2018).

47. Koprinska, I.; Rana, M.; Agelidis, V.G. Correlation and instance based feature selection for electricity load forecasting. *Knowl.-Based Syst.* **2015**, *82*, 29–40. [CrossRef]

48. Nedellec, R.; Cugliari, J.; Goude, Y. GEFCom2012: Electric load forecasting and backcasting with semi-parametric models. *Int. J. Forecast.* **2014**, *30*, 375–381. [CrossRef]

49. Xiao, J.; Li, Y.; Xie, L.; Liu, D.; Huang, J. A hybrid model based on selective ensemble for energy consumption forecasting in China. *Energy* **2018**, *159*, 534–546. [CrossRef]

50. Hall, M.A. Correlation-Based feature selection of discrete and numeric class machine learning. In Proceedings of the Seventeenth International Conference on Machine Learning, San Francisco, CA, USA, 29 June–2 July 2000.

51. Kouhi, S.; Keynia, F.; Ravadanegh, S.N. A new short-term load forecast method based on neuro-evolutionary algorithm and chaotic feature selection. *Int. J. Electr. Power Energy Syst.* **2014**, *62*, 862–867. [CrossRef]

52. Estévez, P.A.; Tesmer, M.; Perez, C.A.; Zurada, J.M. Normalized mutual information feature selection. *IEEE Trans. Neural Netw.* **2009**, *20*, 189–201. [CrossRef] [PubMed]

53. Wang, Z.; Cao, Y. Mutual information and non-fixed ANNs for daily peak load forecasting. In Proceedings of the 2006 IEEE PES Power Systems Conference and Exposition, Atlanta, GA, USA, 29 October–1 November 2006; pp. 1523–1527.

54. Elattar, E.E.; Goulermas, J.; Wu, Q.H. Electric load forecasting based on locally weighted support vector regression. *IEEE Trans. Syst. Man Cybern. Part C* **2010**, *40*, 438–447. [CrossRef]

55. Wi, Y.-M.; Joo, S.-K.; Song, K.-B. Holiday load forecasting using fuzzy polynomial regression with weather feature selection and adjustment. *IEEE Trans. Power Syst.* **2012**, *27*, 596. [CrossRef]

56. Schaffernicht, E.; Gross, H.-M. Weighted mutual information for feature selection. In Proceedings of the International Conference on Artificial Neural Networks, Espoo, Finland, 14–17 June 2011; pp. 181–188.

57. Reis, A.R.; Da Silva, A.A. Feature extraction via multiresolution analysis for short-term load forecasting. *IEEE Trans. Power Syst.* **2005**, *20*, 189–198.

58. Amjady, N.; Keynia, F. Short-term load forecasting of power systems by combination of wavelet transform and neuro-evolutionary algorithm. *Energy* **2009**, *34*, 46–57. [CrossRef]

59. Hu, Z.; Bao, Y.; Xiong, T.; Chiong, R. Hybrid filter–wrapper feature selection for short-term load forecasting. *Eng. Appl. Artif. Intell.* **2015**, *40*, 17–27. [CrossRef]

60. Niu, D.; Wang, Y.; Wu, D.D. Power load forecasting using support vector machine and ant colony optimization. *Expert Syst. Appl.* **2010**, *37*, 2531–2539. [CrossRef]
61. Lin, S.-W.; Ying, K.-C.; Chen, S.-C.; Lee, Z.-J. Particle swarm optimization for parameter determination and feature selection of support vector machines. *Expert Syst. Appl.* **2008**, *35*, 1817–1824. [CrossRef]
62. Hu, Z.; Bao, Y.; Xiong, T. Comprehensive learning particle swarm optimization based memetic algorithm for model selection in short-term load forecasting using support vector regression. *Appl. Soft Comput.* **2014**, *25*, 15–25. [CrossRef]
63. Amjady, N.; Keynia, F.; Zareipour, H. Short-term load forecast of microgrids by a new bilevel prediction strategy. *IEEE Trans. Smart Grid* **2010**, *1*, 286–294. [CrossRef]
64. Sheikhan, M.; Mohammadi, N. Neural-based electricity load forecasting using hybrid of GA and ACO for feature selection. *Neural Comput. Appl.* **2012**, *21*, 1961–1970. [CrossRef]
65. Liang, Y.; Niu, D.; Hong, W.-C. Short term load forecasting based on feature extraction and improved general regression neural network model. *Energy* **2019**, *166*, 653–663. [CrossRef]
66. Santos, P.; Martins, A.; Pires, A. Designing the input vector to ANN-based models for short-term load forecast in electricity distribution systems. *Int. J. Electr. Power Energy Syst.* **2007**, *29*, 338–347. [CrossRef]
67. Ghadimi, N.; Akbarimajd, A.; Shayeghi, H.; Abedinia, O. Two stage forecast engine with feature selection technique and improved meta-heuristic algorithm for electricity load forecasting. *Energy* **2018**, *161*, 130–142. [CrossRef]
68. Hong, W.-C.; Dong, Y.; Lai, C.-Y.; Chen, L.-Y.; Wei, S.-Y. SVR with hybrid chaotic immune algorithm for seasonal load demand forecasting. *Energies* **2011**, *4*, 960–977. [CrossRef]
69. Hu, Z.; Bao, Y.; Chiong, R.; Xiong, T. Mid-term interval load forecasting using multi-output support vector regression with a memetic algorithm for feature selection. *Energy* **2015**, *84*, 419–431. [CrossRef]
70. Swarup, K.S.; Satish, B. Integrated ANN approach to forecast load. *IEEE Comput. Appl. Power* **2002**, *15*, 46–51. [CrossRef]
71. Khotanzad, A.; Afkhami-Rohani, R.; Maratukulam, D. ANNSTLF-artificial neural network short-term load forecaster-generation three. *IEEE Trans. Power Syst.* **1998**, *13*, 1413–1422. [CrossRef]
72. Kalaitzakis, K.; Stavrakakis, G.; Anagnostakis, E. Short-term load forecasting based on artificial neural networks parallel implementation. *Electr. Power Syst. Res.* **2002**, *63*, 185–196. [CrossRef]
73. Zhang, Y.; Wang, J.; Zhao, T. Using Quadratic Programming to Optimally Adjust Hierarchical Load Forecasting. *IEEE Trans. Power Syst.* **2018**. [CrossRef]
74. Sun, X.; Luh, P.B.; Cheung, K.W.; Guan, W.; Michel, L.D.; Venkata, S.; Miller, M.T. An efficient approach to short-term load forecasting at the distribution level. *IEEE Trans. Power Syst.* **2016**, *31*, 2526–2537. [CrossRef]
75. Hong, T.; Shahidehpour, M. *Load Forecasting Case Study*; EISPC, US Department of Energy: Washington, DC, USA, 2015.
76. Capasso, A.; Grattieri, W.; Lamedica, R.; Prudenzi, A. A bottom-up approach to residential load modeling. *IEEE Trans. Power Syst.* **1994**, *9*, 957–964. [CrossRef]
77. Stephen, B.; Tang, X.; Harvey, P.R.; Galloway, S.; Jennett, K.I. Incorporating practice theory in sub-profile models for short term aggregated residential load forecasting. *IEEE Trans. Smart Grid* **2017**, *8*, 1591–1598. [CrossRef]
78. Hyndman, R.J.; Ahmed, R.A.; Athanasopoulos, G.; Shang, H.L. Optimal combination forecasts for hierarchical time series. *Comput. Stat. Data Anal.* **2011**, *55*, 2579–2589. [CrossRef]
79. Gamakumara, P.; Panagiotelis, A.; Athanasopoulos, G.; Hyndman, R.J. *Probabilistic Forecasts in Hierarchical Time Series*; Monash University: Melbourne, Australia, 2018.
80. Nose-Filho, K.; Lotufo, A.D.P.; Minussi, C.R. Short-term multinodal load forecasting using a modified general regression neural network. *IEEE Trans. Power Deliv.* **2011**, *26*, 2862–2869. [CrossRef]
81. Fan, S.; Methaprayoon, K.; Lee, W.-J. Multiregion load forecasting for system with large geographical area. *IEEE Trans. Ind. Appl.* **2009**, *45*, 1452–1459. [CrossRef]
82. Wang, Y.; Chen, Q.; Sun, M.; Kang, C.; Xia, Q. An ensemble forecasting method for the aggregated load with sub profiles. *IEEE Trans. Smart Grid* **2018**, *9*, 3906–3908. [CrossRef]
83. Yang, Y. Combining forecasting procedures: some theoretical results. *Econ. Theory* **2004**, *20*, 176–222. [CrossRef]
84. Hong, T.; Pinson, P.; Fan, S. Global energy forecasting competition 2012. *Int. J. Forecast.* **2014**, *30*, 357–363. [CrossRef]

85. Xie, J.; Chen, Y.; Hong, T.; Laing, T.D. Relative humidity for load forecasting models. *IEEE Trans. Smart Grid* **2018**, *9*, 191–198. [CrossRef]

86. Liu, B.; Nowotarski, J.; Hong, T.; Weron, R. Probabilistic load forecasting via quantile regression averaging on sister forecasts. *IEEE Trans. Smart Grid* **2017**, *8*, 730–737. [CrossRef]

87. Charlton, N.; Singleton, C. A refined parametric model for short term load forecasting. *Int. J. Forecast.* **2014**, *30*, 364–368. [CrossRef]

88. Lloyd, J.R. GEFCom2012 hierarchical load forecasting: Gradient boosting machines and Gaussian processes. *Int. J. Forecast.* **2014**, *30*, 369–374. [CrossRef]

89. Taieb, S.B.; Hyndman, R.J. A gradient boosting approach to the Kaggle load forecasting competition. *Int. J. Forecast.* **2014**, *30*, 382–394. [CrossRef]

90. Fujimoto, Y.; Kikusato, H.; Yoshizawa, S.; Kawano, S.; Yoshida, A.; Wakao, S.; Murata, N.; Amano, Y.; Tanabe, S.-i.; Hayashi, Y. Distributed energy management for comprehensive utilization of residential photovoltaic outputs. *IEEE Trans. Smart Grid* **2018**, *9*, 1216–1227. [CrossRef]

Review

Assessing and Comparing Short Term Load Forecasting Performance

Pekka Koponen [1],*, Jussi Ikäheimo [1], Juha Koskela [2], Christina Brester [3] and Harri Niska [3]

[1] VTT, Technical research Centre of Finland, Smart Energy and Built Environment, P.O. Box 1000, FI-02044 Espoo, Finland; Jussi.Ikaheimo@vtt.fi

[2] Department of Electrical Engineering, Tampere University, P.O. Box 1001, FI-33014 Tampere, Finland; Juha.J.Koskela@tuni.fi

[3] Department of Environmental and Biological Sciences, University of Eastern Finland, P.O. Box 1627, FI-70211 Kuopio, Finland; kristina.brester@uef.fi (C.B.); Harri.Niska@uef.fi (H.N.)

* Correspondence: Pekka.Koponen@vtt.fi; Tel.: +358-20-722-6755

Received: 13 March 2020; Accepted: 17 April 2020; Published: 20 April 2020

Abstract: When identifying and comparing forecasting models, there may be a risk that poorly selected criteria could lead to wrong conclusions. Thus, it is important to know how sensitive the results are to the selection of criteria. This contribution aims to study the sensitivity of the identification and comparison results to the choice of criteria. It compares typically applied criteria for tuning and performance assessment of load forecasting methods with estimated costs caused by the forecasting errors. The focus is on short-term forecasting of the loads of energy systems. The estimated costs comprise electricity market costs and network costs. We estimate the electricity market costs by assuming that the forecasting errors cause balancing errors and consequently balancing costs to the market actors. The forecasting errors cause network costs by overloading network components thus increasing losses and reducing the component lifetime or alternatively increase operational margins to avoid those overloads. The lifetime loss of insulators, and thus also the components, is caused by heating according to the law of Arrhenius. We also study consumer costs. The results support the assumption that there is a need to develop and use additional and case-specific performance criteria for electricity load forecasting.

Keywords: short term load forecasting; performance criteria; power systems; cost analysis

---

1. Introduction

Bessa et al. [1] discussed two different ways of measuring the performance of a forecast. One way is to measure the correspondence between forecasts and observations (forecast quality). Another way is to measure the incremental benefits (economic/or other) when employed by users as an input into their decision-making processes (forecast value). Assessing forecast quality is more straightforward and the standard approach is statistical error metrics, such as:

- Mean absolute error (MAE),
- Mean absolute percentage error (MAPE), and
- Root mean squared error (RMSE).

Typically, these metrics apply some kind of loss function to individual errors and then calculate a summary statistic [2]. For example, Screck et al. [3] surveyed the literature for 681 load forecasts for the residential sector. Altogether, 15 error metrics were used, the most frequently used was mean absolute percentage error (MAPE) with 392 values. The second was normalized root mean squared error (NRMSE) with 209 error values and the third was RMSE. MAPE and NRMSE are relative metrics that aim to be comparable amongst different experiments. In the literature, the meaning of NRMSE varies significantly, because there are many different ways to normalize the RMSE. Here, we use only the most common definition that normalizes RMSE by dividing it by the mean of the measured values. In the alternative definitions, the normalization is done by the difference between the maximum and minimum, by the standard deviation, or be the interquartile range, etc. Some publications, such as [4], even use a NRMSE definition that, similarly to MAPE, normalizes each individual error with the simultaneous measured value before calculating the RMSE. So defined NRMSE and MAPE are much more sensitive than the NRMSE we use to calculate absolute errors when the actual loads are small or very small. MAPE also puts less weight on large deviations than NRMSE. MAPE is based on assumptions that (1) accurate forecasting of small loads is important and (2) one large error is not more significant than an equally large sum of small absolute errors. Both these assumptions are clearly in conflict with the actual consequences of the short-term load forecasting errors that we discuss next.

Often, model identification is easier and computationally more efficient with quadratic criteria such as sum of squared errors (SSE), RMSE and NRMSE. These criteria also reflect the combination of errors from several independent sources. This is important, as typically several different component forecasts are needed for forecasting the total power or power balance. The assumption of independence may not be completely valid, however. For example, the forecasts of loads and generation are often based on the same or mutually correlated weather forecasts or are connected by the behavior of certain groups of humans. Technologies and energy markets also reduce independence of load behavior. For example, demand side responses for electricity markets and ancillary services have very much mutual correlation. Good short-term load forecasting methods utilize such dependencies efficiently and their forecasting errors tend to be rather independent from each other. Correlated forecasting errors may also stem from using the same forecasting methods or methods that have common weaknesses. Mutual correlation of forecasting errors is usually easy to detect, and it is a sign that improving the forecast is possible.

The concept of forecast value, on the other hand, views the forecast user's business process more extensively and is more difficult to assess. Forecast value includes the economic and noneconomic benefits which are available to the user by using the forecast. An example is the reduction in imbalance costs for a balance responsible party. A crucial aspect is that the value is user and problem specific [1]. For example, power market participants and the system operator may measure forecast value differently. For the system operator, the most important issue is the expected maximum forecast error, and not the mean forecast error. An UCTE position paper [5] discussed this issue for wind power forecasts. We will employ a case study below to explore to what extent the consumer and the retailer aggregator have different preferences. Such differences between different actors largely stem from the fact that the electricity markets locally and imperfectly approximate marginal cost changes, but do not perfectly reflect the real costs of the power systems.

It is often infeasible to calculate the benefits of the forecast accurately. When it is possible to accurately model the decision-making process which exploits the forecasts and the resulting costs, the error metric can be selected so that the resulting costs are minimized [6].

Forecast value is also related to the error metrics. According to [6], error metrics should be easy to interpret, quick to compute and reflect the net increase in costs resulting from decisions made based on incorrect forecasts. MAPE also penalizes over-forecasts (where forecast load is greater than realized load) more than under–forecasts [7,8]. In addition, MAPE penalizes relatively lightly such large absolute forecasting errors that occur during load peaks. However, in short-term load forecasting, under–forecasting high loads tends to be especially costly for all the relevant actors, including the system operator and the energy consumer.

The most commonly used error metrics also suffer from a double penalty effect for events which are correctly predicted but temporally displaced [7,9]. There are criteria, such as the parameterized earth mover's distance [10], which do not suffer from this effect. For avoiding this double penalty effect, time shifted error measures such as dynamic time warping and permutated (so called adjusted) errors and their downsides are considered by [9].

Costs of large mutually correlated forecasting errors may behave differently from the SSE and RMSE. MAE assumes that the costs due to forecasting errors depend linearly on the size of the errors. It is often used to avoid the problem that SSE and RMSE put too much weight on large forecasting errors as compared to the assumed real costs. Often this assumption is not valid and the actual costs may grow even faster than the square of the error assumed in SSE and RMSE.

The power systems are changing in an increasing speed. Distributed new energy resources such as active loads and other controllable distributed flexible energy resources make the power flows in distribution grids more variable than before. The power flows to and from the customers of the grid are correlated due to control actions, solar radiation and wind. The traditional load forecasting methods become obsolete. The forecasting performance criteria also need some reconsideration and updating in order to fit to this new situation.

Our aim in the present work is to show that there is still a need to develop the understanding, selection and amendment of criteria for forecasting performance. In forecasting applications, assessing or measuring the incremental benefits from the forecasts is both necessary and difficult. With case studies, we assess how the forecasting errors increase the costs of the competitive electricity market actors, electricity network operators and their customers, the consumers and prosumers that use the power system.

2. Needs to Develop Short Term Load Forecasting Criteria

When studying and developing short term load forecasting methods for active demand, such as [11], we have detected the following challenges:

- Assessing the economic costs of forecasting errors on the electricity markets is more complex than often assumed.
- The cost impacts tend to be asymmetric. Under-forecasting the load peaks typically causes higher costs than a similar amount of over-forecasting.
- The economic costs of forecasting errors tend to concentrate to the load peaks at the network bottleneck and system levels and to the rarely occurring high price peaks on the markets for electricity and the ancillary services of the power system and grids.
- Some popular and useful methods tend to predict higher and shorter load peaks than what actually occur.

The costs of forecasting errors to the competitive electricity market actors mainly stem from the balancing errors caused to their balance responsible party. When forecasting only components of the total balance, the errors relative to the errors in the total balance matter and the errors relative to the component forecast itself are not at all relevant. In addition, we need accurate forecasting during the system peak loads and peak prices and the highest peaks are rare and unpredictable.

Underestimating the load when forecasting critical high load situations can lead to very expensive unwanted measures in managing the grid constraints or peak load reserves at short notice.

A common problem with black box methods, such as machine learning methods, is that although they generally tend to under-forecast load peaks, as shown in [4], for rarely occurring outdoor temperature related peak loads they tend to predict higher and shorter load peaks than what actually occur. This tends to happen because (1) the identification data do not include enough such extreme situations in order to model the load saturation effects or (2) many nonlinear black box methods tend to have large errors when extrapolating outside the situations included in the identification data. There are several methods to deal with this problem. Daizong Dint et al. [12] proposed using (1) a memory network technique to store historical patterns of extreme events for future reference and (2) a new classification loss function called extreme value loss (EVL) for detecting extreme events in the future. Those approaches can improve the deep neural network (DNN) performance regarding only those extreme events that have been included in the learning data. Another approach is to add another model for the power range and use that for limiting out those forecast values that exceed the limit by more than a tolerance [4]. Then, the energy of the peak should be preserved by extending the length of the peak. Physically based model structures describing the existence of constrained maximum power are useful for such peak limitation. Physically based model structures with power constraints can also be used to model the main phenomenon that contributes to the peak load. We have improved forecasting of rarely occurring and extreme situations by combining several different models into hybrid models, see [11] for an example. In order to better assess, compare and develop methods, we need forecasting criteria that adequately reflect the concentration of the economic costs of the forecasting errors to the high load situations.

3. Cases Studied

3.1. Electricity Market Costs Due to Forecasting Errors

The costs relate to the overall forecasting error of the total energy balance of the balance responsible party of the actor in the market. Thus, the load forecast of a consumer group segment is only one component of the total forecast. There are forecasts for different types of load and local generation. The different forecasts are not fully independent, because many of them may use the same weather forecasts and be subject to interactions between consumer group behavior. However, the errors of accurate forecasters tend to be independent and it is easy to check to what extent this assumption holds. Going to such details is complicated and outside the scope of this paper. Thus, for clarity of the analysis, we assume that the errors of different segment forecasts are independent. Then the contribution of the expected individual error component e_1 to the total expected forecast error e is as follows.

$$\mathrm{E}\left[e^2\right] = \mathrm{E}[e_0{}^2 + e_1{}^2] = \mathrm{E}[e_0{}^2] + \mathrm{E}[e_1{}^2] \tag{1}$$

where e_0 is the expected total error of all the other forecast components. Figure 1a shows how the total error e behaves as a function of e_1 when e_0 is set to 1. For small e_1, the increase in e is quadratic.

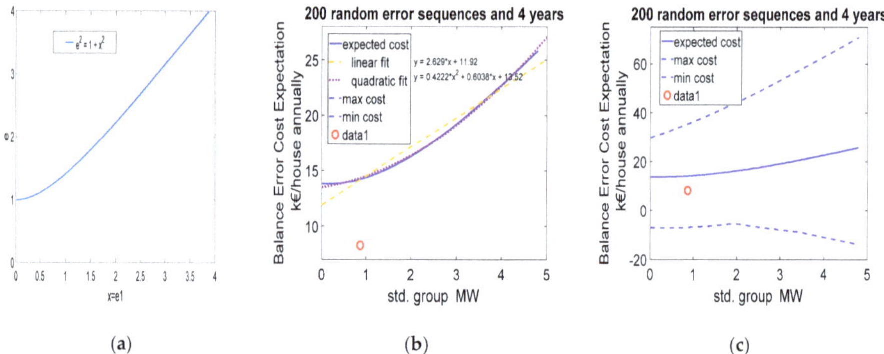

Figure 1. (**a**) Impact of an additional error std. component to the total error std. when assuming normally distributed independent error sequences, (**b**) the behavior of the mean of simulated balancing error costs, (**c**) the range of balancing error cost variation in the simulations.

The monetary cost of the total forecasting error needs to be assessed in the electricity market. For simplicity, we assume that the forecasting errors cause costs via the market balancing errors. The market rules for penalizing the balancing errors of loads vary from country to country. The Nordic power market applies the balancing market price as the cost of the balancing errors. In the Nordic countries, one price system is applied for errors in consumption (load) balance and, in addition, there is a small cost component proportional to the absolute balancing error (volume fee for imbalance power). Power generation plants below 1 MVA are included in the consumption balance.

The price of consumption imbalance power is the price for which Fingrid both purchases balance power from a balance responsible party and sells it to one. In the case of the regulating hour, the regulation price is used. If no regulation has been made, the Elspot FIN price is used as the purchase and selling price of consumption imbalance power. For an explanation, see [13]. All the prices and fees are publicly available at the webpages of Fingrid and NordPool, such as [14]. In addition, imbalance power in the consumption balance is subject to a volume fee. There is also a fee for actual consumption, but that depends only on the actual consumption and not on the forecasting errors. The volume fee of imbalance power for consumption during the whole study period was 0.5 €/MWh. See [15] for the fees.

From an actual or simulated component load forecasting error, the resulting increase the forecasting error of the total energy balance can be calculated, and using this increase in error the resulting balancing cost increase was calculated based on the price of imbalance power and the volume fee of the power imbalance. In this way, we got an estimate for how much electricity market costs the forecasting error causes to the balance responsible actor. We do not know the forecasting errors of the balance responsible party. In the following study, we generated them as a normally distributed sequence that has standard deviation 3 MW. The actual errors may be bigger, but as we see later, that will not affect the conclusions. The price of imbalance price occasionally has very high peaks. Thus, the cost estimate will be very inaccurate, even when a very long simulation period is applied. It is necessary to check the contribution of the very highest price peaks to the cost in order to have a rough idea on the inaccuracy of the results. We avoided this challenge as follows. We made a short-term load forecast using a residual hybrid of physically based load control response models and a stacked booster network as explained in [16] for a four-year-long test period. We found out that the forecasting errors were rather well normally distributed and bounded. We generated 200 random normally distributed bounded error sequences over the four–year period. With each one of these 200 normally distributed bounded random error sequences, we calculated the balancing error costs for the forecast group. Then, the standard deviation of the group errors was varied and the same cost calculation repeated. The variation of the costs between the error sequences was very large. The mean behaved as assumed in Figure 1a and an actual measured and forecast case was clearly in the area where the quadratic dependency dominates (see Figure 1b). The demand response aggregator considers the actual and forecast active loads as trade secrets and does not allow us to make them public information. Except for this one point, all data used for the simulations in Figure 1 are publicly available.

The balancing error cost was very stochastic, because the impact of high balancing market price peaks dominated (see Figure 1c). The balancing error cost over the whole four-year-long period mostly depended on the forecasting error during those few price peaks. The red circle represents the forecasting errors when an experimental short-term demand response forecasting algorithm was applied to the forecasting on the morning of the previous day. There, the aggregated controllable power was slightly over 18 MW. The results support the use of the quadratic error criteria std. and RMSE, rather than linear criteria such as MAPE etc. In this case, data driven forecasters that do not model the dynamic load control responses have so poor accuracy that the cost dependency approaches a linear dependency. Here, we have ignored the fact that especially large forecasting errors can affect the price of imbalance power significantly, thus increasing the balancing error cost much more than linearly.

Another observation is that with the good performance forecasting model, the forecasting errors increased the imbalance costs very little, only 0.53 € per controllable house annually. This suggests that the one price balancing error model gives only very weak incentives to improve the forecasting accuracy in normal situations. The one–price balance error cost model may not adequately reflect the situation from the power system point of view. A further study is needed to find out to what extent, how much and with which market mechanisms the power system can in the future benefit from improving the short-term forecasting accuracy. A conclusion of the H2020 SmartNet project [17] was that improving the forecasting accuracy is critical for getting the benefits from the future ancillary service market architectures for enabling the provision of the ancillary services using distributed flexible energy resources.

Some other countries apply a two–price system for balancing error costs. That means that the price for the balancing errors is separately defined by the balancing market for both directions. Then the costs of load forecasting errors are much higher than in a one–price system. They may even exaggerate the related needs of power system, if the errors of the forecasts of the individual balancing are assumed to be independent from each other. When the share of distributed generation increases, the one price system may become problematic, because the consumption and distributed generation may not have enough incentives to minimize their balancing errors. This increases the need for balancing reserves in the system. The share of distributed generation is expected to increase much during the summertime, which means that also in the Nordic countries there may appear needs to change the market rules regarding the balancing costs somehow. Moving to two–price system may be one such possibility. Thus, it may be worthwhile to repeat the above study using the two–price system of the generation balance.

3.2. Distribution Network Costs Due to Forecasting Errors

Overloading of network components causes high losses that increase the temperature of the components. Overheating reduces the lifetime of the insulator materials in the network components rapidly. If the forecast underestimates the peak load, the operational measures to limit overload are not activated in time, energy losses increase and the component aging increases so much that the expected lifetime is reduced.

The losses in the network components are generally proportional to the square of the current. When the voltage is kept roughly constant and the reactive power, voltage unbalance and distortion are assumed to be negligible, the losses are roughly proportional to the square of the transferred power. In real networks, these assumptions are not accurate enough. Strbac et al. [18] calculated losses using a complete model of three power distribution license areas in UK. The analysis highlighted that 36–47% of the losses are in low voltage (LV) networks, 9–13% are associated with distribution transformer load related losses, 7–10% are distribution transformer no-load losses and the remaining part in higher voltage levels. They [18] (p. 43) showed the marginal transmission losses as a function of system loading. A 1 MWh reduction in load would reduce transmission losses by 0.11 MWh during peak load condition (100%). When system loading is 60%, reducing the load by 1 MWh will reduce transmission losses by 0.024 MWh. This corresponds to the dependency $f(P) = P^{2.98}$. The sample size is small, so the accuracy of this dependency is questionable. Nevertheless, the dependency is clearly different from $f(P) = P^2$ that results from assuming constant voltage at the purely active power load and transmission losses relative to the square of the current. Underestimating the peak loads causes much higher losses and related costs than other load forecasting errors. Thus, the impact of forecasting errors to the energy losses is very nonlinear and depends on the direction of the error and size of the load.

Ref. [19] studied how transformer oil lifetime depends on temperature. Arrhenius' law describes the observed aging rather well. Aging mechanisms of cable polymers were discussed in [20]. The Arrhenius method is often used to predict the aging of cable insulation, although it is valid only in a narrow temperature range. For example, it is applicable only below the melting point of the insulator. For simplicity, we here model the aging using the Arrhenius method. According to it, k the rate of chemical reaction (such as insulator aging) is an exponential function of the inverse of the absolute temperature T.

$$k = Ae^{-E_a/RT} \tag{2}$$

where R is the gas constant, and the pre-exponential factor A and the activation energy E_a are almost independent from the temperature T. In the steady state, the difference between the component or insulator temperature and the ambient temperature is linearly proportional to the losses. Components are normally operated much below their nominal or design capacity and the impact of forecasting errors on the losses and aging is small. When the component during peak load is operated at or above its nominal load and is subject to high ambient temperature and poor power quality high losses, fast component aging, or expensive operational measures result from under-forecasting the load.

Thus, the impact of forecasting errors to the costs is very nonlinear and depends on the direction of the error and size of the load. Most of the time, the network costs from short term load forecasting errors are small or even insignificant. However, the costs of forecasting errors increase rapidly when the load is at or above the nominal capacity of the network bottlenecks, if the actual load is higher than the forecast load.

3.3. A Consumer Cost Case Study: Load Forecasting Based Control of Domestic Energy Storage System

All the costs of the power supply are paid by the users of the electricity grid. The forecasting errors discussed in the other chapters increase the electricity prices and grid tariffs by increasing the costs of the electricity retailers, the aggregators and the grid operators. Here, we consider those costs that the consumer has possibilities to control more directly.

By using energy storage in a domestic building, a customer can get savings in the electricity bill [21]. The amount of the savings depends on many factors. The load profile of the customer and the electricity pricing structure and price levels are the main variables that affect the savings, but the customer has very limited possibilities to change them. The size of the energy storage can be optimized for the customer's load profile, but after that, controlling the energy storage and consumption is the only way to maximize the savings. Energy storage can be used, e.g., to increase the self-consumption of small-scale photovoltaic production, but it can also be used for minimizing costs from different electricity pricing components. If the energy retailer's price is based on the market price of electricity, the customer can get savings by charging the energy storage during low price and discharging during high price as in [21]. If electricity distribution price is based on the maximum peak powers, the customer can get savings by discharging the battery during peak hours, as in [22].

Such electrical energy storage systems are still rare and typically installed to increase the self-consumption of small-scale photovoltaic power production. Although the battery technologies progress all the time, the profitability in such use is still typically poor, especially if there are loads that can be easily shifted in time. One can improve the profitability of the battery system significantly by having several control targets or a more holistic one. Such a possibility is minimizing the costs from different electricity pricing components, but that requires short–term load forecasting.

In this case study, it is assumed that every customer in the study group has a market price-based contract with an electricity retailer and the distribution price is partly based on the maximum peak powers as in [23]. The energy retailer price is based on hourly day-ahead market prices of Finland's area in Nord Pool electricity markets [13]. The electricity retailer adds a margin of 0.25 c/kwh to the hourly prices. Distribution prices are based on the study in [23], where the price components were calculated for the area of the same distribution system operator (DSO) as where the customers' data of this study were measured. The price structure includes two consumption-based components: volumetric charge (0.58 c/kWh) and power charge (5.83 €/kW). The power charge is based on the monthly highest hourly average peak power. When value added tax (24%) is added to these charges, the prices which affect to customers' costs are: volumetric charge 0.72 c/kWh and power charge: 7.23 €/kWh. The same prices were used in [22].

Customers' load data are from one Finnish DSO, whose license area is partly rural and partly small towns. The study group consists of 500 random customers. In simulations, each customer has 6 kWh with a 0.7 C-rate Lithium-ion battery. Battery type is lithium iron phosphate (LFP), because it has good safety for domestic use. The energy storage system is controlled firstly to decrease the monthly maximum peak power and secondly to decrease the electricity costs with market price-based control as in [22]. The battery is discharged when power is forecast to increase over the monthly peak and charged during low prices. The market price-based control algorithm and battery simulation model were presented in [21]. In previous studies, the controlling of energy storage was based on load forecasting. The load forecasts are based on a model, which utilizes customer's historical load data and temperature dependence of load with temperature forecast [24].

In the present study, the dependence between the accuracy of load forecasting and the customers' savings achieved by using the energy storage are studied. The simulations are made for every customer with 11 different load forecasts each having a different load forecasting accuracy. The forecasting accuracy is varied by scaling the forecast error time series. The actual load forecast time series is nominated as the basic level (100%) and the real load data correspond to the ideal forecast (0%). The range of studied error scaling is selected linearly between 0% and 200%, with 20% step size in every hour. Customers' yearly cost savings and different forecast accuracy criteria (SSE, RMSE, NRMSE, MAE and MAPE) have been calculated during simulations. Additionally, because most of the savings come from the decrease in monthly peak powers, the MAE of monthly peak powers (MAE_{max}) was calculated. The monetary value of the cost savings depends on the customer's load profile, so the results are given as percentage values of cost savings. The results of the simulations are shown in Figure 2. From the result points, we calculated least-squares fitted line (R1) and least-squares fitted second-order curve (R2).

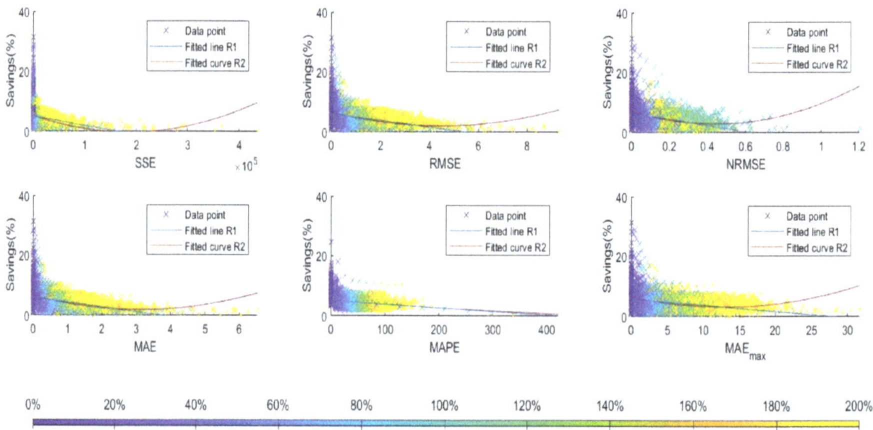

Figure 2. Percentage yearly savings of customers when using energy storage to decrease monthly maximum peak powers and the costs of market priced energy, shown as a function of different forecast error criteria. Color of points shows the used error level (0–200%).

From the results, we see that with ideal forecasts the savings of the customers vary a lot. This stems from the different load profiles of the customers. The customers, whose load profile includes high peaks during several months, can get very high savings. If customer's load profile is very flat, the savings can be low. When the errors in the forecast start to increase, the savings drop very fast at first, but the decrease in the savings slows quickly and the decrease stays low until the end.

From the results of Figure 2 and the results of least-squared fittings, we collected the main results and values for the Table 1, which helps to compare different criteria. In Table 1, data points mean the points (maximum 5500 points) which can be used and logical order means that the data points are in order, such that higher error gives lower savings.

The idea in the comparison is that good criteria predict as accurate as possible the cost savings of a customer. Fitted lines and curves predict the cost savings best with MAPE, but MAPE can be calculated only for a small part of the customers. For this reason, the values of MAPE seem better than they really are. With NRMSE, the order of points is not logical: the savings do not monotonically decrease as the NRMSE value increases. SSE gives the worst values in fittings. RMSE and MAE are almost equal, but MAE is in this case marginally the best criteria. Differences between the criteria are not large and selecting the most suitable criteria in this case requires accurate comparison.

Additionally, the MAE of the monthly peak powers is shown in Figure 2. As we can see, MAE_{max} describe the effect of forecasting error for the savings almost as well as traditional forecast error criteria.

This is logical when most of the savings come from the decrease in monthly peak powers. When the other forecast error criteria take into account all hours during the year (8760 h), this MAE_{max} is calculated only from one hour per month (12 h).

Table 1. Comparison of criteria in a consumer cost case study.

Criteria	Data Points	Logical Order	Mean of Residuals R1	Max Error R1	Median of Residual R1	Mean of Residuals R2	Max Error R2	Median of Residuals R2
SSE	5500	Yes	1.70	26.39	1.32	1.69	26.14	1.31
RMSE	5500	Yes	1.63	25.41	1.25	1.59	24.81	1.23
NRMSE	5500	No	1.61	24.96	1.23	1.57	24.36	1.18
MAE	5500	Yes	1.62	25.43	1.23	1.58	24.83	1.22
MAPE	1310	Yes	1.25	13.20	1.03	1.25	13.17	1.04
MAE_{max}	5500	Yes	1.70	25.62	1.32	1.65	24.89	1.26

3.4. Comparison of Methods across Different Forecasting Cases

When the same method is used to forecast different aggregated loads the values of the same accuracy criteria can be very different. Humeau et al. [25] analyzed the consumption of 782 households and found out how the values of the NRMSE decrease with the increase in the number of sites in clusters. They compared linear regression, support vector regression (SVR) and multilayer perceptron (MLP) in this respect. [26] shows a typical short-term load forecasting accuracy dependence on the prediction time horizon. The weather forecasts and load forecasting methods have improved much so now the accuracy decreases somewhat later but the shape of the dependency is still similar. In addition to the number of sites, the value of criteria depends also on the size and type of sites, type of loads, the presence of active demand, etc. Figure 3 demonstrates some of the dependencies. It summarizes some results from our past publications between the years 2012 and 2020. All these publications can be found via [11,16]. All the most accurate methods in Figure 3 are hybrids that combine several short-term forecasting methods and include both machine learning and physically based models. In the case with about 59,000 customers, the most accurate method includes also a similar day method. All of the most accurate methods use more than one hybridization approach, including residual hybrid, ensemble and range constraining.

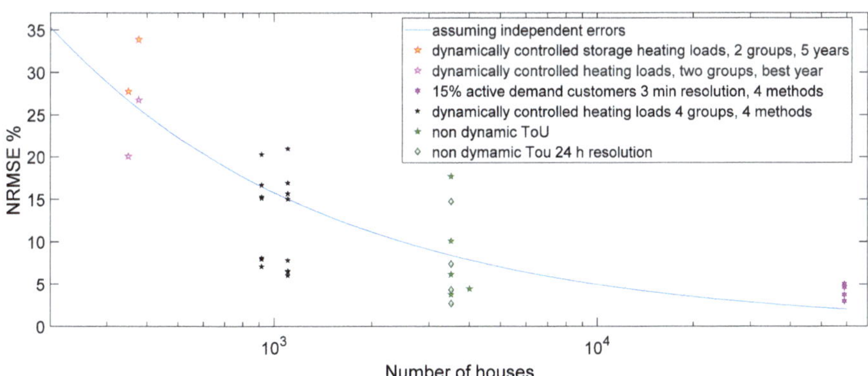

Figure 3. Load forecasting accuracy as a function of the number of aggregated customers as collected from our own published results. The results support the assumption of independent errors except for the largest group size and the other differences reduce the comparability among the cases much more.

In Figure 3, the four markers on the right end represent different forecasting method applied to the exact same case. At about 1000 aggregated customers, there are combinations of four rather similar groups of about 1000 customers and two different methods in all eight combinations. The blue line in the figure shows how the forecasting accuracy measured in std. depends on the number of customers assuming completely independent forecasting errors. The expected behavior of uncorrelated forecasts is like that. The cases are not fully comparable and the amount of them is too small for making any reliable conclusions. Forecasting when active loads are not present is usually more accurate than when dynamic load control is applied. Load control also makes the load behavior strongly correlated, which also tends to increase the correlation of forecasting errors and thus reduce the NRMSE improvement stemming from increasing the number of aggregated customers. In the cases with 59,000 customers at the right side of Figure 3, the forecasting time resolution was 3 min and in the other cases it was 1 h. In the lowest NRMSE case at about 3500 customers, it was 24 h. Using more accurate time resolution causes higher values of criteria. At about 3500 customers, the range of the values shows the impact of the improvement of the methods when the case remains the same. There, the outdated national load profiles had the highest NRMSE. In the four groups that have about 1000 customers, the controllable loads dominate. With two of these four groups, the forecasting performance is very good as compared to the blue line. The two leftmost groups, each having 350 to 380 customers, suffered from communication failures in the first third of the 5-year-long verification period. Selecting only the best year of the verification would have given NRMSE = 20.1% for group 1 and NRMSE = 26.8% for group 2 of that case.

One observation is that the results are not always very well comparable between different groups in the same case, nor between different years for the same group. Thus, meaningful comparison of methods across different individual forecasting cases does not seem feasible. The comparability was even worse when using MAPE instead of NRMSE. In addition, the results support the hypothesis that with the best forecasting methods in the studied cases the assumption of mutually independent forecasting errors may be justified if the number of houses is not very large. The amount of cases is too small for making firm conclusions. When the forecasting cases represent the same time and country the cross correlation can be calculated from the forecasts. We leave that now to further studies. The purpose here is only to show that (1) this kind of analysis may be useful when made with many more cases and (2) the comparison of forecasting methods between different cases is complex and gives only very rough results. Further research with many more cases is needed in order to get reliable quantitative results.

Figure 4 shows how the MAPE and NRMSE depended on each other in 38 short-term forecasts that we have produced in six different forecasting cases. All the forecasts have the same forecasting horizon. The differences in their behavior are rather small. This is the expected result for the errors of accurate forecasts. We expect that the results may be much more different if either low accuracy forecasts or more exceptional situations are included in the comparison. Further studies are needed regarding that.

All those six forecasts that have NRMSE between 15% and 17% are from the same group in the same case but use different forecasting methods. The group behavior in the verification was rather different from the identification. The low MAPE in one of the forecasts may indicate that MAPE there failed to adequately detect the rather large peak load errors caused by the behavior change. That may happen, because the statistical behavior of the errors was not any more normally distributed, as is the case with accurate forecasts.

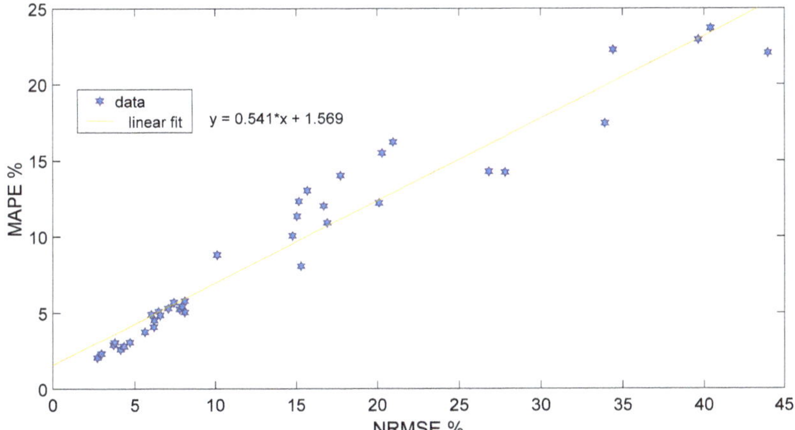

Figure 4. Comparison of normalized root mean squared error (NRMSE) and mean absolute percentage error (MAPE) in 38 forecasts in 6 different short term load forecasting cases comprising together 12 different forecasting methods.

3.5. Load Peak Sensitive Validation

The most valuable forecasts for peak load can be obtained by modeling the actual costs of forecasting errors in the electricity market, in the distribution grid or in both depending on the purpose of forecasting. For most comparisons, this is not practical, because of the complexity and stochastic nature of the costs as shown before. As the cost of the errors is the fundamental reason for the need for accurate forecasts, it is nevertheless important to have at least a general idea on how the costs form and use criteria that reflect the load peak sensitivity of the costs.

Conventional validation statistics cannot solely guarantee the performance of a model in load peak situations. For instance, the recent study on weather forecast-based short-term fault prediction using a neural network (NN) model [27] showed some inherent limitations of the standard MAPE and mean absolute error (MAE) metrics. MAPE does not work due to existence of zero values, i.e., the absence of network faults. MAE does not properly reflect the model performance thoroughly, e.g., in rare peak events. The high fault rate periods are important to predict in order to temporarily increase preparedness to manage them. On the other hand, the results indicate that the index of agreement (IA) [28] may provide a more robust metric for measuring the model performance, including peak events and for model evaluation and comparison in general:

$$IA = 1 - \left(\frac{SSE}{\sum_{i=1}^{n}\left(\left|\hat{y}_i - \overline{y}\right| + \left|y_i - \overline{y}\right|\right)^2} \right) \tag{3}$$

where $SSE = \sum_{i=1}^{n}(y_i - \hat{y}_i)^2$, y_i is the true number of faults, $\hat{y}_i$ is the predictied number of faults, $\overline{y} = \frac{1}{n}\sum_{i=1}^{n} y_i$, $i = 1, n$, n is the sample size, and $IA \in [0, 1]$, higher values of IA indicate better models.

The study [27] also showed that the resampling/boosting of rare faults peaks in the training data can be used to enhance the ability of an NN model to forecast fault events. Figure 5 demonstrates the results of forecasting all types of faults and faults caused by wind, originally presented in [27].

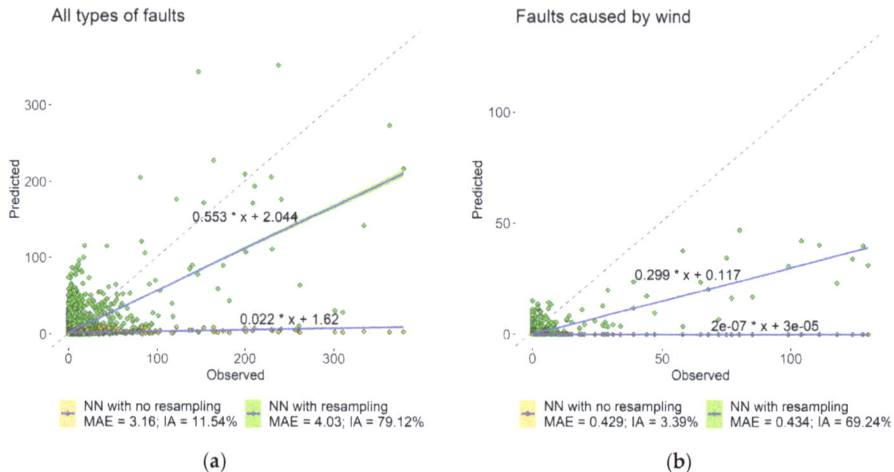

Figure 5. Weather-based fault prediction in the electricity network: comparison of NN models in two experiments with and without resampling. (**a**) All faults, (**b**) Faults caused by wind.

NN models without resampling do not predict peaks but have better MAE as they are more accurate for samples of fewer faults, which are prevalent in the data. On the contrary, NN models with resampling are less accurate for samples of fewer faults, which makes MAE higher. However, the models with resampling are able to predict large peaks, which is more valuable from the problem perspective. For the prediction of all the considered types of faults and wind faults, higher IA values correspond to models with oversampling, which are better at predicting peaks. Based on the given liner regression, we see that there is still a tendency to underestimate peaks. Anyway, the index of agreement may be useful as an additional criterion also in short term load forecasting that has similarly challenges in assessing the performance of forecasting the load peaks.

One option to the aforementioned standard evaluation metrics are categorical statistics i.e., to evaluate model's performance in critical load peak situations by discriminating electric load to a category/class (e.g., low and high) and then apply some conventional index for each class separately or only to the peak loads. Discrimination can also be based on variables that affect the load such as the outdoor temperature and electricity market price. Evaluation of the accuracy of the daily peak load forecast is sometimes used as a peak load sensitive criterion.

Accurate peak load forecasting is so important that short term peak loads are often separately forecast, as in [29,30], for example. Peak load forecasting may be less sensitive to the choice of criterion than the forecasting of the full profile, but absolute maximum error (AME) was used in [30] to complement MAPE when comparing forecasting methods. This seems justified although there both criteria rated the compared four methods clearly in the same order.

Based on the discrimination and the contingency table, a set of different standard metrics can be derived including:

- The fraction of correct prediction (true positive rate, TPR)
- the false positive rate, FPR
- the Success Index (SI): SI = TPR − FPR

where TPR is the true positive rate representing the sensitivity of the model (the fraction of correct predictions) and FPR is the false positive rate, representing the specificity of the model. SI is limited to the range of −1, 1 and for a perfect model SI = 1 [31–33]. With this approach, it is also necessary to have a category for too high predictions, because otherwise too high peak predictions would not be penalized.

Such categorical statistics including probability of detection (POD), critical success index (CSI) and false alarm ration (FAR) are used, e.g., in wind power ramp forecasting as measures of accuracy for forecasts of sudden and large power changes [34,35]. The detection and forecasting of those events is crucial with regard to power production and load balance. However, the ability of forecasting methods to predict those events remains relatively low for short term operation.

- Probability of detection (POD): POD = TP/TP + FN
- Critical success index (CSI): CSI = TP/(TP + FN + FP)
- False alarms ratio (FAR): FAR = FP/(FP + TP).

A possible way to classify the load to suitable categories could be based on the gradient of the load duration curve. Also, proportions of the observed peak load or time could be considered; for example, those times could be used when the load is over 80% of the peak load or 20% of the highest loads measured.

4. Criteria for other Relevant Aspects than Forecasting Performance

4.1. Estimation of Confidence Intervals for Short Term Load Forecasting

Forecasting peak loads too low increases risks and costs as already discussed. Traditionally, these risks are managed by increasing operational margins depending on the situation and based on experience. The operational margin calculation can be the standard deviation (std.) of the short-term load forecast error tuned by a situation dependent coefficient. Forecast error quantiles are also used instead of the std. as the basis of the estimation of the operational margins, because in rarely occurring exceptional situations, such as extreme peak loads, the error distribution may not be normal. For example, Petiau [36] presented a method to estimate the confidence intervals (CI). It is based on the calculation of quantiles of past forecasting errors. CIs quantify the uncertainty of the forecast.

The estimation of the standard errors can be performed, e.g., using the method of bootstrapping. The bootstrapping is a nonparametric approach based on the re-sampling of the data to produce the distribution of re-sampled validation indices [37].

Probabilistic load forecasts provide even more information for risk assessment. Hong and Fan [38] reviewed the methods to produce and evaluate probabilistic forecasts.

Estimation of confidence intervals and the assessment of forecasting performance in exceptional situations are closely related. They both need to focus on exceptional situations where the general assumptions well justified for the errors of accurate forecasts may not be valid.

4.2. Computational Time

In the model identification, many new forecasting methods need much computational resources. This can be problematic, when online learning is applied. For some methods, this is a relevant problem also in off–line model identification, especially because the models need to be updated due to changes in the grid, market aggregation, customer behavior, new priorities, etc. For example, support vector regression (SVR) has so poor computational scalability that it limits the possibilities to exploit it in 1 or 3 min time resolution short-term forecasting needed in electricity grid operation [11]. Based on a brief discussion and comparison of computation times of their models in short term forecasting, Lusis et al. [4] concluded that model run time in model training may be an important factor when choosing between models.

Overly detailed models that have many parameters have poor forecasting performance as the classical textbooks of system identification, such as [39], have shown. In a different application area, a rather recent comparison of several popular statistical and machine learning methods [40] found out that the methods with the best forecasting performance also had the lowest run time computational complexity. Thus, we conclude that if a forecasting method needs excessive computational resources during operation, it is a sign that it also may have both poor performance and unnecessary computations.

That is not always the case, however. Some forecasting methods, such as Gaussian Process Regression models, are more computationally intensive during operation, with no connection to overfitting and produce confidence intervals without the need for bootstrapping, which is a big advantage in the considered short-term load forecasting cases.

Different methods also require very different amounts of working time from different types of experts. Physically based forecasting models require good knowledge of the domain and classical model identification. Many forecasting methods, especially many ML methods, require much expertise and time in the tuning of method parameters and model structure to the specific case. Also, the requirements for preprocessing the data vary.

In order to be able to compare the costs and benefits of the forecasting methods, the computational complexity and the need for expertise and working time need to be assessed in terms of monetary costs. The computational complexity has several dimensions such as the need for computational operations and different types of memory. How easily and efficiently the method suits for parallel processing also affects the cost.

5. Discussion

Electricity market costs caused by the forecast errors are quite stochastic due to the impact of very few very high price peaks. Thus, using the observed market costs as identification and tuning criterion is not feasible as the results of the related case study demonstrate.

The results suggest that improving the short-term forecasting accuracy of the aggregated active demand loads further from our best recent methods [11] now may not give adequately significant savings in the expected value of the imbalance costs to justify much method development investments. It seems more important to mitigate the risks caused by high imbalance price peaks. A possible reason is that the one-price consumption imbalance fees of the case study may not reflect adequately the costs of imbalance to the power system. A more important reason is that the forecasting errors of the aggregated distributed flexible load are rather independent and small as compared to the forecasting errors of the overall balance of the balance responsible party responsible for the flexible load.

In the electricity markets, such forecasting errors that are much smaller than the balancing errors of the actor or its balance responsible party have very small impact on the total balancing error. Thus, it is much more important to focus on large errors of the component forecasts. With them, the balancing error costs depend typically linearly on the size of the balancing error. The balancing error costs depend on the related rules and prices of the particular electricity market but in general they tend to reflect only marginal linearized costs in the powers system, when the balancing errors are assumed not to affect the prices. The actual costs for the system operator grow much faster than linearly and that may be the actual case for the competitive market actors if the impacts on balancing power price and risk–hedging costs are not ignored.

In power distribution networks, the costs of forecasting errors concentrate very much on the load peaks of the network. Network losses depend on the square of the current. The aging of the components is rapidly sped–up by temperature increase caused by the losses. Thus, during low load, the impact of load forecast errors is insignificant. During high load peaks, the load forecasting error requires additional operational margin that is expensive. Thus, it is crucial to focus on the forecasting errors during network peak loads that may be at times that are different from times of the peaks of the component load forecast.

Here our focus is in model verification and validation criteria. In model identification or learning there are more limitations regarding the type of the performance criteria. Some model identification methods or tools may accept only certain criteria types. Some others do not have such strict limitations, but the form and properties of criteria always affect convergence and computational efficiency together with the identification model and method.

It was shown that the actual cost of short-term load forecasting errors is a highly non-linear and asymmetric function of the forecasting error. For example, the penalties for underestimating the load should be higher than for overestimating it. The cost of the errors is also highly load-dependent. Arguably, the right thing to do would be to construct a loss function that is specific to the forecasting problem at hand, and have the machine learning optimize that domain-specific loss function directly. Here we only propose, using domain-specific criteria to complement or refine a common loss function such as RMSE rather than replacing it completely. We do not yet experiment with such domain specific load functions, but only aim to explain what they should include and why they are needed. Many standard forecasting packages may not yet support the possibilities to use domain specific loss functions and adding them there could make them better available to more practitioners. The properties of the loss function affect the properties and number of the solution and the convergence of the method identification. These are difficult to assess if there is no analytic form of the loss function available. We show in Section 3.1 why the prices of the electricity markets and ancillary service markets are not directly applicable in a domain specific loss function, because rarely occurring and very unpredictable price peaks dominate the value of the loss function. Similarly, most of the grid costs tend to concentrate to a very few short overload periods. Thus, domain specific loss functions for the short-term load forecasting model identification may create substantial potential risks. The domain specific loss functions need to be designed and considered carefully. We initially assume that directly using domain specific load functions in identification should be used in parallel with other approaches rather than instead of them. Then the good and weak sides of the approaches can be better detected, and the strengths of the different approaches combined. Experimenting with the use of problem-specific loss functions also in the learning phase is a topic worth considering for future research.

6. Conclusions

The main findings were evident but also way too often ignored. (1) It is important to consider the actual costs and other consequences of the prediction errors when selecting criteria for short term load forecasting. (2) The behavior of the actual costs can be very nonlinear. Even quadratic cost criteria may underestimate the growth of the costs with the size of the error and the load peak. (3) Often, the costs stem mainly from peaks in the loads and in the market prices. For such cases, criteria that normalize each error to the simultaneous measurement, such as MAPE, can be very misleading. (4) The results do not support using any of the commonly used criteria to compare methods across different short-term forecasting cases. (5) The costs in real application cases may be so stochastic that their direct use in validation may not be appropriate. (6) Model development effort, expertise need, and computational complexity are also often relevant when selecting short term forecasting methods.

The research results by the authors suggest that integrating different short-term forecasting methods together into hybrids can combine their mutual strengths, thus improving performance and robustness without excessive development effort. The main challenge of such model integration is that it requires expertise with many different modelling methods.

Our aim was to show that the selection, development and analysis of the performance criteria deserves attention. Blindly using so-called standard criteria may mislead the development, tuning and selection of the short-term forecasting methods in real applications.

Author Contributions: J.I. contributed regarding the literature search, the introduction, and Section 4.1 on confidence intervals, and checked and amended Section 3.1 on electricity market costs. J.K. wrote Section 3.3. on criteria in load forecasting-based control of domestic energy storages. C.B. and H.N. wrote Section 3.5. on peak load sensitive validation. P.K. defined the paper scope and focus, collected most of the data and wrote most of the text. All authors have read and agreed to the published version of the manuscript.

Funding: This research was part of the project Analytics funded by the Academy of Finland.

Acknowledgments: The authors wish to thank the earlier projects, including their funders, partners and supporters for enabling the use of the data and forecasts that were necessary for this study. They also thank Tuukka Salmi for some useful comments.

Conflicts of Interest: The authors declare no conflict of interest.

References

1. Bessa, R.J.; Miranda, V.; Botterud, A.; Wang, J. 'Good' or 'bad' wind power forecasts: A relative concept. *Wind Energy* **2011**, *14*, 625–636. [CrossRef]
2. Frances, P.H. A note on the Mean Absolute Scaled Error. *Int. J. Forecast.* **2016**, *32*, 20–22. Available online: https://doi.org/10.1016/J.IJFORECAST.2015.03.008 (accessed on 5 February 2020). [CrossRef]
3. Schreck, S.; de la Comble, I.P.; Thiem, S.; Niessen, S. A methodological framework to support load forecast error assessment in Local energy markets. *IEEE Trans. Smart Grid* **2020**. [CrossRef]
4. Lusis, P.; Khalipour, K.R.; Andrew, L.; Liebman, A. Short–term residential load forecasting: Impact of calendar effects and forecast granularity. *Appl. Energy* **2017**, *205*, 654–669. [CrossRef]
5. Hodge, B.; Lew, D.; Milligan, M. Short-term load forecast error distributions and implications for renewable integration studies. In Proceedings of the 2013 IEEE Green Technologies Conference (Green Tech), Denver, CO, USA, 4–5 April2013; pp. 435–442. Available online: https://doi.org/10.1109/GreenTech.2013.73 (accessed on 20 February 2020).
6. Abdulla, K.; Steer, K.; Wirth, A.; Halgamuge, S. Improving the On–line Control of Energy Storage via Forecast Error Metric Customization. *J. Energy Storage* **2016**, *10*, 51–59.
7. Haben, S.; Ward, J.; Greetham, D.V.; Singleton, C.; Grindrod, P. A new error measure for forecasts of household-level, high resolution electrical energy consumption. *Int. J. Forecast* **2014**, *30*, 246–256. [CrossRef]
8. Hyndman, R.J.; Koehler, A.B. Another look for measures of forecast accuracy. *Int. J. Forecast* **2006**, *22*, 679–688. [CrossRef]
9. Jakob, M.; Neves, C.; Vukanovic'Greetham, D. Short Term Load Forecasting. In *Forecasting and Assessing Risks of Individual Electricity Peaks*; Mathematics of Planet Earth; Springer: Cham, Switzerland, 2020; pp. 33–34.
10. Lococ, J.; Beeks, C.; Seidl, T.; Skopal, T. Paramterized Eart Movers Distance for Efficient Metric Space Indexing. In Proceedings of the SISAP '11, Lipari, Italy, 30 June–1 July 2011; p. 2.
11. Koponen, P.; Niska, H.; Mutanen, A. Mitigating the Weaknesses of Machine Learning in Short–Term Forecasting of Aggregated Power System Active Loads. In Proceedings of the IEEE INDIN19, Helsinki–Espoo, Finland, 22–25 July 2019; p. 8.
12. Ding, D.; Zhang, M.; Pan, X.; Yang, M.; He, X. Modelling Extreme Events in Time Series Prediction, KDD'19. In Proceedings of the 25th ACM SIGKDD International Conference on Knowledge Discovery & Data Mining, Anchorage, AK, USA, 4–8 August 2019; pp. 1114–1122. Available online: https://doi.org/10.1145/3292500.3330896 (accessed on 11 February 2020).
13. Two-Price and One-Price System, Fingrid. Available online: https://www.fingrid.fi/en/services/balance-service/description-of-balance-model/two-price-and-one-price-system/ (accessed on 11 February 2020).
14. Nord Pool. Elspot Day-a-Head Electricity Prices. Available online: https://www.nordpoolgroup.com/Market-data1/Dayahead/Area-Prices/ALL1/Hourly/?view=table (accessed on 11 February 2020).
15. Fees, Fingrid. Available online: https://www.fingrid.fi/en/services/balance-service/fees/ (accessed on 15 April 2020).
16. Koponen, P.; Salmi, T.; Evens, C.; Takala, S.; Hyttinen, A.; Brester, C.; Niska, H. Aggregated forecasting of the load control responses using a hybrid model that combines a physically based model with machine learning. In Proceedings of the CIRED 2020 Workshop, Berlin, Germany, 22–23 September 2020; p. 4.
17. SmartNet. Available online: http://smartnet-project.eu/ (accessed on 15 April 2020).
18. Strbac, G.; Djapic, P.; Pudjianto, D.; Konstantelos, I.; Moreira, R. *Strategies for Reducing Losses in Distribution Networks*; Imperial College: London, UK, 2018; p. 87.
19. Husnayain, F.; Latif, M.; Garniwa, I. Tranformer Oil Lifetime Prediction Using the Arrhenius Law Based on Physical and Electrical Characteristics. In Proceedings of the IEEE 2015 International Conference on Quality in Research, Lombok, Indonesia, 10–13 August 2015; pp. 115–120.
20. Bowler, N.; Liu, S. Aging mechanisms and monitoring of cable polymers. *Int. J. Progn. Health Manag.* **2015**, *6*, 12.
21. Koskela, J.; Rautiainen, A.; Järventausta, P. Utilization possibilities of electrical energy storages in households' energy management in Finland. *Int. Rev. Electr. Eng.* **2016**, *11*, 607–617. [CrossRef]
22. Koskela, J.; Lummi, K.; Mutanen, A.; Rautiainen, A.; Järventausta, P. Utilization of electrical energy storage with power-based distribution tariffs in households. *IEEE Trans. Power Syst.* **2019**, *34*, 1693–1702. [CrossRef]

23. Lummi, K.; Rautiainen, A.; Järventausta, P.; Heine, P.; Lehtinen, J.; Hyvärinen, M.; Salo, J. Alternative power-based pricing schemes for distribution network tariff of small customers. In Proceedings of the IEEE Innovative Smart Grid Technologies-Asia (ISGT Asia), Singapore, 22–25 May 2018; p. 6.

24. Mutanen, A. *Improving Electricity Distribution System State Estimation with AMR-Based Load Profiles*; Tampere University of Technology: Tampere, Finland, 2018; p. 91.

25. Humeasu, S.; Wijaya, T.K.; Vasirani, M.; Aberer, K. Electricity load forecasting for residential customers: Exploiting aggregation and correlation between households. In Proceedings of the 2013 Sustainable internet and ICT for sustainability, SustainIT, Palermo, Italy, 30–31 October 2013. Available online: http://dx.doi.org/10.1109/SustainIT.2013.6685208 (accessed on 18 April 2020).

26. Koponen, P.; Mutanen, A.; Niska, H. Assessment of some methods for short–term load forecasting. In Proceedings of the IEEE PES ISGT Europe 2014, Istanbul, Turkey, 12–15 October 2014; p. 6.

27. Brester, C.; Niska, H.; Ciszek, R.; Kolehmainen, M. Weather-based fault prediction in electricity networks with artificial neural networks. In Proceedings of the IEEE World Congress on Computational Intelligence (WCCI) 2020, Glasgow, UK, 19–24 July 2020.

28. Willmott, C.J.; Ackleson, S.G.; Davis, R.E.; Feddema, J.J.; Klink, K.M.; Legates, D.R.; O'Donnell, J.; Rowe, C.M. Statistics for the evaluation and comparison of models. *J. Geophys. Res.* **1985**, *900*, 8995–9005. [CrossRef]

29. Sarduy, J.R.G.; Di Santo, K.G.; Saidel, M.A. Linear and non-linear methods for prediction of peak load at University of São Paulo. *Measurement* **2016**, *78*, 187–201. Available online: https://doi.org/10.1016/J.MEASUREMENT.2015.09.053 (accessed on 18 April 2020). [CrossRef]

30. Grant, J.; Eltoukhy, M.; Asfour, S. Short-Term Electrical Peak Demand Forecasting in a Large Government Building Using Artificial Neural Networks. *Energies* **2014**, *7*, 1935–1953. [CrossRef]

31. Youden, W.J. Index for rating diagnostic tests. *Cancer* **1950**, *50*, 32–35. [CrossRef]

32. Van Aalst, R.M.; De Leeuw, F.A. European Topic Centre on Air Quality (RIVM, NILU, NOA, DNMI). In *National Ozone Forecasting Systems and International Data Exchange in Northwest Europe*; Report of the Technical Working Group on Data Exchange and Forecasting for Ozone Episodes in Northwest Europe (TWG-DFO)); European Environment Agency: København, Danmark, 1997.

33. Gajowniczek, K.; Ząbkowski, T. Short Term Electricity Forecasting Using Individual Smart Meter Data. *Procedia Comput. Sci.* **2014**, *35*, 589–597. [CrossRef]

34. Ferreira, C.; Gama, J.; Matias, L.; Botterud, A.; Wang, J. *A Survey on Wind Power Ramp Forecasting*; Technical Report No. ANL/DIS-10-13; Argonne National Laboratory: Lemont, IL, USA, 2010.

35. Zhang, J.; Florita, A.; Hodge, B.-M.; Freedman, J. Ramp forecasting performance from improved short-term wind power forecasting. In Proceedings of the ASME 2014 International Design Engineering Technical Conferences & Computers and Information in Engineering Conference IDETC/CIE 2014, Buffalo, NY, USA, 17–20 August 2014; p. 12.

36. Petiau, R.B. Confidence interval estimation for short-term load forecasting. In Proceedings of the 2009 IEEE Bucharest PowerTech, Bucharest, Romania, 28 June–2 July 2009; pp. 1–6.

37. Efron, B.; Tibshirani, R.J. *An Introduction to the Bootstrap*; Chapman and Hall: New York, NY, USA, 1993.

38. Hong, T.; Fan, S. Probabilistic electric load forecasting: A tutorial review. *Int. J. Forecast.* **2016**, *32*, 914–938. Available online: https://doi.org/10.1016/J.IJFORECAST.2015.11.011 (accessed on 18 April 2020). [CrossRef]

39. Ljung, L. *System Identification, Theory for the User*, 2nd ed.; Prentice Hall: Upper Saddle River, NJ, USA, 1999; p. 640.

40. Madridakis, S.; Spillotis, E.; Assimakopoulos, V. Statistical and Machine Learning forecasting methods: Concerns and ways forward. *PLoS ONE* **2018**, *13*, 26.

Article

Integration of Demand Response and Short-Term Forecasting for the Management of Prosumers' Demand and Generation

María Carmen Ruiz-Abellón [1], Luis Alfredo Fernández-Jiménez [2], Antonio Guillamón [1], Alberto Falces [2], Ana García-Garre [3] and Antonio Gabaldón [3,*]

[1] Department of Applied Mathematics and Statistics, Universidad Politécnica de Cartagena, 30203 Cartagena, Spain; maricarmen.ruiz@upct.es (M.C.R.-A.); antonio.guillamon@upct.es (A.G.)

[2] Department of Electrical Engineering, Universidad de La Rioja, 26004 Logroño, La Rioja, Spain; luisafredo.fernandez@unirioja.es (L.A.F.-J.); alberto.falces@unirioja.es (A.F.)

[3] Electrical Engineering Area, Universidad Politécnica de Cartagena, 30203 Cartagena, Spain; ana.ggarre@upct.es

* Correspondence: antonio.gabaldon@upct.es; Tel.: +34-968-338944

Received: 14 November 2019; Accepted: 13 December 2019; Published: 18 December 2019

Abstract: The development of Short-Term Forecasting Techniques has a great importance for power system scheduling and managing. Therefore, many recent research papers have dealt with the proposal of new forecasting models searching for higher efficiency and accuracy. Several kinds of artificial intelligence (AI) techniques have provided good performance at predicting and their efficiency mainly depends on the characteristics of the time series data under study. Load forecasting has been widely studied in recent decades and models providing mean absolute percentage errors (MAPEs) below 5% have been proposed. On the other hand, short-term generation forecasting models for photovoltaic plants have been more recently developed and the MAPEs are in general still far from those achieved from load forecasting models. The aim of this paper is to propose a methodology that could help power systems or aggregators to make up for the lack of accuracy of the current forecasting methods when predicting renewable energy generation. The proposed methodology is carried out in three consecutive steps: (1) short-term forecasting of energy consumption and renewable generation; (2) classification of daily pattern for the renewable generation data using Dynamic Time Warping; (3) application of Demand Response strategies using Physically Based Load Models. Real data from a small town in Spain were used to illustrate the performance and efficiency of the proposed procedure.

Keywords: short-term load forecasting; demand response; distributed energy resources; prosumers

---

1. Introduction

There is a large literature related to short-term forecasting in the context of electric energy and this topic also has a great interest in many other fields. In fact, the proposal of new forecasting methods is daily increasing because of their applicability to dispatch unit commitment or market operations [1]. In this sense, short-term load forecasting models have always been a key instrument for carrying out such operations, although in recent years, with the increasing integration of power plants based on renewable energy with high variability (mainly wind and solar), forecasting models for these kinds of power plants has gained the attention of researchers and utilities. Photovoltaic (PV) systems are the most widespread renewable based power generation systems (they stand for more than half of the total installed capacity in power plants based on renewable sources) with a large number of small-scale installations on medium or low-voltage grids, right next to residential consumers.

The search for more accurate and faster forecasting methods, both in load and in PV power generation, has resulted in a set of efficient techniques that can be divided into three different categories: time-series approaches, regression based, and artificial intelligence methods (see [2,3]).

In the field of short-term load forecasting there are some examples of classical time-series approaches with auto-regressive integrated moving average (ARIMA), seasonal ARIMA (SARIMA) and ARIMA with exogenous variables (ARIMAX) (see [4–6]). Regression-based models (see [4,7]) are also widely used in the field of short-term load forecasting, including non-linear regression [8] and nonparametric regression [9] methods. Examples of artificial intelligence methods include fuzzy logic [10,11], artificial neural networks (ANN) [12–14], ensemble methods based on regression trees [15,16], and support vector machines (SVM) [17,18]. Recently [19], a new hybrid model was proposed to improve accuracy that performs well in electricity load forecasting.

On the other hand, the development of short-term PV power forecasting models has followed a parallel path. Thus, some published works use classical time-series approaches [20], regression methods [21], fuzzy logic models [22], ANN-based models [23], ensemble methods [24,25], and support vector machines [25]. A comparative study of the forecasting performance of different models of the above-mentioned approaches for the same PV plant is presented in [26], and in which the best model, among those studied, changes according to the data available for the training process.

Undoubtedly, fitting and computation velocity improvements are desirable, but at the same time, it is essential to take advantage of current forecasting methods. The main objective of this paper was not to propose a new short-term forecasting method, but to illustrate how some recent ones can be combined to predict electricity consumption and photovoltaic (PV) generation, in order to propose efficient strategies of Demand Response (DR) for an aggregated load of consumers. On the other hand, DR acts a regulator or damper to correct excursions of net demand of Power System buses with demand and generation (i.e., "prosumers") due to punctual errors of forecasts both in demand, but specially in renewable generation, reducing the own volatility of this last resource.

Political and regulatory scenarios in several countries support the development of the so-called Distributed Energy Resources (DER), i.e., the integration of demand flexibility, energy storage, and generation (mainly Renewable Energy Sources, RES), which facilitates the de-carbonization objectives of power systems by 2030–2050 [27]. For example, the European Commission (EC) is concerned with a necessary increase of flexibility of demand involved with the integration and exploitation of DER possibilities. The Direction General of Energy (DG ENER) reported, in a public dissertation [28], that the theoretical level of Demand Response in 2017 was 100 GW, but only 21 GW were activated (75% of them through incentive based options, i.e., the so-called explicit DR). The policy scenario for 2030 is 160 GW of theoretical demand potential with 52 GW activated (24% of peak load demand, assuming it will reach 570 GW).

To accomplish this forecast, this future scenario makes necessary an increase of 300% in DR resources in a decade, and this seems a difficult task if future forecasts fail again in the trends about the evolution of DR [29]. In this way, it is important to consider and enhance DR. Moreover, the net benefit of the overall deployment of DER and RES could reach €5600 million/year for the EU economy (i.e., generate up to 1% Gross Domestic Product increase over the next decade). The potential in the Spanish case is 17 GW; around 50% of this potential could be explained by DR and RES in small and medium customer segments. For these reasons, DR policies in this paper are centered in these segments. This participation also involves the capability of aggregators and system operators to develop more accurate forecasts both on demand and generation and the necessity of making this information (forecasts) easily accessible to customers in order to increase their participation and engagement in markets (mainly in energy markets but also in Ancillary Service markets). More accurate forecasts could allow customers to take advantage of the retributions of energy markets and avoid possible penalties due to imbalances between generation and consumption. Due to the size of demand and generation, forecasts are more difficult and can represent a barrier for customers, especially in tasks involved with RES forecasting. This fact makes demand flexibility a potential tool to change this scenario.

The proposed methodology can be summarized as follows:

- Firstly, short-term forecasting methods are used to predict hourly load and photovoltaic generation with a horizon of 24 h.
- Secondly, the predicted daily PV generation of the training dataset is grouped into homogeneous clusters according to their shape. Next, a representative PV curve is obtained for each cluster, and a discriminant analysis is developed to assign each predicted PV curve of the test dataset to a cluster.
- Finally, Demand Response strategies are applied to those days with a predicted PV curve in the suitable cluster (the one that provides more accurate predictions).

Among the wide variety of machine learning methods, we have chosen random forest to predict the electricity consumption with a horizon of 24 h due to its proven efficiency in short-term load forecasting [15]: high accuracy, fast computing (even for parameter tuning), and easy understanding of feature importance results. Prediction results obtained with random forest for this dataset showed great accuracy; therefore, other forecasting methods were not really needed.

In the case of forecasting hourly PV generation, several machine learning methods were applied and compared, searching for the most accurate. Specifically, linear regression, neural networks, random forest, gradient boosting, and support vector regression models were developed and tuned choosing the optimal values for their parameters by means of genetic algorithms. Unlike hourly load forecasting, the goodness-of-fit measures of the predicted PV curves showed lower accuracy. Regarding the clustering method applied to the PV curves, the dynamic time warping distance [30] and average linkage were selected for the classification stage.

The rest of the paper is organized as follows: Section 2 describes in detail all forecasting and clustering methods used, as well as DR strategies applied in this paper. Section 3 present the results obtained for load and PV forecasts, explaining the application of clustering and DR to minimize the effects of forecasting errors. Finally, some conclusions and future developments are stated in Section 4.

2. Materials and Methods

2.1. Methodology Overview

Day-ahead markets represent the most active markets in terms of economic value of transactions, but other markets have experienced a noticeable growth (for example Intraday Markets in France and Belgium with growth rates of 54.5% and 82.9%, respectively, in 2018). Real-Time Markets have facilitated the integration of renewables in USA markets in the last decade, and wider-scale markets with later gate closure would facilitate the integration of renewable in other systems (e.g., Balancing Markets, and specifically, Reserve Replacement). The integration of demand-side resources in markets presents both risks and opportunities for Balance Responsible Parties (BRP, responsible for its imbalances) and Balance Service Providers (BSP, i.e., the provision of bids for balancing) and need the development of new and more integrated methodologies. The main idea of this research work is providing new tools to aggregators for a best management of demand and generation in markets, both in the short-term (around 24 h) and in the very short-term (from several minutes to 1 h), to evaluate net demand unbalance, while the aggregator or other parties take into account gate closure times. To perform this task, the proposed methodology takes profit from different databases which should be able for DR (demand, customer, RES, and weather). From these databases, this work applies different methodologies to obtain well fitted 24 h forecasts for demand and PV generation, and with these predictions aggregators evaluate bids and offers to be sent to Intra-Day or Real Time Energy Markets (with the objective of minimizing the energy costs for customers or prosumers). Logically, both models (demand and PV) exhibits errors and these errors can involve penalties in the markets because other agents (BSP, LSE) should change their energy balance and buy or sell energy in the very short term. Considering that PV-forecasts usually have a greater error than demand-forecast, and that a fast model is needed to manage the potential flexibility of demand in the aggregator-side, a simple and very short-term PV forecast model is developed based on historical recordings of PV generation (in the site) and available real-time measurements (Information and Communication Technology, i.e., ICT devices). This model and the results of cassation processes in short-term energy markets provide a reference signal for flexible resources (only DR resources are considered in this paper). With the help of different tools and scripts from a DR-toolbox (e.g., segmentation, classification, disaggregation and modeling), the aggregator can evaluate the "demand baseline" for different end-uses in the short-term and can propose a control signal to change demand according to its requirements. This demand is simulated and evaluated hour by hour with several indices of performance (and modified in some cases) to fit the demand packages offered to energy markets (i.e., net demand). In other cases, when the power system tackles for flexibility, the aggregator can provide additional flexibility to energy and ancillary services markets, agents or Transmission Operators.

Figure 1 presents an overview flowchart, which depicts the methodologies and tools to be used through the paper.

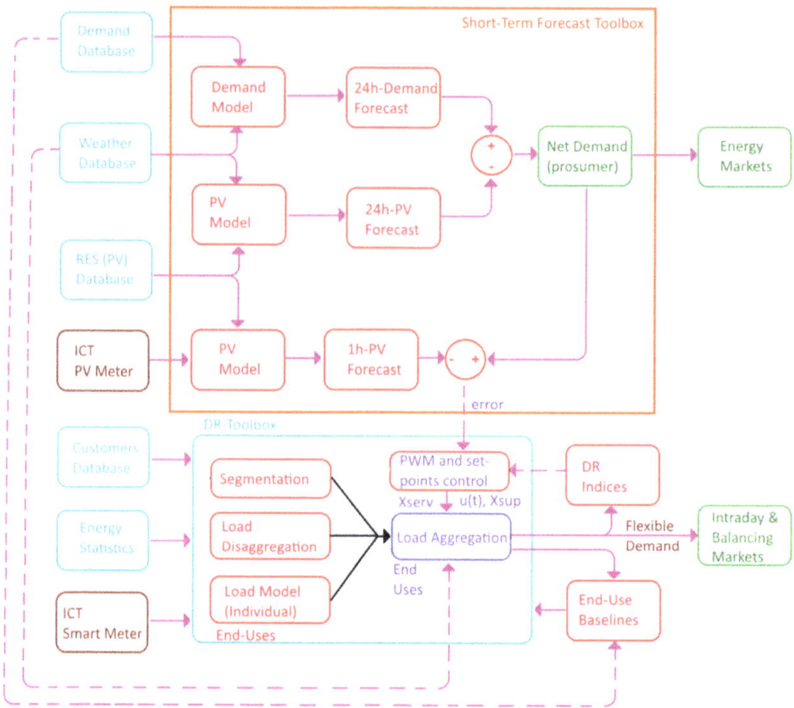

Figure 1. Methodology flowchart.

A quantitative analysis for demand-side flexibility seems necessary thorough the definition and the evaluation of some indicators (i.e., DR indices in Figure 1) that allows to score the flexibility and performance of loads being controlled, basically at the aggregated level. These indicators converge with the idea of some voluntary schemes in the EU that intend to express the "readiness of a building" (in this case the readiness of the load inside buildings). According to these proposals [31], these indicators mean: "readiness to adapt in response to the needs of the occupants, readiness to facilitate maintenance and efficient operation, and readiness to adapt in response to the situation of the energy grid". Taking into account this last requirement, a score is performed through the indicators to be described in Section 3.4.3.

2.2. Characteristics of the Customers: Demand, Photovoltaic Generation, and End-Uses

All electricity customers from a small town (4400 inhabitants), sited in the north Spain, have been selected for simulation purposes in this work. These customers include residential, commercial, and industrial clients, although most of the power consumption is due to the residential ones. Basically, this group corresponds to average residential customers in the south countries of the EU. The rated power per customer ranges from 3 to 8 kW. The climate is continental, and winter temperatures range from 0 to 13 °C and in summer from 13 to 29 °C.

Regional investors built a PV plant in the vicinity of the town, with a significant capacity with respect to its power demand. The PV plant is composed of two-axis solar trackers with a rated power of 1.9 MW, and it is connected to the same power substation that links the town to the grid.

Hourly load and photovoltaic generation data from 1 October 2008 to 31 March 2011 (both included) were available. These data were obtained from the electric utility distributor and corresponded to hourly average power measurements in the substation. It is worth mentioning that it has been very difficult to obtain real data corresponding to a considerable customer group that can act as prosumers (consumers and producers); thus, we had to manage data not as recent as desired.

Figure 2a shows the winter and summer loads for two selected workdays monitored at the distribution level (substation) that supplies power at 13.2 kV to the distribution transformer centers (CT) of customers (basically residential and commercial supply). Figure 2b shows the average temperature in the area for the same two selected workdays. Figure 2c plots the hourly PV power generation on the peak production days (days with the highest energy generation) of January and June 2010. The PV power generation values in the central hours of the day can mean an important fraction of the town consumption (30–40%). The selection of January and June as representative months is due to the fact that June corresponds to the month with the highest PV generation, while January corresponds to the lowest PV generation and the highest energy consumption. In Figure 2, it is also shown the average profiles of demand, temperature and PV generation for the period in which data is available (from 1 October 2008 to 31 March 2011). It is remarkable that the average PV generation profile (Figure 2c) is lower compared to the other ones (June and January). This fact can be explained because months' profiles are representing the peak power days, while the average profile includes days in which there are no PV generation due to adverse weather conditions.

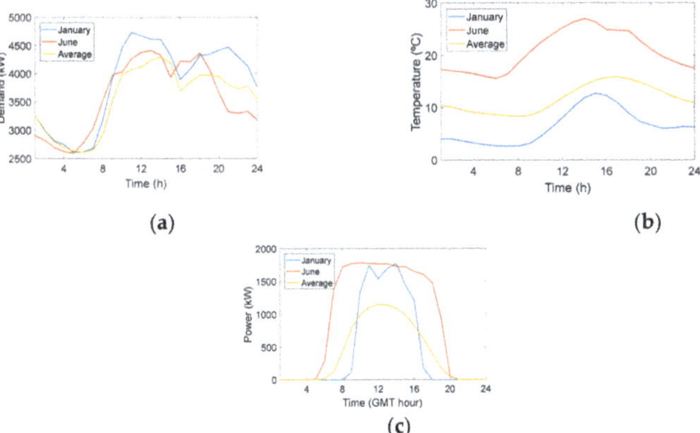

Figure 2. Load, Temperature and PV Generation profiles: (**a**) example of winter and summer Load Curves in the power substation; (**b**) example of winter and summer temperature behavior; (**c**) example of winter and summer power generation in the PV plant on peak production days.

The use of DR portfolio for damping both the errors in the evaluation of demand in short-term and the intrinsic variability of PV generation sources need the evaluation of DR potential and its flexibility in each customer segment. First, this evaluation must be based on the knowledge of end-uses for an average customer. The first alternative to know demand composition behind-the-meter is the use of the information provided by Smart Meters (SM) and then apply some Non-Intrusive Load Monitoring (NILM) methodology, for example [32]. This last approach involves the full development of capacities of available home automation technologies, considering the increasing deployment of Smart Meter in several countries around the world [33]. Some of these methodologies have been reported by authors in previous papers [34] to obtain end-use disaggregation/profiles in residential segments and their real flexibility when DR policies are applied (i.e., DR validation).

In some cases, and from a practical point of view, it is possible that some problems arise for a practical implementation of DR based on end-uses, for instance: small customers do not have yet any SM, confidentiality of data is in question, Data Exchange Platforms (DEO) are not fully developed, and the availability of data is scarce [35] or the aggregator has access to meter data but without the necessary granularity or quality (i.e., it is usual to have data with granularity ranging from 15 min to 1 h which usually makes much more complex the identification of loads through NILM methods). In this way, an alternative access to demand data should also be considered by aggregators to accomplish the evaluation of DR potential. This alternative is based on periodic surveys performed by international or national Energy Agencies, for example EIA (data from 2015, [36]) in the United States or the Joint Research Centre (data from 2016, [37]) in the European Union. In this way, a residential "average" EU or USA customer (and its end-use share) can be estimated according to these data. In the case under study, available reports from the Institute for Energy Diversification and Energy Savings (IDAE, Spain) and the Spanish Government [38] have been analyzed. Table 1 depicts the main end-uses for Spain, EU-28 countries and the USA. Notice that in European Mediterranean countries, the Air Conditioning load represents a higher percent (66% of households have this appliance and the increasing trend is quite solid). A similar trend can be reported in the USA, because 87% of homes use air conditioning. It accounts for 12% of annual residential energy expenditures and is a large factor in fluctuations in residential electricity use. Heat Pumps (HP) exhibit similar trends according to EIA estimations [36]: the share of heated homes using HP increased from 8% to 12% in a decade (from 2005 to 2015). At the same time, the share of homes using electricity for water heating (WH) increased by 7% (to 46%). Due to this fact, both loads (HP and WH) have been considered to evaluate demand flexibility in this work. Moreover, winter period has been selected for simulation purposes in the following paragraphs, because demand in winter is higher than in summer and the climate in this Spanish area is more restrictive for PV generation possibilities.

Table 1. Main end-uses share in the residential sector according to some estimates in the USA [36], EU-28 [37], and Spain [38] adapted and updated from [39].

Type of End-Use	USA (2015) All Fuels	USA (2015) Electricity	EU (2016) All Fuels	Spain (2014) All Fuels	Spain (2014) Electricity
Space Heating	43	14.76	64.7	42.9	7.36
Water Heater	19	13.65	14.5	17.9	7.47
Air Conditioning	6.24	16.89	0.3	0.98	2.33
Refrigerators	4.75	7.02	-	7.94	-
Other *	29.8	47.67	20.5	39.22	82.84

*: Appliances and Electronics.

To obtain some representative profiles, it seems necessary to evaluate load dynamics, and the service the customer obtains from them. Figure 3a,b shows some real end-use load profiles for a household belonging to the overall customer demand, previously shown in Figure 2. In this case, feedback from everyday activities [40] of the customer is important to refine profiles, improve DR&EE (Demand Response and Energy Efficiency) success and gain customer interest in energy concerns.

Regulations can help aggregators to establish load profiles. Figure 3a shows an average HP consumption profile in winter, as in this study, DR simulations to balance generation are focused on this period. Figure 3b shows an average water flow use to determine WH requirements extracted from EN 15316-3-1, Section 5 (EU normative). Figure 3c shows the proposed end-use profiles for an average customer.

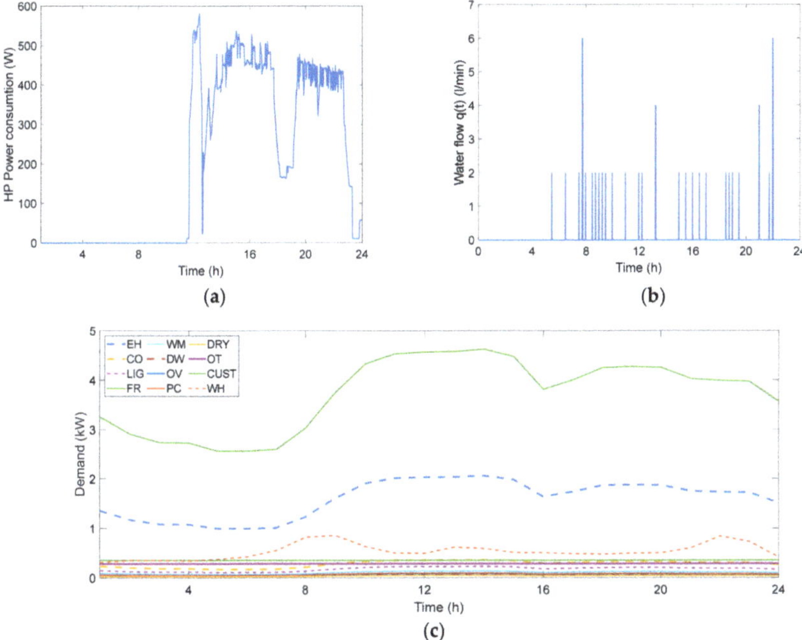

Figure 3. Daily end-uses and customer service: (**a**) HP profile; (**b**) daily use of residential Water Heater. (**c**) Daily profiles for main end-uses (winter). Acronyms: EH: Heating (Electric heaters and Heat pumps), WH: Water Heating; CO: cooking; LIG: lighting; FR: fridges; WM: washing machines; DW: dishwasher; OV: Oven; PC: computers; DRY: dryers; OT: Others; and CUST: overall customer demand.

The procedure for obtaining end-uses profiles (Figure 3) could be explained as follows. In the first place, the aggregator needs to recover basic information about customer daily overall demand (aggregated or not) through Smart Metering (Figure 1, left bottom side). This information, alongside weather databases and public reports of energy household demand and share of end-uses in terms of energy, allows the aggregator the calculation of "household baselines". At this point, the aggregator is able to run and refine Physically Based Load Models (PBLM) both at elemental and aggregated levels (i.e., include inputs/outputs for these models). With PBLM and average weather inputs the aggregator obtains "end-use baselines" for each end-use with relevant potential for DR (e.g., HVAC, space heater, WH, or thermal ceramic heaters, Figure 1), and their average daily demand in each season/month. Finally, "elemental baselines" (kWh and profiles) are modulated through coefficients (considering weather conditions and customer behavior) to fit the "overall baseline" for the customer.

2.3. Short-Term Forecasting Methods

In this section, the forecasting methods used to predict hourly load and PV generation are described.

2.3.1. Random Forest

Random forest is an ensemble method based on regression trees whose efficiency in load forecasting has been widely illustrated [15]. Being based on regression trees makes random forest a flexible method in case of complex and non-linear relationships, and as an ensemble method, it can provide low bias and reduces the variance of predictions.

Some other ensemble methods based on regression trees are bagging, conditional forest or boosting, whose efficiency in load forecasting has been shown in different papers (see for instance [41]).

Random forest is a generalization of bagging (bootstrap aggregating), but only a random sample of *"mtry"* predictors can be chosen at each split of the regression tree. This approach will reduce the variance of the estimations more than bagging, mainly in the case of correlated predictors. In this paper we have decided to use random forest to predict the electricity consumption instead of conditional forest of boosting due to its simplicity in parameter calibration and fast computation.

The efficiency of random forest depends on a suitable selection of the number of trees N and the number of predictors *"mtry"* tested at each split. However, random forests will not overfit when N increases, thus, we can focus on calibrating only the parameter *"mtry"*. The calibration of the parameter and the random forest method have been implemented throughout the R package *"caret"*, see [42].

2.3.2. Stochastic Gradient Boosting (SGB)

Another ensemble method based on regression trees is the stochastic gradient boosting. It has been successfully used in short-term load [43] and PV power [44] forecasting applications. The SGB method is based on the sequential construction of additive regression models, usually in the form or regression trees with a maximum tree size. At each iteration, the models are fitted, by least squares, to a random sample of pseudo-residuals of the previous stage. Thus, the SGB method applies a gradient descent algorithm, reducing the error (difference between output value and expected value) at each iteration.

The develop of an SGB forecasting model requires the selection of the values for a set of tuning parameters which include the number of trees (also called as iterations), the interaction depth, the shrinkage or learning rate, the minimum number of observations in terminal nodes of the trees, and the bagging or sampling fraction. Unlike the random forest method, SGB models with many trees are prone to overfitting; thus, that number must be carefully chosen. The complexity of the SGM model is related to the interaction depth which corresponds to the maximum size of each tree. The shrinkage parameter manages the influence of each sequential tree on the final value provided by the SGM model. The minimum number of observations in terminal nodes or leaves of the trees limits the observations used to provide their mean value as the response of that terminal node. Finally, the bagging fraction corresponds to the fraction of the training dataset observations randomly selected to build each tree (small values reduce the possibility of overfitting, but increase the model uncertainty). For a detailed description of the SGB method, see [45].

2.4. Time-Series Clustering

Clustering is an unsupervised technique whose objective is to separate objects (represented by a multivariate dataset) into homogeneous groups (called clusters), such that objects in the same cluster have high similarity among them, but low similarity with the objects in a different cluster. It is considered an exploratory technique very useful by itself or as a previous step for other kind of data analysis. Depending on the way the clusters are generated, clustering methods can be divided in two big sets: hierarchical methods and divisive methods. In addition, the resulting clusters are determined by the distance or similarity measure and the linkage method selected.

A special case is time-series clustering, where each object to be grouped corresponds to a sequence of values as a function of time. One of the main advantages of clustering time-series is that it allows the discovery of hidden patterns in time-series datasets. Generally, three different objectives can be considered in this context: finding similar time series in time, in shape, or in change (structural similarity). The selection of a suitable distance measure is essential and depends on the objective pursued. Interesting surveys in the field can be found in [46,47].

In this paper, we have focused on similarity in shape to cluster the daily curves of photovoltaic generation. Dynamic Time Warping (DTW) distance, described below, was chosen for that purpose.

Regarding the nature of the clustering, hierarchical technique together with average linkage were selected. These clustering methods were developed by means of the R package TSclust [48].

DTW distance, introduced by [30], is commonly used for measuring shape-based similarity between two time series, which may vary in timing. The main advantage against other shape-based distances such as Euclidean or Wavelet Transform is its invariance to warping. In our context, daily curves of PV generation are conditioned by sunlight hours, which vary along the different seasons. That makes DTW distance suitable for clustering a set of daily PV curves along different years.

Given two time series (x_i) $I = 1, \dots, m$ and (y_j) $j = 1, \dots, m$, it starts calculating a nxm matrix $D = (Dij)$ with the distance between every possible pair of point x_i and y_j in the two time series, $Dij = d(x_i,y_j)$, $I = 1, \dots, n, j = 1, \dots, m$, where $d(x_i,y_j)$ can represents the Euclidean or the absolute distance. According to [30], a warping path w is a contiguous set of matrix elements which defines a mapping between (x_i) and (y_j) that satisfies the following conditions:

- Boundary conditions: $w_1 = (1; 1)$ and $w_k = (m; n)$, where k is the length of the warping path.
- Continuity: if $w_i = (a, b)$ then $w_{i-1} = (a', b')$, where $a - a' \leq 1$ and $b - b' \leq 1$.
- Monotonicity: if $w_i = (a, b)$ then $w_{i-1} = (a', b')$, where $a - a' \geq 0$ and $b - b' \geq 0$.

The objective in DTW distance is to find the shortest warping path. Due to its high computational cost, different approaches have been proposed to optimize the calculation (see [49]).

2.5. Demand Response Strategies

Demand Response policies have been used by ISOs since the early 1980s. In the first years of DR, the objective was to achieve a more rational planning and operation of resources. In recent years, with the development of energy markets and the increasing share of RES in the generation mix, DR becomes more centered in the customer and in the integration of the available RES potential. Demand Response can be divided into explicit and implicit DR. Implicit DR means the change of demand due to prices whereas explicit DR involves the change of demand when System or Distribution Operators (i.e., ISO or DSO) forecast and declare an event into the system in the short-term.

To respond to these events or prices, the most common policy is to limit demand. This reduction can be performed through the cycling of power supply (the supply is switched ON and OFF alternatively following a rectangular wave $u(t)$). If the natural "cycling" of the end-use being controlled, $m(t)$ (the operating state of the control device), is greater than $u(t)$, the DR action is effective (notice that an appliance can describe cycles or not, for example a fridge or an inverter heat pump, but every load has its operating state $m(t)$ with respect to rated power). Considering that, in practice, demand measurements are discrete (every 5, 15 or 60 min) and it is necessary to define average values in a time period $[t, t + k]$. Mathematically, Equation (1):

$$m(t) = \frac{P_{eu}(t)}{RP_{eu}}; \quad \overline{m}(t, t+k) = \frac{1}{kRP_{eu}} \int_t^{t+k} P_{eu}(t)dt; \quad \overline{u}(t, t+k) = \frac{t_{ON}}{k} \tag{1}$$

where $\overline{m}(t, t + k)$ and $\overline{u}(t, t + k)$ are mean values of $m(t)$ and $u(t)$ in the time period k, respectively, and t_{ON} is the time in this period where a "representative" (average) load remains switched *ON* and demands power.

The models to be used (to obtain $m(t)$ and apply $u(t)$) are PBLM, a methodology proposed first to solve problems such as cold load pickup. The main reason for this choice is that these models are "white" [50] or "grey" [51] models which allow physically explaining the dynamics of the appliance and its environment and consequently foreseeing its changes. In this work, PBLM "grey" models for HVAC (Heating, Ventilation and Air Conditioning) and WH loads (heating and ventilation) previously proposed in [39] have been used. Figure 4 shows an electrical-thermal equivalent for this model for heating loads (a broader explanation of parameters can be found in several references [52,53]).

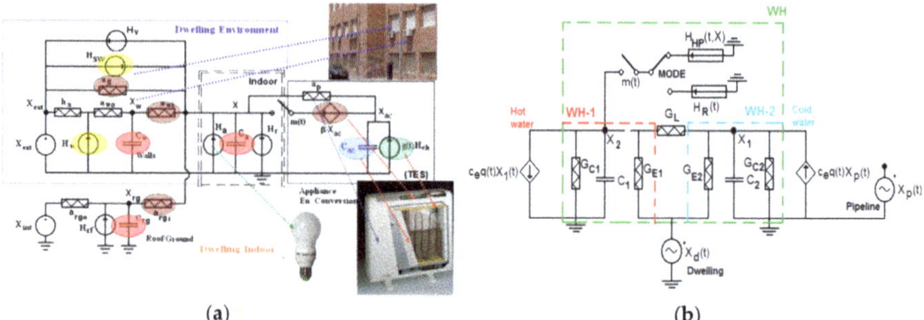

Figure 4. Example of PBLM for: (**a**) HVAC (Heating, Ventilation and Air Conditioning); (**b**) WH (Water Heater).

The main features of these models are: they consider heat gains and losses, for instance solar radiation (H_{sw}, H_w) or internal gains due to inhabitants (H_r) or appliances (H_a) (Figure 4a); the model takes into account heat storage from the specific heat of external walls (C_w), indoor masses (C_a, C_1 and C_2, especially important for WH) or roof/ground (C_{rg}); and it considers the control mechanisms which drive appliances (for instance thermostats $m(t)$ and DR policies $u(t)$). Moreover, their state variables are temperatures: indoor (X_i), walls (X_w), roof/ground (X_{rg}) for HVAC loads (Figure 4a), and X_1, X_2 for the stratification of water in the reservoir ("hot" WH1 and "cold" WH2 sub-tanks, Figure 4b), that is to say, characteristics that allow the evaluation of energy flows and storage capabilities (i.e., the indirect capacity of storage in the envelope of buildings), the direct storage in WH or the loss of customer service due to the application of DR (i.e., internal or hot water temperature).

These models are individual ones and need a further aggregation to reach a minimum demand level (size of reduction packages) established by specific regulations of electricity markets to bid or offer into these markets (e.g., from 100 kW to 1 MW depending on each specific service or market [54,55]). This task is often developed by energy aggregators.

To achieve these packages, the aggregator needs to rise ON-time to increase demand of each specific end-load whereas a decrease of demand requires a reduction of ON-time. The second alternative (the traditional one) is easier because the aggregator only needs to manage the rate of switching-OFF and switching-ON times of the power supply to load. This is easy to perform through hardware by classical methods (e.g., ripple control of WH in Germany, [56,57]) or home automation methods (e.g., controlled plugs through Z-Wave protocols [58] and universal software platforms for control [59]).

An important concern for the practical implementation of modern DR policies is the Automated Demand Response (ADR), because customer manual control does not fit the requirements both of accuracy and reliability of response. It is imperative for the success of ADR the development of standards for the communications between operators, aggregators and their customers' automation equipment [60] and the feedback from them. Open automated demand response protocol (OpenADR [61]) represents a good example of such a standard. Every day, more and more devices are certified to use OpenADR 2.0 protocols, and especially Smart Thermostats, but this certification is not necessary if some gateway assumes the role of "last-mile" controller and is compliant to receive and transmit OpenADR protocols. For example, home automation platforms such as Universal Devices ISY994 Series [62] allows the communication of residential customers with OpenADR, sending consigns and commands to home automation technologies working with different protocols (Zwave+, Insteon/X10, Zigbee Pro, Amazon Echo or Google Home). Other platforms, from well-known manufacturers, such as ABB SACE's Emax 2 Power Controller, develop similar functions [63] but at building scale. Examples of ADR capabilities of grid-integrated buildings and building microgrids, architecture, and standards can be found in [60].

The rate of change is defined to the PBLM software by PWM waves: the carrier wave being tried has a frequency of 0.833 mHz (i.e., 1 cycle every 20 min) and the modulating waveform is the desired decrease in the average value in $m(t, t + 20 \text{ min})$ according to deviations between 24-h PV forecasts (see Section 3.2) and 1-h PV forecasts used in markets (Figure 1).

The reasons for choosing 20 min have physical and technical senses. The first can be explained from the point of view of load service in the case of HVAC: if a harsh control is needed, switch-OFF times greater than 20 min can cause thermal fluctuations in the dwellings, easily noticeable by consumers (this can produce a lack of customer engagement in DR). The second reason is the so-called "lock-out" or mechanical delay of heating and air conditioning units. This mechanism is used to prevent a rapid recycling of the compressor avoiding mechanical damages. From the point of view of DR, it can cause an additional delay when applying ON/OFF and thermostat control signals. To evaluate the effect and characterize (from a statistical point of view) this process, some residential HP appliances (rated power from 1 kW to 3 kW) were monitored by authors. Changes in customer demand due to control actions have been recorded by an electronic meter and several Z-wave wall plug switches which send data to PCs using an USB gateway. Results depict that ON latency time ranges from 20–60 s and OFF latency times range from 10–40 s [64]. That is to say, the minimum ON-time should be in the range of one minute to limit inherent errors due to latency.

The first alternative (i.e., the increase of demand) is a less traditional option for DR [65]. Several reasons explain both the lack of use of these alternatives and their real interest. For instance, the increase of demand requires the control of thermostats. This control is more expensive than the supply control because smart thermostats are expensive. The cheapest option (e.g., Z-Wave) cost around € 150–200, whereas a remote switch costs around € 40–60. Fortunately, modern appliances include control of temperature though mobile-phones or PC, and these alternatives can be used for control (notice that some of them are compliant with well-known DR standard protocols [61]). In other cases, where the control device is intrusive (this is the case of WH), the cost of thermostat is similar, but the same maintenance (labor cost) is needed to include this option in the appliance. Over the last few years, some HPWH manufacturers in the USA have included these options for large units (200–500 L reservoir/storage tanks), for example [66].

The control of the thermostat (up or down, according to season, and usually used for pre-heating or pre-cooling policies in the dwelling being conditioned) has been proposed as a "virtual-storage" resource [67] for customers to take profit from Time of Use (ToU) tariffs or to "prepare" loads to face to DR events policies and maintain customer service (i.e., internal temperatures of houses or dwellings).

Usually, these policies have been used before the load is controlled, but the proposal in this paper is to use them continuously to adapt demand to changes in the forecasted PV generation. The control of the thermostat is evaluated and changed, if necessary, $\pm 0.5\,°C$ every 20 min. The reason for selecting this value is that $0.5\,°C$ is a usual value for the change of temperature settings in home smart-thermostats.

In this way, the proposed control strategy $u(t, t + k)$ for heating is done by Equation (2):

$$u(t,t+k)\begin{cases} > m(t,t+k) \rightarrow X^{sup} = \begin{cases} X^{sup} + \Delta X; \; u(t,t+k) > u(t-k,t) + \frac{db}{2} \text{ and } X^{sup} < X^{lim} \\ X^{sup}; \; u(t,t+k) > u(t-k,t) + \frac{db}{2} \text{ and } X^{sup} \geq X^{lim} \\ X^{sup}; \; u(t-k,t) - \frac{db}{2} < u(t,t+k) < u(t-k,t) + \frac{db}{2} \\ X^{sup} - \Delta X; \; u(t,t+k) < u(t-k,t) - \frac{db}{2} \end{cases} \\ < m(t,t+k) \rightarrow u(t,t+k) = \begin{cases} PWM(\Delta PV gen); \; X_i(t-k,t) > X_i^{serv} \\ u(t-k,t); \; X_i(t-k,t) < X_i^{serv} \end{cases} \end{cases} \tag{2}$$

where X_{sup} is the upper temperature of load's thermostat, which is set as a simple hysteresis cycle with dead-band *db* (usually ranging from 0.01–0.03 pu), and X_{lim} is the maximum reasonable temperature inside the dwelling (for example 22–23 °C in the case of HVAC, in winter) or the maximum temperature of water inside the tank (68 °C), which avoids risk of burns if a mixing valve is not used for the control of hot water pipeline. X_i^{serv} is a minimum service level for the appliance (a minimum comfort temperature inside the dwelling, for example 16 °C, or a minimum temperature of hot water inside the WH, for example 36 °C).

Basically, Equation (2) means that load control is done by a double control. In the case of heating (electric heaters or HP) if the load must go up, the thermostat goes up until it reaches the maximum allowable value (X_{lim}). Otherwise, if demand must fall to balance a decrease in PV generation (with respect to 24 h forecast), the thermostat or the supply is controlled to reduce demand. Notice that a "baseline", (i.e., load demand evaluated without control $m(t, t + k)$) is also needed as reference for controlled load. This baseline also comes from PBLM models (Figure 1).

3. Results and Discussion

3.1. Prediction Results for the Electricity Consumption

In this subsection, we provide 24-h-ahead predictions for the electricity consumption of the Spanish town in order to apply them to the context of Demand Response. For that, the ensemble method random forest described above was applied and some other aspects such as predictors importance or parameter selection was also developed.

On the one hand, it is well known that hourly loads of the previous days are the most important factors in short-term load forecasting. On the other hand, temperature is a factor that might affect the electricity consumption (cooling and heating of buildings). Therefore, prediction of hourly temperature for the location of the town under study was also used as an input in the load forecasting model, obtained as explained in the Section 3.2. In addition, several calendar variables such as the hour of the day, the day of the week, the month of the year and holidays have been taken into account in the design of the load forecasting model. Table 2 depicts the 49 predictors used for the load forecasting model: 23 dummies for the hour of the day, six dummies for the day of the week, 11 dummies for the month of the year, one dummy for holidays, one predictor of the forecasting temperature and seven predictors of historic loads (lags 24 h, 48 h, 72 h, 96 h, 120 h, 144 h, and 168 h).

Table 2. Description of the predictors.

Predictors	Description
H2, H3, … H24	Hourly dummy variables corresponding to the hour of the day
WH2, WH3, … WH7	Hourly dummy variables corresponding to the day of the week
MH2, MH3, … , MH12	Hourly dummy variables corresponding to the month of the year
FH1	Hourly dummy variable corresponding to national, regional or local holidays
Temperature	Predicted hourly external temperature.
LOAD_lag_i	Hourly load lagged "*i*" hours, with i = 24, 48, … ,168.

Before any analysis, a previous data filtering has been developed in order to detect and substitute missing cases or measurement errors. Moreover, in all cases the training period for model fitting ranges from 1 October 2008 to 31 November 2010, whereas the test period ranges from 1 January 2011 to 31 March 2011.

Three different measurements, given in Equations (3)–(5), were used to obtain the accuracy of the forecasting models: the root mean square error (RMSE), the R-squared (percentage of the variability explained by the forecasting model), and the mean absolute percentage error (MAPE).

The root mean square error is defined by:

$$\text{RMSE} = \sqrt{\sum_{t=1}^{n} \frac{(y_t - \hat{y}_t)^2}{n}} \tag{3}$$

The R-squared is given by:

$$\text{R} - \text{squared} = 1 - \frac{\sum_{t=1}^{n} (y_t - \hat{y}_t)^2}{\sum_{t=1}^{n} (y_t - \bar{y})^2} \tag{4}$$

The mean absolute percentage error is defined by:

$$\text{MAPE} = \frac{100}{n} \sum_{t=1}^{n} \left| \frac{y_t - \hat{y}_t}{y_t} \right| \tag{5}$$

where n is the number of data, y_t is the actual load at time t, $\hat{y}_t$ is the forecasting load at time t, and $\overline{y}_t$ is the mean value of the actual load. A slightly variants of this measure are the mean absolute error (MAE) and the cumulative absolute error (CAE).

Taking into account that the accuracy for special days (weekends and holidays) is usually lower than for regular days, the above goodness-of-fit measurements were obtained separately for each group of the test data.

Parameter tuning in random forest mainly refers to select an optimal number of trees (*ntree*) and an optimal number of predictors considered at each split (*mtry*). In fact, the selection of parameter *ntree* is quite easy because higher values do not lead to overfitting; thus, only a high enough value is needed. The optimal *mtry* = 12 was obtained by means of 10-fold cross-validation with three repeats and using a random grid with nine values (among the 49 possible).

Table 3 shows the goodness-of-fit measures for the training and test datasets after applying random forest with *ntree* = 200 and *mtry* = 12, and even for regular and special days separately. Furthermore, the importance of each predictor in the forecasting model has been obtained through the node impurity, getting that the electricity consumption at the same hour of the previous week (predictor *LOAD_lag_168*) is the most important predictor and that the following five most important ones are also historical loads. Temperature was in the 11th position of importance.

Table 3. Goodness-of-fit measures for regular and special days in the training and test datasets.

Measure	Regular Days	Special Days	All Days
Error_mean_train (kW)	6.88	−11.87	0.83
Error_mean_test (kW)	35.39	−4.94	22.84
Error_sd_train (kW)	114.29	107.18	112.39
Error_sd_test (kW)	173.84	154.95	169.19
Error_skewness_train	−0.16	−0.19	−0.15
Error_skewness_test	0.37	0.45	0.42
Error_kurtosis_train	10.93	8.48	10.21
Error_kurtosis_test	4.05	5.74	4.44
RMSE_train (kW)	114.49	107.83	112.39
RMSE_test (kW)	177.34	154.92	170.68
R-squared_train	0.98	0.94	0.98
R-squared_test	0.95	0.81	0.95
MAPE_train	2.05	2.45	2.18
MAPE_test	3.36	3.63	3.44

Figure 5 represents the evolution of the goodness-of-fit measures for each hour of the day, where the best accuracy is reached early in the morning.

As an example, Figure 6 shows the actual and predicted load for a complete week in the test dataset.

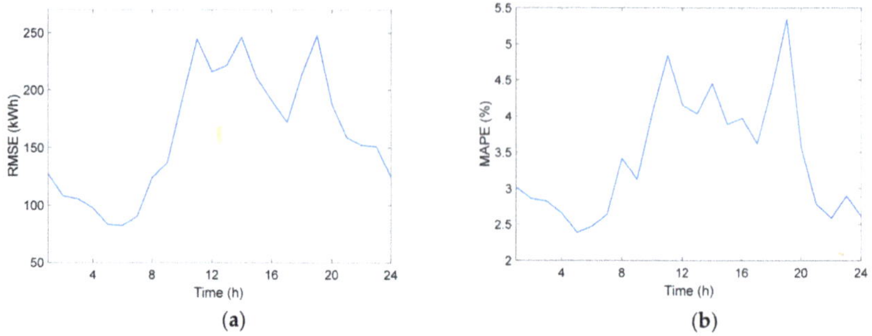

Figure 5. Goodness-of-fit measures by hour of the day in the test dataset: (**a**) using RMSE; (**b**) using MAPE.

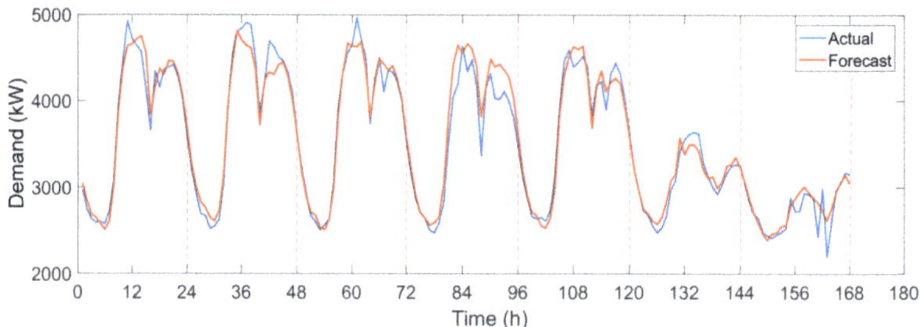

Figure 6. Actual and forecasting load for a week (14–20 February 2011).

3.2. Prediction Results for the Photovoltaic Generation

In this subsection, we describe the short-term PV power forecasting model able to offer 24-h-ahead predictions for the PV plant placed in the town under study in the context of Demand Response. As mentioned above, this forecasting model is based on the SGB method, which allowed the creation of the forecasting model with the lowest RMSE from a set of models developed with techniques such as linear multivariate regression, artificial neural networks, random forest, and support vector machines. The SGB method was selected because it achieved the lowest RMSE with a five-fold cross-validation procedure with the training dataset. All the forecasting models used the same training dataset, and their parameters were optimized following a similar procedure to the described, for the SGB, in the following paragraphs.

Table 4 shows the explanatory variables used to develop the PV power forecasting model. The dependent or output variable was the hourly power generation in the PV plant for each hour of the day (only daylight hours were considered). The explanatory variables include hourly weather predictions obtained with the Weather and Research Forecast (WRF) mesoscale model [68], a numerical weather prediction (NWP) model able to produce forecasts for a geographical region with the desired spatiotemporal resolution. In order to produce the weather forecasts, the WRF model is started every day with the initial and boundary conditions provided by the forecasts of the GFS model, an NWP model with global coverage, run and maintained by the National Centers for Environmental Prediction (NCEP) from USA. From the values provided by the GFS model with a $1° \times 1°$ (latitude–longitude) spatial resolution for the 00:00 Universal Time Coordinated (UTC) cycle, the WRF model provided predictions of weather variables over the region where the town under study is located with a time resolution of one hour, and a spatial resolution (distance between points of the grid of analysis) around 12 km. The forecasts of the desired weather variables for the location of the PV plant, or for the location of the town, were calculated by bilinear interpolation from the forecasts for the four nearest grid points.

For a real operation, these weather predictions can be available for a new day before dawn and include all the forecasts for the 24 h ahead.

Table 4. Explanatory variables for the PV power forecasting model.

Name	Description
swflx	Surface downwelling shortwave flux (W·m^{-2})
temp	Temperature at 2 m (Kelvin)
pres	Surface sea level pressure (hPa)
mod	Wind speed at 10 m (m/s)
dir	Wind direction at 10 m (degrees)
rh	Relative humidity at 2 m (per unit)
cft	Global cloud cover (per unit)
cfl	Cloud cover at low levels (per unit)
cfm	Cloud cover at medium levels (per unit)
cfh	Cloud cover at high levels (per unit)
prec	Accumulated rainfall in the hour (kg·m^{-2})
vis	Visibility (m)
clear	Clear-sky global horizontal irradiance (W·m^{-2})
aghi	Average global horizontal irradiance (W·m^{-2})
aip	Average irradiance on panel (W·m^{-2})
h1	Cosine of the day fraction for the hour
h2	Sine of the day fraction for the hour

The WRF model provided the hourly values of most of the 17 explanatory variables of Table 4 (variables from *swflx* to *clear*). The variable *swflx* corresponded to the global horizontal irradiance. Wind speed and direction were included because of the effect they could have on the temperature of the PV panels and, therefore, in their efficiency. The *aghi* variable matched to the average value of the forecasts for two consecutive hours of the *swflx* variable, that is, the average value of the global horizontal irradiance throughout the last hour. The *aip* variable corresponded to the average value of the irradiance on the PV panel throughout the last hour and it was calculated considering the characteristics of the PV panel with two-axis trackers and the solar geometry, as the aggregation of the direct normal and total diffuse irradiances on the tilted surface of the PV panels. The direct normal irradiance was obtained with the Erbs model [69] using the values of *aghi* and the total diffuse irradiance was calculated by means of the King model [70]. The variables *h1* and *h2* were used to code the hour (on UTC hour basis).

In order to select the best structure of the forecasting model, an optimization methodology was used based on the genetic algorithm (GA) with advanced generalization capabilities. This methodology is the GA-PARSIMONY [71], which allows the selection of parsimonious models. The main difference of this methodology with respect to the conventional GAs is a rearrange in the ranking of the individuals based on their complexities, so that individuals with less complexity (in this case, models with a less complex structure) are promoted to the best position of each generation. The promotion of less complex models with respect to the rest of individuals in the same generation with comparable fitness, allows the obtainment of models with improved generalization capability.

The GA-PARSIMONY methodology is implemented in the R package GAparsimony [72], which was the tool used to optimize the PV power forecasting models. In the case of the SGB model, the optimization process could choose the number of trees (in the range 20–250), the maximum interaction depth (range 3 to 8), the shrinkage value (range 0.001 to 0.25), and the minimum number of observations per terminal node (range 2 to 8), as well as select the input variables among those available (Table 4). The bagging fraction was fixed in 0.5. The fitness function corresponded to the negative value of the average RMSE obtained with five-fold cross-validation and three repeats with the training dataset. The complexity of the forecasting models evaluated in the optimization process was ten times the number of input variables used by the model plus the square of the maximum interaction depth. The number of individuals per generation was 50, the maximum number of generations 100, and the re-rank error value 0.1 (individuals with lower complexity and difference in the fitness value lower than the re-rank error were promoted to top positions in the ranking of each generation). The final model, achieved after the optimization process, used only 11 input variables (*swflx*, *hr*, *pres*, *prec*, *mod*, *clear*, *cfm*, *temp*, *dir*, *cfl*, and *h2*), had 182 trees, its maximum interaction depth was 6, its shrinkage value was 0.1197, and the number of minimum observations in a terminal node was 3. Table 5 shows the goodness-of-fit measures for the training and test datasets, where the RMSE, R-Squared and MAPE values are calculated using Equations (3)–(5), taking as y_t the actual PV generation power at time t and $\hat{y}_t$ the forecasted value for such hour. The high MAPE values are mainly due to very low actual PV generations in early and late daylight hours, when a small forecasting error can correspond to a very high absolute percentage error value.

Table 5. Goodness-of-fit measures in the training and test datasets for the PV power forecasting model.

Measure	Value
Error_mean_train (kW)	4.96
Error_mean_test (kW)	−19.52
Error_sd_train (kW)	308.60
Error_sd_test (kW)	362.12
Error_skewness_train	−0.021
Error_skewness_test	−0.173
Error_kurtosis_train	0.906
Error_kurtosis_test	0.994
RMSE_train (kW)	302.52
RMSE_test (kW)	350.34
R-squared_train	0.78
R-squared_test	0.70
MAPE_train	237.31
MAPE_test	310.06

Figure 7 plots the actual and forecasted hourly PV power generation values for a week in the testing dataset. Notice that the forecast is carried out each day before dawn and, up to now, no correction is applied along the day.

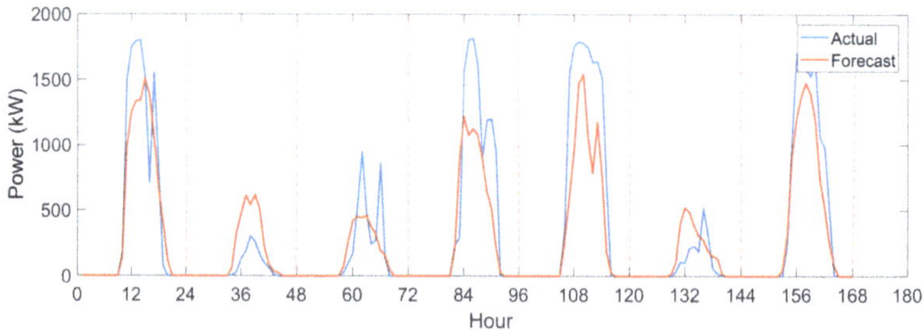

Figure 7. Actual and forecasting PV power for a week (14–20 February 2011).

3.3. Classification Results of Photovoltaic Curves

The classification stage of the proposed method has been carried out in three steps: firstly, the predicted daily curves of PV generation corresponding to the training period (from 1 October 2008 to 31 December 2010) are clustered into homogenous groups using DTW distance and average linkage; secondly, the "desired" cluster is selected (the one whose predicted PV curves better fit the real PV curves) and a centroid curve for each resulting cluster is obtained; finally, each predicted daily curve in the test period (from 1 January 2011 to 31 March 2011) is classified into the nearest cluster by computing its DTW distance with each centroid curve. Those days in the test dataset of which the predicted PV curves are classified into the "desired" cluster will be the ones selected for applying DR policies.

First step described above implies the hierarchical clustering of 822 time series (number of days in the training dataset) of length 24 (hourly data). The resulting dendrogram provided five possible groups, whose centroid curves are given in Figure 8 (observe that the main difference among the curves falls on the magnitude of the generated energy, except for the fifth cluster). For each of the five resulting clusters, we computed the median of the percentage error for all days in the clusters, obtaining 19.36% for Cluster 1, 39.95% for Cluster 2, 77.41% for Cluster 3, 57.09% for Cluster 4, and 48.15% for Cluster 5. Therefore, Cluster 1 provided lower fitting errors than the rest of clusters, and hence, it was considered the "desired" cluster for our purpose. As the desired cluster provides lower errors, those forecasting PV curves of the test dataset that are classified into the desired cluster are expected to better fit the real PV curves than the ones that are classified into any of the not desired clusters.

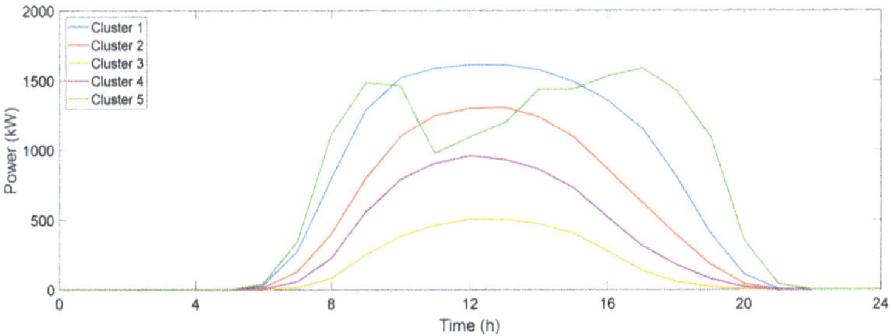

Figure 8. PV centroid curves of each cluster in the training dataset.

The results obtained in the final step of the classification stage were the following. A total of 31 days in the test period were classified into Cluster 1 (the "desired" cluster), whose dates are given in Table 6 (recall that all days refer to year 2011) and some examples comparing the real PV and predicted PV (24 h ahead) are given in Figure 9.

Table 6. Dates of the test period (2011) classified into Cluster 1 (and therefore selected for DR actions).

Day (Number)	Date (dd/mm)	Day (Number)	Date (dd/mm)	Day (Number)	Date (dd/mm)
1	4 February	11	10 March	21	20 February
2	5 February	12	11 March	22	20 March
3	5 March	13	12 February	23	21 March
4	6 February	14	14 January	24	22 March
5	6 March	15	14 February	25	3 January
6	7 February	16	16 January	26	23 March
7	7 March	17	18 February	27	24 January
8	8 February	18	18 March	28	25 February
9	9 February	19	19 March	29	28 March
10	10 February	20	20 January	30	29 March
				31	31 March

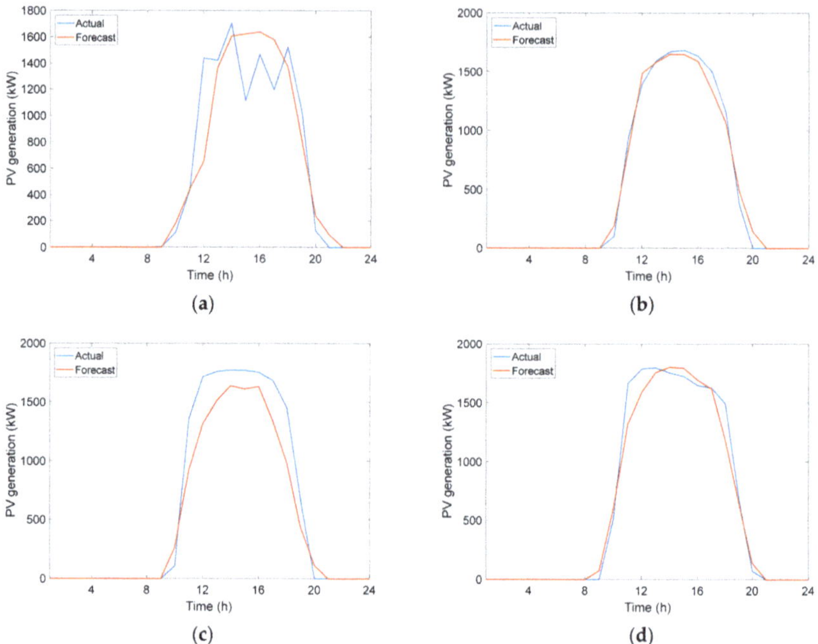

Figure 9. Comparisons of actual PV and 24 h forecast PV generation for some days in Cluster 1: (**a**) day 3; (**b**) day 4; (**c**) day 9; (**b**) day 28.

3.4. Results for Demand Response Strategies

3.4.1. Very Short-Term PV Adjusted Forecasting

As was seen in Sections 3.1 and 3.2, the main accuracy problem when obtaining a 24 h-ahead forecast of net demand is due to errors in the prediction of PV generation (compares Figures 6 and 7), because this forecast is carried out each day before dawn, and no correction is applied during the day. The problem of balancing net load with respect to 24 h predictions should be taken into account by aggregators, because any imbalance can produce an important money flow from aggregators to Balance Service Providers (BSP) or Load Serving Entities (LSE). For this reason, a very short-term correction has been proposed. The idea is similar to the procedure used by ISOs to correct demand when some power event is declared in the system taking into account measurements of demand before the correction (i.e., the generation of customer's baselines or CBL, for example [73]). In these methods (used for the retribution of demand response in reliability programs), the adjustment factor is obtained by means of the first two hours of the four-hour period prior to the commencement of the reliability event. In our case, the methodology to correct the forecasting values should balance accuracy and fast computation (it takes less than a minute to have enough time for the load control processing). In this stage, we propose the determination of an adjustment factor by means of the first 60 min of actual PV records of the one-and-a-half-hour period prior to the forecast window (it is assumed that PV generation has an SM able to record every minute and that sends this information to the aggregator). This adjustment factor is evaluated through the Equation (6):

$$af(d,t) = \frac{\sum_{k=30}^{k=90} PVA(d, t-k)}{\sum_{k=30}^{k=90} PVF(d, t-k)} \tag{6}$$

where $af(d, t)$ denotes the adjusted factor for the time t of the current day d used to fix generation forecasts, $PVA(d, t - k)$ is the actual PV generation k min before, and $PVF(d, t - k)$ is the predicted 24 h PV generation for day d at time $t–k$. Then, a first approximation of the forecasted PV baseline (*PVBL_aux*) is computed in the time interval $[t - 30, t]$ to obtain $[t, t + 30]$ values (that is, predictions are corrected in a time-window corresponding to very short-term according to the reduction of imbalances. For the participation of customers in other markets, the window can be enlarged according to gate closure times) through Equation (7):

$$PVBL\_aux(d, t + k) = af(d, t) \times PVF(d, t + k); k = 1, 2 \ldots 30 \; min \tag{7}$$

To improve the goodness of this 1 h-ahead forecast, $PVBL$ is compared with the average of historical values of PV generation in the last two weeks:

$$PV_{upl}(d, t) = \frac{1}{450} \sum_{j=1}^{15} \sum_{k=-15}^{k=15} PVA(d - j, t + k) \tag{8}$$

where PV_{upl} is the upper limit considered as acceptable for any correction through the adjusted factor $af(d, t)$. Therefore, Equation (7) is improved by Equation (9):

$$PVBL(d, t + k) = \begin{cases} PVBL\_aux(d, t + k), \; if \; PVBL\_aux(d, t + k) < PV_{upl}(d, t + k) \\ PV_{upl}(d, t + k), \; otherwise \end{cases} ; k = 1, ..30 \tag{9}$$

Results showed that the proposed correction by Equation (9) suits well its objective (Table 7 depicts the MAPE of the 24 h-ahead forecasts (*PVF*) and the 1 h-ahead forecasts (*PVBL*) for some representative days in Cluster 1). As expected, 1 h-ahead forecasts outperform 24 h-ahead forecasts, but achieve a significant improvement on days when the most serious errors took place (days 14, 16, 17, and 23). In some cases, a small increase of errors is reported (days 5 and 26).

Table 7. Improvement in PV forecast attributable to the adjustment given in Equation (9).

Day	MAPE (%) of PVF (24 h-Ahead Forecast)	MAPE (%) of PVBL (1 h-Ahead Forecast)
1	12.5	7.7
2	13.9	9.1
4	18.6	8.7
5	4.8	5.6
14	28.2	11.1
16	128.1	30.6
17	39.2	11.9
20	22.9	5.8
23	52.6	21.7
26	9.7	9.9
28	8.2	6.7

3.4.2. Balancing Net Demand through DR

Once the predictions (for customers' demand and PV generation) were calculated, and the clustering process selected the "desired" days for applying DR, the next objective was trying to adapt the net demand (difference between customers' real demand and real PV generation) to the predicted net demand made for the day ahead (Figure 1).

In order to achieve this aim, DR policies were applied to two flexible loads: WHs and HVACs. The reason of choosing these loads is their facility for implementing control strategies by changing the thermostat temperature and their ability to act as thermal energy storage systems (by doing preheating of water in the case of WH and precooling and preheating of rooms/walls in the case of HVAC), see Section 2.5. DR policies were applied through PBLM and aggregation models [52], with the aim of ensuring that the final net consumption is adapted in a significant extent to the profile of net energy demand predicted the day before, so that it was not necessary to trade additional resources into the wholesale electricity market or to pay BSP.

The planning of DR actions that have to be performed was obtained hour-by-hour. That is to say that DR actions for each next hour were planned, taking into account the differences between predictions made for 24 h ahead and predictions made for 1 h ahead. As forecasts for electricity consumption are more accurate than PV forecasts, in this study, the predictions made for 1 h ahead only estimated the PV generation, thus, DR strategies only acted in order to manage the PV forecasting error. This fact means that the control was performed in the period in which there was PV generation, that is, approximately, from 8 a.m. to 9 p.m.

The PV variations between 24-h predictions and 1-h predictions must be compensated with changes in the consumption of WHs and HVACs. As WHs and HVACs represent the 17.9% and 42.9% of the total consumption, respectively (see Table 1); 25% of the PV variations was managed by WHs loads, and the rest (75%) was assigned to HVACs.

To demonstrate the ability of the loads (HVAC and WH) to adapt their consumption and the capacity of minimizing variability between predictions and real consumption and generation, the 31 days obtained from the "desired" cluster of PV curves have been simulated. Two different examples (days 14 and 23) have been selected to illustrate the application of DR strategies and its results, being explained in detail. Later, in Tables 8 and 9, overall results and indicators for a set of representative days in Cluster 1 will be shown.

Table 8. Results for the net energy consumption (day 23).

Energy 24-h (MWh)	Energy w/o DR (MWh)	Energy w/DR (MWh)	CAE 24 h-w/o DR (MWh)	CAE 24 h-w/DR (MWh)	Error w/o DR (%)	Error w/DR (%)	Max. ΔP w/o DR (kW)	Max. ΔP w/DR (kW)
42.16	45.96	40.37	6.02	2.96	14.27	7.02	1256.93	579.32

Table 9. Results for net energy consumption (day 14).

Energy 24-h (MWh)	Energy w/o DR (MWh)	Energy w/DR (MWh)	CAE 24 h-w/o DR (MWh)	CAE 24 h-w/DR (MWh)	Error w/o DR (%)	Error w/DR (%)	Max. ΔP w/o DR (kW)	Max. ΔP w/DR (kW)
50.13	47.23	49.43	4.13	1.79	8.25	3.58	805.45	451.54

Figure 10a,b presents the results for day 23 (21 March 2011) in which the PV generation forecasts made 24 h in advanced overestimate the real PV energy generated. As can be seen in Figure 10a, the forecasts made 1 h in advance are much more accurate; thus, they can be used as a baseline curve to plan the DR actions (Table 7 presents overall results). Figure 10b shows the effects on the net consumption forecasts. In order not to trade additional resources, the objective is to reduce the consumption in the way that the final actual net consumption matches the 24-h-ahead forecasts.

Taking into account that there is PV generation only from 8 a.m. to 9 p.m., control actions will be applied to WH and HVAC only during this period. Figure 11 presents the variations in each predicted load demand that have to be performed to obtain the desired net consumption profile.

Figure 12 depicts the load consumption after DR control strategies and the desired consumption profile from 24-h forecasts. As can be seen, DR actions work properly and the differences between the final and the "desired" load consumption have been significantly reduced.

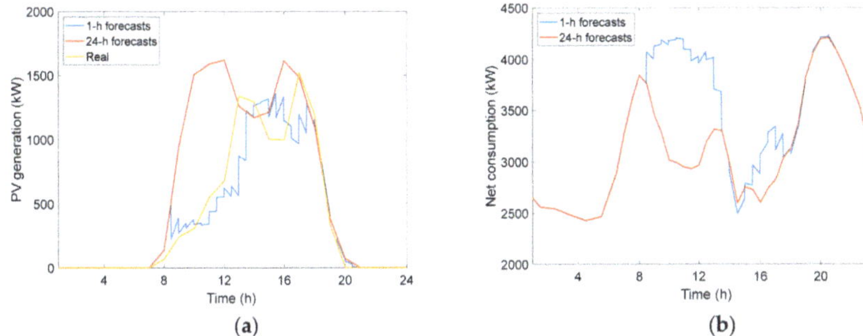

Figure 10. Differences between 24-h-ahead and 1-h-ahead forecasts (day 23): (**a**) PV generation; (**b**) Net consumption.

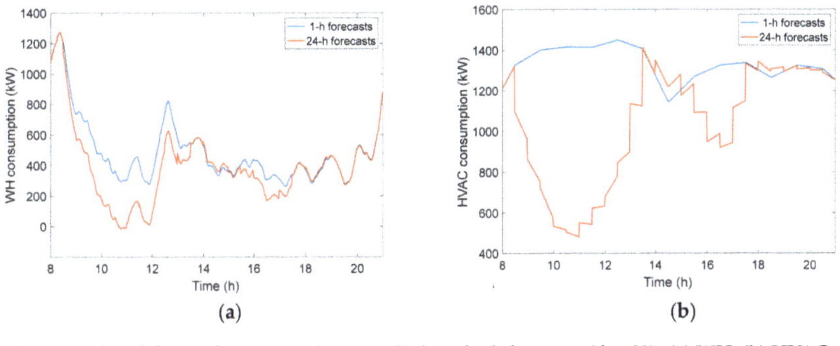

Figure 11. Load demand variations between 24-h and 1-h forecasts (day 23): (**a**) WH; (**b**) HVAC.

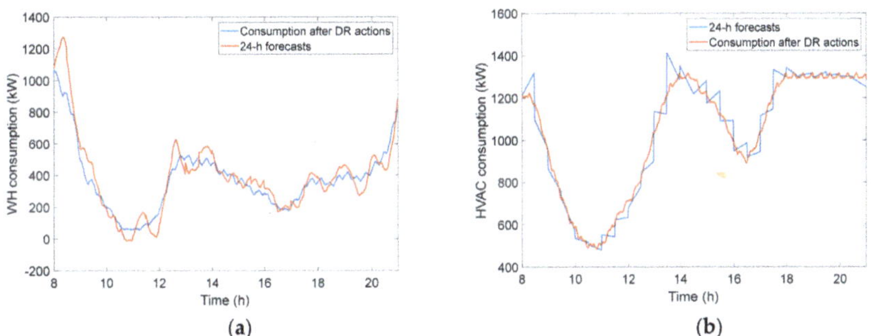

Figure 12. Load consumption after DR actions and 24-h forecasts (day 23): (**a**) WH; (**b**) HVAC.

By modifying the WH and HVAC load consumption, it is also changed the temperature of water and rooms respectively. These variations can cause some comfort problems for users, thus, aggregators should assure that there are no large temperature variations and that the customer comfort is always guaranteed. This effect is considered in double control strategies defined in Section 2.5. by Equation (2).

Figure 13 shows the temperature profile of the loads. In the case of WHs (Figure 13a), the temperature is always above 45 and 50 °C in the cold and hot sub-tank, respectively [74]. Figure 13b shows the temperature of the rooms, the temperature of the internal walls, the temperature of the external walls, and the external ambient temperature. As can be seen, the internal temperature was always above 16 °C, while the maximum internal and ambient temperature (external) were 19.5 and 13 °C, respectively; the difference between internal and external temperature was always above 5.5 °C.

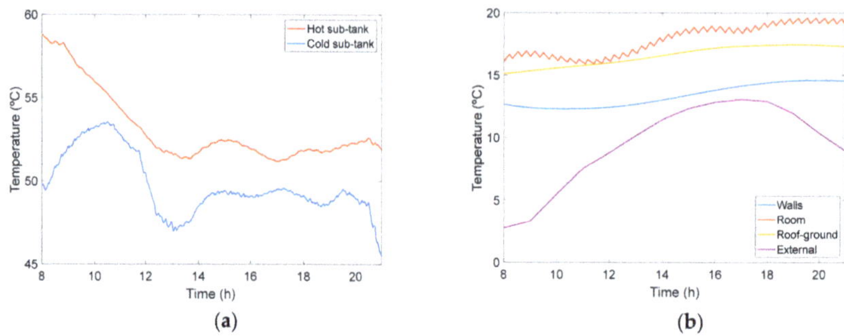

Figure 13. Loads' temperature profile after DR actions (day 23): (**a**) WH (state variables X_1 and X_2, Figure 4b); (**b**) HVAC (state variables X_i, X_w, X_{rg}, and input X_{ext}, Figure 4a).

Finally, Figure 14 shows the net profiles for the 24-h-ahead forecasts compared with final net consumption after DR actions and the real net consumption if no DR action was performed.

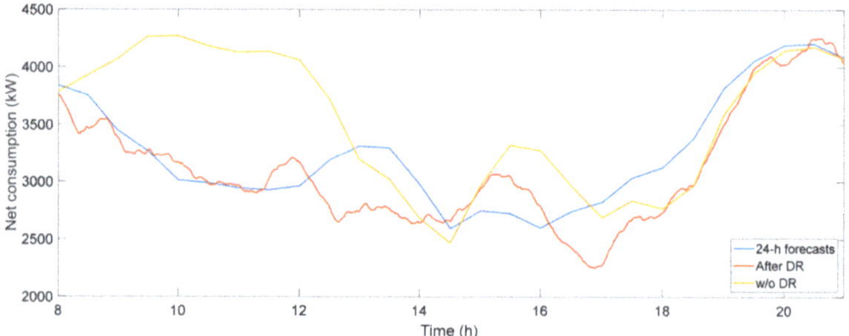

Figure 14. Net consumption profiles (day 23): 24-h forecasts, after DR actions and without DR.

It is clearly demonstrated that DR actions significantly reduce the differences with 24-h-ahead net profile, efficiently compensating the forecasting errors and balancing the final net load consumption. Table 8 presents some numerical results from this example. All variables were calculated during the control period (from 8 a.m. to 9 p.m.).

As can be deduced from the Table 8, the total net consumption of the day is reduced from 45.96 MWh if no DR action is taken to 40.37 MWh if DR actions are applied. The differences between the 24-h-ahead forecasts and the final net consumption are also shortened: cumulative absolute error (CAE) reduces from 6.02 MWh without applying DR to 2.96 MWh with DR actions. In the same way, the percentage of error, understood as the rate between the CAE and the net consumption for 24-h-ahead forecasts, decreased from 14.27% (without DR) to 7.02% (with DR), which is a 51% of reduction.

In the next paragraphs, a different example of the DR strategy will be presented. In this case, day 14 (14 January 2011) was analyzed, where the 24-h-ahead forecast predicted less PV energy than the final real PV generation; thus, the aggregator has bought more energy than is necessary in Electricity Markets required. In order to not waste this energy, it was used to increase the temperature of WH and HVAC, exploiting their capacities to act as thermal energy storage systems. Figure 15a presents the 24-h-ahead and 1-h-ahead forecasts compared with real PV generation data. As can be seen, the 24-h-ahead forecasts have underestimated the PV generation, while 1-h-ahead forecasts have much more precision. Figure 15b shows that it is necessary to increase the final net demand to adapt the 1-h-ahead profile to 24-h-ahead forecasts, and in this way, consume the "excess" of PV generation.

Figure 16 depicts the results from the application of DR policies to the flexible loads (WH and HVAC). In both cases, the consumption after DR actions follow the target, to match its energy consumption with the 24-h-ahead predicted energy consumption for each load. Thus, it is fair to say that the DR control was working efficiently.

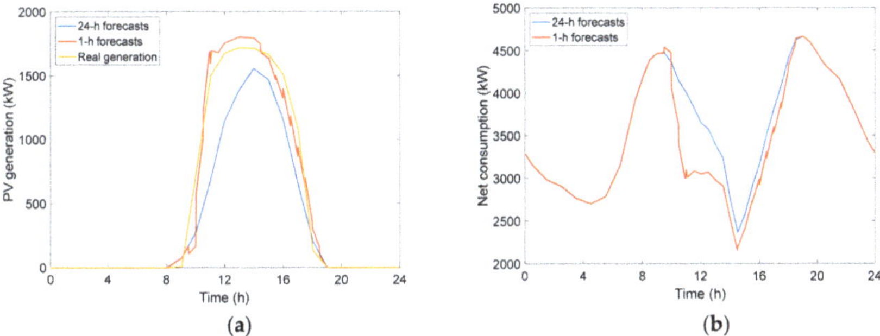

Figure 15. Differences between 24-h-ahead and 1-h-ahead forecasts (day 14): (**a**) PV generation; (**b**) Net consumption.

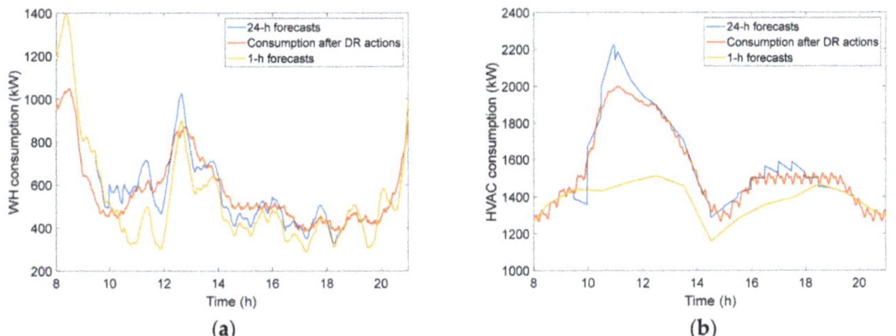

Figure 16. Load demand variations among 24-h forecast, 1-h forecasts and final load consumption after DR (day 14): (**a**) WH; (**b**) HVAC.

As mentioned before, the increase in the energy consumption of both WH and HVAC was used to rise the temperature of the water inside the tank (in the case of WHs) and the temperature inside the rooms (HVAC). In the case of the WHs (Figure 17a), the temperature increases in the cold sub-tank from 49 to 59 °C and in the hot sub-tank from 58 to 63 °C (always ensuring that, for health security issues, the temperature is not above 68 °C in any WH), while in the HVACs (Figure 17b), the internal temperature of the rooms increased from 17 to 20.5 °C, whereas the maximum ambient temperature (external) was 11.5 °C.

Figure 18 depicts the 24-h-ahead forecasts for the net energy consumption. The final net energy consumption profile after the application of DR strategies is also shown, as is its comparison with the net energy that will be consumed if no DR action is taken. The graph shows that DR actions reduced the differences between the 24-h forecasts and final net consumption, minimizing the necessity of selling back to Electricity Markets the PV generation surplus or to pay additional charges (or penalties) for unbalance in markets.

Table 9 presents some numerical results from this example. All variables were calculated only during the control period (8 a.m.–9 p.m.). Results show that the net energy consumption with DR strategies increased to 49.43 MWh compared with the 47.23 MWh consumed without the application of DR, and reached 50.13 MWh for 24-h forecasts. The variations between forecasts and final net consumption were reduced from 4.13 MWh to 1.79 MWh, reducing the percentage of error by 57% (from 8.25 to 3.58%). The maximum power peak difference was also reduced by 44%.

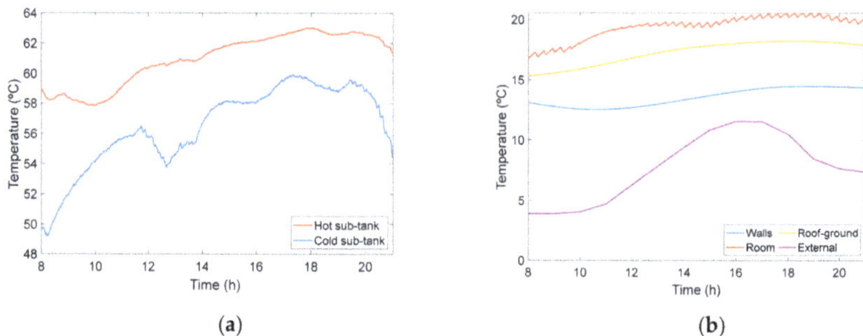

(a)

(b)

Figure 17. Loads' temperature profile after DR actions (day 14): (**a**) WH (state variables X_1 and X_2, Figure 4b); (**b**) HVAC (state variables Xi, X_w, Xrg, and input X_{ext}, Figure 4a).

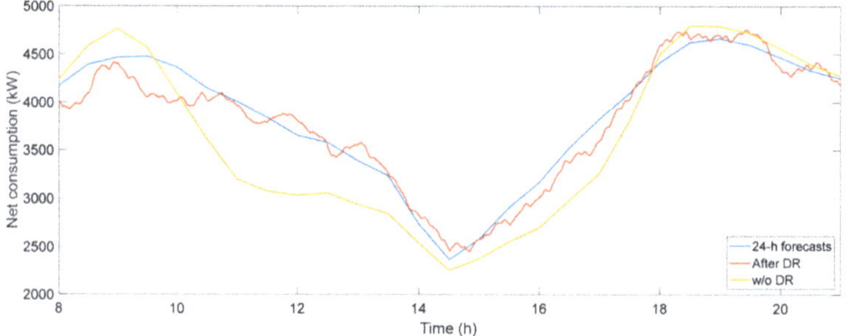

Figure 18. Net consumption profiles (day 14): 24-h forecasts, after DR actions and without DR.

Table 10 shows overall results for net energy consumption of a representative part of the 31 days in Cluster 1. Day 16 (Figure 19a) obtained the greatest reduction in the error (from 31.63 to 6.7%), whereas day 28 (Figure 19b) was the one in which the percentage of error increased most (from 6.55 to 9.99%). Notice that this rise in the error was not due to DR actions, but because of the lack of accuracy in the prediction of customers' load profile of that day (day 28), because only deviations of PV generations were corrected by means of DR (see also Figure 9).

Table 10. Values for net energy consumption.

Day	Energy 24-h (MWh)	Energy w/o DR (MWh)	Energy w/DR (MWh)	CAE 24 h-w/o DR (MWh)	CAE 24 h-w/DR (MWh)	Error w/o DR (%)	Error w/DR (%)	Max. ΔP w/o DR (kW)	Max. ΔP w/DR (kW)
1	44.84	43.41	43.09	3.09	3.04	6.90	6.78	629.51	558.59
2	32.52	30.93	31.22	2.39	2.00	7.37	6.14	732.47	449.26
4	27.06	25.19	26.42	3.24	1.71	11.99	6.32	581.92	411.86
5	22.34	22.07	21.86	1.32	1.44	5.94	6.48	419.74	380.21
14	50.13	47.23	49.43	4.13	1.79	8.25	3.58	805.45	451.54
16	29.70	38.64	30.09	9.39	1.99	31.63	6.70	1622.69	472.04
17	47.15	41.59	45.25	6.40	2.85	13.58	6.05	1108.13	538.99
20	48.31	47.04	48.74	2.89	1.68	5.99	3.49	622.92	342.69
21	27.59	23.97	25.27	3.86	2.59	13.99	9.40	776.20	534.47
23	42.16	45.96	40.37	6.02	2.96	14.27	7.02	1256.93	579.32
26	41.01	42.38	42.08	2.01	1.90	4.92	4.63	487.96	572.66
28	41.97	41.25	39.32	2.75	4.19	6.55	9.99	792.76	1124.70

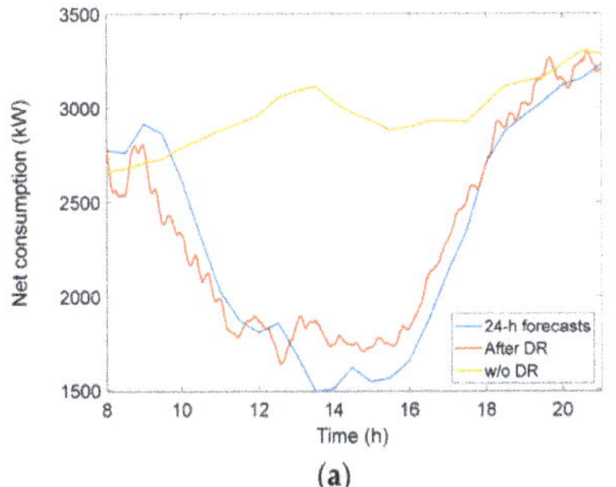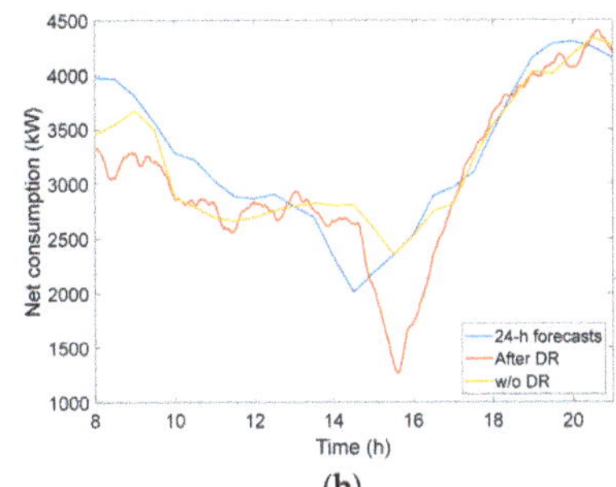

(a)   (b)

Figure 19. Net consumption profiles: 24-h forecasts, after DR and without DR: (**a**) day 16; (**b**) day 28.

Finally, Figure 20 presents the cumulative absolute error for the 31 days in Cluster 1 when comparing the target (the 24-h-ahead forecast) with the final net consumption in two different scenarios (without DR and after DR actions). It can be seen that, in general, errors were reduced after DR, except in some particular days where they increased slightly. Notice that the set of days selected in Tables 10 and 11 (among the 31 days belonging to Cluster 1) represent different scenarios (low, medium, and high reduction of the errors as well as increasing of them after DR policies).

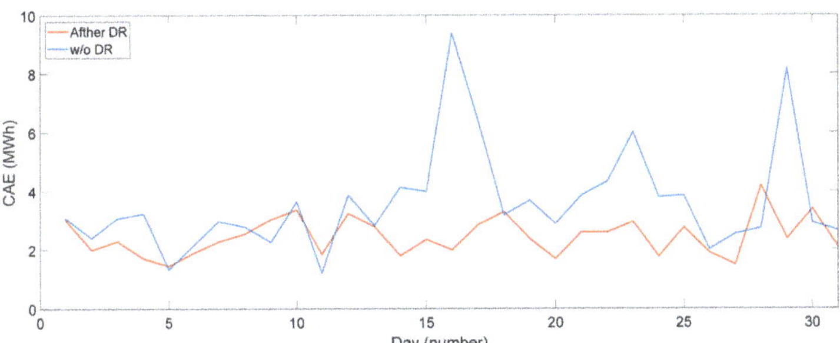

Figure 20. Cumulative absolute error (CAE) between the final net consumption (without DR and after DR actions) and the target (the 24 h-ahead forecast of the net consumption).

Table 11. DR performance and flexibility indicators.

Day	Balance Mileage (MWh)	Demand Mileage (MWh)	Mileage Ratio (%)	Symmetry Equation (12)	Performance Equation (13)
1	2.64	5.54	47.5	0.95	0.27
2	2.62	4.52	57.5	1.59	0.34
4	2.86	4.06	70.17	4.24	0.05
5	1.52	3.92	38.89	0.83	1.89
14	3.12	5.72	54.50	7.06	0.047
16	3.06	4.12	74.20	0.008	0.036
17	4.32	5.34	80.9	17.92	0.064
20	1.66	5.54	28.9	5.24	0.268
23	4.14	5.28	78.13	0.038	0.073
26	1.88	5.16	36	0.953	0.467
28	2.08	5.22	39.8	0.224	2.34

3.4.3. Analysis of DR Flexibility

As mentioned in Section 2.1, a quantitative analysis for demand-side flexibility has been performed thorough some indicators defined and calculated at an aggregated level.

The first indicator of flexibility refers to signals' dynamic. This is done through an indicator that gives an idea about the variation of a signal and it is very close to the "mileage" score used for the verification of assets in Ancillary Services. In this way, the "signal_mileage" is the absolute sum of movement of the analyzed signal in a given time period with respect to the average value of the signal, in our case daily:

$$\text{signal\_mileage} = \frac{\sum_{kk=ini+1}^{end} \text{abs}(\text{signal}(kk) - \text{signal}(kk-1))}{\sum_{kk=ini+1}^{end} \text{signal}(kk)} \tag{10}$$

where signal refers to the target foreseen for balancing PV generation through load (variable "balance") or the load demand (with DR or without DR, i.e., the variable "baseline" of demand or the new demand with DR, in this case the variable "demand_DR"). These indicators give, respectively, an idea of the variability of demand (hourly, daily,...) for the segment under study and the "amount of work" involved through "balance" signals to match PV/demand forecast errors.

A second indicator is the "mileage_ratio", which measures the relation of the value of mileage of the balance signal sent to flexible demand versus the value of changes in demand in the steady state (without DR). This indicator gives the aggregator a first insight with respect the effort that demand is forced to yield to match PV generation in the short-term:

$$\text{mileage\_ratio} = \frac{\text{balance\_mileage}}{\text{baseline\_mileage}} \tag{11}$$

The third indicator represents the symmetry of the effort required from demand to follow the energy balance signal (i.e., the overall increase of flexible demand versus the reduction of demand). As has been discussed in previous paragraphs, in some cases, it is more difficult for the load to increase in demand than achieve a reduction in demand (for instance, electric heating in winter). For these reasons, the positive changes in demand were evaluated with respect to negative changes of demand. Mathematically:

$$\text{symmetry} = \frac{\sum_{kk=ini}^{end} \text{abs}((\text{demand\_DR}(kk) - \text{baseline}(kk)) > 0)}{\sum_{kk=ini}^{end} \text{abs}((\text{demand\_DR}(kk) - \text{baseline}(kk)) < 0)} \tag{12}$$

Finally, the aggregator calculated a daily performance score that reflects the load resource's accuracy in increasing or decreasing its demand to provide balance in response to balance dispatch signal. The performance score calculation evaluates each resource's accuracy in following the balance signal, that is to say:

$$\text{performance} = \text{abs}\left(1 - \frac{\sum_{kk=ini}^{end} \text{abs}(\text{demand\_DR}(kk) - \text{baseline}(kk))}{\sum_{kk=ini}^{end} \text{abs}(\text{balance}(kk) - \text{baseline}(kk))}\right) \tag{13}$$

These indicators have been evaluated through and Table 11 presents the main results.

A brief explanation of the results can help the reader to better understand the physical meaning of these indicators. It is also interesting to consider the results of Section 3.4.2 for days 14 and 23. Table 11 shows that the days 14 and 23 require a noticeable effort from flexible demand (3.12 and 4.14 for "balance_mileage" values that are above the average effort). Steady state fluctuations of demand are similar for both days (5.72 and 5.28, see demand mileage column). This index (mileage_demand) is of interest to reflect unusual changes of demand pattern.

Column five in Table 11 presents the symmetry of the effort. Day 14 requires a net increase of demand (symmetry = 7.06), whereas day 23 basically requires a strong shaving of demand (symmetry = 0.038). The score for the performance of flexible demand depicts that flexible loads follow with enough accuracy their targets (performance indicator is around zero). Moreover, from Table 11, the aggregator can deduce that flexible demand fails in days 5 and 28 (performance has the greatest values and over 1), but these days do not represent a big problem, because the effort required from demand (balance_mileage) is low (1.52 and 2.08). These results and the results previously discussed in Section 3.4.2 help both the customer and the aggregator to familiarize themselves with the demand response and the usefulness of short-term predictions to manage their new role as prosumers or energy aggregators.

4. Conclusions

Energy issues are a main concern for the sustainability of our society. This sustainability is based on the integration of renewable sources and in the development of new energy markets, which should be more customer-centered than in the past. These objectives need the development and validation of additional tools to facilitate this change and to contribute to the effective engagement of customers in these markets, as has happened in telecommunications markets. Technological aspects, such as the forecast of demand and renewable sources or the management of energy, appeared as significant barriers to the effective deployment of new markets in small and medium customer segments, i.e., benefits usually do not balance the complexity for new responsibilities and tasks in these scenarios. Moreover, forecasts get more complex when the level of assets' aggregation decreases, and this makes the above-mentioned objectives more difficult. For these reasons, this work developed and validated both demand and renewable generation forecasting methods at low aggregation levels (in the order of some MW) of the power system (distribution), but focused the methodological effort on the application of these methods to demonstrate the possibility of participation of "prosumers" in markets rather than in the achievement of small improvements in forecasts accuracy (MAPE, RMSE, CAE) through exponential complexity.

The interaction of demand, generation and management models demonstrated that this feedback or linkage among models can balance errors through a "closed control loop" that drove the net demand of "prosumers". In the analyzed scenarios, results showed that 50% of prediction errors can be balanced with naïf correction models (very short term) and a "reduced" portfolio of loads (HVAC and WH) and policies. The paper also demonstrated the ability of these small and medium customers, through demand aggregators, to exhibit in the market the necessary flexibility in demand (up and down) to manage the volatility of renewables and build new power systems and new markets in the horizon 2030–2050.

Further developments are necessary to advance in this approach, for instance: the consideration of the participation of customers in new markets and more complex services (mixed participation in two or more markets or services), the refinement of very short-term models (both in demand and generation), the introduction and synthesis of new end-use PBLM and their further aggregation, the integration and deployment of ICTs in the models and the validation, the hybridization of ESS and demand models, both following PBLM philosophy, to provide more capabilities for flexibility, and the adjustment and improvement of these models in actual customers through pilots. In the medium term, and with these tools, the potential flexibility of these small and medium customer segments could be exploited and used to balance the integration of renewable both in Smart Grids and in conventional Power Systems.

Author Contributions: L.A.F.-J., M.C.R.-A., A.G. (Antonio Guillamón) and A.G. (Antonio Gabaldón) conceived and designed the experiments and structure of the paper. A.G.-G. and A.G. (Antonio Gabaldón) programmed the algorithms and developed and wrote the part concerning HVAC and WH modeling, the technical evaluation of prosumers, and the simulation and evaluation of DR targets. M.C.R. and A.G. (Antonio Guillamón) managed the customer database of demand/end-uses and supervised some of the mathematic algorithms for short-term load models. L.A.F.-J. and A.F. managed PV generation database and defined and supervised the mathematic algorithms for short-term PV generation models. All authors have approved the final manuscript. All authors have read and agreed to the published version of the manuscript.

Funding: This work was supported by the supported by the AEI/10.13039/501100011033, (Ministerio de Ciencia, Innovacióna y Universidades, Spanish Government) under research projects ENE-2016-78509-C3-2-P, ENE-2016-78509-C3-3-P, RED2018-102618-T, and EU FEDER funds. The authors have also received funds from these grants for covering the costs to publish in open access.

Acknowledgments: This work was supported by the Ministerio de Economía, Industria y Competitividad (Spanish Government) under research projects ENE-2016-78509-C3-2-P, ENE-2016-78509-C3-3-P, RED2018-102618-T, the Ministerio de Educación (Spanish Government) under doctorate grant FPU17/02753, and the special support of EU FEDER funds.

Conflicts of Interest: The authors declare no conflict of interest.

References

1. Catalão, J.P.S. *Electric Power Systems: Advanced Forecasting Techniques and Optimal Generation Scheduling*; CRC Press: Boca Raton, FL, USA, 2012.
2. Hahn, H.; Meyer-Nieberg, S.; Pickl, S. Electric load forecasting methods: Tools for decision making. *Eur. J. Oper. Res.* **2009**, *199*, 902–907. [CrossRef]
3. Antonanzas, J.; Osorio, N.; Escobar, R.; Urraca, R.; Martinez-de-Pison, F.J.; Antonanzas-Torres, F. Review of photovoltaic power forecasting. *Sol. Energy* **2016**, *136*, 78–111. [CrossRef]
4. Alfares, H.K.; Nazeeruddin, M. Electric load forecasting: Literature survey and classification of methods. *Int. J. Syst. Sci.* **2002**, *33*, 23–34. [CrossRef]
5. Yang, H.T.; Huang, C.M.; Huang, C.L. Identification of ARMAX model for short term load forecasting: An evolutionary programming approach. *IEEE Trans. Power Syst.* **1996**, *11*, 403–408. [CrossRef]
6. Taylor, J.W.; de Menezes, L.M.; McSharry, P.E. A comparison of univariate methods for forecasting electricity demand up to a day ahead. *Int. J. Forecast.* **2006**, *22*, 1–16. [CrossRef]
7. Massana, J.; Pous, C.; Burgas, L.; Melendez, J.; Colomer, J. Short-term load forecasting in a non-residential building contrasting models and attributes. *Energy Build.* **2015**, *92*, 322–330. [CrossRef]
8. Bruhns, A.; Deurveilher, G.; Roy, J.S. A non-linear regression model for mid-term load forecasting and improvements in seasonality. In Proceedings of the 15th Power Systems Computation Conference, Liege, Belgium, 22–26 August 2005.
9. Charytoniuk, W.; Chen, M.S. Nonparametric regression based short-term load forecasting. *IEEE Trans. Power Syst.* **1998**, *13*, 725–730. [CrossRef]
10. Li, K.; Su, H.; Chu, J. Forecasting building energy consumption using neural networks and hybrid neuro-fuzzy system: A comparative study. *Energy Build.* **2011**, *43*, 2893–2899. [CrossRef]
11. Liao, G.C.; Tsao, T.P. Application of a fuzzy neural network combined with a chaos genetic algorithm and simulated annealing to short-term load forecasting. *IEEE Trans. Evol. Comput.* **2006**, *10*, 330–340. [CrossRef]
12. Tso, G.K.F.; Yau, K.K.W. Predicting electricity energy consumption: A comparison of regression analysis, decision tree and neural networks. *Energy* **2007**, *32*, 1761–1768. [CrossRef]
13. Dong, Y.; Ma, X.; Ma, C.; Wang, J. Research and application of a hybrid forecasting model based on data decomposition for electrical load forecasting. *Energies* **2016**, *9*, 1050. [CrossRef]
14. Chen, K.; Chen, K.; Wang, Q.; He, Z.; Hu, J.; He, J. Short-Term Load Forecasting with Deep Residual Networks. *IEEE Trans. Smart Grid* **2019**, *10*, 3943–3952. [CrossRef]
15. Dudek, G. Short-term load forecasting using random forests. *Adv. Intell. Syst. Comput.* **2015**, *323*, 821–828. [CrossRef]
16. Lin, Y.; Luo, H.; Wang, D.; Guo, H.; Zhu, K. An Ensemble Model Based on Machine Learning Methods and Data Preprocessing for Short-Term Electric Load Forecasting. *Energies* **2017**, *10*, 1186. [CrossRef]
17. Niu, D.; Wang, Y.; Wu, D.D. Power load forecasting using support vector machine and ant colony optimization. *Expert Syst. Appl.* **2010**, *37*, 2531–2539. [CrossRef]
18. Zhang, X.; Wang, J.; Zhang, K. Short-term electric load forecasting based on singular spectrum analysis and support vector machine optimized by Cuckoo search algorithm. *Electr. Power Syst. Res.* **2017**, *146*, 270–285. [CrossRef]
19. Zhang, J.; Wei, Y.M.; Li, D.; Tan, Z.; Zhou, J. Short term electricity load forecasting using a hybrid model. *Energy* **2018**, *158*, 774–781. [CrossRef]
20. Li, Y.; Su, Y.; Shu, L. An ARMAX model for forecasting the power output of a grid connected photovoltaic system. *Renew. Energy* **2014**, *66*, 78–89. [CrossRef]
21. Li, Y.; He, Y.; Su, Y.; Shu, L. Forecasting the daily power output of a grid-connected photovoltaic system based on multivariate adaptive regression splines. *Appl. Energy* **2016**, *180*, 392–401. [CrossRef]

22. Abdullah, N.A.; Abd Rahim, N.; Gan, C.K.; Nor Adzman, N. Forecasting Solar Power Using Hybrid Firefly and Particle Swarm Optimization (HFPSO) for Optimizing the Parameters in a Wavelet Transform-Adaptive Neuro Fuzzy Inference System (WT-ANFIS). *Appl. Sci.* **2019**, *9*, 3214. [CrossRef]

23. Chu, Y.; Urquhart, B.; Gohari, S.M.I.; Pedro, H.T.C.; Kleissl, J.; Coimbra, C.F.M. Short-term reforecasting of power output from a 48 MWe solar PV plant. *Sol. Energy* **2015**, *112*, 68–77. [CrossRef]

24. Galicia, A.; Talavera-Llames, R.; Troncoso, A.; Koprinska, I.; Martínez-Álvarez, F. Multi-step forecasting for big data time series based on ensemble learning. *Knowl. Based Syst.* **2019**, *163*, 830–841. [CrossRef]

25. Antonanzas, J.; Pozo-Vázquez, D.; Fernandez-Jimenez, L.A.; Martinez-de-Pison, F.J. The value of day-ahead forecasting for photovoltaics in the Spanish electricity market. *Sol. Energy* **2017**, *158*, 140–146. [CrossRef]

26. Ferlito, S.; Adinolfi, G.; Graditi, G. Comparative analysis of data-driven methods online and offline trained to the forecasting of grid-connected photovoltaic plant production. *Appl. Energy* **2017**, *205*, 116–129. [CrossRef]

27. Australian Energy Market Commission Integrating Distributed Energy Resources for the Grid of the Future. Economic Regulatory Framework Review. Available online: https://www.aemc.gov.au/sites/default/files/2019-09/Finalreport-ENERFR2019-EPR0068.PDF (accessed on 6 November 2019).

28. Sánchez Jiménez, M. Regulatory Proposal for deployment of flexibility. In Proceedings of the India SMART GRIDS Week, New Delhi, India, 8–10 March 2017; p. 15.

29. EURELECTRIC Designing Fair and Equitable Market Rules for Demand Response Aggregation. Available online: http://www.eurelectric.org (accessed on 4 September 2019).

30. Bemdt, D.J.; Clifford, J. Using Dynamic Time Warping to find patterns in time series. In Proceedings of the KDD Workshop, Seattle, WA, USA, 31 July–1 August 1994; pp. 359–370.

31. Smart Energy Europe. SmartEn White Paper: A Vision for Smart and Active Buildings. Available online: https://www.smarten.eu/wp-content/uploads/2019/07/FINAL-smartEn-white-paper-Smart-Buildings.pdf (accessed on 6 November 2019).

32. Zeifman, M.; Roth, K. Nonintrusive appliance load monitoring: Review and outlook. *IEEE Trans. Consum. Electron.* **2011**, *57*, 76–84. [CrossRef]

33. Federal Energy Regulatory Commission (FERC). Assessment of Demand Response and Advanced Metering: Staff Report. Available online: https://www.ferc.gov/legal/staff-reports/2016/DR-AM-Report2016.pdf (accessed on 14 June 2019).

34. Gabaldón, A.; Molina, R.; Marín-Parra, A.; Valero-Verdú, S.; Álvarez, C. Residential end-uses disaggregation and demand response evaluation using integral transforms. *J. Mod. Power Syst. Clean Energy* **2017**, *5*, 91–104. [CrossRef]

35. Jenssen, Å.; Borsche, T.; Wolst, J. *Data Exchange in Electric Power Systems: European State of Play and Perspectives*; THEMA Consulting: Oslo, Norway, 2017; ISBN 9788283680133.

36. Residential Energy Consumption Survey (RECS)—Data—U.S. Energy Information Administration (EIA). Available online: https://www.eia.gov/consumption/residential/data/2015/ (accessed on 6 November 2019).

37. Bertoldi, P.; López Lorente, J.; Labanca, N. *Energy Consumption and Energy Efficiency Trends in the EU-28 2000–2014*; Publication Office of the European Commission: Luxembourg, 2016.

38. IDAE Consumo por usos y Energías del Sector Residencial (2010–2017). Available online: https://www.idae.es/estudios-informes-y-estadisticas (accessed on 6 November 2019).

39. García-Garre, A.; Gabaldón, A.; Álvarez-Bel, C.; Ruiz-Abellón, M.; Guillamón, A. Integration of Demand Response and Photovoltaic Resources in Residential Segments. *Sustainability* **2018**, *10*, 3030. [CrossRef]

40. Ellegård, K.; Palm, J. Visualizing energy consumption activities as a tool for making everyday life more sustainable. *Appl. Energy* **2011**, *88*, 1920–1926. [CrossRef]

41. Del Carmen Ruiz-Abellón, M.; Gabaldón, A.; Guillamón, A. Load forecasting for a campus university using ensemble methods based on regression trees. *Energies* **2018**, *11*, 2038. [CrossRef]

42. Caret R Package. Available online: https://cran.r-project.org/web/packages/caret/caret.pdf (accessed on 6 November 2019).

43. Ben Taieb, S.; Hyndman, R.J. A gradient boosting approach to the Kaggle load forecasting competition. *Int. J. Forecast.* **2014**, *30*, 382–394. [CrossRef]

44. Wang, J.; Li, P.; Ran, R.; Che, Y.; Zhou, Y. A Short-Term Photovoltaic Power Prediction Model Based on the Gradient Boost Decision Tree. *Appl. Sci.* **2018**, *8*, 689. [CrossRef]

45. Friedman, J.H. Stochastic gradient boosting. *Comput. Stat. Data Anal.* **2002**, *38*, 367–378. [CrossRef]

46. Warren Liao, T. Clustering of time series data—A survey. *Pattern Recognit.* **2005**, *38*, 1857–1874. [CrossRef]

47. Aghabozorgi, S.; Seyed Shirkhorshidi, A.; Ying Wah, T. Time-series clustering—A decade review. *Inf. Syst.* **2015**, *53*, 16–38. [CrossRef]

48. Montero, P.; Vilar, J.A. TSclust: An R package for time series clustering. *J. Stat. Softw.* **2014**, *62*, 1–43. [CrossRef]

49. Dau, H.A.; Silva, D.F.; Petitjean, F.; Forestier, G.; Bagnall, A.; Mueen, A.; Keogh, E. Optimizing dynamic time warping's window width for time series data mining applications. *Data Min. Knowl. Discov.* **2018**, *32*, 1074–1120. [CrossRef]

50. Wilson, E.; Christensen, C. Heat Pump Water Heater Modeling in EnergyPlus. In Proceedings of the Building America Residential Energy Efficiency Stakeholder Meeting, Austin, TX, USA, 29 February–1 March 2012; pp. 1–36.

51. Li, X.; Wen, J. Review of building energy modeling for control and operation. *Renew. Sustain. Energy Rev.* **2014**, *37*, 517–537. [CrossRef]

52. Gabaldón, A.; Álvarez, C.; Ruiz-Abellón, M.; Guillamón, A.; Valero-Verdú, S.; Molina, R.; García-Garre, A. Integration of Methodologies for the Evaluation of Offer Curves in Energy and Capacity Markets through Energy Efficiency and Demand Response. *Sustainability* **2018**, *10*, 483. [CrossRef]

53. Demand Response (DR) Web Page. Available online: http://www.demandresponse.eu/ (accessed on 14 June 2019).

54. NYISO Emergency Demand Response Program Manual. Available online: https://www.nyiso.com/demand-response (accessed on 4 September 2019).

55. NYISO Day-Ahead Demand Response Program Manual. Available online: http://online.fliphtml5.com/qzli/zqaf/#p=1 (accessed on 6 November 2019).

56. PROLAN Ripple Control Receiver and Tone Frequency Receiver (HKV-RKV). Available online: https://www.prolan.hu/en/solutions/HKV-RKV (accessed on 6 November 2019).

57. TenneT und Bayernwerk: Dezentrale Flexibilität aus Bayern für die Energiewende—TenneT. Available online: https://www.tennet.eu/de/news/news/tennet-und-bayernwerk-dezentrale-flexibilitaet-aus-bayern-fuer-die-energiewende/ (accessed on 6 November 2019).

58. Wireless Smart Home and Home Automation | FIBARO. Available online: https://www.fibaro.com/en/ (accessed on 6 November 2019).

59. Symcom. IP-Symcon Integrators. Automation Software. Available online: https://www.symcon.de/en/product/integrators/ (accessed on 6 November 2019).

60. Samad, T.; Koch, E.; Stluka, P. Automated Demand Response for Smart Buildings and Microgrids: The State of the Practice and Research Challenges. *Proc. IEEE* **2016**, *104*, 726–744. [CrossRef]

61. OpenADR Alliance. Available online: https://www.openadr.org/ (accessed on 2 December 2019).

62. Universal Devices Web Page. Available online: https://www.universal-devices.com/ (accessed on 2 December 2019).

63. Cui, T.; Carr, J.; Brissette, A.; Ragaini, E. Connecting the Last Mile: Demand Response in Smart Buildings. In Proceedings of the 8th International Conference on Sustainability in Energy and Buildings, SEB-16, Turin, Italy, 11–13 September 2016; Volume 111, pp. 720–729.

64. Gabaldón, A.; Álvarez, C.; Moreno, J.I.; Matanza, J.; López, G.; Ruiz-Abellón, M.C.; Valero-Verdu, S. Evaluation of the performance of Aggregated Demand Response by the use of Load and Communication Technologies Models. In Proceedings of the EEDAL Conference, Irvine, CA, USA, 13–15 September 2017.

65. Monitoring Analytics LLC. 2019 Quarterly State of the Market Report for PJM: January through September. Available online: https://www.monitoringanalytics.com/reports/PJM_State_of_the_Market/2019/2019q3-som-pjm.pdf (accessed on 2 December 2019).

66. Steffes, P. The Path to Grid-Interactive Water Heating (GIWH), Opportunities & Challenges. Available online: https://www.peakload.org/assets/36thConf/PLMASteffesPresentation11-13-17.pdf (accessed on 6 November 2019).

67. Lueken, R.; Hledik, R.; Chang, J. The Hidden Battery. Opportunities in Electric Water Heating. In *Proceedings of the PLMA Grid Interactive Behind the Meter Storage Interest Group Meeting*; The Brattle Group; NRECA; NRDC; PLMA: Boston, MA, USA, 2015.

68. Skamarock, W.C.; Klemp, J.B.; Dudhia, J.; Gill, D.O.; Barker, D.M.; Duda, M.G.; Huang, X.-Y.; Wang, W.; Powers, J.G. A Description of the Advanced Research WRF Version 3 (No. NCAR/TN-475+STR). Available online: https://opensky.ucar.edu/islandora/object/technotes%3A500/datastream/PDF/view (accessed on 6 November 2019).

69. Erbs, D.G.; Klein, S.A.; Duffie, J.A. Estimation of the diffuse radiation fraction for hourly, daily and monthly-average global radiation. *Sol. Energy* **1982**, *28*, 293–302. [CrossRef]

70. Lave, M.; Hayes, W.; Pohl, A.; Hansen, C.W. Evaluation of global horizontal irradiance to plane-of-array irradiance models at locations across the United States. *IEEE J. Photovolt.* **2015**, *5*, 597–606. [CrossRef]

71. Sanz-Garcia, A.; Fernandez-Ceniceros, J.; Antonanzas-Torres, F.; Pernia-Espinoza, A.V.; Martinez-De-Pison, F.J. GA-PARSIMONY: A GA-SVR approach with feature selection and parameter optimization to obtain parsimonious solutions for predicting temperature settings in a continuous annealing furnace. *Appl. Soft Comput. J.* **2015**, *35*, 13–28. [CrossRef]

72. 'GAparsimony' R Package. Available online: https://cran.r-project.org/web/packages/GAparsimony/GAparsimony.pdf (accessed on 13 November 2019).

73. Lake, C. PJM Empirical Analysis of Demand Response Baseline Methods. Available online: https://www.pjm.com/-/media/markets-ops/demand-response/pjm-analysis-of-dr-baseline-methods-full-report.ashx?la=en (accessed on 4 September 2019).

74. Gabaldon, A.; Valero-Verdu, S.; Garcia-Garre, A.; Senabre, C.; Alvarez-Bel, C.; Lopez, M.; Penalvo, E.; Sanchez, E.P. A physically-based model of heat pump water heaters for demand respose policies: Evaluation and testing. In Proceedings of the 2018 International Conference on Smart Energy Systems and Technologies, Sevilla, Spain, 10–12 September 2018; pp. 1–6.

Article

Deep Learning-Based Short-Term Load Forecasting for Supporting Demand Response Program in Hybrid Energy System

Sholeh Hadi Pramono *, Mahdin Rohmatillah *, Eka Maulana, Rini Nur Hasanah and Fakhriy Hario

Department of Elctrical Engineering, University of Brawijaya, Veteran Road, Lowokwaru, Malang 65113, Indonesia

* Correspondence: sholehpramono@ub.ac.id (S.H.P.); rohmatillahmahdin1994@gmail.com (M.R.); Tel.: +62341-554166 (S.H.P.)

Received: 8 August 2019; Accepted: 28 August 2019; Published: 30 August 2019

Abstract: A novel method for short-term load forecasting (STLF) is proposed in this paper. The method utilizes both long and short data sequences which are fed to a wavenet based model that employs dilated causal residual convolutional neural network (CNN) and long short-term memory (LSTM) layer respectively to hourly forecast future load demand. This model is aimed to support the demand response program in hybrid energy systems, especially systems using renewable and fossil sources. In order to prove the generality of our model, two different datasets are used which are the ENTSO-E (European Network of Transmission System Operators for Electricity) dataset and ISO-NE (Independent System Operator New England) dataset. Moreover, two different ways of model testing are conducted. The first is testing with the dataset having identical distribution with validation data, while the second is testing with data having unknown distribution. The result shows that our proposed model outperforms other deep learning-based model in terms of root mean squared error (RMSE), mean absolute error (MAE), and mean absolute percentage error (MAPE). In detail, our model achieves RMSE, MAE, and MAPE equal to 203.23, 142.23, and 2.02 for the ENTSO-E testing dataset 1 and 292.07, 196.95 and 3.1 for ENTSO-E dataset 2. Meanwhile, in the ISO-NE dataset, the RMSE, MAE, and MAPE equal to 85.12, 58.96, and 0.4 for ISO-NE testing dataset 1 and 85.31, 62.23, and 0.46 for ISO-NE dataset 2.

Keywords: short-term load forecasting; deep learning; wavenet; long short-term memory; demand response; hybrid energy system

---

1. Introduction

Nowadays, the hybrid energy system has become more popular in the electricity industry. The main reason of this trend is the exponential reduction of energy storage cost and the development of digital connection, enabling real time monitoring and smarter grid establishment. Moreover, the hybrid energy system is considered as one of the best solutions in tackling intermittency experienced by most renewable energy schemes including solar- and wind-based energy. For example, in solar photovoltaic, the energy is only delivered when obtaining sufficient solar irradiance. As a consequence, a lot of research has been conducted in order to provide the best scheme of this hybrid system [1].

Among provided hybrid energy system schemes, the most possible way to be implemented in the majority of countries is the integration between renewable and fossil energy, because fossil energy has already well established. In case of sustainability, this scheme is very good, because fossil energy is obviously able to supply adequate power to the grid when the alternative sources cannot handle users' load demand. The major concern of this scheme is the cost to be borne by users once the renewable sources cannot supply adequate power to the grid in which they will pay an expensive fossil-based electric price. Moreover, fossil energy tends to gradually increase leading to economic conflict in society [2]. Therefore, in order to tackle this issue, an appropriate demand response scheme can be applied.

Demand response is the change of electric usage by users due to change of electric price or maybe an incentive as a reward of lowering their power consumption [3]. Applying demand response to this hybrid system is very beneficial for shaving peak load demand [4,5], leading to the reduction of fossil energy consumption. Moreover, it can provide short-term impact and economic benefit for both consumer and utility.

In order to support this demand response, short-term load forecasting (STLF) is very important for predicting whether the energy storage from renewable sources is able to handle the forthcoming power consumption or not. If the prediction states that the storage is not adequate to support the future load, then the electricity utility can announce this situation to the users, which eventually triggers them to reduce their electric usage, because users do not only want to pay more for conventional energy source but also want to get incentives from the authorities.

Fortunately, with the help of developed infrastructures like smart meters equipped with a lot of sensors and the Internet of Things (IoT), a robust STLF method is feasible to be implemented. Broadly speaking, research in load forecasting can be categorized into two research classes, traditional and advanced model. Traditional model uses simple statistics method for example regression models [6] and Kalman filtering model [7]. Nevertheless, among proposed traditional models, autoregressive integrated moving average (ARIMA) and generalized autoregressive conditional heteroscedascity (GARCH) are two of the most popular techniques in regression function that were used in several precedent research studies [8,9]. Unfortunately, these traditional models only provide good accuracy if the electrical load and other parameters have a linear relationship. Meanwhile, the advanced model is a data-driven model implementing the machine learning technique for instance support vector machine (SVM) [10], K-nearest neighbor (KNN) [11], and others [12–14].

However, based on the recent publications [15–18], the deep learning-based methods show the most convincing performance by outperforming other machine learning-based solutions. The main reason of the deep learning superiority is first, deep learning does not highly rely on feature engineering and the hyperparameters tuning is relatively easier compared to other data-driven models. The second is the availability of huge datasets, where deep learning can precisely map the inputs to the certain output by making complex relations among layers in the network based on that huge training data. Moreover, since the availability of Graphics Processing Unit (GPU) parallel computation and methods providing weights sharing like convolutional neural network (CNN) [19], the computational speed of deep learning models become significantly faster.

Because of the superiority of the deep learning, this research proposes a method in load forecasting task, specifically STLF, to predict the hourly power consumption by using deep learning algorithms which is the combination of the advanced version of the convolutional neural network (CNN), dilated causal residual CNN, and long short-term memory (LSTM) [20].Dilated causal residual CNN is inspired by the Wavenet model [21], which is very famous for audio generation, and the residual network [22] with gated activation function. This model will learn the trend based on long sequence input while the LSTM layer works as a model's output self-correction which relates the output of the wavenet-based model with the recent load demand trend (short sequence).

The main contribution of this research is that we propose a novel model utilizing a combination of dilated causal residual CNN and LSTM utilizing long and short sequential data and fine tuning technique. External feature extraction or feature selection data are not included in this research. Moreover, this research only takes into account time index information as the external factor data, making it easy to be compared as a benchmark model for future research.

In order to prove the generality of our proposed model performance, two different scenarios of model testing are conducted. The first scenario is using the testing dataset having identical distribution with the validation dataset, while the second is using dataset having unknown distribution. As a comparison, our proposed model results are compared with the performance of the model from [15,16] and the standard wavenet [21]. The simulation result shows that our proposed model outperforms other deep learning models in root mean squared error (RMSE), mean absolute error (MAE), and mean absolute percentage error (MAPE).

Therefore, due to the accuracy of our model performance, this model can be used for supporting utilities in applying demand response program since it can help the utilities to obtain accurate prediction about the future load demand that eventually providing precise information to the users whether the future load demand can be supplied by the renewable source or not.

The rest of the paper is organized as follows. Section 2 describes the dataset used in detail including preliminary data analysis and data preprocessing. Section 3 explains the model architecture and its parameter, training, and testing stage. Section 4 provides information about results obtained by using the proposed models compared with other deep learning-based models and clear explanation about the reason why the proposed model can achieve the result. Lastly, Section 5 summarize the findings discussed in this paper and also possible future works.

2. Dataset

In order to prove the generality of our proposed method, two datasets are used as the model's input which are the ENTSO-E (European Network of Transmission System Operators for Electricity) [23] and ISO-NE (Independent System Operator New England) dataset [24]. The ENTSO-E dataset is the dataset obtained from load demand in every country in Europe. In this research we only take into account data gathered from Switzerland. Meanwhile, the ISO-NE dataset is data of hourly load demand in New England.

Those datasets have different kinds of characteristics, especially in the case of load demand range and complexity. In the ENTSO-E dataset, the lowest and highest value of load are 1483 KWh and 18,544 KWh, respectively, while in the ISO-NE dataset they are 9018 KWh and 26,416 KWh, respectively. Another different characteristic is the fluctuation trend in a single day load demand. In the ENTSO-E dataset, the load demand is more oscillated compared to the ISO-NE dataset. As proof, Figures 1 and 2 show the average daily power consumption and example of load trend in a day based on those datasets and example of demand trend of each dataset.

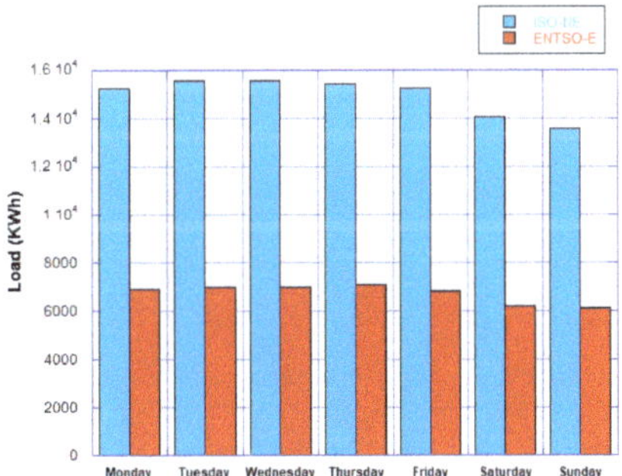

Figure 1. Average daily power consumption of each dataset.

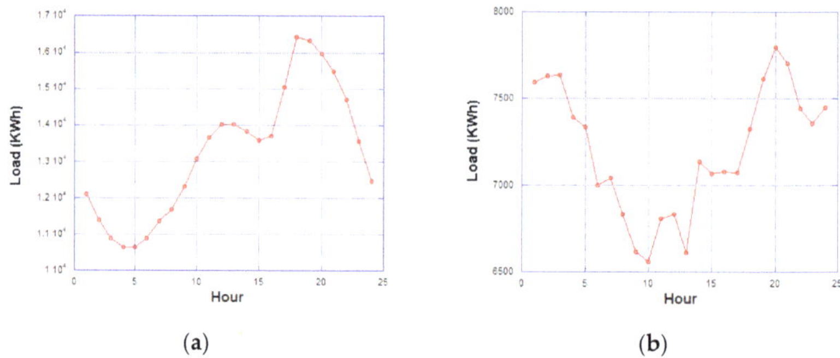

(a)   (b)

Figure 2. Example of one-day load demand of each dataset (**a**) ISO-NE dataset; (**b**) ENTSO-E dataset.

In building a solution for load forecasting, deep understanding of the load demand trend is very necessary. Figure 3 shows data in both datasets over a year period. From Figure 3, broadly speaking, the trend of load demand has a periodicity that will be repeated over the next weeks. However, this trend is highly affected by a lot of external factors causing a fluctuation over a certain period, for example, the economic, weather, and time index. Unfortunately, obtaining those external factors are difficult, the easiest external data that can be gathered is time index data containing information about the date and clock. Therefore, in this research we not only fed the model by load demand trend but also time index data represented by one-hot vector. One-hot vector is a sparse vector that maps a feature with M categories into a vector which has M elements where only the corresponding element is one while the remaining are zero.

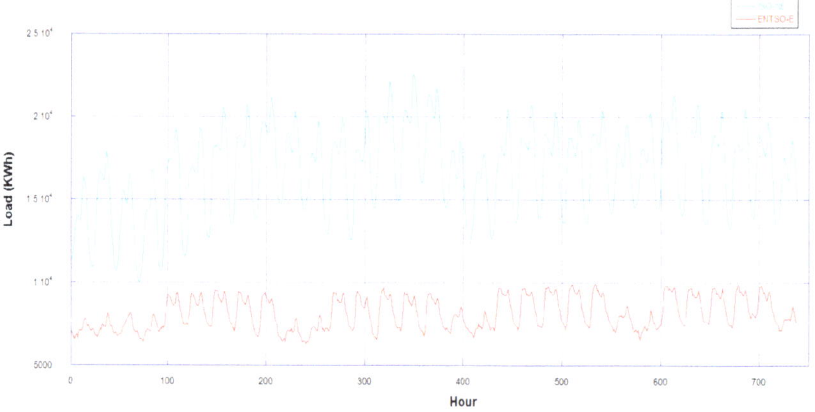

Figure 3. Load demand trend over a year.

Before fed to the model, the datasets must be pre-processed, which consists of checking for null values, splitting dataset into training, validation and testing dataset, and eventually data normalization. The data used for this research is limited only to data taken from 1 January 2015 until 30 May 2017 for the ENTSO-E dataset and from 1 January 2004 until 30 May 2006 for the ISO-NE dataset. The first two years of data are used for the training stage, while the rest of the 3600 data are split into 3000 and 600 data. The first 3000 data are randomly taken for validation and testing data with proportion of nearly 0.65 and 0.35 to be used for validation and the testing stage, respectively. In other words, the total of validation data and first testing data are 1900 and 1100, respectively, while another 600 data are used for the second testing data. We conduct two testing stage, because the first testing data have a nearly identical distribution with the validation data which clearly make the model provide good accuracy in the testing stage. Meanwhile, the second testing data is clearly new data that their distribution is never experienced by the model both in the training and validation stage. This kind of testing process is appropriate for proving the generality of the model.

In the data normalization process, min-max scaling method as expressed in Equation (1) is implemented.

$$z_i = \frac{x_i - \min(x)}{\max(x) - \min(x)} \tag{1}$$

$x = x_1, \ldots, x_n$ and z_i is the ith normalized data. The parameters in the normalization process must come from training dataset only, because we assume that the future data (validation and test set) have different distribution with the training dataset.

3. Model Design

In this research, the cores of the proposed model are wavenet architecture implementing both of the dilated causal CNN and residual network and LSTM, which have proven very well in time-series data prediction.

Our proposed model consists of two stages that have a function to learn long and short sequence data. Inspired by the success of wavenet architecture and LSTM in handling time series data, the long sequence taken from the 32 time steps before the target is learned using wavenet while the short sequence taken from 4 time steps before target is learned using LSTM.

3.1. Wavenet

Wavenet consists of deep generative models utilizing the dilated causal convolutional neural network of audio waveform. Causal convolution means that the output of the recent time step is only affected by the previous time step. Meanwhile, the dilated convolutional neural network is a modified convolutional neural network where the filter weight alignment is expanded by a dilation factor that eventually results in a broader receptive field that can be expressed as follows:

$$(F *_l k)(p) = \sum_{s+lt=p} F(s)k(t) \tag{2}$$

while the standard convolution is expressed as follows:

$$(F * k)(p) = \sum_{s+t=p} F(s)k(t) \tag{3}$$

The dilated convolution is denoted by $*_l$ notation. The difference between dilated convolution with standard convolution is the notation l representing the dilation factor which causes the filter to skip one or several points during the convolution process.

Figure 4 shows the dilated convolution applied in one dimensional data. The blue, white, and orange circle are input data, hidden layer output, and output layer output, respectively. There are 32 input data taken from $t = 1$ until $t = 32$ that are convoluted with filter with the size of two. The dilation rate is increased by one in every hidden layer that causes a broader receptive field. This dilation rate is repeated twice. In the output layer, only the last value is taken, which we assume represents the feature of load at $t = 33$.

In order to optimize the usage of the dilated convolutional neural network, the residual technique [22] is applied to the model. The implementation of the residual network will take into account lower levels outputs which have features that will help in predicting the future power demand, especially in the case of a network implementing a sparse filter which has the potential to lose several information from the previous layers.

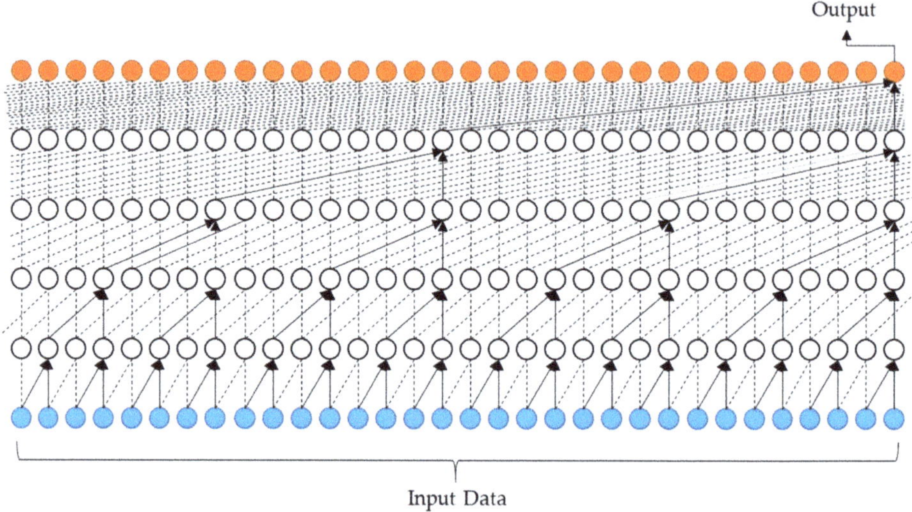

Figure 4. Dilated causal convolutional neural network with filter size 2.

The residual model is a famous way to build the deep neural network that was firstly proposed for the image recognition task. Using this way, instead of only mapping input data x to a function $H(x)$ that outputs $\hat{y}$, the mapping scenario from the previous residual block $f(x, \{W_i\})$ where W_i is the learned weights and biases from the residual block i is also considered. Therefore, the output of the residual block can be expressed as:

$$H(x) = f(x, \{W_i\}) + x \tag{4}$$

Moreover, since we use the stacked residual block, then the output of the residual block can be represented as:

$$x_K = x_0 + \sum_{i=1}^{K} f(x_{i-1}, W_{i-1}) \tag{5}$$

x_K is the output of residual block K, x_0 is the input of the residual network and $f(x_{i-1}, W_{i-1})$ is the output and associated weight of the previous residual blocks. As a result of several summation between the previous and final residual block, then the back propagation of the network to x_0 can be calculated using the following equation:

$$\frac{\partial \mathcal{L}}{\partial x_0} = \frac{\partial \mathcal{L}}{\partial x_K} \frac{\partial x_K}{\partial x_0} = \frac{\partial \mathcal{L}}{\partial x_K} \left(1 + \frac{\partial}{\partial x_0} \sum_{i=1}^{K} f(x_{i-1}, W_{i-1}) \right) \tag{6}$$

$\mathcal{L}$ is the total loss of the network and constant 1 indicates that the gradient of the network output can be directly back-propagated without considering layers' parameters (weights and biases). This formulation ensures the layers do not suffer of vanishing gradient, even the weights are small. Figure 5 shows the basic residual learning process.

Moreover, skip connection and gated activation are applied to the network for speeding up the convergence and avoiding overfitting. The process of residual and skip connection with gated activation is shown in Figure 6.

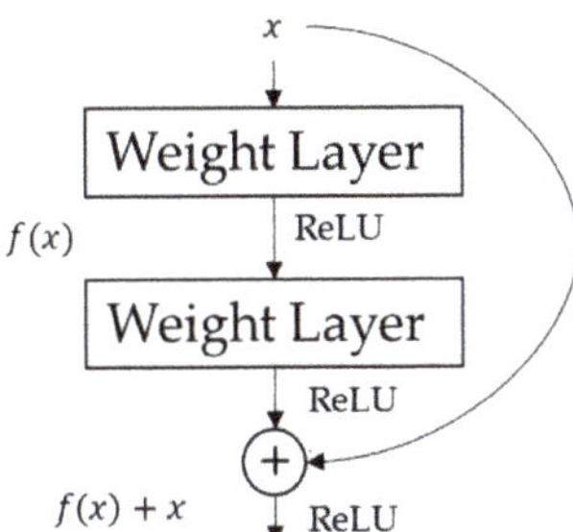

Figure 5. Residual learning process.

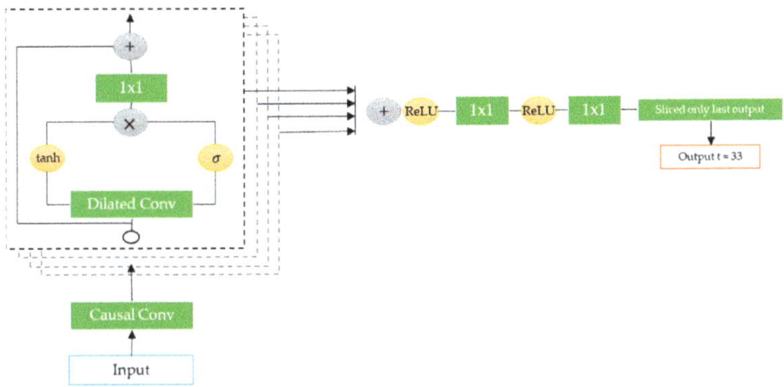

Figure 6. Overview of residual dilated convolutional block and gated activation function.

The gated activations are inspired by the LSTM layer where tanh and sigmoid (σ) work as learned filter and learned gate, respectively. The use of gated activations has been proved to work better compared to using ReLU activation in time series data [21]. The output of dilated convolution with gated activations can be expressed as:

$$z = \tanh\left(w_f * x\right) \odot \sigma\left(w_g * x\right) \tag{7}$$

where w_f and w_g are learned filter and learned gate, respectively.

3.2. LSTM

In the case of forecasting future data, the knowledge of the recent trend is very essential. As an illustration, in predicting future data, we mostly start to figure out a long sequence trend. After we have already known the pattern of the trend based on the long sequence of previous data, then in order to provide better forecast, we also try to relate our understanding of long sequence data with the recent trend. The same concept is applied to our proposed method. We fine tune the wavenet-based model with one LSTM layer assigned to help the network to relate the output of dilated CNN with the recent trend. This step can also be considered as a correction step of the dilated CNN output, as we assert a fix input to be concatenated with dilated CNN output which are then fed to the LSTM layer that also work as output layer.

Brief Explanation of LSTM

LSTM is the developed version of the standard recurrent neural network (RNN) where instead of only having a recurrent unit, LSTM has "LSTM cells" that have an internal recurrence consisting of several gating units controlling flow of information [25]. Comparison between the simple RNN and LSTM layer using tanh as the activation function is depicted in Figure 7.

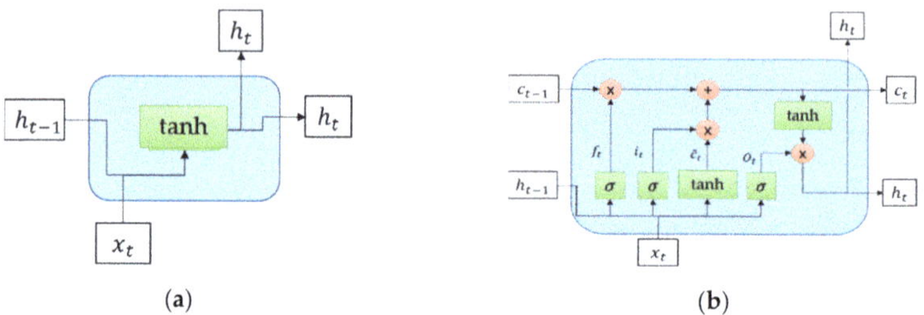

Figure 7. Comparison between simple recurrent neural network (RNN) and long short-term memory (LSTM) layer: (**a**) simple RNN layer, (**b**) LSTM layer.

From Figure 7, the difference between the simple RNN and LSTM layer is clear. The LSTM layer is more complex than the simple RNN, because LSTM not only takes into account input (x) and hidden state (h) at a certain time step, but also LSTM cells (c) that will replace the hidden state to prevent older signals from vanishing during the process. Three control gates ruling the LSTM cells, forget gate, input gate, and output gate are represented by f_t, i_t, O_t, respectively. Those gates use sigmoid activation function having an output range between 0 and 1 represented by σ. Meanwhile, $\widetilde{c}_t$ is the input node that works identical to the simple RNN layer.

Mathematically explained, forget gate and input gate, respectively can be expressed as:

$$i_t = \sigma(W_i \times [h_{t-1}, x_t] + b_i) \tag{8}$$

$$f_t = \sigma\left(W_f \times [h_{t-1}, x_t] + b_f\right) \tag{9}$$

$[h_{t-1}, x_t]$ is the concatenation between input and hidden state value, while W and b are weight matrices, respectively. On the other hand, the cell state is updated with the formulation:

$$c_t = f_t \times c_{t-1} + i_t \times \widetilde{c}_t \tag{10}$$

where $\widetilde{c}_t$ is expressed as:

$$\widetilde{c}_t = \tanh(W_c \times [h_{t-1}, x_t] + b_c) \tag{11}$$

The last, the output gate o_t and hidden state h_t is calculated by using the following equation:

$$o_t = \sigma(W_o \times [h_{t-1}, x_t] + b_o) \tag{12}$$

$$h_t = o_t \times \tanh(c_t) \tag{13}$$

3.3. Detailed Model Setup

Complete representation of our model is depicted in Figure 8. The wavenet-based model is fed by 32 data sequences containing information of load demand and time index information (clock, day, and month). This wavenet model is used for the initial forecasting algorithm based on long sequence data. Before fed to dilated CNN, long sequence data are prepossessed using standard 1D-CNN with filter size equal to one. Next, the prepossessed data is convoluted by dilated causal CNN with filter size of 2. All of the convolutional layers have ReLU activation function expressed as:

$$f(x) = \max(0, x) \tag{14}$$

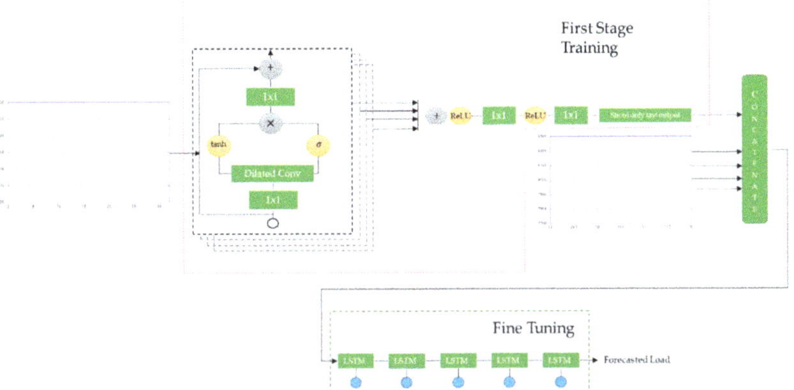

Figure 8. Proposed model architecture.

In the case of the residual network, the total residual block in our model is 10 with a dilation rate set to be a repetition of a sequence dr = [1, 2, 4, 8, 16]. Dilation rate is a hyperparameter that represents how large the gap between the element of the filter is, as shown in Figure 4 and indicated by the arrows. All of the residual blocks are summed followed by the ReLU activation function. The post-processing is conducted before the last convolutional layer which works similar with time distributed fully connected layer in which it is assigned for normalization of the residual output.

Because the length of input and output in the dilated causal CNN are identical, then the customize layer is built for taking only the last output's neuron representing load at t = 33. After completing the wavenet-based network training, the output of the last convolutional layer is concatenated with recent data sequences (t = 29 until t = 32) that work as LSTM input. Here, by using the fine tuning technique, LSTM with linear activation function is assigned to make self-correction of convolutional output in order to provide more accurate prediction based on short sequence. This self-correction method is nearly identical to how humans make a prediction based on data sequences. For example, in predicting the environment temperature, humans will relate the understanding between the temperature trend from the previous day with the recent temperature trend in order to make an accurate prediction for the next hour temperature.

Table 1 shows the summary of our model's parameter where f, ks, s, and dr are the number of filters, kernel size, stride, and dilation rate, respectively. Loss function and optimizer are mean absolute error and adaptive and momentum (ADAM) [26], respectively, and batch size is 512. The model was trained using Nvidia GTX 1070, Tensorflow 1.13.1 [27], CUDA 10, and CuDNN 7.6.2 with 500 epochs and the final model is chosen based on the validation accuracy.

Table 1. Forecasting result obtained using dataset 1.

Layer	Parameters
Input Layer (1)	32, 44
Conv1D (1)	$[f, ks, s]$ = [16, 1, 1]
Dilated Causal Conv1D	$[f, ks, s, dr]$ = [32, 2, 1, dr]
Conv1D (2)	$[f, ks, s]$ = [16, 1, 1]
Conv1D (3)	$[f, ks, s]$ = [128, 1, 1]
Conv1D (4)	$[f, ks, s]$ = [1, 1, 1]
Input Layer (2)	5, 1
LSTM	1 output node

4. Result and Discussion

4.1. Model Performance Evaluation

In order to evaluate our model performance, we mainly compared this model with the two previous deep learning-based models that work very well in the case of STLF. Those models are inspired by [15] (Model 1), which utilizes stacked LSTM and [16] (Model 2) which combines the stacked CNN and LSTM layer with the feature fusion layer. The configuration of each comparison model is identical to the published papers. Model 1 consists of two stacked LSTM layers consisting of 20 units followed by fully connected layers as an output layer. Model 2 is built with a combination between the LSTM and CNN layer where their outputs are concatenated, which is eventually fed to the fully connected layer called a feature fusion layer. All of the models are trained with the same input data. Moreover, the original wavenet model is also used for a model comparison in order to prove the benefit of LSTM as a self-correction layer in our proposed model.

This section reports on the performance of hourly load forecasting by using our proposed method compared to other forecasting methods. In the testing stage, all of the models are evaluated with three difference commonly used metrics, root mean squared error (RMSE), mean absolute error (MAE), and mean absolute percentage error (MAPE). MAE is the average of absolute difference values between predicted load and actual load consumption. MAPE is just identical to MAP but it uses a ratio between the difference with the actual load while RMSE is another metric that tends to have a higher value compared to other metrics. The higher value which results from the metrics, the worse performance of the model. Those metrics are defined as follow:

$$\text{RMSE} = \sqrt{\frac{1}{N} \sum_{i=1}^{N} (\hat{y}_i - y_i)^2} \tag{15}$$

$$\text{MAE} = \frac{1}{N} \sum_{i=1}^{N} |\hat{y}_i - y_i| \tag{16}$$

$$\text{MAPE} = \frac{1}{N} \sum_{i=1}^{N} \left| \frac{\hat{y}_i - y_i}{y_i} \right| \tag{17}$$

4.2. ENTSO-E Load Prediction

Tables 2 and 3 show all of the models' performance on dataset 1 and dataset 2, respectively. Overall, our proposed model outperforms other models, especially in the case of dataset 1 where the data distribution is nearly identical with the validation data, while the worst performance is shown by [16]-based model. In Table 2, our proposed model clearly shows its excellence over other methods which shows the success of the combination between deep residual causal CNN and LSTM using fine tuning technique in understanding long sequence and short sequence data, respectively. Moreover, compared to the standard wavenet model, our model performs better accuracy indicating the usefulness of the LSTM layer in making self-correction for initial load forecasting output by dilated causal residual CNN.

Table 2. Forecasting result obtained using European Network of Transmission System Operators for Electricity (ENTSO-E) testing dataset 1.

Model	RMSE	MAE	MAPE (%)
Tian et al.	240.57	171.71	2.45
Kong et al.	222.44	155.83	2.22
Wavenet	217.98	157.28	2.24
Our Model	203.23	142.23	2.02

Table 3. Forecasting result obtained European Network of Transmission System Operators for Electricity (ENTSO-E) testing dataset 2.

Model	RMSE	MAE	MAPE (%)
Tian et al.	306.77	216.19	3.42
Kong et al.	304.07	209.22	3.29
Wavenet	305.04	212.99	3.36
Our Model	292.07	196.95	3.1

However, all of models exhibit downgraded performance in dataset 2. It indicates that all of the model still cannot understand data which has slightly different distribution with training, validation, and testing data 1. The failure of all of the models in testing using dataset 2 is highly affected by the quality of the ENTSO-E dataset where the inconsistency of hourly power usage or unpredicted trends occur several times. Figures 9 and 10 show the result of STLF on dataset 1 and dataset 2, respectively.

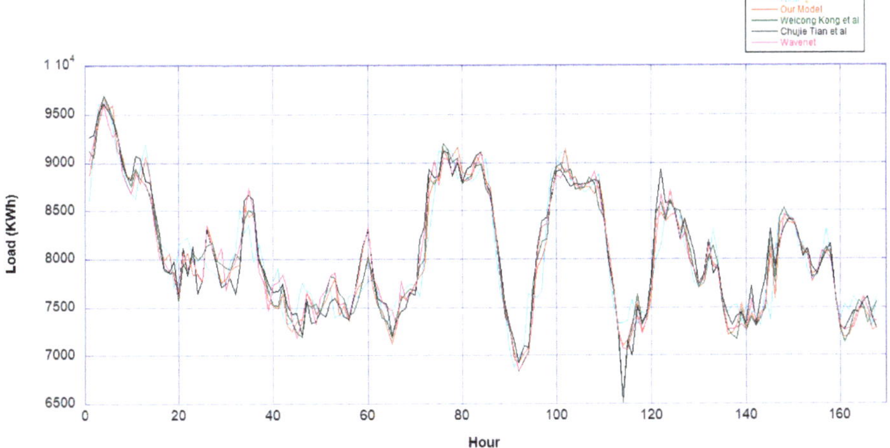

Figure 9. Forecasting result using ENTSO-E dataset 1.

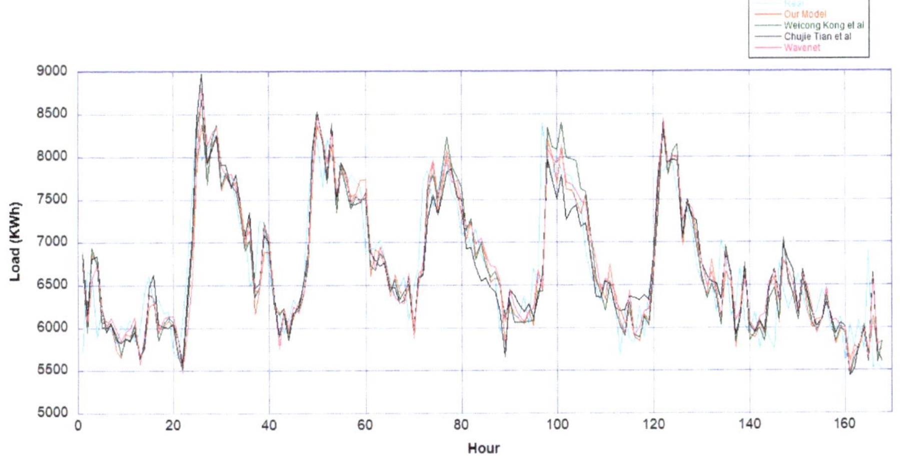

Figure 10. Forecasting result using ENTSO-E dataset 2.

4.3. ISO-NE Load Prediction

Tables 4 and 5 show all of models performance on ISO-NE dataset 1 and dataset 2 respectively. Different with performances shown in the previous subsection, all of the models perform very well both in known and unknown data distribution. However, the model based on [16] still exhibits the worst accuracy, while our model still performs the best although its excellence is not absolute compared to the model based on [15]. Same with the result obtained using the ENTSO-E dataset, the implementation of the LSTM layer for tuning the wavenet-based model, is proven to help the network in making more accurate load predictions.

Table 4. Forecasting result obtained using Independent System Operator New England (ISO-NE) testing dataset 1.

Model	RMSE	MAE	MAPE (%)
Tian et al.	114.33	82.18	0.56
Kong et al.	92.7	62.55	0.42
Wavenet	109.76	77.69	0.52
Our Model	85.12	58.96	0.4

Table 5. Forecasting result obtained using Independent System Operator New England (ISO-NE) testing dataset 2.

Model	RMSE	MAE	MAPE (%)
Tian et al.	141.97	89.07	0.66
Kong et al.	100.5	65.12	0.48
Wavenet	125.11	78.02	0.57
Our Model	88.31	62.23	0.46

The success of all models in understanding an unknown data distribution in this dataset is mainly because of the ISO-NE data property, which is simpler compared to the ENTSO-E dataset in which more fluctuations are experienced in certain ranges of time due to external factors. This simplicity results in high accuracy both in validation and testing data, enabling all models to handle forthcoming data sequences. Figures 11 and 12 show the forecasting result using input from dataset 1 and dataset 2, respectively.

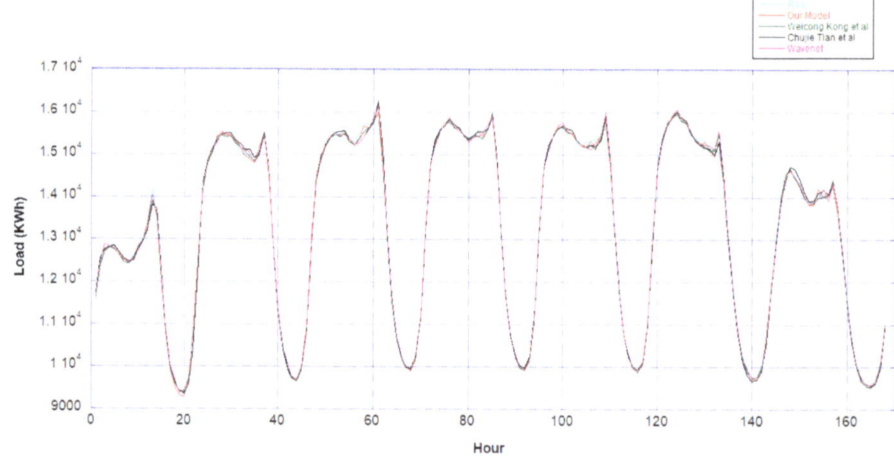

Figure 11. Forecasting result using ISO-NE dataset 1.

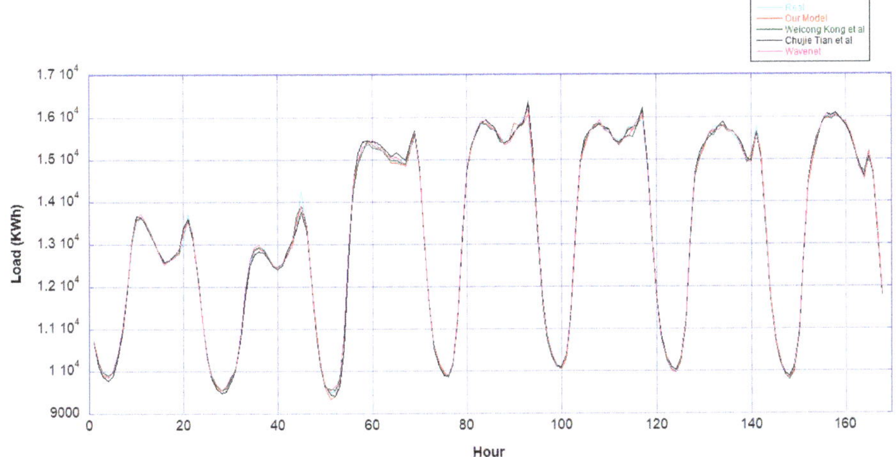

Figure 12. Forecasting result using ISO-NE dataset 2.

4.4. Discussion

Overall, our proposed model shows the best performance compared to the other deep learning-based models. It indicates that each network in the proposed model works very well. The wavenet-based network provides understanding of long sequence data and the LSTM layer helps the model do self-correction by relating the output of the first network with the recent or short sequence data.

However, as suggested by [12,28,29], in order to show our proposed model significance over the other models, both the Wilcoxon signed rank test [30] and Friedman test [31] are conducted using all the models' forecasting error given input from those testing datasets. The Wilcoxon signed rank test will compare the $W_{statistic}$ with the Wilcoxon critical value W which are expressed in Equations (18) and (19) (for huge number of data), respectively.

$$W_{statistic} = \min\{r^+, r^-\} \tag{18}$$

$$W = \frac{N(N+1)}{4} \tag{19}$$

r^+ and r^- are the sum of the positive and negative rank, respectively, while N is the number of data. If $W_{statistic}$ is less than W, and the p-value is less than α, then the null hypothesis is rejected, and it indicates the superiority of our model.

On the other hand, the Friedman test is applied to show the significant differences of our proposed models over all comparison models. The statistic F is expressed by:

$$F = \frac{12N}{k(k+1)}\left[\sum_{j=1}^{k} R_j{}^2 - \frac{k(k+1)^2}{4}\right] \tag{20}$$

where N is the number of forecasting results, k is the number of compared models, and R_j is the average rank sum based on forecasting error r for jth compared model expressed by:

$$R_j = \frac{1}{N}\sum_{1}^{N} r_i{}^j \tag{21}$$

If the p-value of F is less than the critical value, than the null hypothesis is not accepted, indicating the superiority of our model.

Tables 6–9 show the significance test in testing dataset 1 and dataset 2 ENTSO-E and testing dataset 1 and 2 ISO-NE, respectively. In the Wilcoxon signed-rank test, significant levels are set to $\alpha = 0.025$ and $\alpha = 0.05$ while in the Friedman test it is conducted only under $\alpha = 0.05$. From results showed in the tables, our proposed model shows significant contribution in forecasting accuracy improvement and superiority over the other models, except in the ISO-NE dataset, where in the Wilcoxon signed rank test between our proposed model and Kong et al. model the results indicate no significance, although the results in terms of RMSE, MAE, and MAPE show that our model performs better.

Table 6. Result of the Wilcoxon signed-rank test and Friedman test using ENTSO-E testing dataset 1.

Compared Models	Wilcoxson Signed-Rank Test				Friedman Test
	$\alpha = 0.025$ $W = 285{,}423$	p-value	$\alpha = 0.05$ $W = 285{,}423$	p-value	$\alpha = 0.05$
Our Model vs Kong et al.	243,792	3.64×10^{-5}	243,792	3.64×10^{-5}	$H_0 : e_1 = e_2 = e_3 = e_4$
Our Model vs Tian et al.	211,190.5	1.81×10^{-13}	211,190.5	1.81×10^{-13}	$F = 47.4618$
Our Model vs Wavenet	246,095	9.60×10^{-5}	246,095	9.60×10^{-5}	$p = 2.77 \times 10^{-10}$ (Reject H_0)

Table 7. Result of the Wilcoxon signed-rank test and Friedman test using ENTSO-E testing dataset 2.

Compared Models	Wilcoxson Signed-Rank Test				Friedman Test
	$\alpha = 0.025$ $W = 80{,}792$	p-value	$\alpha = 0.05$ $W = 80{,}792$	p-value	$\alpha = 0.05$
Our Model vs Kong et al.	68,149.5	0.001227	68,149.5	0.001227	$H_0 : e_1 = e_2 = e_3 = e_4$
Our Model vs Tian et al.	66,885	3.77×10^{-4}	66,885	3.77×10^{-4}	$F = 17.59$
Our Model vs Wavenet	69,587	0.004169	69,587	0.004169	$p = 0.00053$ (Reject H_0)

Table 8. Result of the Wilcoxon signed-rank test and Friedman test using ISO-NE testing dataset 1.

Compared Models	Wilcoxson Signed-Rank Test				Friedman Test
	$\alpha = 0.025$ $W = 285{,}423$	p-value	$\alpha = 0.05$ $W = 285{,}423$	p-value	$\alpha = 0.05$
Our Model vs Kong et al.	274,482	2.78×10^{-1}	274,482	2.78×10^{-1}	$H_0 : e_1 = e_2 = e_3 = e_4$
Our Model vs Tian et al.	179,311	6.68×10^{-26}	179,311	6.68×10^{-26}	$F = 140.032$
Our Model vs Wavenet	206,826.5	6.43×10^{-15}	206,826.5	6.43×10^{-15}	$p = 3.72 \times 10^{-30}$ (Reject H_0)

Table 9. Result of the Wilcoxon signed-rank test and Friedman test using ISO-NE testing dataset 2.

Compared Models	Wilcoxson Signed-Rank Test				Friedman Test
	$\alpha = 0.025$ $W = 80{,}792$	p-value	$\alpha = 0.05$ $W = 80{,}792$	p-value	$\alpha = 0.05$
Our Model vs Kong et al.	80,556	0.950686	80,556	0.950686	$H_0 : e_1 = e_2 = e_3 = e_4$
Our Model vs Tian et al.	57,955	5.29×10^{-9}	57,955	5.29×10^{-9}	$F = 40.198$
Our Model vs Wavenet	66,618.5	0.00029	66,618.5	0.00029	$p = 9.67 \times 10^{-9}$ (Reject H_0)

5. Conclusions and Future Works

This paper proposes a novel method for hourly load forecasting case which is very important in the case of demand response for hybrid energy systems, especially for system use of both renewable and fossil energy in order to reduce fossil energy usage. The proposed method is mainly inspired by the wavenet-based model utilizing dilated causal residual CNN and LSTM. In this approach, two different data sequences are fed to the model. The long data sequences are fed to the wavenet-based model while the short data sequences are fed to the LSTM layer assigned for model self-correction using fine-tuning technique.

In order to prove the generality of our model, two different datasets, which are the ENTSO-E and ISO-NE dataset, are used with two different testing scenarios. The first testing scheme uses the dataset having nearly identical distribution with the validation dataset, while the second uses a dataset from slightly different data distribution. Based on the obtained result, our proposed model exhibits the best performance compared to other deep learning-based models in terms of RMSE, MAE, and MAPE. In detail, our model achieves RMSE, MAE, and MAPE equal to 203.23, 142.23, and 2.02 for ENTSO-E testing dataset 1 and 292.07, 196.95, and 3.1 for ENTSO-E dataset 2. Meanwhile, in the ISO-NE dataset, the RMSE, MAE, and MAPE equal to 85.12, 58.96, and 0.4 for ISO-NE testing dataset 1 and 85.31, 62.23, and 0.46 for ISO-NE dataset 2. However, there are several findings that can be improved in future work. The first is in the ENTSO-E dataset testing result; all models cannot provide high accuracy forecasting if they are fed using slightly different data distribution. It indicates that all models face difficulties in understanding fluctuated or unpredicted data like the ENTSO-E dataset. The second is although RMSE, MAE, and MAPE show our model exhibits better accuracy compared to the Kong et al. model in the ISO-NE dataset, our model cannot provide a significant improvement.

Therefore, for future work, additional external factors data like information of holidays and weather conditions can be fed as the models' input in order to improve our findings. In addition, building a new model can also be conducted since this research area and artificial intelligence (AI) algorithms are developed very quickly, making a new idea come up very fast. All of the codes and datasets used in this research are available on Github.

Author Contributions: Conceptualization, S.H.P.; methodology, S.H.P. and M.R.; software, M.R. and E.M.; validation, S.H.P. and M.R.; formal analysis, S.H.P., M.R., E.M. and R.N.H.; investigation, S.H.P., M.R. and E.M.; resources, S.H.P.; data curation, M.R. and F.H.; writing—original draft preparation, M.R. and E.M.; writing—review and editing, S.H.P. and R.N.H.; visualization, M.R., R.N.H. and F.H.; supervision, S.H.P.; project administration, F.H.; funding acquisition, S.H.P.

Funding: This research was funded by Indonesia Ministry of Research and Technology (KEMENRISTEKDIKTI) (Grant No.332.51/UN10.C19/PN/2019).

Conflicts of Interest: The authors declare no conflict of interest.

References

1. Krishna, K.S.; Kumar, K.S. A review on hybrid renewable energy systems. *Renew. Sustain. Energy Rev.* **2015**, *52*, 907–916. [CrossRef]

2. Olson, C.; Lenzmann, F. The social and economic consequences of the fossil fuel supply chain. *MRS Energy Sustain.* **2016**, *3*. [CrossRef]

3. Siano, P. Demand response and smart grids—A survey. *Renew. Sustain. Energy Rev.* **2014**, *30*, 461–478. [CrossRef]

4. Nguyen, D. Demand Response for Domestic and Small Business Consumers: A New Challenge. In Proceedings of the IEEE PES T&D 2010, New Orleans, LA, USA, 19–22 April 2010; pp. 1–7.

5. Marwan, M.; Kamel, F.; Xiang, W. Mitigation of electricity price/demand using demand side response smart grid model. In *Proceedings of the 1st International Postgraduate Conference on Engineering, Designing and Developing the Built Environment for Sustainable Wellbeing (eddBE 2011), Brisbane, Australia, 27–29 April 2011*; Queensland University of Technology: Brisbane, Australia, 2011; pp. 128–132.

6. Vu, D.H.; Muttaqi, K.M.; Agalgaonkar, A. A variance inflation factor and backward elimination based robust regression model for forecasting monthly electricity demand using climatic variables. *Appl. Energy* **2015**, *140*, 385–394. [CrossRef]

7. Al-Hamadi, H.; Soliman, S. Short-term electric load forecasting based on Kalman filtering algorithm with moving window weather and load model. *Electr. Power Syst. Res.* **2004**, *68*, 47–59. [CrossRef]

8. Cho, M.; Hwang, J.; Chen, C. Customer Short Term Load Forecasting by Using ARIMA Transfer Function Model. In Proceedings of the 1995 International Conference on Energy Management and Power Delivery EMPD'95, Singapore, 21–23 November 1995; pp. 317–322.

9. Hao, C. A new method of load forecasting based on generalized autoregressive conditional heteroscedasticity model. *Autom. Electr. Power Syst.* **2007**, *15*, 12–13.

10. Niu, D.; Wang, Y.; Wu, D.D. Power load forecasting using support vector machine and ant colony optimization. *Expert Syst. Appl.* **2010**, *37*, 2531–2539. [CrossRef]

11. Fan, G.-F.; Guo, Y.-H.; Zheng, J.-M.; Hong, W.-C. Application of the Weighted K-Nearest Neighbor Algorithm for Short-Term Load Forecasting. *Energies* **2019**, *12*, 916. [CrossRef]

12. Dong, Y.; Zhang, Z.; Hong, W.-C. A hybrid seasonal mechanism with a chaotic cuckoo search algorithm with a support vector regression model for electric load forecasting. *Energies* **2018**, *11*, 1009. [CrossRef]

13. Zhang, R.; Dong, Z.Y.; Xu, Y.; Meng, K.; Wong, K.P. Short-term load forecasting of Australian National Electricity Market by an ensemble model of extreme learning machine. *IET Gener. Transm. Distrib.* **2013**, *7*, 391–397. [CrossRef]

14. Ghofrani, M.; Ghayekhloo, M.; Arabali, A.; Ghayekhloo, A. A hybrid short-term load forecasting with a new input selection framework. *Energy* **2015**, *81*, 777–786. [CrossRef]

15. Kong, W.; Dong, Z.Y.; Jia, Y.; Hill, D.J.; Xu, Y.; Zhang, Y. Short-term residential load forecasting based on LSTM recurrent neural network. *IEEE Trans. Smart Grid* **2017**, *10*, 841–851. [CrossRef]

16. Tian, C.; Ma, J.; Zhang, C.; Zhan, P. A Deep Neural Network Model for Short-Term Load Forecast Based on Long Short-Term Memory Network and Convolutional Neural Network. *Energies* **2018**, *11*, 3493. [CrossRef]

17. Han, L.; Peng, Y.; Li, Y.; Yong, B.; Zhou, Q.; Shu, L. Enhanced deep networks for short-term and medium-term load forecasting. *IEEE Access* **2018**, *7*, 4045–4055. [CrossRef]

18. Park, K.; Yoon, S.; Hwang, E. Hybrid load forecasting for mixed-use complex based on the characteristic load decomposition by pilot signals. *IEEE Access* **2019**, *7*, 12297–12306. [CrossRef]

19. LeCun, Y.; Bengio, Y. Convolutional networks for images, speech, and time series. *Handb. Brain Theory Neural Netw.* **1995**, *3361*, 1995.

20. Hochreiter, S.; Schmidhuber, J. Long short-term memory. *Neural Comput.* **1997**, *9*, 1735–1780. [CrossRef]

21. Oord, A.V.D.; Dieleman, S.; Zen, H.; Simonyan, K.; Vinyals, O.; Graves, A.; Kalchbrenner, N.; Senior, A.; Kavukcuoglu, K. Wavenet: A generative model for raw audio. *arXiv* **2016**, arXiv:1609.03499.

22. He, K.; Zhang, X.; Ren, S.; Sun, J. Deep residual learning for image recognition. In Proceedings of the IEEE Conference on Computer Vision and Pattern Recognition, Las Vegas, NV, USA, 26 June–1 July 2016; pp. 770–778.

23. Handbook-Glossary, U.O. European network of transmission system operators for electricity. Available online: https://transparency.entsoe.eu/load-domain/r2/totalLoadR2/show (accessed on 30 April 2019).

24. Shamsollahi, P.; Cheung, K.; Chen, Q.; Germain, E.H. A neural network based very short term load forecaster for the interim ISO New England electricity market system. In Proceedings of the PICA 2001. Innovative Computing for Power-Electric Energy Meets the Market. 22nd IEEE Power Engineering Society. International Conference on Power Industry Computer Applications (Cat. No. 01CH37195), Sydney, Australia, 20–24 May 2001; pp. 217–222.

25. Goodfellow, I.; Bengio, Y.; Courville, A. *Deep Learning*; MIT Press: Cambridge, MA, USA, 2016.

26. Kingma, D.P.; Ba, J. Adam: A method for stochastic optimization. *arXiv* **2014**, arXiv:1412.6980.

27. Abadi, M.; Barham, P.; Chen, J.; Chen, Z.; Davis, A.; Dean, J.; Devin, M.; Ghemawat, S.; Irving, G.; Isard, M. Tensorflow: A system for large-scale machine learning. In Proceedings of the 12th {USENIX} Symposium on Operating Systems Design and Implementation ({OSDI} 16), Savannah, GA, USA, 2–4 November 2016; pp. 265–283.

28. Diebold, F.X.; Mariano, R.S. Comparing predictive accuracy. *J. Bus. Econ. Stat.* **2002**, *20*, 134–144. [CrossRef]

29. Derrac, J.; García, S.; Molina, D.; Herrera, F. A practical tutorial on the use of nonparametric statistical tests as a methodology for comparing evolutionary and swarm intelligence algorithms. *Swarm Evolut. Comput.* **2011**, *1*, 3–18. [CrossRef]

30. Wilcoxon, F. Individual comparisons by ranking methods. In *Breakthroughs in Statistics*; Springer: New York, NY, USA, 1992; pp. 196–202.

31. Friedman, M. A comparison of alternative tests of significance for the problem of m rankings. *Ann. Math. Stat.* **1940**, *11*, 86–92. [CrossRef]

Article

Classification of Special Days in Short-Term Load Forecasting: The Spanish Case Study

Miguel López *, Carlos Sans, Sergio Valero and Carolina Senabre

Electrical Engineering Area, University Miguel Hernández, Av. de la Universidad, s/n, 03202 Elche, Spain; carsantr@gmail.com (C.S.); svalero@umh.es (S.V.); csenabre@umh.es (C.S.)
* Correspondence: m.lopezg@umh.es; Tel.: +34-965-222-407

Received: 4 February 2019; Accepted: 27 March 2019; Published: 1 April 2019

Abstract: Short-Term Load Forecasting is a very relevant aspect in managing, operating or participating an electric system. From system operators to energy producers and retailers knowing the electric demand in advance with high accuracy is a key feature for their business. The load series of a given system presents highly repetitive daily, weekly and yearly patterns. However, other factors like temperature or social events cause abnormalities in this otherwise periodic behavior. In order to develop an effective load forecasting system, it is necessary to understand and model these abnormalities because, in many cases, the higher forecasting error typical of these special days is linked to the larger part of the losses related to load forecasting. This paper focuses on the effect that several types of special days have on the load curve and how important it is to model these behaviors in detail. The paper analyzes the Spanish national system and it uses linear regression to model the effect that social events like holidays or festive periods have on the load curve. The results presented in this paper show that a large classification of events is needed in order to accurately model all the events that may occur in a 7-year period.

Keywords: load forecasting; special days; regressive models

1. Introduction

Short-term load forecasting (STLF) is a determining factor for operation of an electric system. It is a necessary process in order to ensure the balance between generation and demand. The system operator needs to know the expected load to make decisions and to perform an optimal control of the electrical system. Many countries have liberalised electrical markets, which promotes the participation of multiple agents. This participation yields a competitive system, which leads to reduced costs to the final consumer. The accuracy of load forecasting leads to an optimization of the power generation and of operation of the system and a consequent reduction of the costs. In addition, a good STLF leads to a better share of renewable energy in the electric system, therefore reducing CO_2 emitted to the atmosphere conforming with the Paris Agreement [1], which has been ratified by several countries. This reduction helps to avoid the emission of excess CO_2 for countries like those of EU, Iceland, Liechtenstein and Norway, which must comply with the EU Emissions Trading System [2]. The optimization of the electric generation reduces its cost, improving competitiveness among companies and subsequently, the economical and industrial development of a country.

Several works have been published about STLF models in the last decades [3]. These methods split into three large groups: artificial intelligence [4–19], statistical [20,21] and hybrid models [22–24]. Regarding artificial intelligence, these techniques have been successfully applied for STLF, such as artificial neural networks (ANN) [4,5], extreme learning machine (ELM) [6,7], support vector machine (SVM) [8,9,11], adaptive neuro-fuzzy inference system (ANFIS) [10–14], fuzzy logic (FL) [14–16], genetic algorithm (GA) [16] and self-organizing map (SOM) [17–19]. On the other hand, statistical models such as autoregressive (AR) model [20,21], autoregressive integrated moving average (ARIMA) [21] and exponential smoothing (ES) [25,26] have been extensively used in STLF. Other forecasting methods are considered hybrid models [23,24], which use the combination of various techniques (statistical models and/or artificial intelligence) to obtain forecasting load.

One of the most used methods of AI for the STLF is Neural Networks (NNs), and the amount of research work on this topic found in reference databases like SCOPUS is much higher than for the rest of artificial intelligence techniques. This technique has been used over the last decades obtaining successful results. In the early days, a review [4] of different works published between 1991 and 1999, reports the use of different models of NNs for STLF describing the doubts suggested by some authors about adopting this technique for STLF. This study concludes that many research groups used small data sets with only one NN for STLF. However, other groups have over-parameterized, increasing the complexity of the forecasting model and decreasing the accuracy. In conclusion, further investigations are needed to perform an NN prediction model that clears the controversy in the scientific community.

In recent years, different techniques based on NNs have provided good results forecasting load due to the increase of the historical data used. In [5], different NN methodologies are compared, multilayer perceptron (MLP) [27], radial basis function neural networks (RBFNN) [28], generalized regression neural network (GRNN) [29] and counter-propagation neural networks (CPNN) [30] which learn from patterns that represent the daily load curves. The results showed that the STLF of a GRNN model is more accurate than the rest of NN models analysed in this work.

The ELM technique shown in [6] is a different forecasting model of AI. The method shown in [7] is used to prove the accuracy for STLF versus other NN models. An ELM model provides a better efficiency of the training and a better accuracy of the predictions.

Another technique based on AI is SVM, which is described in [8]. This method is used in [9] to forecast load by combining four SVM models according to certain values of temperature and demand. In [11], a comparison between SVM and ANFIS [12,13] is presented. The comparison uses essential information about the days of a week, achieving much closer results to the actual load by the SVM model.

The performance of the FL model was also compared to that of the ANFIS model using the same parameters and data [14]. The predictions obtained in both cases were successful, although the ANFIS model was more accurate than the FL model. The latter technique was used to obtain the STLF by using different parameters in [15] (e.g., weather, time, historical data), demonstrating that this method can provide a more accurate prediction than conventional models.

In most cases, the artificial intelligence technique called genetic algorithm was used to select the most important parameters for forecasting electricity demand. A genetic algorithm like simulated annealing is used with other technique like FL to obtain the optimal parameters [16] by means of the back propagation method. This method improves the accuracy of the predictions.

Kohonen's Self-Organizing Map (SOM) [17] was also used to successfully obtain forecasting loads [18]. Moreover, the SOM model was a very useful tool for classifying the data of the parameters used to forecast the load. The classification of the meteorological data [19] by using the SOM model to cluster the data provided a prediction of demand through nonlinear autoregressive network with exogenous inputs (NARX) [31].

The autoregressive models are the most commonly used statistical models. Many research articles employ these statistical methods for STLF. In [20], Baharudin et al. analised autoregressive (AR) model and autoregressive–moving-average (ARMA) [32] to obtain forecasting load, concluding that the performance of AR model was more accurate.

A comparison of different statistical models was studied by Taylor et al. [21]. It compared methods such as autoregressive (AR) models, autoregressive integrated moving average (ARIMA) [33], a regression method with principal component analysis (PCA) [34], exponential smoothing (ES) [25] and the Holt-Winters exponential smoothing method. The best performing method was double seasonal Holt-Winters exponential smoothing.

Regarding the hybrid models, which use the techniques from two different methods to obtain the demand prediction, Fan et al. [23] analysed a SOM Neural Network to cluster each data set into subsets. In addition, 24 SVMs are used to adjust each subset to the next day's load profile. Hybrid models can use several forecasting models, where the final forecast is provided by the combination of both models. In previous research [24], the authors stated that, AR and NN methods provide separate forecasts and a final result given by the linear combination of both methods. A linear combination was implemented in order to enhance forecasting accuracy.

One issue that has not received as much attention as the forecasting engine is the forecasting of special days. Several days throughout the year show a profile that does not match the expected profile for its weekday. These differences may be caused by temperature [26] but, the more extreme cases are caused by socio-economic effects of the calendar. In Figure 1, we can see how the load profile for a special day (1 November, a national holiday in Spain) has lower demand values with respect to any normal day. This and other special days are harder to forecast and incur higher forecasting errors that increase forecasting losses. This is especially important because a low average forecasting error with large peak errors may be costlier than a slightly less accurate forecast that has smaller peak errors. This research focuses on how to provide the proper information about these special days so that a forecasting engine may be able to forecast them more accurately.

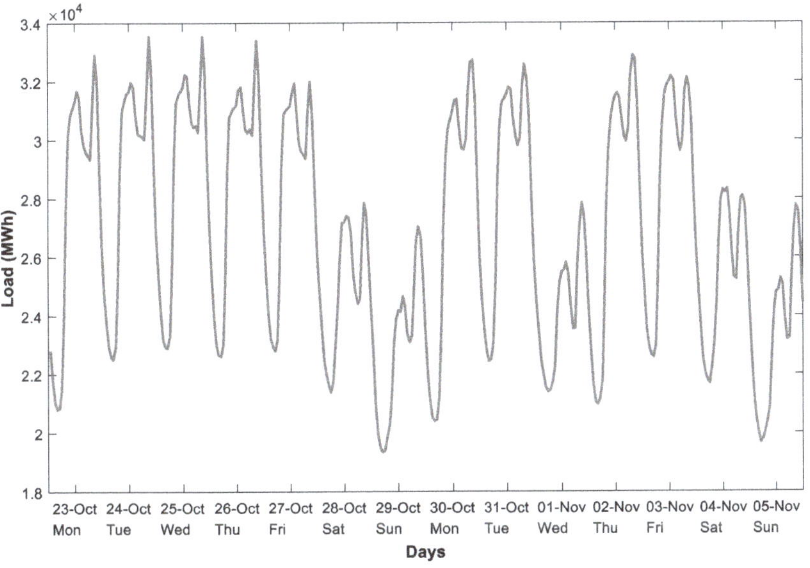

Figure 1. Hourly electricity demand in Spain from 23 October 2017 to 5 November 2017.

There are several ways in which the calendar affects load profiles. Figure 1 already shows that a national holiday has a profile more similar to a Sunday than to a typical weekday. However, this does not mean that all national holidays share the same profile. In addition, the demand pattern of normal days can be altered by the proximity of special days (see Figure 2) such as the Monday before a holiday or the Friday after a holiday. The load profiles of special days do not have the same demand pattern, it can be seen in Figure 2 that the load profile of 12 October is different to the load profiles of normal days. In addition, the demand pattern on 13 October is lower than the demand pattern of a normal Friday, which is because it happens after a holiday. If we consider all special days belonging to only one kind of day, the accuracy of the demand prediction will be affected.

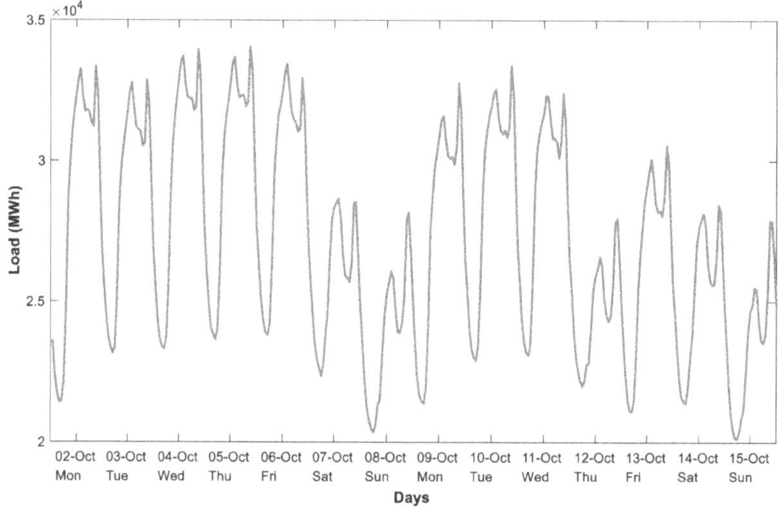

Figure 2. Hourly electricity demand in Spain from 2 October 2017 to 15 October 2017.

Each day of the week may have its own profile and, in many cases, there are interactions causing the type of special day to be conditioned by its weekday. The same special day may have different demand patterns depending on the day of the week. Figure 3 displays the profile load of the special day 7 December for the period 2010 to 2017. The day before and after 7 December is a holiday. However, the demand load for Monday, Saturday and Sunday are very different from the other days, as well as the first hours of Thursday and the period between 9 am and 7 pm on the same day. Special days are classified into several types depending on the day of the week.

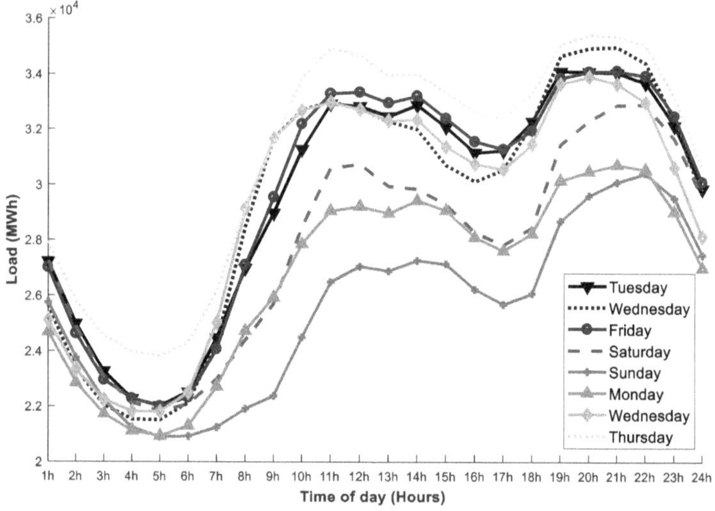

Figure 3. Hourly electricity demand for December 7th, which fell on Tuesday (2010), Wednesday (2011), Friday (2012), Saturday (2013), Sunday (2014), Monday (2015), Wednesday (2016) and Thursday (2017).

However, this classification of special days is not enough to obtain all the information that lies within the profiles of the special days. A deeper classification of the special days is necessary to improve forecasting accuracy, which has been the subject of this work. All demand patterns different from the demand patterns of normal days should be analyzed, taking into account the influences on demand described in this paper. Special days with similar demand patterns are grouped to reduce the complexity and computing times of the forecasting model.

The contribution of this research line in:

- This paper proposes a classification for special days that includes more than 40 types of day, while most published articles limit their classification to 15 or less.
- The performance of this classification is tested against less extensive classifications, proving that its use provides a more accurate forecast for each type of day.
- Non-linearities present in special days load profiles are overcome by using separate models for each hour and categorical variables for each type of day.
- A detailed analysis of special days in Spain is provided, which can help to replicate this classification in other systems.
- The benefits of this approach stem not only from the reduction of the average forecasting error but especially on the reduction of error peaks which usually happen on these special days.

This paper is organized as follows: Section 2 describes the state of the art and previous work related to forecasting special days. Section 3 shows the available data for the analysis, the characteristics of the mathematical model used, the treatment to the input variables in order to be processed by the model and the different experiments carried out. Section 4 includes the results of these experiments, analyzing how each classification of days affects modeling accuracy. Section 5 is a brief conclusion expressing the relevance of using complex classification schemes for the different types of days in order to achieve accurate forecasts. Appendices A and B include a more extensive review of the results.

2. State of the Art and Related Work

The aim of this study is to improve STFL accuracy for special days by defining a more detailed classification. Load profiles of the different days do not follow the same pattern and if we group the most similar demand patterns, the accuracy of the prediction will increase. Demand patterns of special days and demand patterns of normal days are very different. If our forecasting model does not differentiate between normal and special days, the results obtained will be inaccurate on these special days.

In most of the research works related to STFL, all days are classified into two or three large groups such as weekdays, weekends and holiday periods [23,35–37]. Some research articles, where special days are classified according to the demand pattern, they are grouped into 5 [38] and 15 [39] different days, obtaining better results than choosing only three types of days. The profile loads on special days do not have the same demand pattern (i.e., days adjacent to holidays, period of Christmas, Easter, national holidays, week before Christmas) [26,38], consequently the forecasting uncertainty is greater for these days. The day of the week is also an important factor in the load profiles of special days [40–42]: the same special day may have different demand patterns depending on the day of the week. In addition, demand pattern of normal days can be altered by the proximity of special days [38,40,41,43].

Several works have been published for anomalous load forecasting [26,38–44]. In the case of research [42], it only takes into account the days with the greatest errors in the prediction. These days are the holidays that fall on a Saturday or a Monday. This research was done in Korea where Sundays are holidays. Special days are classified into four categories (Tuesday, Wednesday, Thursday and Friday; Saturday; Monday; Sunday). This research only classifies special days depending on the day of the week they fall on. This method reduces the highest prediction errors. However, the accuracy of the prediction can be improved if a deep classification of special days is performed. In [43], the different types of day are classified based on the shape of the load curve into three categories (weekdays (Monday to Friday); Saturday; Sunday and Holidays). Due to the application of special rules, the proximity of the forecast day to a holiday is taken into account. This classification does not differentiate special days from each other. In [26], data are classified according to the type of day and month, to capture the effects of seasonality on the load profile. In addition, three variables are added to check the impact on electricity consumption of holidays, days following a holiday and Easter. In [38], special days are classified into five different types of day (weekday, Saturday, Sunday, Monday and special holidays), but a neural network is necessary for each type of special day. In addition, a fuzzy inference model forecasts the maximum and the minimum loads of a special day. Therefore, if the number of types of special days increases, the forecasting model will be more complex. In [39], SOM is used to group the days with similar load profiles and STFL is performed by means of an NN. SOM performs the classification of the different types of days into groups that can vary between 11 and 15. This technique has been discarded, because the separation of the different types of days requires prior knowledge that is difficult to assemble and whose result is not clear. The classification proposed in this research is similar to the classification described in [40]. Special days are treated according to whether they fall on the same date, the same day of the week, the day of the week is weekday or weekend. However, the classification of special days must be greater as well as the number of days considered as special days. In [41], a variation of the forecasting model described in [40], is used, increasing training period to 8 years and formulating a specific rule to be applied in France. However, the classification of special days into seven categories is still insufficient. In [44], special days are classified into four categories such as common holidays (some national holydays and all local and regional holidays) and three special national holidays.

The use of categorical variables to formulate such classifications in linear regression models is very common [26,41,44–46] and it is the same approach used in this paper. These categorical variables define the type of day and are translated into dummy binary categories that allow the regression model to estimate each type of day individually without any linear assumption among normal or special days.

3. Methodology

The starting point of this study is a load forecasting system currently in use at the Red Electrica de España (REE) headquarters. REE is the Transport System Operator for the Spanish system. The following paragraphs aim to describe the available data, the variables actually introduced in the system and the mathematical aspects of the model. In addition, this section will describe the different experiments carried out to determine how the variety of special days can be classified and the type of information used to characterize each type of day.

3.1. Available Data

The input data can be grouped as load, temperature and calendar data:

- Load: The load data are hourly values from the whole Spanish inland system. Data covers the period 2010 through 2017. The data have been filtered to discard incoherent outliers and fill in missing data. Only 0.01% of the data points were affected.
- Temperature: The temperature data consists of maximum and minimum daily values from 59 stations scattered throughout the country. As described in [24], only the most relevant locations are actually included in the model.
- Calendar: The information on special days is initially extracted from the B.O.E. (National Official Gazette) [47]. This document includes all national and regional holidays for each year. However, even though these dates present a specific load profile, they are not the only ones that need special treatment and, therefore, additional information will be needed to establish a definite classification of special days. This additional information can be extracted from the characteristics of the load but in this case, the forecasting model was designed by an expert in the field.

3.2. Mathematical Model

The forecasting model used as starting point is thoroughly described in [24]. It includes a neural network and an autoregressive model whose output is combined to provide a single forecast. The combination of both outputs is more accurate than both of them, therefore, both techniques have advantages and are useful as forecasters. However, the black-box characteristics of the neural network makes it less useful to extract conclusions about the model and it is consequently discarded for this study. In addition, the limitation that regression models impose of linearity between each variable and the output can be overcome by linearization methods (temperature) or other techniques. The abnormalities that special days cause in the daily profiles are non-linear because each hour is affected differently and, therefore, the resulting profile is not a scaled copy of the profile of a regular day. In addition, there is no linear relation between the nature of each special day and its effect on the load. The regression model used allows us to include these abnormalities by using individual models for each hour, whose coefficients, therefore, are not restricted in any way and by using separate binary categories for each type of day. The model then provides specific profiles for each category without any relation between the hours within a profile or between profiles of different categories.

The autoregressive model is described by Equation (1), where a given output y_t is a linear combination of the model's p previous known errors (e_{t-i}), a number of exogenous variables included in X_t and a random shock ε_t:

$$y_t = \sum_{i=1}^{p} \varphi_i \cdot e_{t-i} + X_t \cdot \theta + \varepsilon_t \tag{1}$$

This type of process is useful for characterizing time series which are self-correlated to some extent. However, the autoregressive part may also cover up the effect of the exogenous variables. Therefore, in our study, the autoregressive part of the model has been eliminated as shown in Equation (2):

$$y_t = X_t \cdot \theta + \varepsilon_t \tag{2}$$

The effect that any exogenous variable may have on the load may vary throughout the day, which means that the coefficient for each variable may take different values at different times. In order to meet this requirement, the 24 h profile is obtained by using one model for each hour. The input structure for each model is the same. In addition, the output used is the natural logarithm of the load, which experimentally shows a lower modeling error than the actual load.

3.3. Input for the Model

The exogenous variables used in the model stem from the available data described above. However, due to the non-linearities present in most load-variable relations, a pretreatment is necessary to conform the definite variables going into the exogenous variables matrix:

- Load: The initial model contained two variables used to model the long-term trend as a quadratic function of time. This approach may be valid for shorter periods of time (3 years) but for longer periods, the long-term behavior is not reproduced by a quadratic function. Therefore, these variables are substituted by a 52-week moving average of the previous load for each model. In addition, the initial model included the last known load value at the time of the forecast. This variable, as mentioned above about the autoregressive terms, may hide the effect of other variables like temperature or special days. Therefore, it is also removed from our study.
- Temperature: The effect of temperature is non-linear as both hot or cold temperatures cause an increase in electricity consumption. Moreover, it has a certain inertia and temperature during previous days also has an influence on current demand. In addition, for large regions with a diversity of climates, it is not desirable to use a calculated average temperature for the whole region as it masks the extreme temperatures that may trigger high demands locally. To account for all these factors, the temperature variables are first selected from all available locations: Only 5

(Madrid, Barcelona, Seville, Bilbao and Zaragoza) of the 59 available series are used representing the different climate areas in Spain. Linearity is achieved by using the Hot and Cold Degree Days (HDD, CDD) method [44].This technique splits the data series into two different series each one accounting for demand increases for hot and cold temperatures. It requires defining two thresholds splitting the temperature range into three parts: cold days below the cold threshold, neutral zone in between thresholds and hot days above the hot threshold. Therefore, this method models the load-temperature relationship as a piecewise linear function that calculates different slopes for cold and hot days while it sets the slope for the neutral zone to zero. Figure 4 illustrates this methodology. In order to model temperature inertia, the model includes the current value and lagged variables from all locations for the last two days. To sum up, the temperature variables include HDD and CDD series from five locations with the current and two lagged values. This treatment adds up to 30 variables.

- Calendar data: The information about the type of day is critical as load profiles vary greatly from regular Mondays, to Fridays, Sundays, and even among holidays and special periods throughout the year. Therefore, a detailed classification is key to forecasting these special days accurately. In the starting model, 53 variables are used to classify the type of day along with eleven more to assign the month. The variables included in the model from these 53 variables are described in Tables 1–5.

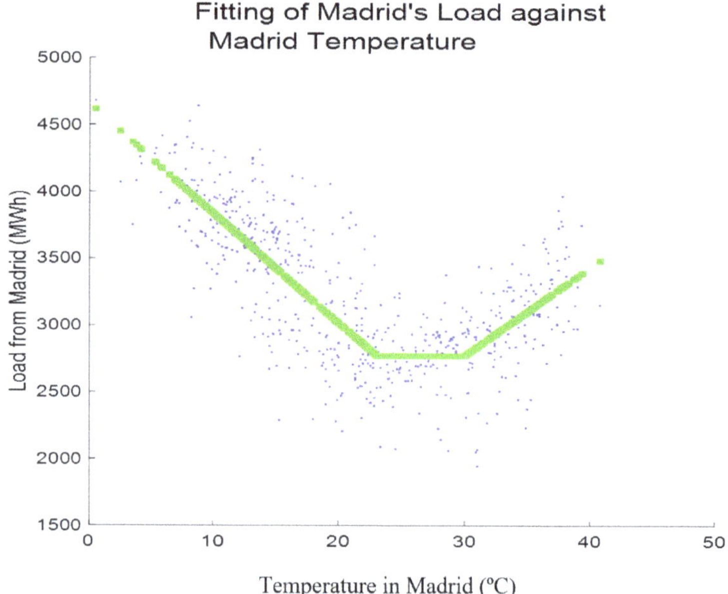

Figure 4. Scatter plot of load from the region of Madrid against temperature from the Madrid weather station. The linearization through HDDs and CDDs is shown in green.

A set of 24 variables is used to identify 24 specific days that are considered to have a profile of their own, incompatible with any other day. These cases are described in Table 1:

Table 1. National special days.

Type 1 (6)		Type 2 (5)	Type 3 (13)	
All instances are included regardless of the day of the week. (6 variables)	1 Jan 6 Jan 1 May 24 Dec 25 Dec 31Dec	Only whenever the day of the week is Mon-Fri. [1] (5 variables)	2 Jan 5 Jan 7 Dec 26 Dec 30 Dec	Each day from Mon. before Good Friday to Sat. after Easter Monday (13 variables)

[1] Weekend instances may qualify into other categories.

The three variables used to classify the rest of national holidays from the B.O.E. are described in Table 2:

Table 2. Regular national holidays.

Weekend	Monday	Tuesday-Friday
Days included as holidays in BOE:		Typically: 15 August 12 October 1 November 6 December 8 December

Days adjacent to a holiday or special day may show a different load profile depending on the day of the week. Table 3 shows the four variables used to model this phenomenon.

Table 3. Days before and after a holiday.

Before		After	
Tuesday to Friday	Monday [1]	Monday to Thursday	Friday [1]

[1] Typically, the profile is more affected if the day is in between the weekend and the holiday.

The classification of regular days is done through six variables for days Monday to Saturday. Variables from Tables 1–3, described categories defined as exclusive: each day belonging to any of these categories may not belong to any other. However, the following categories are thought of as modifiers to the day of the week and may be active at the same time as the day of the week.

Religious holidays are widely observed in Spain and the work and school calendar includes two periods besides the summer-time in which people concentrate their vacation days. The effect of Easter was included in the exclusive variables because, by definition, it happens on the same weekday every year, however, the period around Christmas presents a complexity that forces the use of the following modifiers shown in Table 4:

Table 4. Periods affected by Christmas.

Week before Christmas Day	Week after Christmas Day	Week after New Year Day
Four variables [1] each for one day: 20 December 21 December 22 December 23 December	One variable for days only Monday to Friday 27–29 December	One variable for days only Monday to Friday 2–5 January

[1] Days from the week before seem to be differently affected among them while the days in other periods are all equally affected within their own period.

The summer period is affected by a lower demand from industry due to holidays but a higher demand from services due to tourism. This fluctuation is not constant throughout the summer, but may vary weekly. The inclusion of the month variables helps modeling this behavior but, in the case of August, three additional variables have been added to model differences among weeks.

Finally, regional or local holidays are published in regional gazettes and identified by a variable which, in this case, is not binary but equivalent to the fraction of the National Gross Product that the particular region represents.

The effect of Daylight Savings Time is also considered in the initial forecasting model in two ways. Firstly, it relies on the autoregressive part to phase out the error from the time shifts in March and October and secondly, in order to ease the transition, the first three days from each season are considered as special. These special days are not considered in this study because the autoregressive part is removed. To sum up, Table 5 summarizes all variables for the type of day.

Table 5. Types of day category summary.

Type of Category	General Description	Specifics	Number of Variables
EXCLUSIVE	Special days with specific profiles	Easter	13
		Any day of the week	6
		Only weekdays	5
	National holidays with generic profile	Weekend	1
		Monday	1
		Tuesday to Friday	1
	Days adjacent to holidays	Before	2
		After	2
	DAYS OF THE WEEK		6
MODIFIERS	Christmas periods	Before Christmas day	4
		After Christmas day	1
		After New Year's day	1
	Summer vacation (August)	Week number	3
	Regional holidays	GNP fraction	1

In summary, the model includes one long-term load variable, 30 temperature variables, 11 binary variables for the month and 53 binary variables for the type of day. The output variable is the natural logarithm of the load.

3.4. Experiments and Results

The aim of this study is to determine the different load profiles that a given day may have depending on social- and labor-related characteristics. To determine whether the classification above is adequate, simplistic or overly complex, this study proposes a series of tests to determine how the accuracy of the model varies as the complexity of the classification increases. These experiments are based on eight different classifications starting from the most simplistic, in which only the day of the week is observed and finishing with the most complex, in which all categories described above are considered. The different models are incrementally defined in Table 6, in which each incremental change is described.

Special days are scarce and certain types may only happen every two or three years. This causes a problem when splitting the data set into training data and out-of-sample testing data. In order to solve this, all experiments have been carried out using each one of the 8 years as the testing period and the other 7 as training data. Therefore, the results provided for each model correspond to an 8-year period for which every year is obtained as an out-of-sample test of the model trained with the other 7 years.

Table 6. Summary of incrementally complex models.

Number of Model	General Description		
0	Only six variables are included which are used to describe the day of the week.		
1	Three variables are added	Holidays	Type 1 of special holidays, Maundy Thursday, Good Friday, Easter Monday and all regular national holidays
		Days before holidays	Jan 5th, Dec 7th, Dec 30th, Good Wednesday and all days from the before-holiday final category
		Days after holidays	Jan 2nd, Dec 26th and all days from the after-holiday final category
2	Five variables are added	Regional	As described in the final classification
		Week before Christmas	Dec 20th to Dec 23rd in one single category
		Week after Christmas	Dec 27th to Dec 29th in one single category
		Week after New Year's	Jan 2nd to Jan 5th in one single category
		Easter	Monday and Tuesday prior to Good Friday and Wednesday to Saturday from the next week
3	One variable added	Summer holiday	First three weeks of August
4	Four variables are added	Holidays is split into three categories	Holidays I.P.M. that happen on weekend Holidays I.P.M. that happen on Monday. Holidays I.P.M. that happen on Tuesday to Friday
		Days before holidays is split into two categories	Days before holidays I.P.M. that happen on Monday. Days before holidays I.P.M. that happen on Tuesday to Friday
		Days after holidays is split into two categories	Days after holidays I.P.M. that happen on Friday. Days after holidays I.P.M. that happen on Monday to Thursday
5	Thirteen variables are added	Easter	Easter I.P.M. each day gets its own category (6) Saturday and Sunday after Good Friday each get their own category (2).
		Week before Christmas August	Dec 20th to Dec 23rd each get their own category (3) Each week gets its own category (2)
6	Eleven variables are added	Specific holidays	All type one (6) and type 2 (5) national special days are assigned their own category. (11)
7	Four variables are added	Easter	Wednesday, Thursday, Good Friday and Easter Monday each get their own category (4)

Each model provides a different output which differs especially on days for which other models use a different classification. It is obvious that when a model includes a new category to better suit a type of day, the days falling under said category are expected to be more accurately modeled. However, it is worth mentioning that by "cleaning up" the category in which these days were before, the remaining days in the category also experience a change in their profile that should be for the best. Therefore, even among models that apparently have the same definition of a category, i.e., all models have a regular Monday category, there may be differences in these categories among these models.

The measure of reported error is the Mean Average Percentage Error (MAPE) and it is categorized by type of day to focus on the specific changes among models. For each general category of type of days, the error from each model that introduces a significant change in the definition of said category is reported. Models that treat a category in the same way and that may only experience collateral changes are not reported for clarity reasons.

In addition to the accuracy of a given classification, it is important to understand how each classification models the load profile. This will help us in determining not only which classification is more accurate but also, and more importantly, how each special day's profile is different from each other and learn the expected load from each of them.

Since the output of each model is not the actual load but the natural logarithm, the effect of each coefficient on the load can be considered not as an addend but as a multiplier to the expected load. Considering the way that the type of day categories are defined, the base profile is that of a regular January Sunday. Therefore, a category coefficient higher than 1 means that the typical load at that particular hour for that particular category is larger than the load expected for a regular January Sunday controlling for other factors like temperature.

The coefficient profile calculated for each category by each model is the second result of this study and it provides a useful tool for understanding the nature of each type of special day and the behavioral changes of the consumer on such types of day. These profiles are included in the results section if they are relevant but all of them are also included in Appendix A for the reader's reference.

4. Results

The following section describes the results obtained by each model for every category of type of day. These results include the MAPE results for each model to assess the accuracy improvement that the refining of each category adds to the model. In addition to the error, it also includes the coefficient profile so that the accuracy change can be interpreted based on how the category is differently modeled.

4.1. Regular Days

Even though the aim of this study is not regular days, it will help understand the rest of the results if this category is analyzed. The definition of the category remains unchanged through all eight models; however, more and more special days are removed from the category from model 1 to model 7 and, therefore, the modeling of regular days is more accurate. Figure 5 shows the changes in MAPE on each day of the week through all models. The largest improvement occurs from model 0 to model 1. This model is the first one to acknowledge the existence of special days. Nevertheless, accuracy increases continuously through model 7, especially on Sundays, Mondays and Saturdays.

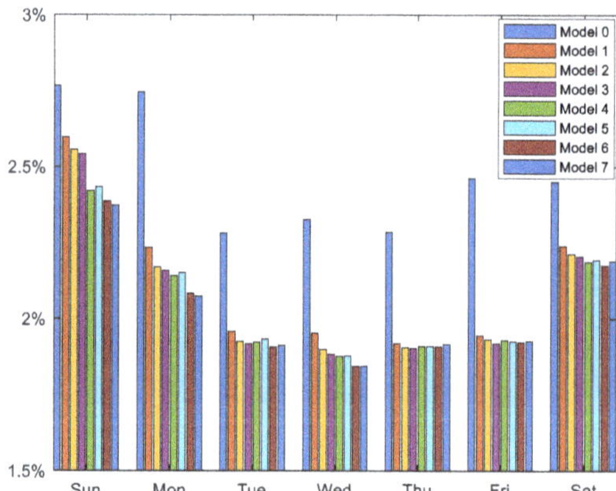

Figure 5. MAPE error of all seven models for the seven days of the week.

The change in the models can be seen in Figure 6a, in which the load profile and the reported MAPE for each model is shown for regular Mondays. The graph shows how the model 0 profile is lower than the other two, which are essentially equivalent. This is because model 0 models regular Mondays after all Mondays in the data set while models 4 and 7 exclude most or all special days, which have a lower profile. The rest of the regular days' results are shown in Appendix A (Figure A3). In Figure 6b the profile for all days of the week from model 7 can be seen and how Monday has a lower start while Friday has a lower finish. Saturday has a unique profile and Tuesday, Wednesday and Thursday are very similar.

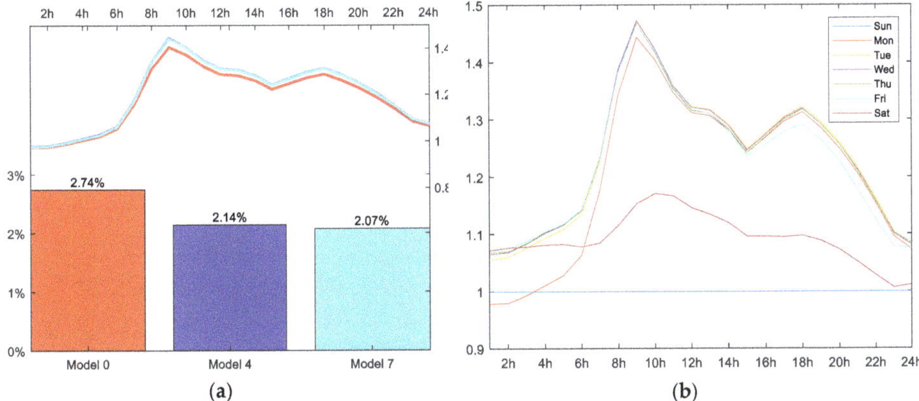

Figure 6. (**a**) Coefficient profile and error chart for regular Mondays.; (**b**) Coefficient profiles for all regular days in model 7. Sunday is a straight line because it is the reference.

It is also important to quantify whether two profiles are similar enough to be considered the same type of day. In order to do so, Table 7 presents the maximum difference between each profile and its most similar pair. This measure expresses the difference between the profiles for two types of days. Specifically, it is the maximum percentage difference between the two 24 h profiles. It is a measure of how two profiles may or may not be equivalent and represent the same type of day. Both type-of-day categories are named at the top and bottom of the table. In this case, there is less than 1% difference between Tuesday, Wednesday and Thursday and, therefore these three categories are candidates for a joint category.

Table 7. Differences between days of the week.

Date	Sun	Mon	Tue	Wed	Thur	Fri	Sat
Difference	17.11%	8.27%	0.93%	0.57%	0.57%	3.26%	17.11%
Most similar	Sat	Tue	Wed	Thur	Wed	Thur	Sun

4.2. Special Days Type 1 and Type 2

Special day types 1 and 2 include 11 fixed dates (see Table 1) that are considered to have specific load profiles. The models that introduce differences into these categories are 0 (all days are regular days), 3 (includes holidays and days before and after), 5 (distinguishes profiles for different days of the week) and 7 (each day is considered unique). Figure 7a shows the model accuracy of the category in these four models. Including these categories as general holidays or before/after holidays lowers the error from 15.6% to around 7%, but then considering each individual holiday yields a much better result of 2.8%. For these special days, assigning different profiles according to the day of the week (model 3 vs. 5) does not improve the result.

In addition, Figure 7b shows the example of Christmas day and how the coefficient profile changes in each model for this particular day. It is interesting how Christmas day shows a much lower load profile than most other special days and so it needs to be considered apart. The coefficient profile and error graphs for the rest of days within the category are included in Appendix A (Figure A3). The maximum difference between the most similar profiles is detailed in Table 8 and it shows that all profiles are independent.

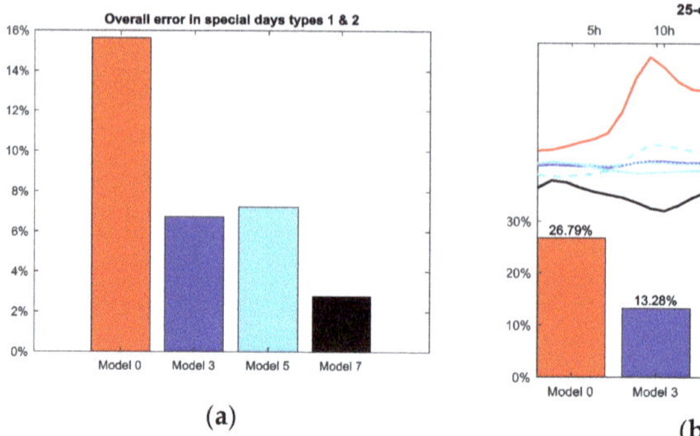

(a)   (b)

Figure 7. (**a**) MAPE for models 0, 3, 5 and 7 for all types 1 and 2 days. (**b**) Coefficient profile and MAPE for models 0, 3, 5, and 7 for Christmas Day.

Table 8. Differences between special days type 1 and 2.

Date	1 Jan	2 Jan	5 Jan	6 Jan	1 May	7 Dec	24 Dec	25 Dec	26 Dec	30 Dec	31 Dec
Difference	5.79%	11.2%	10.2%	6.65%	6.65%	10.2%	3.43% [1]	5.79%	11.9%	10.8%	3.43%
Most similar	25 Dec	7 Dec	7 Dec	1 May	6 Jan	5 Jan	31 Dec	1 Jan	31 Dec	7 Dec	24 Dec

[1] The most similar profiles belong to 24 Dec and 30 Dec, but the peak difference is almost 3.5%.

4.3. Special Days Type 3

Special days type 3 include the Easter period. Each day is considered to have its own profile but differently from types 1 and 2, these days do not happen on the same date each year. The models that affect this type of days are 0, 3, 5 and 7 and the overall accuracy for these days is shown in Figure 8.

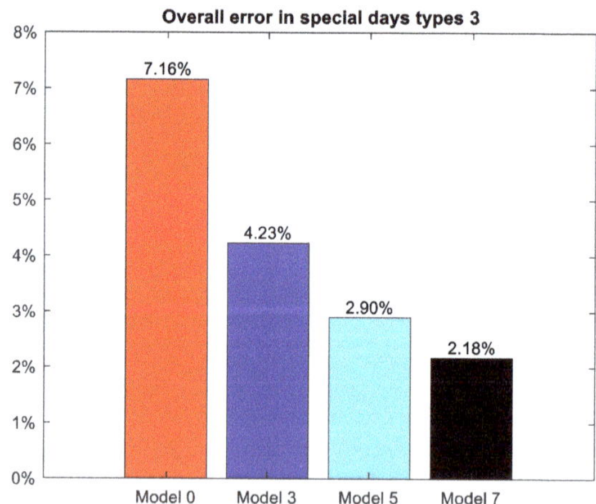

Figure 8. MAPE for models 0, 3, 5 and 7 for all type 3 days.

This type of special days includes some whose profile is deeply affected, like Good Friday, and others, like the Wednesday after Easter Monday, that are only slightly different to a regular day. Figure 9 shows the coefficient and error graph for these two examples while the rest are included in Appendix A (Figure A3). For clarity reasons, days that happen in the same week as Good Friday are from now on referred to as *A*, while those that happen on the next week will be known as *B*.

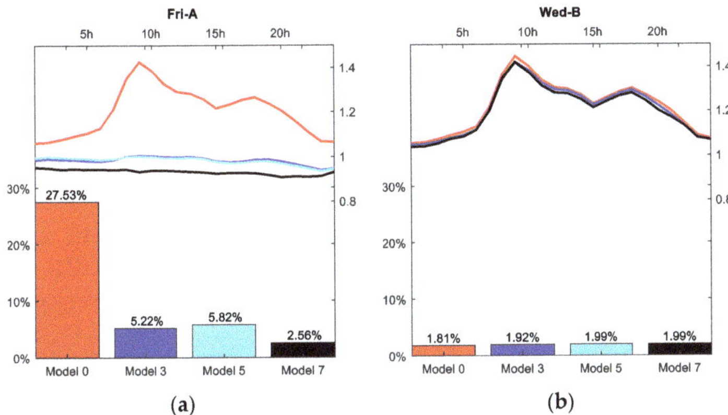

Figure 9. (**a**) MAPE for models 0, 3, 5 and 7 for Good Friday. (**b**) Coefficient profile and MAPE for models 0, 3, 5, and 7 for the Wednesday after Good Friday.

While Figure 9a shows how accuracy on Good Friday increases as the classification is more complex, it does not happen the same way for the next Wednesday, as it is shown in Figure 9b. This may lead to the conclusion that such a Wednesday should not be considered special, as model 0 yields the most accurate result. However, as aforementioned, the profiles for regular weekdays from model 0 are lower than actual regular days due to the presence of holidays that, in that model, are considered regular. The MAPE for such Wednesday if it was considered as a regular day in model 7 would go as high as 3.3%. Figure 10 shows these profiles and illustrates how the profile for that specific Wednesday is lower than both the profile of a regular Wednesday in models 0 and 7. This justifies the use of these specific categories.

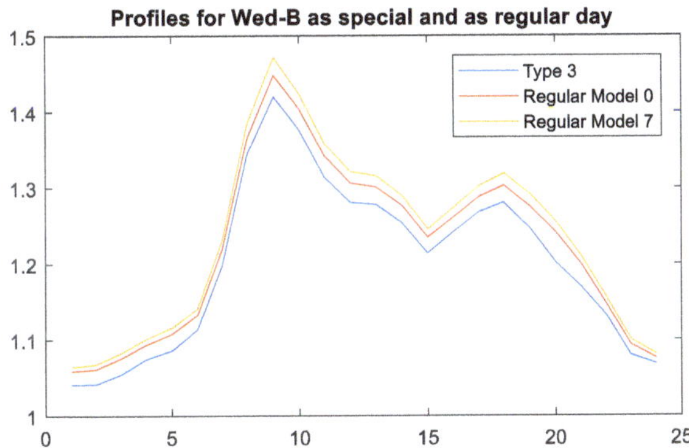

Figure 10. The profile for Wednesday B is lower than any of the regular day profiles calculated in models 0 or 7.

Table 9 shows the similarities among the coefficient profiles within the category. In order to illustrate the fact that days like Wednesday B actually differ from regular days, the profiles for regular days from model 7 are included as candidates for most similar. The results show that Friday and Saturday from week B have the most similarities with regular Fridays and Saturdays, but this difference is larger than two percent.

Table 9. Differences between special days type 3.

Date	Mon-A	Tue-A	Wed-A	Thu-A	Fri-A	Sat-A	Sun-A	Mon-B	Tue-B	Wed-B	Thu-B	Fri-B	Sat-B
Similarity	2.55%	4.24%	4.24%	8.02%	3.15%	9.55%	3.15%	11.27%	2.55%	2.81%	3.22%	2.17%	2.64%
Most similar	Tue-B	Wed-A	Tue-A	Sat-B	Sun-A	Fri-A	Fri-A	Thu-A	Mon-A	Fri-B	Wed-B	Reg Fri	Reg Sat

4.4. Regular Holidays and Days Before and After

Regular holidays are those that are considered to have a generic holiday profile. The days before and after include days which happen before or after a special day, not necessarily a regular holiday. The models that affect these days are 0, 3, 5 and 7. The overall accuracy for these dates and models is shown in Figure 11. In this case it is clear that taking into account the day of the week improves the accuracy (from model 3 to model 5) but also that removing the previously studied special days (some of which were holidays, but others were days before and after) also improves the generic ones (from model 5 to 7).

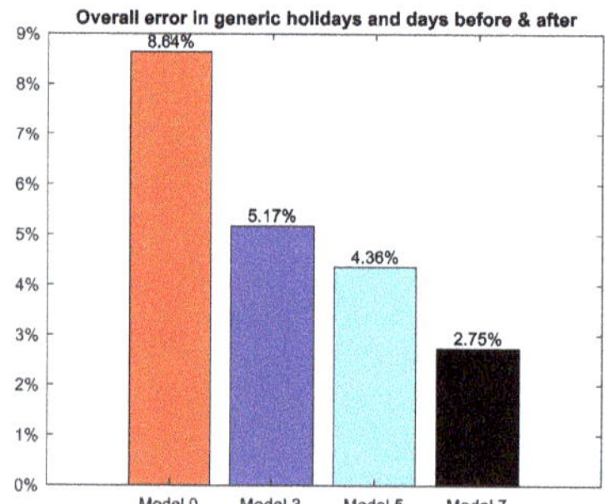

Figure 11. MAPE for models 0, 3, 5 and 7 for generic holidays and days before or after a holiday.

The coefficient profiles and errors graphs for all days in these categories are included in A4. In the case of holidays, the category is split into holidays on weekends, on Mondays and during the rest of the week. Figure 12 shows the coefficient profiles for these categories in the final model.

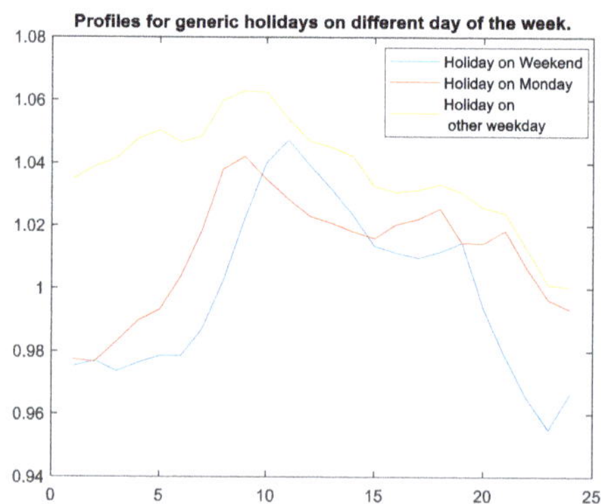

Figure 12. The different profiles that a generic holiday may have depend on the day of the week.

The difference between them can be explained by considering the next and previous days. Holidays whose previous day was a weekday start with a higher profile and holidays whose next day is a holiday finish with a lower profile than a regular Sunday.

A similar phenomenon happens regarding days before and after a holiday. Their coefficient profiles are shown in Figure 13a,b. In both cases, it is clear how, if the day is adjacent to both the holiday and a weekend, then the profile is lower and it is closer to a holiday's profile. However, Figure 13a shows that if the day before a holiday happens Tuesday to Friday then it is more similar to a Friday than to the corresponding weekday. Similarly, Figure 13b shows that if the day after a holiday happens Monday to Thursday, then its profile is more similar to a Monday than to its corresponding profile. Nevertheless, in both cases the profile is slightly lower than a regular day. Table 10 shows these similarities that justify the use of these variables.

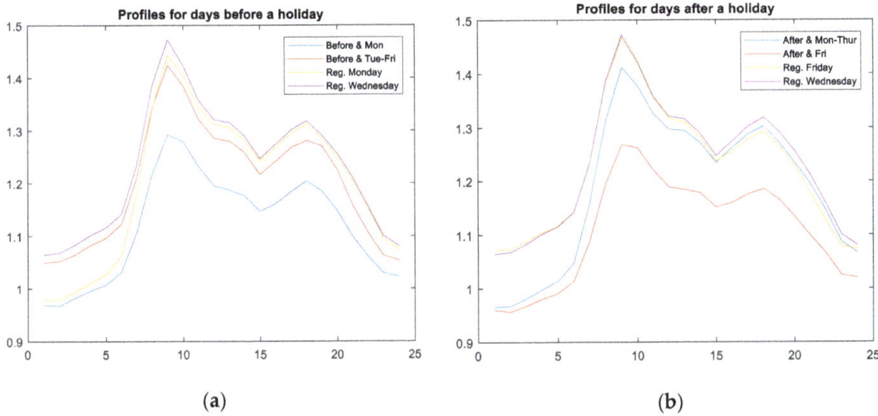

Figure 13. (**a**) Coefficient profiles for days before a holiday with regular Mondays and Wednesday as references. (**b**) Coefficient profiles for days after a holiday with regular Fridays and Wednesday as references.

Table 10. Differences between special days type 1 and 2.

Date	Holiday Sun & Sat	Holiday Mon	Holiday Tue-Fri	Before Holiday & Monday	Before Holiday & Tue-Fri	After Holiday & Mon-Thu	After Holiday & Friday
Difference	4.14%	4.14%	6.2%	2.38%	4.26%	3.17%	2.38%
Most similar	Holiday Mon	Holiday Sun & Sat	Holiday Mon	After & Friday	Reg Fri	Reg Mon	Before & Monday

4.5. Christmas Periods

Three different periods are included in this category as described in Table 5. The models that affect the classification of these periods are 0, 1, 4 and 7. Model 1 only affects the last period because it considers Jan 2nd and 5th as special days, as described in Table 6. The overall accuracy for all the periods is shown in Figure 14. The error increases from model 0 to 1 because both consider all these days as regular, but model 0 assigns a lower profile because it includes more real holidays. Model 4 introduces one category for each period and model 7 assigns one category for each of the four days of the first period.

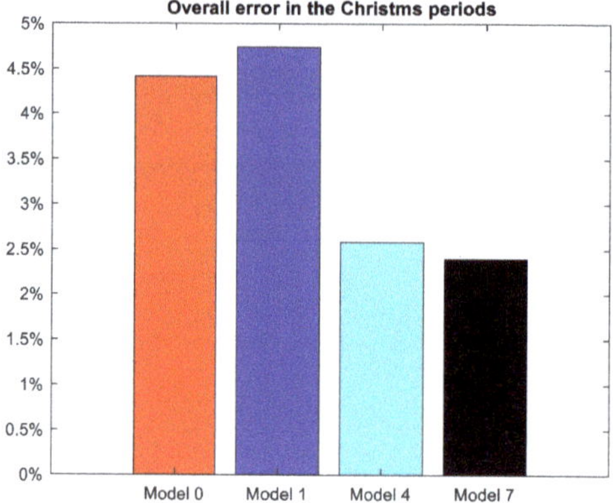

Figure 14. MAPE for models 0, 1, 4 and 7 for all days included in the three Christmas periods.

The first period (20–23 Dec) is defined by one variable for each day. These variables are modifiers, which means that, in addition to them, the day of the week is also active. The meaning of the coefficient profile of these variables is how a regular day is modified by having this variable active. Figure 15 shows the coefficient profiles associated with these four modifiers. It shows how, especially after noon, all days are different to a regular day (which would equal a straight line with a value of 1). In addition, the effect of Christmas is incremented gradually: each day closer to Christmas gets a lower profile than the one before.

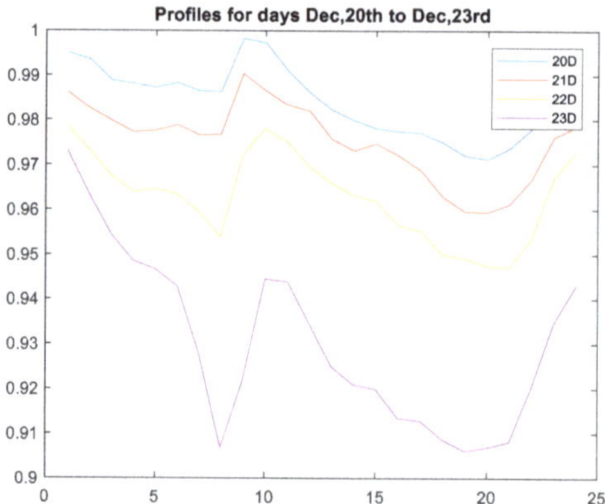

Figure 15. Coefficient profiles for the modifying variables of 20–23 Dec. Each day closer to Christmas day lowers the demand.

The other two periods are defined by only one variable for each of them. Figure 16 shows the coefficient profiles and error for these two periods. It can be seen in Figure 16a how model 1 yields a worse result because regular days from model 1 have a higher, more realistic profile than in model 0. Nevertheless, the even lower profiles from models 4 and 7 provide a much better model. Panel (b) shows that the last period is affected by the definition of special days in model 1. The lower error in this period from model 0 to model 1 is not obtained through a different general profile (the graph shows very similar profiles) but because two days in this period are affected by another variable.

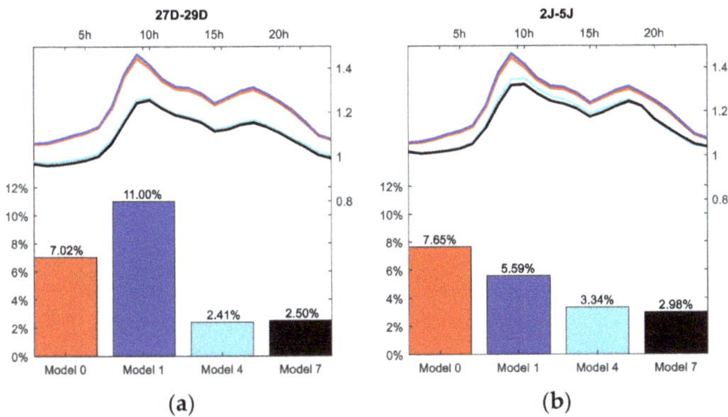

Figure 16. (**a**) Coefficient profile and MAPE for models 0, 1, 4, and 7 for the Christmas period 27–29 Dec. (**b**) Coefficient profile and MAPE for models 0, 1, 4, and 7 for the Christmas period 2–5 Jan.

To determine whether any of these profiles can be joined together, or if any of them is sufficiently similar to a regular day, Table 11 shows the most similar profiles.

Table 11. Similarities and differences among Christmas periods profiles.

Date	27 D–29 D	2 J–5 J	20 D	21 D	22 D	23 D
Difference	9.10%	5.46%	1.61%	1.61%	3.15%	5.46%
Most similar	2 J–5 J	23 D	21 D	20 D	21 D	2 J–5 J

The coefficient profiles and error graphs for all variables are included in Table A5.

4.6. Summer Vacation

The last period is the summer vacation during the month of August. As aforementioned, because vacations are usually assigned by full weeks it is possible that each week will have a specific profile. The results for models 0, 2, 4 and 7 are shown in Figure 17. Model 2 introduces the possibility of distinguishing between the first three weeks and the fourth while model 4 allows the distinction among all four weeks. Model 7 uses the same classification but it benefits from improvements from the definition of other special days (probably related to the national holiday of Aug 15th.).

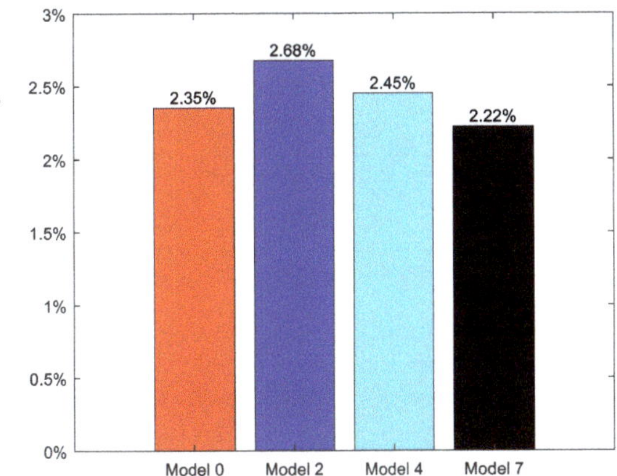

Figure 17. MAPE for models 0, 2, 4 and 7 for all days included in the summer vacation periods.

The different profiles for each week are shown in Figure 18. Weeks 2 and 3 show a profile around 4% lower than weeks 1 and 4, indicating that the two middle weeks concentrate vacations more than the first and last. The coefficient profile and error graphs are included in A6.

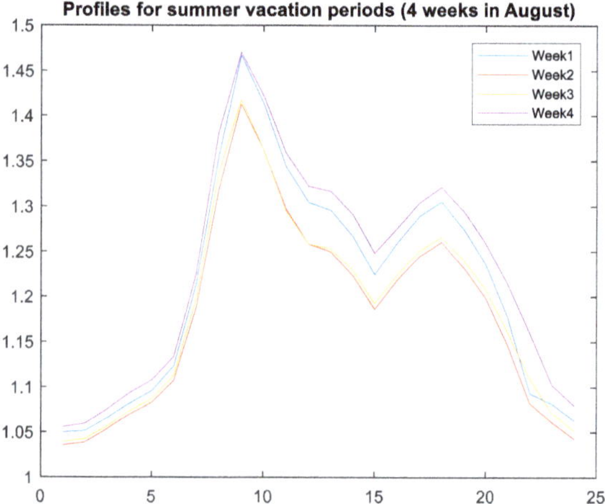

Figure 18. Coefficient profiles for days within the summer vacation period. Central weeks in August experience a lower demand than weeks 1 and 4.

4.7. Overall Result

The overall results for all the days that are considered special are shown in Figure 19. The boxplot shows the 5th, 15th, 50th, 85th and 95th percentile of the model errors for regular and special days from model 0 to model 7.

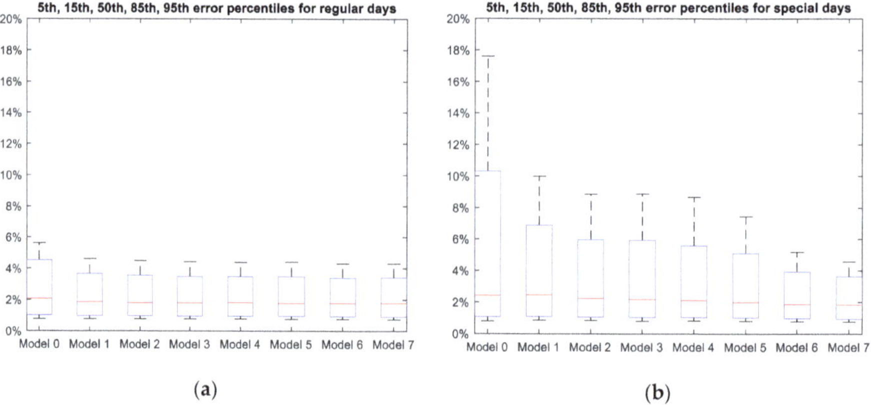

(a)   (b)

Figure 19. (a) Boxplot of the distribution of the error on regular days (b) Boxplot of the distribution of the error on special days.

The main conclusions that can be drawn from this are that the proposed special day classification yields an average error for special days of 1.84%, which is very near to the average error for regular days (1.78%). In addition, the 95th percentile for special days drops from 17.6% in model 0 to only 4.56% in model 7. This means that only 5% of special days have a modeling error larger than 4.56%. This number is even lower for regular days (4.33%). These results, shown in Table 12, suggest that the proposed classification is valid for modeling the special days present in the Spanish system almost as accurately as regular ones.

Table 12. Differences between special days type 1 and 2.

	%ile.	Model 0	Model 1	Model 2	Model 3	Model 4	Model 5	Model 6	Model 7
SPECIAL DAYS	5th	0.80%	0.85%	0.84%	0.81%	0.82%	0.80%	0.78%	0.76%
	15th	1.17%	1.18%	1.13%	1.13%	1.11%	1.10%	1.03%	1.01%
	50th	2.43%	2.47%	2.24%	2.19%	2.11%	1.99%	1.88%	1.84%
	85th	7.89%	5.85%	5.01%	4.95%	4.54%	4.31%	3.50%	3.32%
	95th	17.61%	9.99%	8.86%	8.88%	8.68%	7.45%	5.18%	4.56%
REGULAR DAYS	5th	0.77%	0.76%	0.77%	0.77%	0.76%	0.75%	0.74%	0.73%
	15th	1.09%	1.04%	1.03%	1.01%	1.01%	1.01%	0.99%	0.98%
	50th	2.07%	1.85%	1.80%	1.80%	1.81%	1.79%	1.78%	1.78%
	85th	4.17%	3.34%	3.25%	3.18%	3.17%	3.16%	3.09%	3.13%
	95th	5.62%	4.61%	4.48%	4.42%	4.38%	4.41%	4.31%	4.32%

The extensive categorization presented may be exported to other electrical systems although not all categories may prove to be relevant and some others may be needed. Categories that are not relevant can be identified by assessing the difference between their profile and those of other categories. If two categories present similar profiles, then they may be joined together. However, in order to define new variables, it would be necessary to study the error profiles of the least accurately modeled days and search for a similar pattern in days that can be jointly described in their own category.

5. Conclusions

Load forecasting is a key activity to any electric system and a lack of accuracy leads to an increase in operating costs. These costs grow exponentially as the error increases which leads to high costs on days for which load is hard to anticipate. These special days are those on which working or social habits differ from the ordinary like on holidays, vacation periods or days adjacent to them. The importance of correctly modeling the behavior of consumers in these special dates is key to reducing the maximum errors of a forecasting system. This paper has tested the validity of a special day classification system with more than 40 variables. This large number of variables may seem excessive as most reported models use much fewer categories. However, the results of this system compared to simpler versions of itself show that in order to model accurately the extensive variety of effects that the calendar has on consumer behavior it is necessary to implement a complex classification system like the one tested in this research. The methodology described can be transferred to other electrical systems with some adjustments to the category definition, but it is within reason that it would prove useful at least in similar systems like France, Portugal or Italy.

Author Contributions: M.L. conceived and designed the experiments; C.S. and M.L. performed the experiments; M.l. analyzed the data. M.L., C.S., C.S. and S.V. wrote the paper.

Funding: This research has financial support from "Subvenciones para Grupos de Investigación Consolidables AICO/2018/102", from: Consellería de Educación, Investigación, Cultura y Deporte de la GVA (Generalitat Valenciana). Dirección general de Universidad, Investigación y Ciencia.

Conflicts of Interest: The authors declare no conflict of interest.

Appendix A

The coefficient profiles and error charts for all categories are included here:

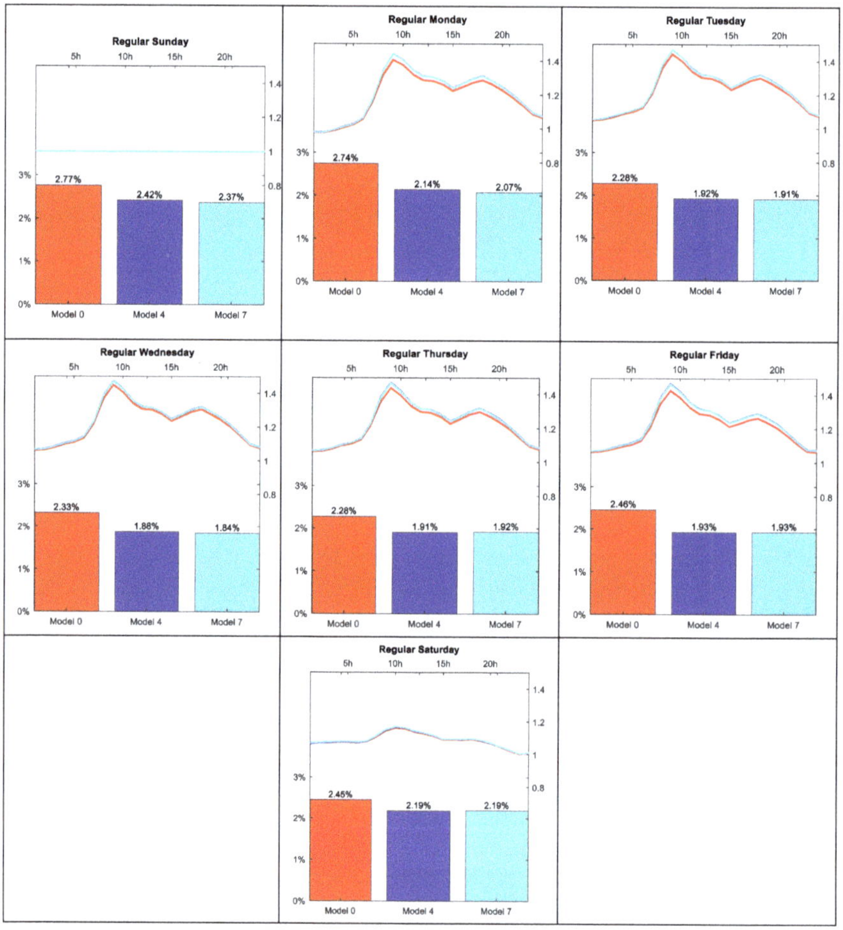

Figure A1. Regular days.

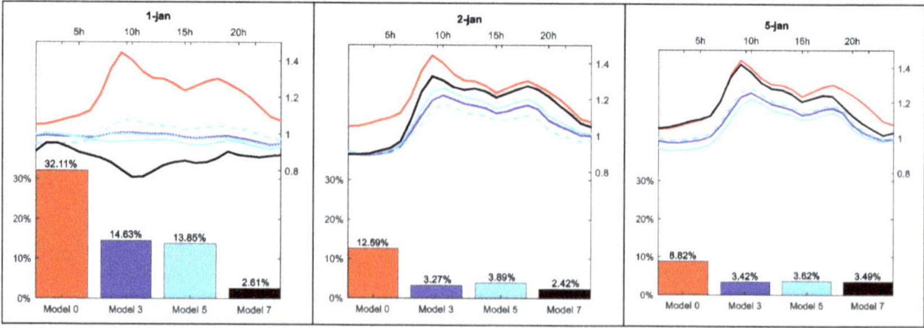

Figure A2. *Cont.*

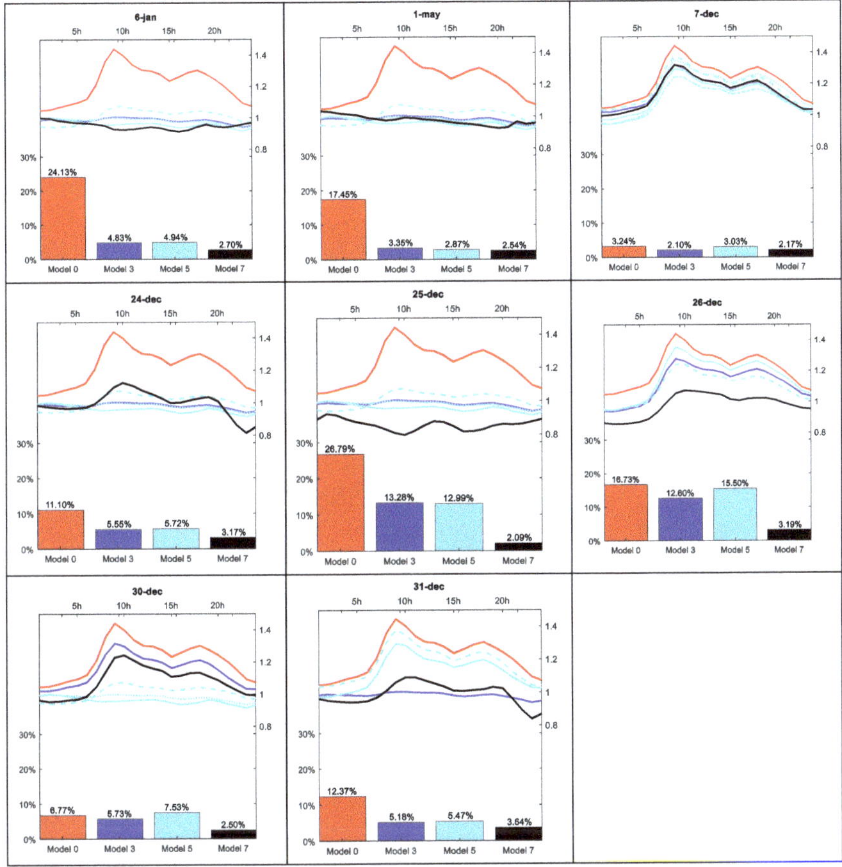

Figure A2. Special days type 1 and 2.

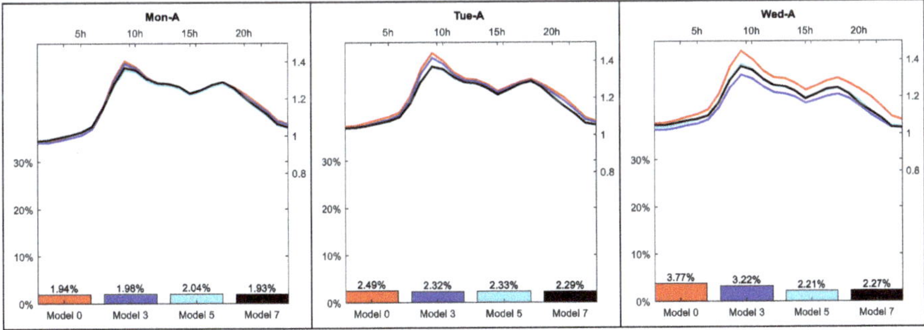

Figure A3. *Cont.*

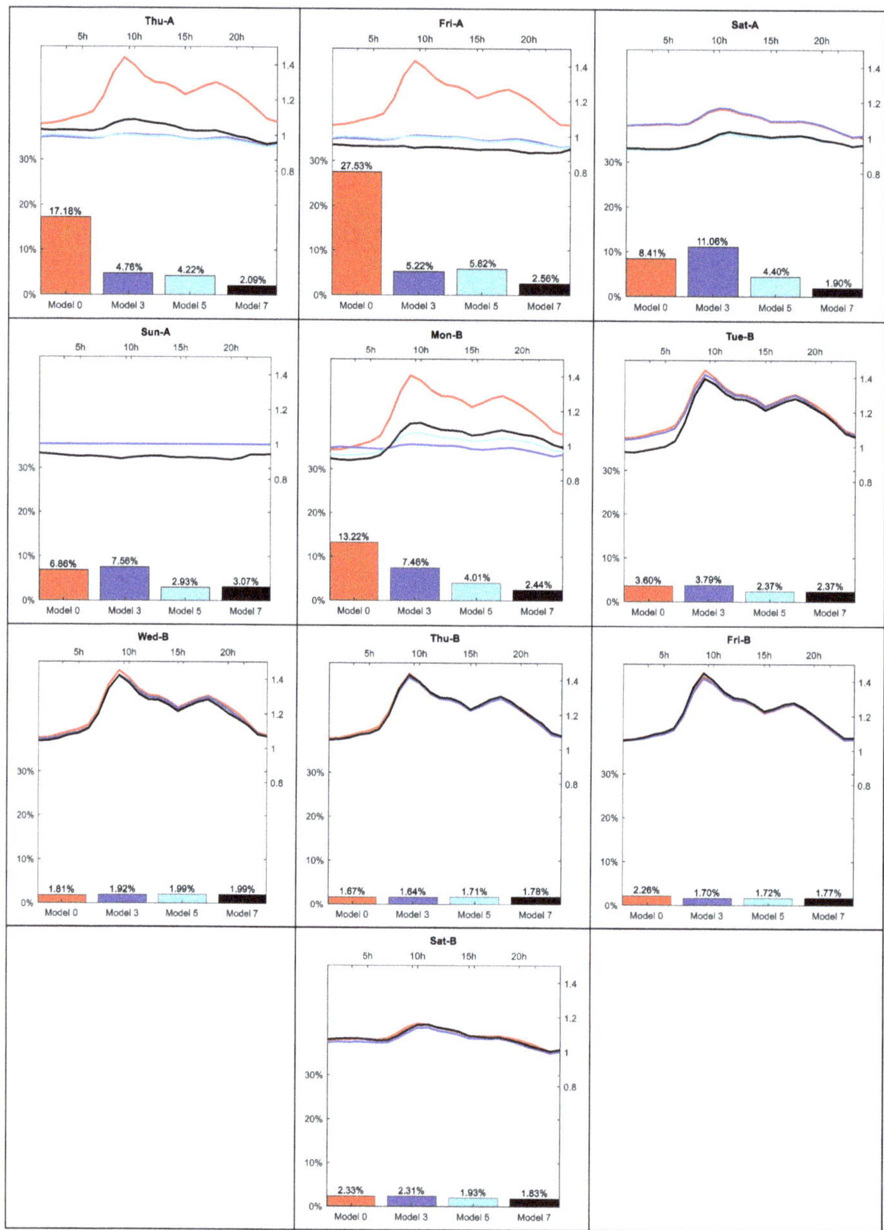

Figure A3. Special days type 3.

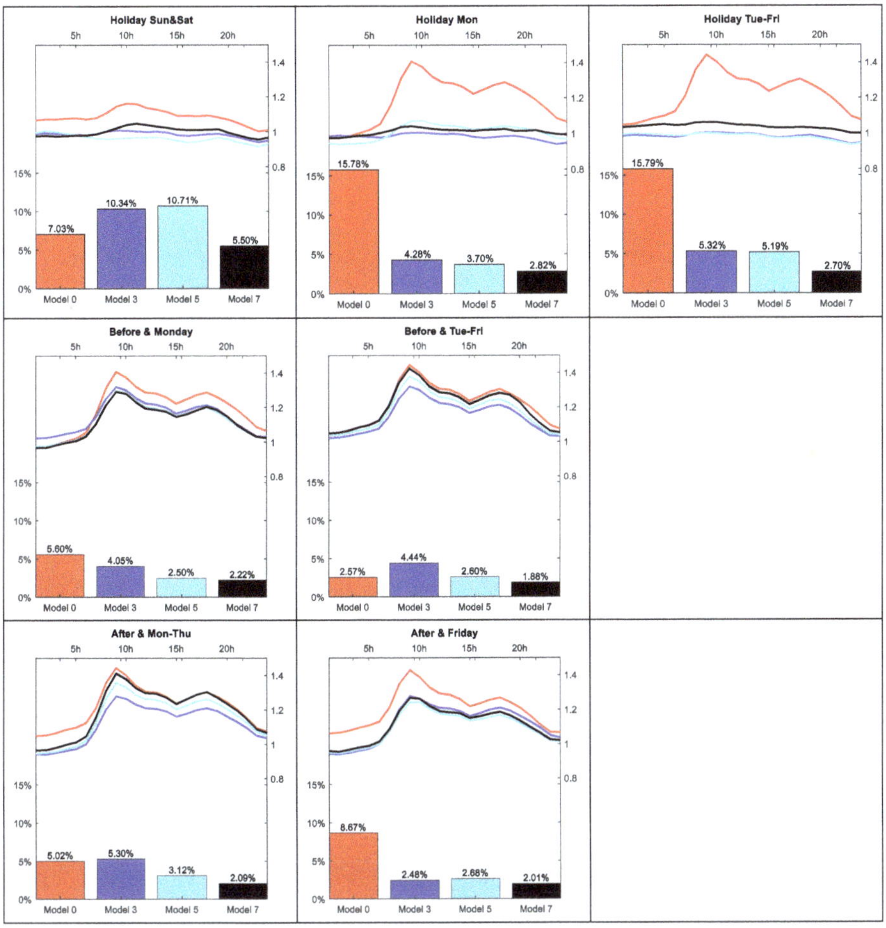

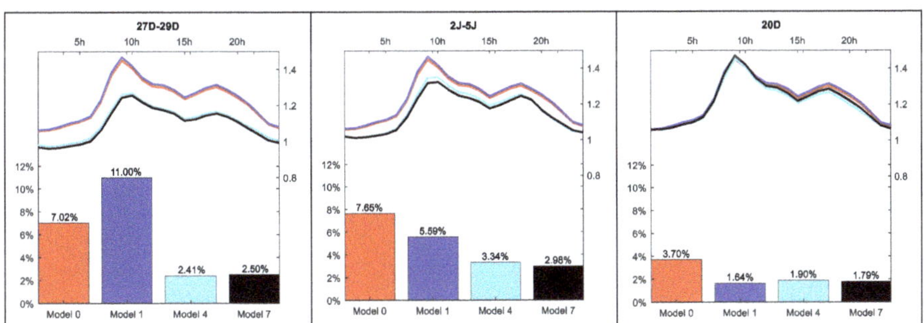

Figure A4. Regular holidays and days before and after.

Figure A5. *Cont.*

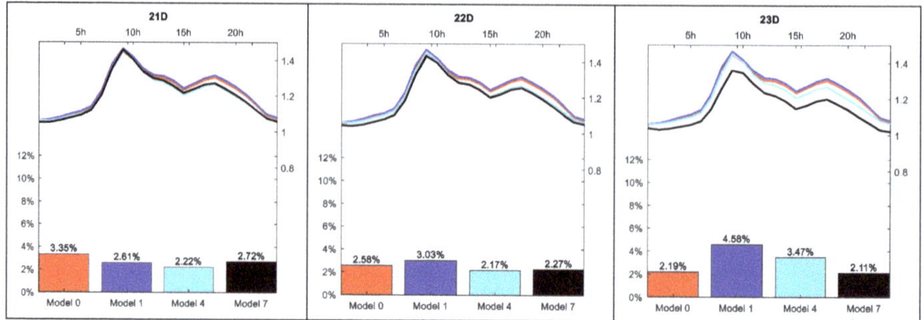

Figure A5. Christmas periods.

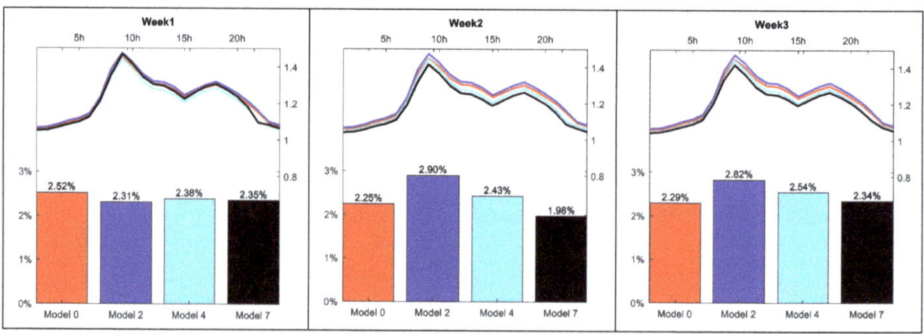

Figure A6. Summer vacation.

Appendix B

All calculated coefficients from the final model for each variable are detailed here:

Table A1. Regular days.

CAT	H1	H2	H3	H4	H5	H6	H7	H8	H9	H10	H11	H12	H13	H14	H15	H16	H17	H18	H19	H20	H21	H22	H23	H24
M	0.98	0.98	0.99	1.01	1.03	1.06	1.18	1.34	1.44	1.40	1.35	1.31	1.31	1.28	1.24	1.27	1.30	1.31	1.29	1.25	1.20	1.15	1.10	1.07
T	1.06	1.06	1.08	1.09	1.11	1.13	1.23	1.38	1.47	1.42	1.36	1.32	1.32	1.29	1.25	1.28	1.30	1.32	1.29	1.26	1.21	1.16	1.10	1.08
W	1.06	1.07	1.08	1.10	1.12	1.14	1.23	1.39	1.47	1.42	1.36	1.32	1.32	1.29	1.2 5	1.27	1.30	1.32	1.29	1.26	1.21	1.16	1.10	1.08
Th	1.07	1.07	1.08	1.10	1.12	1.14	1.23	1.38	1.47	1.42	1.35	1.32	1.31	1.28	1.24	1.27	1.30	1.32	1.29	1.26	1.21	1.16	1.10	1.08
F	1.07	1.07	1.09	1.10	1.12	1.14	1.23	1.38	1.47	1.42	1.36	1.32	1.31	1.28	1.24	1.26	1.28	1.29	1.26	1.23	1.18	1.12	1.08	1,07
Sa	1.07	1.08	1.08	1.08	1.08	1.08	1.08	1.12	1.15	1.17	1.17	1.15	1.13	1.12	1.10	1.10	1.09	1.10	1.09	1.07	1.05	1.03	1.01	1.01

Categories: (M) Regular Mondays, (T) Regular Tuesdays, (W) Regular Wednesdays, (Th) Regular Thursdays, (F) Regular Fridays, (Sa) Regular Saturdays.

Table A2. Special days types 1 and 2.

CAT	H1	H2	H3	H4	H5	H6	H7	H8	H9	H10	H11	H12	H13	H14	H15	H16	H17	H18	H19	H20	H21	H22	H23	H24
J1	0.90	0.95	0.95	0.93	0.90	0.88	0.87	0.84	0.80	0.76	0.77	0.80	0.83	0.85	0.85	0.84	0.84	0.87	0.90	0.88	0.88	0.87	0.88	0.88
J2	0.89	0.89	0.90	0.91	0.93	0.97	1.08	1.24	1.33	1.31	1.27	1.25	1.26	1.24	1.21	1.23	1.25	1.27	1.25	1.22	1.17	1.12	1.07	1.05
J5	1.05	1.06	1.07	1.09	1.10	1.13	1.21	1.35	1.42	1.38	1.31	1.28	1.27	1.25	1.20	1.22	1.24	1.24	1.18	1.13	1.08	1.05	1.01	1.03
J6	1.00	1.00	0.98	0.97	0.97	0.97	0.96	0.95	0.93	0.93	0.93	0.94	0.94	0.94	0.92	0.91	0.92	0.94	0.96	0.94	0.94	0.95	0.96	0.97
M1	1.03	1.03	1.02	1.01	1.01	0.99	0.98	0.98	0.98	0.99	0.99	0.98	0.98	0.97	0.96	0.95	0.95	0.94	0.93	0.93	0.93	0.97	0.95	0.96
D7	1.00	1.00	1.01	1.03	1.04	1.06	1.14	1.25	1.32	1.31	1.26	1.22	1.21	1.20	1.17	1.19	1.21	1.22	1.19	1.15	1.11	1.07	1.03	1.03
D24	0.99	0.98	0.97	0.97	0.97	0.97	0.99	1.04	1.10	1.12	1.11	1.08	1.06	1.03	1.00	1.00	1.02	1.03	1.03	1.01	0.94	0.86	0.81	0.85
D25	0.89	0.92	0.92	0.89	0.88	0.87	0.85	0.83	0.81	0.80	0.82	0.85	0.88	0.87	0.85	0.82	0.82	0.83	0.85	0.86	0.86	0.86	0.88	0.89
D26	0.86	0.86	0.86	0.86	0.87	0.89	0.94	1.00	1.05	1.07	1.07	1.06	1.05	1.04	1.02	1.01	1.02	1.02	1.02	1.01	0.99	0.97	0.96	0.95
D30	0.96	0.95	0.96	0.96	0.97	0.99	1.05	1.14	1.23	1.24	1.21	1.18	1.16	1.15	1.11	1.12	1.13	1.13	1.11	1.09	1.05	1.02	1.00	0.99
D31	0.96	0.95	0.95	0.94	0.94	0.95	0.97	1.01	1.06	1.09	1.09	1.07	1.06	1.04	1.01	1.01	1.01	1.02	1.03	1.02	0.96	0.89	0.84	0.87

Categories: (J1) Jan 1st, (J2) Jan 2nd, (J5) Jan 5th, (J6) Jan 6th, (M1) May 1s, (D7) Dec 7th, (D24) Dec 24th, (D25) Dec 25th, (D26) Dec 26th, (D30) Dec 30th, (D31) Dec 31st.

Table A3. Special days type 3.

CAT	H1	H2	H3	H4	H5	H6	H7	H8	H9	H10	H11	H12	H13	H14	H15	H16	H17	H18	H19	H20	H21	H22	H23	H24
MA	0.97	0.98	0.99	1.01	1.02	1.05	1.15	1.28	1.37	1.36	1.31	1.28	1.28	1.26	1.23	1.25	1.28	1.29	1.26	1.21	1.16	1.12	1.06	1.05
TA	1.04	1.04	1.05	1.06	1.08	1.10	1.17	1.29	1.37	1.36	1.31	1.29	1.28	1.26	1.22	1.25	1.27	1.29	1.26	1.20	1.16	1.11	1.06	1.05
HW	1.05	1.05	1.06	1.07	1.08	1.10	1.17	1.29	1.36	1.35	1.30	1.26	1.25	1.23	1.19	1.21	1.24	1.25	1.21	1.16	1.12	1.08	1.03	1.03
MT	1.03	1.03	1.03	1.03	1.03	1.02	1.03	1.07	1.08	1.09	1.08	1.07	1.06	1.05	1.03	1.02	1.03	1.03	1.01	0.99	0.99	0.97	0.95	0.96
GF	0.95	0.95	0.94	0.95	0.94	0.94	0.94	0.94	0.93	0.94	0.94	0.94	0.93	0.93	0.93	0.93	0.93	0.93	0.92	0.91	0.91	0.91	0.91	0.93
HS	0.94	0.94	0.94	0.94	0.94	0.94	0.95	0.97	0.99	1.02	1.03	1.03	1.02	1.01	1.00	1.01	1.01	1.01	1.00	0.99	0.98	0.97	0.95	0.96
ES	0.95	0.94	0.94	0.93	0.93	0.93	0.93	0.92	0.91	0.92	0.93	0.93	0.93	0.92	0.92	0.92	0.92	0.92	0.92	0.91	0.92	0.94	0.94	0.94
EM	0.92	0.92	0.91	0.92	0.92	0.94	1.00	1.08	1.13	1.13	1.11	1.09	1.09	1.08	1.06	1.07	1.08	1.09	1.08	1.06	1.06	1.04	1.01	0.99
TB	0.97	0.97	0.98	0.99	1.00	1.03	1.14	1.30	1.39	1.36	1.31	1.28	1.27	1.25	1.21	1.24	1.27	1.28	1.26	1.22	1.19	1.14	1.08	1.06
WB	1.04	1.04	1.05	1.07	1.09	1.11	1.20	1.34	1.42	1.38	1.31	1.28	1.28	1.25	1.21	1.24	1.27	1.28	1.25	1.20	1.17	1.13	1.08	1.07
ThB	1.06	1.06	1.07	1.08	1.09	1.11	1.20	1.35	1.43	1.39	1.33	1.30	1.30	1.27	1.23	1.26	1.29	1.31	1.28	1.23	1.19	1.15	1.10	1.08
FB	1.06	1.06	1.07	1.09	1.10	1.13	1.22	1.37	1.45	1.40	1.34	1.30	1.30	1.27	1.23	1.24	1.27	1.28	1.25	1.21	1.16	1.12	1.08	1.07
SB	1.07	1.08	1.08	1.08	1.07	1.07	1.07	1.09	1.13	1.15	1.16	1.14	1.13	1.12	1.09	1.09	1.08	1.09	1.07	1.06	1.03	1.02	1.00	1.01

Categories: (MA) Monday before Good Friday, (TA) Tuesday before Good Friday, (HW) Holy Wednesday, (MT) Maundy Thursday, (GF) Good Friday, (HS) Holy Saturday, (ES) Easter Sunday, (EM) Easter Monday, (TB) Tuesday after Easter Monday, (WB) Wednesday after Easter Monday, (ThB) Thursday after Easter Monday, (FB) Friday after Easter Monday, (SB) Saturday after Easter Monday.

Table A4. Generic holidays and days before or after a holiday.

CAT	H1	H2	H3	H4	H5	H6	H7	H8	H9	H10	H11	H12	H13	H14	H15	H16	H17	H18	H19	H20	H21	H22	H23	H24
HW	0.98	0.98	0.97	0.98	0.98	0.98	0.99	1.00	1.02	1.04	1.05	1.04	1.03	1.02	1.01	1.01	1.01	1.01	1.01	0.99	0.98	0.97	0.96	0.97
HM	0.98	0.98	0.98	0.99	0.99	1.00	1.02	1.04	1.04	1.03	1.03	1.02	1.02	1.02	1.02	1.02	1.02	1.03	1.01	1.01	1.02	1.01	1.00	0.99
HR	1.03	1.04	1.04	1.05	1.05	1.05	1.05	1.06	1.06	1.06	1.05	1.05	1.05	1.04	1.03	1.03	1.03	1.03	1.03	1.03	1.02	1.01	1.00	1.00
BM	0.97	0.97	0.98	1.00	1.01	1.03	1.10	1.22	1.29	1.28	1.23	1.19	1.19	1.18	1.15	1.16	1.18	1.20	1.19	1.15	1.10	1.06	1.03	1.02
BR	1.05	1.05	1.06	1.08	1.10	1.12	1.21	1.34	1.42	1.39	1.32	1.29	1.28	1.26	1.22	1.24	1.27	1.28	1.27	1.23	1.16	1.11	1.06	1.05
AR	0.97	0.97	0.98	1.00	1.01	1.05	1.16	1.31	1.41	1.38	1.33	1.30	1.29	1.27	1.23	1.26	1.29	1.30	1.27	1.23	1.19	1.14	1.09	1.07
AF	0.96	0.96	0.97	0.98	0.99	1.01	1.09	1.20	1.27	1.26	1.22	1.19	1.18	1.18	1.15	1.16	1.18	1.19	1.17	1.13	1.10	1.07	1.03	1.02

Categories: (HW) Holiday on Weekend, (HM) Holiday on Monday, (HR) Holiday on Tuesday to Friday, (BM) Before a holiday and Monday, (BR) Before a holiday and Tuesday to Friday, (AR) After a holiday and Monday to Thursday, (AF) After a holiday and Friday.

Table A5. Christmas periods.

CAT	H1	H2	H3	H4	H5	H6	H7	H8	H9	H10	H11	H12	H13	H14	H15	H16	H17	H18	H19	H20	H21	H22	H23	H24
P1	0.91	0.90	0.89	0.88	0.88	0.88	0.86	0.83	0.85	0.88	0.90	0.90	0.89	0.90	0.90	0.88	0.88	0.88	0.88	0.88	0.89	0.90	0.91	0.92
P2	0.96	0.95	0.94	0.93	0.93	0.92	0.91	0.89	0.90	0.93	0.94	0.94	0.94	0.94	0.94	0.94	0.94	0.94	0.95	0.93	0.93	0.94	0.96	0.96
D20	1.00	0.99	0.99	0.99	0.99	0.99	0.99	0.99	1.00	1.00	0.99	0.99	0.98	0.98	0.98	0.98	0.98	0.98	0.97	0.97	0.97	0.98	0.98	0.98
D21	0.99	0.98	0.98	0.98	0.98	0.98	0.98	0.98	0.99	0.99	0.98	0.98	0.98	0.97	0.97	0.97	0.97	0.96	0.96	0.96	0.96	0.97	0.98	0.98
D22	0.98	0.97	0.97	0.96	0.96	0.96	0.96	0.95	0.97	0.98	0.98	0.97	0.97	0.96	0.96	0.96	0.96	0.95	0.95	0.95	0.95	0.95	0.97	0.97
D23	0.97	0.96	0.95	0.95	0.95	0.94	0.93	0.91	0.92	0.94	0.94	0.93	0.92	0.92	0.921	0.91	0.91	0.91	0.91	0.91	0.91	0.92	0.93	0.94

Categories: (P1) Dec 27th to Dec 29th, (P2) Jan 2nd to Jan 5th, (D20), Dec 20th, (D21) Dec 21st, (D22) Dec 22nd, (D23) Dec 23rd.

Table A6. Summer vacation.

CAT	H1	H2	H3	H4	H5	H6	H7	H8	H9	H10	H11	H12	H13	H14	H15	H16	H17	H18	H19	H20	H21	H22	H23	H24
W1	0.99	0.98	0.98	0.98	0.98	0.98	0.99	0.98	1.00	0.99	0.99	0.99	0.98	0.98	0.98	0.99	0.99	0.99	0.99	0.98	0.97	0.95	0.98	0.98
W2	0.97	0.97	0.97	0.97	0.97	0.97	0.96	0.95	0.96	0.96	0.96	0.95	0.95	0.95	0.95	0.96	0.96	0.96	0.95	0.95	0.95	0.94	0.96	0.96
W3	0.98	0.98	0.97	0.97	0.97	0.97	0.97	0.96	0.96	0.96	0.95	0.95	0.95	0.95	0.96	0.96	0.96	0.96	0.96	0.96	0.96	0.96	0.97	0.97

Categories: (W1) 1st week of August, (W2) 2nd week of August, (W3) 3rd week of August.

References

1. Paris Agreement - Status of Ratification | UNFCCC. Available online: https://unfccc.int/process/the-paris-agreement/status-of-ratification (accessed on 9 November 2018).
2. EU Emissions Trading System (EU ETS). Available online: https://ec.europa.eu/clima/policies/ets_en (accessed on 26 November 2018).
3. Gross, G.; Galiana, F.D. Short-term load forecasting. *Proc. IEEE* **1987**, *75*, 1558–1573. [CrossRef]
4. Hippert, H.S.; Pedreira, C.E.; Souza, R.C. Neural networks for short-term load forecasting: a review and evaluation. *IEEE Trans. Power Syst.* **2001**, *16*, 44–55. [CrossRef]
5. Dudek, G. Neural networks for pattern-based short-term load forecasting: A comparative study. *Neurocomputing* **2016**, *205*, 64–74. [CrossRef]
6. Huang, G.-B.; Zhu, Q.-Y.; Siew, C.-K. Extreme learning machine: Theory and applications. *Neurocomputing* **2006**, *70*, 489–501. [CrossRef]
7. Zhang, R.; Dong, Z.Y.; Xu, Y.; Meng, K.; Wong, K.P. Short-term load forecasting of Australian National Electricity Market by an ensemble model of extreme learning machine. *Trans. Distrib. IET Gene.* **2013**, *7*, 391–397. [CrossRef]
8. Cortes, C.; Vapnik, V. Support-vector networks. *Mach. Learn* **1995**, *20*, 273–297. [CrossRef]
9. Jain, A.; Satish, B. Integrated architecture for short term load forecasting using support vector machines. In Proceedings of the 2008 40th North American Power Symposium, Calgary, AB, Canada, 28–30 September 2008; pp. 1–8.
10. Jang, J.-R. ANFIS: adaptive-network-based fuzzy inference system. *IEEE Trans. Syst. Man Cybern.* **1993**, *23*, 665–685. [CrossRef]
11. M, A.E.; Perez, L.P. Application of support vector machines and ANFIS to the short-term load forecasting. In Proceedings of the 2008 IEEE/PES Transmission and Distribution Conference and Exposition: Latin America, Bogota, CO, USA, 13–15 August 2008; pp. 1–5.
12. Lee, S.C.; Lee, E.T. Fuzzy Neural Networks. *Math. Biosci.* **1975**, *23*, 151–177. [CrossRef]
13. Takagi, T.; Sugeno, M. Fuzzy identification of systems and its applications to modeling and control. *IEEE Trans. Syst. Man Cybern.* **1985**, *SMC-15*, 116–132. [CrossRef]
14. Çevik, H.H.; Çunkaş, M. Short-term load forecasting using fuzzy logic and ANFIS. *Neural Comput & Applic* **2015**, *26*, 1355–1367.
15. Mamlook, R.; Badran, O.; Abdulhadi, E. A fuzzy inference model for short-term load forecasting. *Energy Policy* **2009**, *37*, 1239–1248. [CrossRef]
16. Liao, G.C.; Tsao, T.P. Application of a fuzzy neural network combined with a chaos genetic algorithm and simulated annealing to short-term load forecasting. *IEEE Trans. Evolut. Comput.* **2006**, *10*, 330–340. [CrossRef]
17. Kohonen, T. *Self-Organization and Associative Memory*, 3rd ed.; Springer Series in Information Sciences; Springer: Berlin, Heidelberg, 1989; ISBN 978-3-540-51387-2.
18. López, M.; Valero, S.; Senabre, C.; Aparicio, J.; Gabaldon, A. Application of SOM neural networks to short-term load forecasting: The Spanish electricity market case study. *Electr. Power Syst. Res.* **2012**, *91*, 18–27.
19. Li, H.; Zhu, Y.; Hu, J.; Li, Z. A localized NARX Neural Network model for Short-term load forecasting based upon Self-Organizing Mapping. In Proceedings of the 2017 IEEE 3rd International Future Energy Electronics Conference and ECCE Asia (IFEEC 2017-ECCE Asia), Kaohsiung, Taiwan, 3–7 June 2017; pp. 749–754.
20. Baharudin, Z.; Kamel, N. Autoregressive method in short term load forecast. In Proceedings of the 2008 IEEE 2nd International Power and Energy Conference, Johor Bahru, Malaysia, 1–3 December 2008; pp. 1603–1608.
21. Taylor, J.W.; McSharry, P.E. Short-Term Load Forecasting Methods: An Evaluation Based on European Data. *IEEE Transactions on Power Systems* **2007**, *22*, 2213–2219. [CrossRef]
22. Hanmandlu, M.; Chauhan, B.K. Load Forecasting Using Hybrid Models. *IEEE Transactions on Power Systems* **2011**, *26*, 20–29. [CrossRef]
23. Fan, S.; Chen, L. Short-term load forecasting based on an adaptive hybrid method. *IEEE Transactions on Power Systems* **2006**, *21*, 392–401. [CrossRef]
24. López, M.; Valero, S.; Rodriguez, A.; Veiras, I.; Senabre, C. New online load forecasting system for the Spanish Transport System Operator. *Electric Power Systems Research* **2018**, *154*, 401–412. [CrossRef]
25. Brown, R.G.; Meyer, R.F.; D'Esopo, D.A. The Fundamental Theorem of Exponential Smoothing. *Oper. Res.* **1961**, *9*, 673–687. [CrossRef]

26. Pardo, A.; Meneu, V.; Valor, E. Temperature and seasonality influences on Spanish electricity load. *Energy Econ.* **2002**, *24*, 55–70. [CrossRef]

27. Pal, S.K.; Mitra, S. Multilayer perceptron, fuzzy sets, and classification. *IEEE Trans. Neural Netw.* **1992**, *3*, 683–697. [CrossRef]

28. Park, J.; Sandberg, I.W. Universal Approximation Using Radial-Basis-Function Networks. *Neural Comput.* **1991**, *3*, 246–257. [CrossRef]

29. Rutkowski, L. Generalized Regression Neural Networks in a Time-Varying Environment. In *New Soft Computing Techniques for System Modeling, Pattern Classification and Image Processing*; Rutkowski, L., Ed.; Studies in Fuzziness and Soft Computing; Springer Berlin Heidelberg: Berlin, Heidelberg, 2004; pp. 73–134, ISBN 978-3-540-40046-2.

30. Kuzmanovski, I.; Novič, M. Counter-propagation neural networks in Matlab. *Chem. Intell. Lab. Syst.* **2008**, *90*, 84–91. [CrossRef]

31. Huo, F.; Poo, A.-N. Nonlinear autoregressive network with exogenous inputs based contour error reduction in CNC machines. *Int. J. Mach. Tools Manuf.* **2013**, *67*, 45–52. [CrossRef]

32. Huang, S.J.; Shih, K.R. Shih Short-term load forecasting via ARMA model identification including non-Gaussian process considerations. *IEEE Trans. Power Syst.* **2003**, *18*, 673–679. [CrossRef]

33. Box, G. Box and Jenkins: Time Series Analysis, Forecasting and Control. In *A Very British Affair: Six Britons and the Development of Time Series Analysis During the 20th Century*; Mills, T.C., Ed.; Palgrave Advanced Texts in Econometrics; Palgrave Macmillan UK: London, UK, 2013; pp. 161–215. ISBN 978-1-137-29126-4.

34. Jolliffe, I.T. *Principal Component Analysis*; Springer Series in Statistics; Springer: New York, NY, USA, 1986; ISBN 978-1-4757-1904-8.

35. Lee, K.Y.; Cha, Y.T.; Park, J.H. Short-term load forecasting using an artificial neural network. *IEEE Trans. Power Syst.* **1992**, *7*, 124–132. [CrossRef]

36. Rahman, S.; Bhatnagar, R. An expert system based algorithm for short term load forecast. *IEEE Trans. Power Syst.* **1988**, *3*, 392–399. [CrossRef]

37. Christiaanse, W.R. Short-Term Load Forecasting Using General Exponential Smoothing. *IEEE Trans. Power Appar. Syst.* **1971**, *PAS-90*, 900–911. [CrossRef]

38. Kim, K.-H.; Youn, H.-S.; Kang, Y.-C. Short-term load forecasting for special days in anomalous load conditions using neural networks and fuzzy inference method. *IEEE Trans. Power Syst.* **2000**, *15*, 559–565.

39. Lamedica, R.; Prudenzi, A.; Sforna, M.; Caciotta, M.; Cencellli, V.O. A neural network based technique for short-term forecasting of anomalous load periods. *IEEE Trans. Power Syst.* **1996**, *11*, 1749–1756. [CrossRef]

40. Arora, S.; Taylor, J.W. Short-Term Forecasting of Anomalous Load Using Rule-Based Triple Seasonal Methods. *IEEE Trans. Power Syst.* **2013**, *28*, 3235–3242. [CrossRef]

41. Arora, S.; Taylor, J.W. Rule-based autoregressive moving average models for forecasting load on special days: A case study for France. *Eur. J. Oper. Res.* **2018**, *266*, 259–268. [CrossRef]

42. Song, K.B.; Baek, Y.S.; Hong, D.H.; Jang, G. Short-term load forecasting for the holidays using fuzzy linear regression method. *IEEE Trans. Power Syst.* **2005**, *20*, 96–101. [CrossRef]

43. Chang, C.S.; Liew, A.C. Demand forecasting using fuzzy neural computation, with special emphasis on weekend and public holiday forecasting. *IEEE Trans. Power Syst.* **1995**, *10*, 1897–1903.

44. Cancelo, J.R.; Espasa, A.; Grafe, R. Forecasting the electricity load from one day to one week ahead for the Spanish system operator. *Int. J. Forecast.* **2008**, *24*, 588–602. [CrossRef]

45. Ramanathan, R.; Engle, R.; Granger, C.W.J.; Vahid-Araghi, F.; Brace, C. Short-run forecasts of electricity loads and peaks. *Int. J. Forecast.* **1997**, *13*, 161–174. [CrossRef]

46. Kim, M.S. Modeling special-day effects for forecasting intraday electricity demand. *Eur. J. Oper. Res.* **2013**, *230*, 170–180. [CrossRef]

47. Gobierno de España - Agencia Estatal Boletín Oficial del Estado. Available online: https://www.boe.es/ (accessed on 10 December 2018).

Article

HousEEC: Day-Ahead Household Electrical Energy Consumption Forecasting Using Deep Learning

Ivana Kiprijanovska [1,2,†], Simon Stankoski [1,2,†], Igor Ilievski [3], Slobodan Jovanovski [3], Matjaž Gams [1,2] and Hristijan Gjoreski [4,*]

[1] Department of Intelligent Systems, Jožef Stefan Institute, 1000 Ljubljana, Slovenia; ivana.kiprijanovska@ijs.si (I.K.); simon.stankoski@ijs.si (S.S.); matjaz.gams@ijs.si (M.G.)
[2] Jožef Stefan Postgraduate School, Information and Communication Technologies, 1000 Ljubljana, Slovenia
[3] ITS Iskratel, ITS Softver Centar, 1000 Skopje, North Macedonia; igor.ilievski@its-sk.com.mk (I.I.); jovanovski@iskratel.si (S.J.)
[4] Faculty of Electrical Engineering and Information Technologies, Ss. Cyril and Methodius University, 1000 Skopje, North Macedonia
* Correspondence: hristijang@feit.ukim.edu.mk
† The first two authors have contributed equally to this work.

Received: 31 March 2020; Accepted: 20 May 2020; Published: 25 May 2020

Abstract: Short-term load forecasting is integral to the energy planning sector. Various techniques have been employed to achieve effective operation of power systems and efficient market management. We present a scalable system for day-ahead household electrical energy consumption forecasting, named HousEEC. The proposed forecasting method is based on a deep residual neural network, and integrates multiple sources of information by extracting features from (i) contextual data (weather, calendar), and (ii) the historical load of the particular household and all households present in the dataset. Additionally, we compute novel domain-specific time-series features that allow the system to better model the pattern of energy consumption of the household. The experimental analysis and evaluation were performed on one of the most extensive datasets for household electrical energy consumption, Pecan Street, containing almost four years of data. Multiple test cases show that the proposed model provides accurate load forecasting results, achieving a root-mean-square error score of 0.44 kWh and mean absolute error score of 0.23 kWh, for short-term load forecasting for 300 households. The analysis showed that, for hourly forecasting, our model had 8% error (22 kWh), which is 4 percentage points better than the benchmark model. The daily analysis showed that our model had 2% error (131 kWh), which is significantly less compared to the benchmark model, with 6% error (360 kWh).

Keywords: short-term load forecasting; day ahead; feature extraction; deep residual neural network; multiple sources; electricity

---

1. Introduction

Electrical energy (EE) is one of the most significant driving forces of economic development, and is considered essential to daily life. Although EE is a clean form of energy when it is used, the production and transmission of electricity can have a negative effect on the environment. Additionally, overproduction of EE is problematic, because storing excess electricity is challenging and difficult even with today's technological advances. Hence, a system that can accurately predict EE consumption can be used for electricity production planning, and significantly reduce the problems with storage and overproduction.

In recent years, with the introduction of deregulation and liberalization of the energy markets, EE consumption forecasting has become even more relevant. An accurate short-term load forecasting

(STLF) system can play a crucial role in effective power system operation and efficient market management. Such a system has multiple benefits: (i) it can optimize the production process, thus reducing the cost of overproduction and improving equipment utilization; (ii) it is eco-friendly, with fewer resources used to produce electricity; (iii) it can help in optimizing power grid load and strengthening reliability; (iv) it can potentially decrease EE consumption costs for households by better planning the production/buying of EE in advance; and (v) it emphasizes EE trading possibilities.

The massive development of smart grid technologies in the residential sector brings many challenges to the load forecasting community. It allows EE consumption to be obtained in close to real time, and allows extraction of valuable data that both the supply and demand side can use for efficient management of the electricity load network.

In recent years, there have been various data-driven approaches for modeling and forecasting EE consumption. Most of them focus on industrial objects, factories, and companies, and some are more focused on households. Furthermore, some focus on short-term forecasts (hourly, daily) with a small prediction horizon (an hour in advance), and some focus on long-term forecasts (weekly, monthly). The studies that focus on STLF with a large prediction horizon (at least one day ahead) are quite limited. Therefore, in this paper, we present the household electrical energy consumption (HousEEC) forecast system, which provides day-ahead household electrical energy (EE) consumption forecasts, using a deep residual neural network (DRNN) that combines multiple sources of information. The key contributions of the paper are as follows:

- A review of the existing EE consumption approaches and a highlight of their current limitations (Section 2).
- An extensive analysis and evaluation of the Pecan Street dataset, the largest and richest household EE consumption dataset (Section 3).
- A novel deep learning (DL) method with a scalable architecture that can work with different numbers of households. It is based on DRNN and includes multisource feature extraction, regression learning, and forecasting of hourly EE consumption of multiple households one day in advance. The proposed DRNN uses pre-activation residual blocks and separate input branches for different types of features (Section 4).
- Novel domain-specific historical time series, from which numerous time and frequency features are extracted (Section 4). These features give new insight into the time-series dynamics and significantly increase the performance of the forecast models.
- An extensive evaluation of the method, including: (i) a comparison of our proposed method with seven machine learning (ML) algorithms, five deep learning (DL) approaches, and three benchmark/reference approaches; (ii) error analysis of different application scenarios (hourly, daily and monthly EE consumption); and (iii) a comparison of achieved results for household STLF with results from other state-of-the-art approaches (Sections 5 and 6).
- A practical implementation of the system in a prototype web application, where ML models are deployed and execute the forecasts on a daily basis (Section 7).
- A discussion about the results, the forecasting efficiency and its significance, and potential use of the model in a commercial EE monitoring system (Section 8).

2. Related Work

Selecting a forecasting method depends on multiple factors, including the availability and relevance of historical data, desired prediction accuracy, the forecast horizon, and so forth. In recent years, the STLF problem has been tackled by utilizing various methods, each one characterized by different advantages and disadvantages in terms of training complexity, prediction accuracy, limitations in the forecasting horizon, etc. In general, the related work in STLF can be divided into two categories, depending on the type of user (industrial entities or households) and method used (e.g., statistical, ML, DL).

2.1. Related Methods

With the advent of statistical software packages and artificial intelligence techniques, numerous methods have been proposed to model future EE consumption and improve forecasting performance. These methods can be divided into two categories: conventional statistical methods and methods based on artificial intelligence (AI).

Statistical methods provide explicit mathematical models where the load is represented as a function of several input factors. These were the first used methods, and for years represented the benchmark among systems for STLF. All of these methods, which include smoothing techniques, data extrapolation and curve fitting, assume that the load data have an internal structure. Autoregressive moving average (ARMA) models were among the first used in STLF [1–3]. Soon they were replaced by autoregressive integrated moving average (ARIMA) models [4] and seasonal ARIMA models [5] to deal with time variance often exhibited by load consumption profiles. Other examples of statistical methods used in STLF are multiple regression [6], exponential smoothing [7], adoptive load forecasting [8,9] and Kalman filtering [10,11]. The major weakness of these approaches is their assumption of the linearity of the observed system. EE forecasting is a complex multivariable and multidimensional estimation problem, and these methods are not always suitable for finding the nonlinear relationship between the independent influencing variables and the EE consumption.

On the contrary, advanced ML methods are suitable for finding patterns and regularities in the data and use them to forecast future EE consumption. ML based methods have shown great performance in the field of STLF. The most commonly used ML algorithms for STLF are support vector machines (SVM) [12,13], random forest [14,15] and artificial neural networks (ANNs) [16]. However, as shown in numerous studies and in the benchmark Global Energy Forecasting Competition 2012 (GEFCom2012) [17], very often, simple ML methods applied to manually crafted complex features (polynomial and exponential interaction features combining multiple variables) achieve better and more robust performance [18]. These features often use the lagged and recency effect, first introduced in [19]. One of the winning teams [20] at GEFCom2012 used lagged hourly and average daily temperature variables in the competition. They applied a gradient boosting algorithm to learn the dependencies between features and target variables. Another winning team at GEFCom2012 [21] used exponentially smoothed temperature variables. They used generalized additive models and kernel regression for long-term load and medium-term forecasting, and random forests for short-term load forecasting.

Over the past few years, DL has been a subject of intense study in many fields, especially in time-series prediction. Deep neural networks (DNNs) have shown the capability to approximate any complex function with arbitrary precision. In [22], the authors showed that some DNN architectures are able to outperform classical ML approaches in the load forecasting task. The authors of [23] proposed convolutional neural network (CNN), as an effective and accurate approach for household-level load forecasting. They showed that CNN is able to capture short-term trends in load data and that a data-augmentation technique can improve the load forecasting accuracy. Compared with conventional feedforward neural networks, recurrent neural networks (RNNs) have the particular advantage of coping with historical data through a feedback connection. In [24], the authors presented a deep RNN to predict electricity consumption for commercial and residential buildings. As an extension of RNN, long short-term memory (LSTM) networks have been used in the load forecasting field in the last few years [25]. The authors of [26] utilized two types of LSTM networks (standard and encoder-decoder architecture) to make predictions for one household. The authors of [27] proposed enhanced-LSTM for EE consumption forecast of a metropolitan power system in France. Their method takes into account the periodicity characteristic of the load consumption by using multiple sequences of input time lags, and achieves higher performance than a single-sequence LSTM. Moreover, different hybrid architectures have been explored in order to avoid the limitations of individual models. A hybrid approach for STLF is presented in [28], where the authors processed the load signal in parallel with a LSTM and CNN. The features generated by the two networks were then used as input in a fully connected network in charge of forecasting the day-ahead load. The authors of [29] proposed a hybrid model which combines general regression neural network (GRNN), minimal redundancy maximal relevance technique and empirical model decomposition. The efficiency of the model is validated on aggregated load data from a power system in China. It shows higher forecasting accuracy than single GRNN and SVM. In [30], a hybrid method is proposed, which combines LSTM, empirical mode decomposition and similar-days selection to build a prediction architecture for short-term load forecasting. The authors concluded that the robustness of individual methods in the hybrid scheme can be an advantage for the forecasting model.

2.2. Related Studies According to User Type

According to the type of user, EE consumption forecasting approaches can be divided into those that focus on industrial entities (industrial consumption) and those that focus on households (residential consumption). The industrial approaches focus on entities such as factories, enterprises and companies, and have substantial commercial potential because industry consumes significant amounts of EE. STLF for industrial entities in Spain is discussed in [31]. The authors presented a neuro-fuzzy system with a backpropagation learning algorithm and compared the results achieved with those of other techniques, such as multilayer perceptron and statistical ARIMA processes. In [32], the authors present a model for STLF for a hospital in China. They combined LSTM and CNN and explored the network performance by considering coupling of electrical loads, gas and heating. The authors of [33] introduced an ARMA model for load forecasting of industrial companies, with focus on EE consumption profiles where stochastic changes in the regime can be observed. In [34], a set of multiple linear regression models are developed for modeling industrial loads. The data used in the study were collected from an Italian factory. In this study, the authors showed how few qualitive variables characterize the production schedule. In [35], the authors develop different models for forecasting the next hour load using data from a Spanish industrial pole. With an optimized model for single-hour prediction, a hybrid strategy was applied to build a complete day-ahead hourly load forecasting model. In general, the studies related to industrial EE consumption provide more accurate models compared to households, probably because industrial entities have strict regulations (i.e., shifts and working time), which makes the forecasting less challenging.

On the other hand, residential EE consumption is more challenging to forecast. Each household has its own pattern and electricity consumption profile, which are determined by the number of occupants, their lifestyle, the household area, electrical appliances present in the household, etc. Additionally, household-level EE consumption can vary considerably from one day to the next due to work schedules, holidays, weather conditions, etc. Therefore, most of the approaches in this field tend to avoid such uncertainty by using load aggregation: they focus on forecasting EE consumption of clusters of households, usually grouped by location (i.e., buildings and neighborhoods). Load aggregation usually reduces the inherent variability in load consumption, which results in smoother load shapes that are more predictable. This effect is illustrated in Figure 1.

In [36], the authors used clustering method to divide different types of households. For each cluster, a neural network is fitted, and their forecasts are added together to form predictions for the aggregated load. The authors demonstrate that clustering significantly increases forecast accuracy. Similarly, in [37], the authors propose a three-step process, consisting of clustering approaches, load forecasting for each cluster, and aggregating the forecasts to obtain results at a system level. The authors of [38] also show that aggregating more households improves the relative forecasting performance. They compare load forecasting accuracy at various levels of aggregation for many forecasting methods. In [39], the load consumption forecasting problem is addressed using random forest and support vector regression (SVR). Predictions are made on three spatial scales, and the obtained results show that combination of K = 32 clusters and random forest yields highest forecasting accuracy.

The systems that focus on neighborhoods lose vital information about each household; thus, they have lower commercial value, i.e., such systems cannot monitor and learn the behavior of individual households. Therefore, they cannot offer personalization and planning of EE consumption, which will be useful for cost reduction. There are just a handful of recent studies covering short-term load forecasts (e.g., day-ahead, hourly) for individual households, since they are still very challenging. The authors of [40] present a pooling-based deep recurrent neural network (PDRNN), which batches groups of customer EE consumption profiles into a pool of inputs. The authors of [41] applied Kalman filtering to single household data for a sampling period and forecast horizon of one hour. In [42], an approach is proposed to model the load of individual households based on daily schedule pattern analysis and context information.

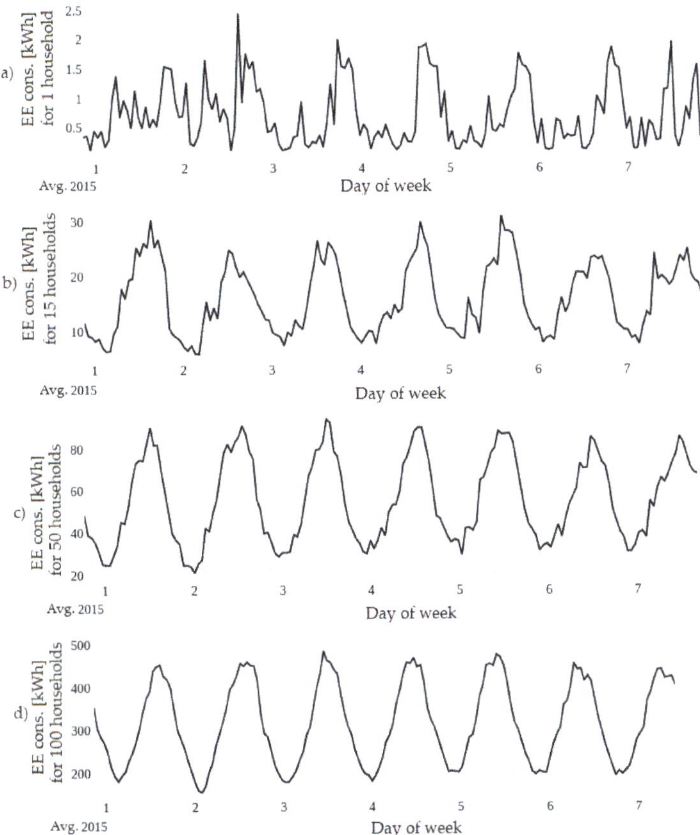

Figure 1. Weekly electrical energy (EE) consumption of: (**a**) 1 household, (**b**) 15 households, (**c**) 50 households, (**d**) 100 households.

However, the authors focus on predicting consumption with a prediction horizon shorter than one day, which does not have the same economic value as one-day-ahead hourly forecasts. Typically, the results of day-ahead forecast are used as a baseline for planning of the 24 h period of the next day, while forecasts with forecast horizon shorter than one day (intraday forecast) are mostly used for adjustment of day-ahead purchases [43]. Accurate day-ahead forecast minimizes the possibility of overproduction and underproduction, and satisfies load requirements in a more economical way, thus reducing the total operation costs [44].

Our proposed solution for EE consumption forecasting includes short-term forecasting (day-ahead forecast, for each hour of the day separately) for household consumption, which has significant economic and industrial value. In our study, we focus on STLF of individual households, which we believe is very specific and challenging due to the variability in consumption and randomness of households.

3. Dataset

3.1. Pecan Street Dataset

In order to develop a model that can accurately and reliably forecast the EE consumption, we performed a thorough analysis of the existing datasets. We analyzed most of the datasets in this domain and then selected Pecan Street dataset as the most appropriate one for our study. An extensive analysis of other relevant datasets and their characteristics can be found in Appendix A.

The Pecan Street dataset is one of the richest datasets related to residential EE consumption. It consists of EE consumption data, obtained from approximately 1000 households in the USA, mainly Austin, Texas. The dataset contains the actual EE consumption values from each household in one-minute intervals, collected by eGauge devices [45]. Our analysis is based on hourly household EE consumption, given in kilowatt-hours (kWh). Descriptive statistics of the EE consumption are provided in Table 1.

Table 1. Descriptive statistics of EE consumption.

Number of Samples	Minimum	Maximum	Mean	Standard Deviation	25th Percentile	50th Percentile	75th Percentile
4,832,504	0.001	35.19	1.28	1.32	0.43	0.82	1.66

Figure 2 shows the average daily EE consumption, i.e., each line in the figure represents average EE consumption for one day in the dataset. Each line is obtained by averaging the load consumption values for each hour in the day separately. The dashed line represents the mean EE consumption at hourly intervals.

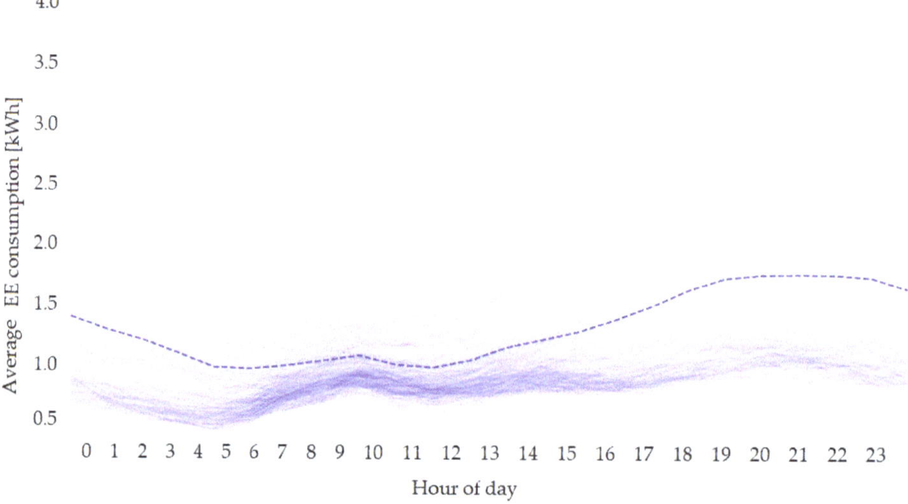

Figure 2. Average daily EE consumption.

Additionally, the Pecan Street dataset contains extensive weather data for the observed region. STLF is mainly influenced by weather parameters, because heating, ventilation and air-conditioning (HVAC) are highly dependent on outdoor temperature, humidity, wind speed, etc. Figure 3 shows a two-dimensional heatmap of EE consumption. The heatmap represents average hourly consumption in appropriate time intervals with predefined colors, where warmer colors represent higher consumption. Figure 3 shows that there is a noticeable increase in average electricity consumption in the summer months. This is specific to this dataset, i.e., it is collected in Texas, USA, where the summer temperature is significantly high, and there is increased use of air-conditioning. Therefore, the steady increase in EE consumption during the summer months can be attributed to the use of air-conditioners.

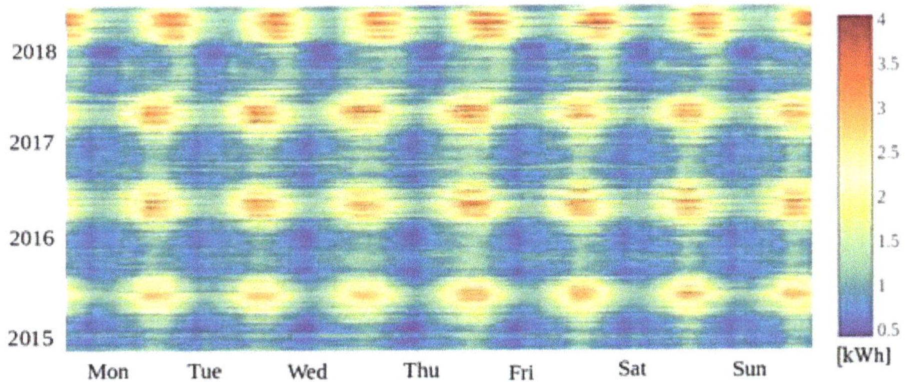

Figure 3. Heatmap of average EE consumption in a week.

The data used in this study were collected from 925 households for a period of almost four years (2015, 2016, 2017, and nine months of 2018). In order to accurately evaluate the proposed forecasting model's performance, we divided the data into three parts: (i) 27 months were used for training data (6 even-numbered months of 2015, all of 2016, and the first 9 months of 2017). (ii) Six months (odd-numbered months) from 2015 were taken for validation data. (iii) The last 12 months were chosen for test data (last 3 months of 2017 and 9 months of 2018).

3.2. Dataset Preprocessing

One of the most important steps towards developing an accurate ML model is data preprocessing. This process prepares the data for analysis by dealing or removing the data that is incorrect, incomplete, irrelevant, duplicated or improperly formatted. The preprocessing of the dataset included the following steps:

- Handling incorrect values for certain variables—In particular, we encountered instances with negative values for consumed electricity, which is impossible and indicates a mistake. In this case, we were able to calculate the value from other variables available in the dataset. For instance, we calculated the total EE consumption as the sum of consumed power from the solar grid and power drawn from the electrical grid.
- Handling outliers (instances which greatly deviate from the expected range) and missing values—If the outliers or missing values pertained to weather-related variables, the true value could be extracted from other instances referring to the same moment in time. However, in the case that the reported load consumption was incorrect from the start, the particular instance was omitted from the dataset entirely.
- Handling sequential values for EE consumption that are identical—In some situations, the sensors in certain households reported a constant value over a prolonged period of time. In this case, we assumed there was a fault with the sensor. Due to the large number of distinct households in the dataset, we could remove these instances.

4. Methodology

In the day-ahead electricity market, generation companies and retailers submit supply and demand orders for every hour of the following day. Therefore, the focus of our work was to create a model that can forecast electricity consumption one day ahead, at 10:00, for every hour of the following day (shown in Figure 4) [46].

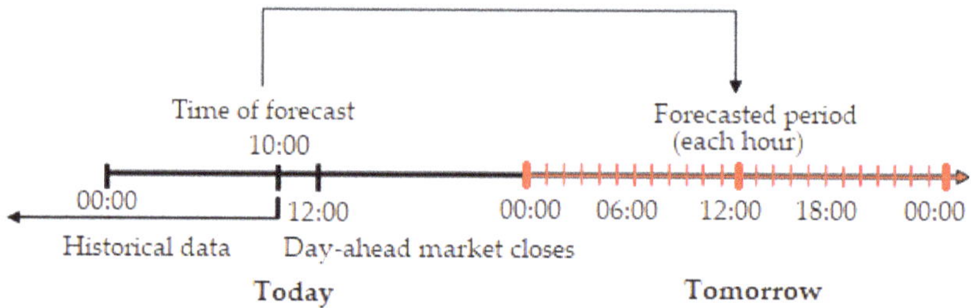

Figure 4. Day-ahead forecast timeline.

This timeline allows planning of the production for the following day in accordance with the day-ahead electricity market. According to this timeline, we developed two models that make predictions for different hours of the next day: one for the hours from midnight to 09:00, and one for the rest of the day. The main reason for developing two models is that we want to include the 24-h-before load consumption value for the hours from midnight to 09:00, which, at the time when the predictions are made (at 10:00), are only available for these hours. We considered this as valuable additional information that can improve forecasting for the first nine hours of the following day, because the periodic nature of EE consumption makes the most recent EE consumption values the dominant factor in STLF [47].

4.1. Feature Engineering

EE load forecasting is a complex multivariable and multidimensional estimation problem. The impacts of many influencing factors that affect load consumption need to be studied in order to develop a precise load forecasting model. Thus, we extracted several features from multiple sources, which can mainly be grouped into two categories: contextual and historical load features.

4.1.1. Contextual Features

- Weather features

The weather is a crucial driving factor in EE consumption. That is why it is a common EE consumption forecasting practice to include weather variables, such as wind speed, humidity and precipitation intensity, in forecasting models. The factor that has the most influence on EE consumption is temperature. Several weather-related features were extracted, and the main focus was on the temperature-related features.

- Calendar features

The social element is part of the reason for the hourly, daily and weekly patterns in EE consumption [48]. To allow forecast models to take into account the EE consumption variations which are tied to days, times of the day and seasons, we included some calendar data as nominal features. We also included information about the special days according to the area of interest, Austin, Texas.

- Interaction features

We also used interaction features, i.e., combinations of two existing features [49]. The hours in the days of a week may result in different loads due to human activities. For instance, there may be a smaller load on weekend mornings than weekday mornings, because people usually do not get up as early as when they have to go to work. This results in lower EE consumption values. The implementation of this group of features was simply done by multiplying two features.

For a full table with all extracted contextual features, see Appendix B.

4.1.2. Historical Load Features

Load consumption is highly related to historical load, due to its periodic nature. Thus, in this study, historical loads of up to one week were used to predict the day-ahead hourly load.

- Standard features

Due to the strong daily patterns of EE consumption, it is highly correlated to consumption at the same hour of previous days [50–52]. That is why the following lagged values were used in the training process of the forecasting model:

1. Historical EE consumption values by individual household for particular hours: $load_{t-24h}$, $load_{t-25h}$, $load_{t-26h}$, $load_{t-48h}$, $load_{t-49h}$, $load_{t-50h}$, $load_{t-72h}$, $load_{t-96h}$, $load_{t-120h}$, $load_{t-144h}$, and $load_{t-168h}$
2. Average historical load consumption values from all households for particular hours: avg_load_{t-24h}, avg_load_{t-25h}, avg_load_{t-26h}, avg_load_{t-48h}, avg_load_{t-49h}, avg_load_{t-50h}, avg_load_{t-72h}, avg_load_{t-96h}, avg_load_{t-120h}, avg_load_{t-144h}, and avg_load_{t-168h}

Features $load_{t-24h}$, $load_{t-25h}$ and $load_{t-26h}$ are used only for the first model, for the hours from midnight to 09:00 (see Section 4).

- Domain-specific historical load features

Based on the fact that future EE consumption is highly related to historical load, we additionally analyzed four types of time series. The first two take into account the strong daily pattern of EE consumption, and consist of historical load data from the day previous to the day when the predictions are made (all 24 h): one refers to the average load consumption in each hour, calculated from all households present in the system, and the other refers to the load consumption of each household in the same hours. The other two types of time series take into account the significance of the lagged values of EE consumption related to the same hours of previous days. More specifically, one of these time series consists of average values for load consumption (from all households) from hour 24, 48, 72, 96, 120, 144 and 168 prior to the forecasted hour, and the other consists of load consumption of each household in the same hours. As mentioned before, the 24-h-before EE consumption value is only used for instances referring to the first defined interval (midnight to 09:00). It should be noted that in the previous section, the lagged values of EE consumption were used as actual features, but in this section they are used for constructing time series from which additional time and frequency features will be extracted.

To include valuable characteristics about the manner of EE consumption in the feature vector, for each instance we generated a comprehensive set of features based on these four types of time series. The features were extracted using the TSFRESH (https://tsfresh.readthedocs.io/en/latest/text/list_of_features.html) Python package, which offers extraction of time and frequency domain features from time-series. We generated 400 new features for each instance. These features include minimum, maximum, variance, correlation, covariance, skewness, kurtosis, number of times the signal is above/below its mean, signal mean change, its autocorrelations (correlations for different delays), etc. These new features give new insight into time-series dynamics, and we believe that they can be significant in improving forecast accuracy. Figure 5 shows how the four time series are constructed for a forecast for a particular household at 08:00.

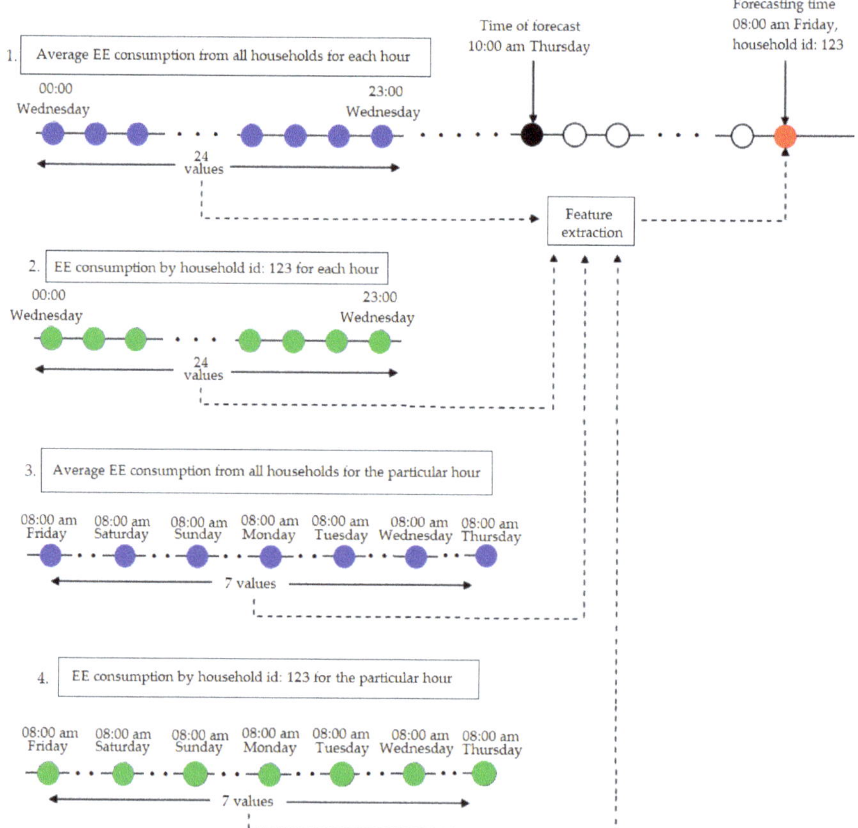

Figure 5. Domain-specific historical load feature extraction procedure.

4.2. Deep Residual Neural Network

DL is part of ML, and is based on artificial neural network architecture [53]. DL allows models comprised of numerous processing layers to learn data representations with multiple levels of abstraction. DL architectures have been applied to many fields, where they have produced results comparable, or in some cases superior, to those of human experts.

One type of DNN that was recently proposed is the deep residual neural network (DRNN). This type of deep network has performed extremely well on natural language processing tasks [54,55] and has emerged as a state-of-the-art architecture in computer vision, image segmentation and object detection [56,57] More recently, architectural variants of DRNN have also been used in load forecasting, where they have shown improvement in aggregated load forecast compared to conventional regression models [58,59]. Therefore, in this work we further explore the effectiveness of DRNN architecture in day-ahead load forecasting for single households. A DRNN can easily be constructed by stacking several residual blocks (Figure 6a). In the residual block, a mapping from x to Θ is learned, where Θ is a set of weights related to the residual block. Accordingly, the general representation of the residual block can be written as shown in Equation (1):

$$H(x) = F(x, \Theta) + x. \tag{1}$$

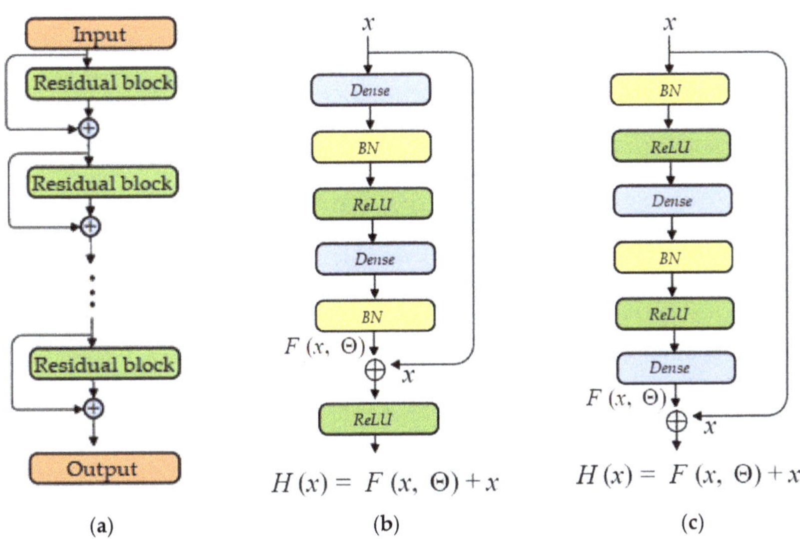

Figure 6. (a) Deep residual neural network (DRNN) structure; (b) original residual block; (c) pre-activation variant of residual block.

The forward propagation of the structure, where k residual blocks are stacked, can be represented as shown in Equation (2):

$$x_K = x_0 + \sum_{i=1}^{K} F(x_{i-1}, \Theta_{i-1}) \tag{2}$$

where x_0 and x_K are the input and the output of the residual network, respectively, and $\Theta_i = \{\Theta_{i,l} | 1 \leq l \leq L\}$ is the set of weights related to the ith residual block, L being the number of layers within the block. Basically, x has no parameters and only adds the output from the previous layer to the layer ahead. The original structure of a residual block used for building a DRNN is shown in Figure 6b.

As DRNNs gain more and more popularity in the research community, their architecture is more intensely studied. There are many proposed interpretations of DRNN architecture and variants of residual blocks. For our DRNN architecture, we used a pre-activation variant of the residual block, proposed in [60]. In this residual block, the activation function rectified linear unit (ReLu) and batch normalization (BN) are used as pre-activation of the weight layers, in contrast to the conventional approach of post-activation. The residual block used for building the DRNN architecture is shown in Figure 6c. In our case, instead of using convolutional layers as weight layers within the block, we used dense layers, making the network more applicable for feature-based input.

4.3. Proposed Architecture for Household Electrical Energy Consumption Forecast (HousEEC)

In this section, we present our proposed architecture for STLF, which is based on a deep residual neural network. First, we collect daily EE consumption, weather and calendar data. Weather and calendar data are used for extracting contextual features (see Section 4.1.1). From the daily EE consumption data, we extract standard historical load-related features referring to a particular household, or average values for all households in the system. Additionally, we define four time series (see Section 4.1.2) to extract domain-specific historical load features. The values of the extracted contextual and load-related features are then transformed in such a way that their distribution is centered around 0 (has a mean value 0) with a standard deviation of 1. This is done feature-wise, i.e., independently for each feature.

The structure of the DRNN for load forecasting is illustrated in Figure 7. The input features are separated into two groups, and each group is used as input in a separate branch. One branch uses contextual features in combination with the classical historical load features as input, and the other uses only domain-specific features. The left branch starts with a residual block containing 32 neurons in the fully connected layers, while the right branch starts with a residual block containing 64 neurons in the fully connected layers. The use of fully connected layers instead of the original convolutional layers in the residual blocks makes the network more applicable for feature-based input and regression [61]. The output of the first two branches is then concatenated with the raw input features, and as such is fed to a DRNN with five additional residual blocks. Each residual block consists of two fully connected layers, activation function and batch normalization. The fully connected layers in the blocks consist of 64, 32, 16, 16 and 8 neurons, consecutively. All such layers in the residual blocks use ReLu as the activation function. Mathematically, it is defined as $f(x) = \max(0,x)$, which makes it suitable for the STLF problem, since the forecasted consumption cannot have negative values. Additionally, we used a dropout rate of 0.1 in order to reduce the chances of overfitting. A total of 6 levels of residual blocks are stacked (1 input level with 2 residual blocks and an additional 5 levels after the concatenation block), forming a 12 layer DRNN.

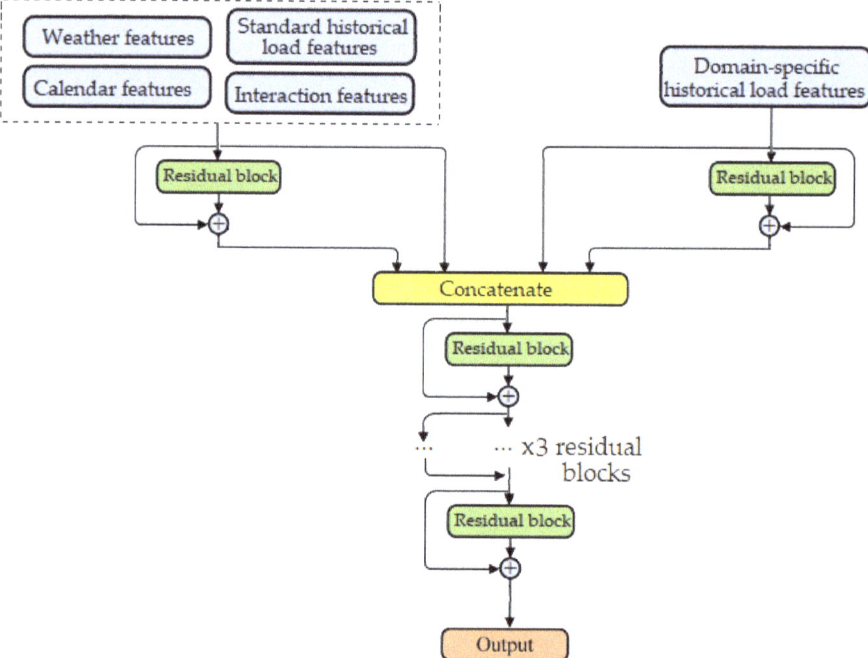

Figure 7. Proposed deep residual neural network (DRNN) architecture for short-term load forecasting (STLF).

5. Experimental Setup

5.1. Evaluation Metrics

In order to evaluate and compare the models, several evaluation metrics were used: root-mean-square error (RMSE) [62], mean absolute error (MAE) [63], and R^2 score [64], which are well-known metrics used to measure performance on regression tasks.

MAE and RMSE are directly interpretable in terms of the used measurement unit (kWh in our case). RMSE is a measure that shows how much the residuals are spread out. Residuals are the difference between actual and predicted values. The definition of RMSE indicates that large errors have higher weight. Since in our regression problem the forecasted values are in a small range, large errors are particularly undesirable. Since we want to penalize large errors more, we focused more on RMSE. MAE shows how close the forecasted values are to actual values. It is calculated as a mean of the absolute values of each prediction error on all instances of the test dataset. R^2 expresses how well the model replicates the observed outcomes, based on the proportion of total variation of outcomes explained by the model. This metric is positively oriented, and its highest value can be 1. RMSE, MAE and R^2 scores are calculated as shown in Equation (3)–(5):

$$\text{RMSE} = \sqrt{\frac{1}{n}\sum_{i=1}^{n}\left(y_{predicted} - y_{true}\right)^2}, \tag{3}$$

$$\text{MAE} = \frac{1}{n}\sum_{i=1}^{n}\left|y_{predicted} - y_{true}\right|, \tag{4}$$

$$R^2 = 1 - \frac{\sum_{i=1}^{n}\left(y_{predicted} - y_{true}\right)^2}{\sum_{i=1}^{n}\left(y_{true} - y_{average}\right)^2}, \tag{5}$$

where n is the number of data samples.

5.2. Reference Models

A reference model or benchmark uses simple summary statistics to create predictions. These predictions are used to measure the benchmark performance, and then this result becomes what we compare our ML model results against. For this study, we implemented three baseline models. One model provides the amount of consumed EE by a specific user 24 or 48 h before the hour of prediction. The 24-h-before value is used for prediction of instances in the first interval, midnight to 09:00, and the 48-h-before value is used for prediction of instances in the second interval, 10:00 to 23:00. Another baseline model is the Vanilla multiple regression Benchmark model [19]. This model uses multiple sources of data to predict future load; in particular, polynomials of temperature and their interaction with calendar variables. To enhance the accuracy of STLF, we augmented the Vanilla multiple regression model by adding some lagged load variables, as well as other combinations of variables that enhance the interaction effect. The last benchmark model is seasonal autoregressive integrated moving average (SARIMA) [65].

For more detailed explanation of the reference models, see Appendix C.

6. Experimental Results

To explore the performance of our proposed model in EE consumption forecast, we did a series of experiments. Sections 6.1–6.5 present numerous comparisons of results for disaggregated hourly load forecast, and Section 6.6 presents the efficiency of the proposed method in aggregated load forecast. Section 6.7 presents a general model to overcome the cold start issue.

6.1. Comparison of Forecasting Techniques

To verify the predictive performance of our STLF model, we made comparison with the previously mentioned benchmarks (see Section 5.2), as well as other ML algorithms—linear regression [66], K-nearest neighbors (KNN) [67], decision tree regressor [68], random forest [69], linear SVR [70], gradient boosting [71,72] and xgboost [73] (see Appendix D). We also considered a classic DRNN, comprising five residual blocks, that takes all features together as input for the first residual block.

Table 2 shows RMSE, MAE and correlation R^2 score for each model and the two benchmarks. A comparison of the performance of the models using different sets of features was also conducted. In the first scenario, only contextual features and standard historical load features were used as input. In the second scenario, the proposed domain-specific historical load features were also included. From the results, the benefit of including domain-specific historical load features can be seen. In almost all cases, the proposed domain-specific historical load features significantly improved the model performance. In addition, the results show that our proposed input structure of the DRNN significantly improves the forecasting accuracy. Our proposed model outperformed all other models in both scenarios, achieving RMSE of 0.44 kWh, MAE of 0.23 kWh and R^2 score of 0.90.

Table 2. Performance of methods using different feature sets.

Method	Contextual + Standard Historical Load Features			Contextual + Standard Historical Load Features + Domain-Specific Historical Load Features		
	RMSE $\downarrow$	MAE $\downarrow$	$R^2 \uparrow$	RMSE $\downarrow$	MAE $\downarrow$	$R^2 \uparrow$
Linear SVR	1.89	1.27	−0.80	1.91	1.28	−0.84
KNN	1.05	0.59	0.44	1.04	0.59	0.46
Decision tree	1.35	0.80	0.09	0.89	0.47	0.60
Linear regression	0.89	0.52	0.60	0.81	0.48	0.67
Gradient boosting	0.89	0.50	0.60	0.72	0.4	0.74
XGBoost	0.89	0.49	0.60	0.71	0.4	0.74
Random forest	0.96	0.59	0.54	0.64	0.33	0.79
DRNN	0.88	0.49	0.61	0.51	0.28	0.87
Statistical benchmark				1.13	0.60	0.35
SARIMA [65]				1.21	0.75	0.28
Vanilla benchmark [19]				1.00	0.58	0.50
HouseEEC (ours)				**0.44**	**0.23**	**0.90**

Computation time for execution of models' training and testing is important for practical implementation in a system, in the case of models retraining with new data and making daily predictions. The training and testing times of the models used in the experiments are shown in Table 3. In all, 2,544,962 instances were used for training and 1,654,499 for testing the models. The deep learning models were trained and tested on NVIDIA Titan X GPU, with 12 GB GDDR5X memory and memory bandwidth of 480 GB/s, while the conventional ML models were trained and tested on AMD Ryzen 7 2700 CPU with 8 cores and maximum clock frequency of 4.1 GHz.

Table 3. Computation time for model training and testing.

Method	Training Time (s)	Testing Time (s)
Linear SVR	2.428	3
KNN	239	10.053
Decision tree	1.849	6
Linear regression	204	3
Gradient boosting	8.383	12
XGBoost	12.232	22
Random forest	12.222	32
DRNN	948	331
HouseEEC	1.016	336

6.2. Error Analysis of Application Scenarios

- Hourly forecast

Figure 8 shows the RMSE score for each hour of the day. The results are obtained by averaging the errors for all users for each hour. Larger error can be observed for 03:00, 23:00, and the afternoon hours when most people return from work and perform different activities at home.

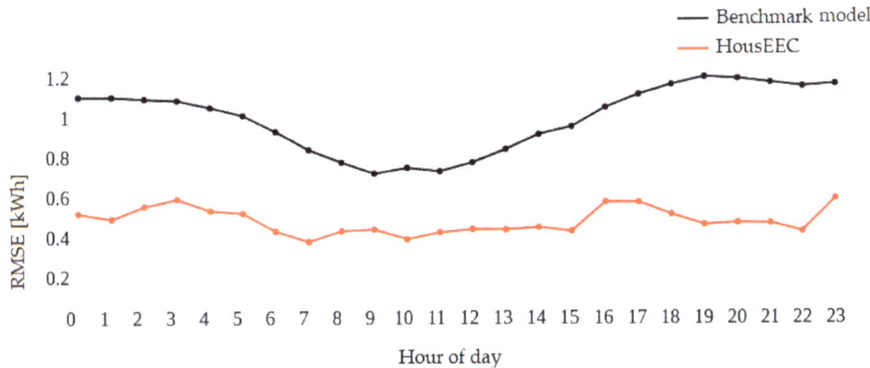

Figure 8. Root-mean-square error (RMSE) for each hour of the day.

However, our model reports quite low error for the morning hours, which is significant because morning hours are related to increased EE consumption, especially on workdays. Overall, there is no significant difference in the reported error for any specific part of the day. Our model significantly outperforms the benchmark model for each hour of the day.

- Weekday forecast

Figure 9 shows the RMSE score for each day of the week. The results are obtained by averaging the errors for all users for each day. The benchmark makes a larger error for weekend days; they are more challenging to forecast due to vacations, trips and irregularities in peoples' lives. However, our model performs similarly for each day of the week, regardless of the uncertainties that are usually present on weekend days.

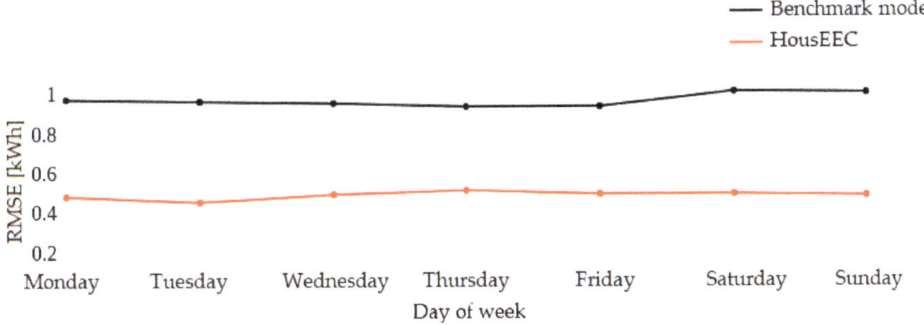

Figure 9. Root-mean-square error (RMSE) for each day of the week.

- Monthly forecast

Figure 10 shows the RMSE score for each month of the year. The results are obtained by averaging the errors for all users for each month. Both the benchmark and our proposed method follow a similar trend in terms of the prediction error; the RMSE score is lowest for the spring months when there is no need for heating or cooling. The largest error made by our model can be observed for May, when the cooling season starts. However, after some time, the increased trend of EE consumption is incorporated into the extracted features, so the prediction errors start decreasing. This is a very important characteristic of our model, since the rest of the summer months are also characterized by increased EE consumption. This is certainly not the case with the benchmark model, which reports the largest error when EE consumption is at its peak.

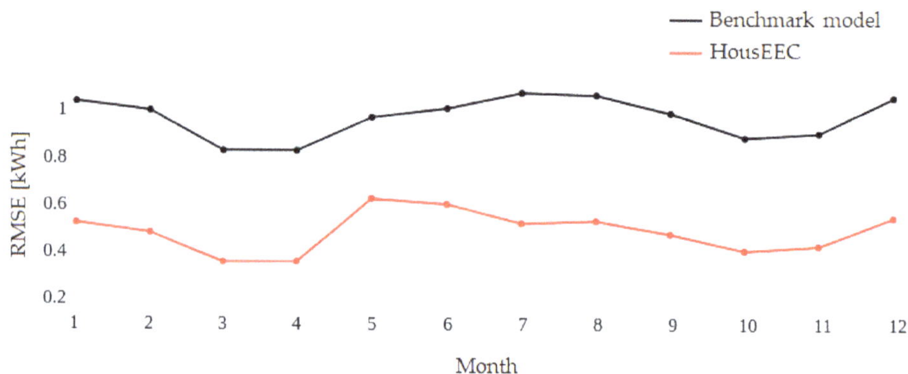

Figure 10. Root-mean-square error (RMSE) for each month of the year.

6.3. Comparison with Other Deep Learning Approaches that Use only Time Series

We additionally made performance comparisons between our method and the most recent DL architectures relevant to load forecasting, described in [74]. The authors present seven architectures designed for 24 h prediction and evaluate them using the individual household electric power consumption (IHEPC) [75] dataset, which contains 47 months of EE consumption data of single households. Based on their results, we chose the five best architectures and evaluated them using the Pecan Street dataset with classical feedforward neural network (FFNN), deep residual neural network (DRNN), temporal convolutional network (TCN), long short-term memory (LSTM) and gated recurrent unit (GRU). This DRNN uses different residual blocks compared to the one in our proposed model. All mentioned networks are described in detail in the paper. For this evaluation, we used seven week-long time series as input for the networks, two related to the historical load and five related to weather data. The first time series is actual EE consumption by a specific household in the past week, and the second is average load consumption by all households in the past week. The weather-related time series are temperature, humidity, apparent temperature, wind speed and precipitation. The Pecan Street dataset contains weather and load measurements for each hour, resulting in 168-hourly-measurements long input and 24-hourly-measurements long output for the networks. Since the results in the paper showed that including calendar information improves prediction accuracy, we additionally included the following information: hour of the day, day of the week, month and work-/non-workday. For training, we used the multiple input–multiple output (MIMO) strategy, meaning that a single predictor is trained to forecast a whole 24 values-long output sequence in a single shot.

Table 4 shows the results: HousEEC shows better results in terms of RMSE, MAE and R^2 compared to end-to-end DL-based methods for load forecasting on the household level. The main conclusion that can be drawn from these results is that the time-series consisting of 168 historical load values does not contain enough information for proper training of DL end-to-end architectures. However, one-week historical load appears to be enough for proper training of the feature-based DRNN, especially when it is trained with extensive feature sets consisting of the domain-specific features which give new insights into the load time-series dynamics.

Table 4. Performance of end-to-end deep learning (DL) approaches.

Method	RMSE	MAE	R^2
FNN	1.05	0.70	0.47
DFNN	0.84	0.55	0.53
TCN	0.78	0.50	0.59
LSTM	0.81	0.54	0.54
GRU	0.80	0.54	0.54
HousEEC	**0.44**	**0.23**	**0.90**

6.4. State-of-the-Art STLF on Household Level

The STLF field lacks a unified comparison between conducted studies. There are many studies in this field that address different segments related to load forecasting, and most of them are not directly comparable. Nevertheless, we believe that a summary of the results achieved with state-of-the-art methods might be informative and useful for new studies in a few ways. Authors can select the most commonly used dataset for their work in order to produce comparable results, and it can help researchers to avoid selecting nonrepresentative data for evaluation of their methods. In this section, selected studies relevant to STLF on the household level are presented. The two criteria for study selection were the forecasting horizon (up to 24 h) and the evaluation metric (RMSE). In order to include more relevant studies, we additionally considered studies reporting normalized root mean squared error (NRMSE), calculated as shown in Equation (6):

$$ \text{NRMSE} = \frac{\sqrt{\frac{1}{n} \sum_{i=1}^{n} \left(y_{predicted} - y_{true} \right)^2}}{\frac{1}{n} \sum_{i=1}^{n} y_{true}}. \tag{6} $$

We ended up with 12 relevant studies, including ours. Table 5 presents a summary of the studies in terms of forecasting horizon, number of households used for evaluation, duration of the test data, and results achieved in terms of RMSE (NRMSE). One parameter that should be considered in this comparison is the size of the data used for evaluation. EE consumption is highly affected by the weather; a lot of electricity is used for cooling in summer and heating in winter. This leads to the conclusion that studies that use shorter periods for their evaluation might present unreliable results without checking model performance in different seasons. Only one of the selected studies evaluated their methods using data collected in a period of 12 months. In order to show how robust the proposed methods are, more households are needed for evaluation. This is because there are different types of users, such as elderly people who spend most of the time at home, people who go to work, students who have a dynamic lifestyle, etc. Only four studies considered datasets with fewer than 100 households.

Table 5. Summary of state-of-the-art STLF studies.

Authors	Forecasting Horizon	Number of Households	Duration of Evaluation	RMSE	NRMSE
Shi et al. [40]	1 h	920	1 month	0.45	–
Lusis et al. [76]	30 min	27	28 days	0.52	–
Muralitharan et al. [77]	24 h	–	–	0.62	–
Gasparin et al. [74]	24 h	1	12 months	0.75	–
Yildiz et al. [78]	24 h	14	–	0.80	–
Ali et al. [79]	1 h	34	6 months	0.80	–
Ganz et al. [80]	24 h	74	2 months	0.85	–
Gerossier et al. [81]	24 h	226	2 months	–	0.43
Vos et al. [82]	24 h	200	6 months	–	0.53
Wijaya et al. [83]	24 h	782	6 months	–	0.61
Humeau et al. [50]	24 h	782	6 months	–	0.80
HousEEC	24 h	297	12 months	0.44	0.34

For our work, we addressed the previously mentioned challenges; our model provides forecasting one day ahead, and the results are evaluated on 297 households over a period of one year. We believe that our results are very promising, considering that they show great performance of the model for a large number of households evaluated for a period of 12 months.

6.5. Analysis of Different Lengths of Training Set

Over time, new households with different EE consumption patterns can appear in the forecasting system. Therefore, it is a common practice for forecasting models to be trained with new data after a certain time. This section presents the HousEEC model's performance for three subsets of the initial test set, when additional data is used for training. For comparison, we used the final HousEEC model (trained with 27 months) to predict the EE consumption of the three new test subsets. Table 6 shows the RMSE, MAE and R^2 scores for different train-test splits.

Table 6. Performance of models with different train-test splits.

	Train-Test Splits (M—Months).					
	27 M vs. 9 M	**30 M vs. 9 M**	**27 M vs. 6 M**	**33 M vs. 6 M**	**27 M vs. 3 M**	**36 M vs. 3 M**
RMSE	0.43	0.46	0.47	0.48	0.45	0.45
MAE	0.28	0.27	0.27	0.28	0.24	0.25
R^2	0.92	0.9	0.9	0.9	0.9	0.92

Even though it is expected that constant inclusion of new data expands the knowledge of the existing model, the results from this analysis showed that there was no significant benefit of it when there were no changes in the dataset in terms of new households.

6.6. Aggregated Consumption Forecasting

Forecasting of aggregated load can be implemented by the standard strategy of direct forecast of the aggregated load, or by aggregating the forecasts for individual households. In Figure 11, we observe the curve of aggregated forecasts for all households and compare it to the actual aggregated load curve. The observed period is the first week of July. It is one of the hottest months in the year, characterized with increased EE consumption due to air-conditioning (Figure 3). Since the peak EE consumption is the most challenging to forecast, we more closely observed the model's performance for a whole week in July—the month during which the EE consumption is the highest in our dataset. It is obvious that the forecast successfully follows the trend of actual consumption, even for July 5, when a significant drop of EE consumption is noted, which is not typical for the time period observed.

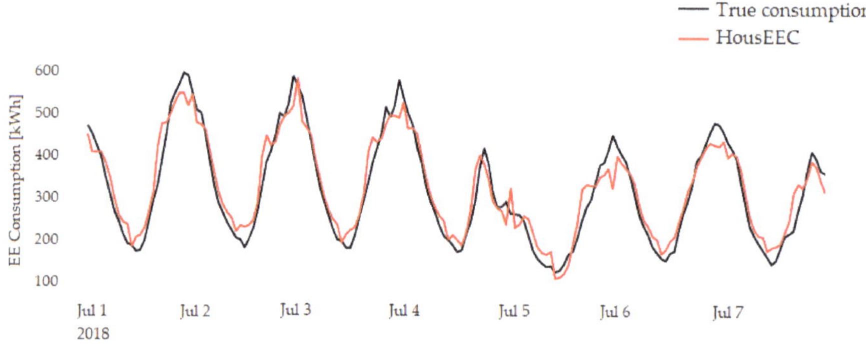

Figure 11. Weekly aggregated EE consumption.

Finally, we calculated total consumption of all households and total error for hourly and daily analysis. The hourly analysis showed that, on average, the aggregated EE consumption of the households is 263 kWh. Our model makes 8% error on average per hour forecast (22 kWh); the Vanilla multiple regression benchmark makes 12% error (32 kWh). The daily analysis showed that, on average, the aggregated EE consumption of all households is 6140 kWh. Our model makes 2% error on average per day forecast (131 kWh); the baseline makes 6% error (360 kWh).

6.7. Cold Start Issue

In order to predict next-day EE consumption of a new household in the system, the HousEEC model requires the household's historical EE consumption of the previous week. This means that it suffers from a one-week cold start, which is a technical limitation of the model. To overcome this limitation, we trained an additional general model that does not use household-specific standard historical load features and domain-specific historical load features that are extracted from the third type of time-series (see Figure 5). This model will to serve as a model for prediction of household EE consumption for only the first week. We used the same HousEEC architecture, and the only difference was the number of input features. The performance of this model on the same test set used in the previous experiments can be seen in Table 7. As expected, this general model provided less precise results compared to the final HousEEC model. However, we consider these results as acceptable, given that this general model would be used for only a short period of time in an actual implementation of the system. The presented results are also additional evidence of the significance of domain-specific historical load features for a particular household.

Table 7. Comparison of performance of general and HousEEC models.

Method	RMSE	MAE	R^2
General model	0.50	0.26	0.87
HousEEC	0.44	0.23	0.90

7. HouseEEC System Prototype

This section presents the practical implementation of the HousEEC system and deployment of the ML model in a prototype web application. The system enables end users to quickly and easily access a service that allows different analyses. One of the most important features of this system is that it can be easily implemented in larger systems that have different monitoring devices for electricity consumption in households. The only prerequisite for implementing such a system for analyzing and forecasting electricity consumption is access to the measured values of household EE consumption. The system contains three main modules:

- Graphical user interface (GUI), through which forecasts of EE consumption can be accessed.
- Back end, which provides the functionality that is served to users through the graphical interface. This section is also responsible for communication with the database, deployment and launch of the forecast module, and similar functions. It also provides application program interfaces (APIs) for interconnection with the ML module and the GUI.
- ML module, which is responsible for deployment of the ML model and its practical use. It contains all the steps required for an ML pipeline: preprocessing data and dealing with missing data; extracting features; and predicting with the ML model. For the implementation, we used libraries including Pandas, Sklearn, NumPy, Tsfresh, SciPy, Keras and Tensorflow.

A visualization of the system and its households is shown in Figure 12. For better visualization, multiple households that are very close geographically are presented as a group (blue circles on the map). Note that this is a simulation, and the geographic locations are for illustration purposes only; the dataset does not contain location information about the households. Next, the application provides access to a table of measured EE consumption of all households for the last 24 h. In addition, there is an option to search for a specific household, which can provide insight into its individual time series of EE consumption. This table also enables easy control of the accuracy of household measuring devices: whether they show values or whether there are erroneous values in the metered data (negative values for consumed electricity or values outside the expected range). If unexpected data are spotted in the table, they can be deleted from the database, preventing them from affecting prediction by the ML model in the following days.

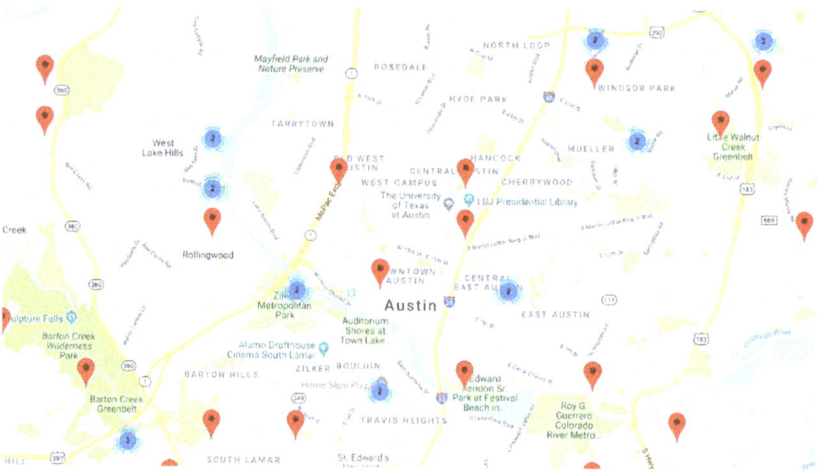

Figure 12. Map of simulated households' locations.

The next service of the system represents daily predictions of EE consumption for each hour of the next day. These forecasts are obtained by executing the ML model at 10:00 every day. This allows sufficient time for planning the actions of the day-ahead electricity market, which, as mentioned before, closes at 12:00. Although forecasts are obtained at the household level, they are presented at the aggregated level for all households. Figure 13 shows three lines, representing predicted electricity consumption achieved by the three chosen models: random forest, benchmark and our final HousEEC.

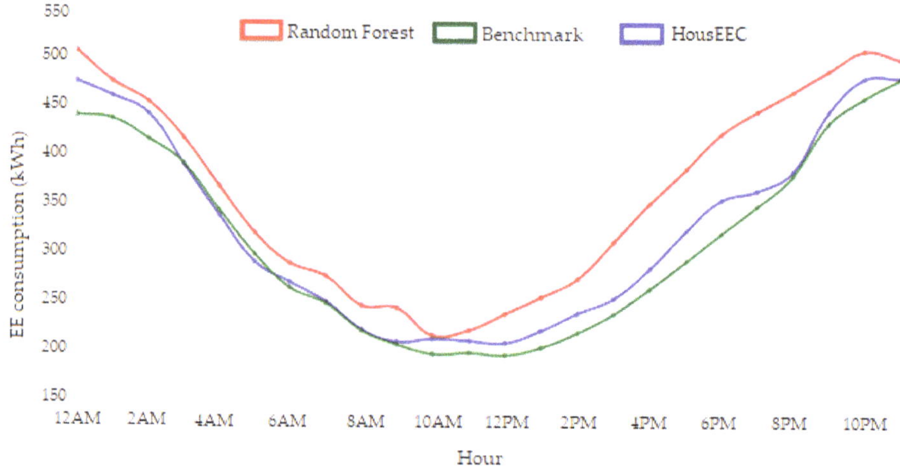

Figure 13. Aggregated hourly predictions of EE consumption for the next day.

The final service offered by the system is the performance analysis of the ML models (Figure 14). With this service, users can load predictions for the past period and compare them to actual consumption values. First, the user selects the interval of interest and the models. Then the system outputs the predictions and true consumption. For example, Figure 14 shows predictions of the random forest and HousEEC models and the true consumption for the randomly selected period from 1–15 June 2018. The user can visually inspect the model errors.

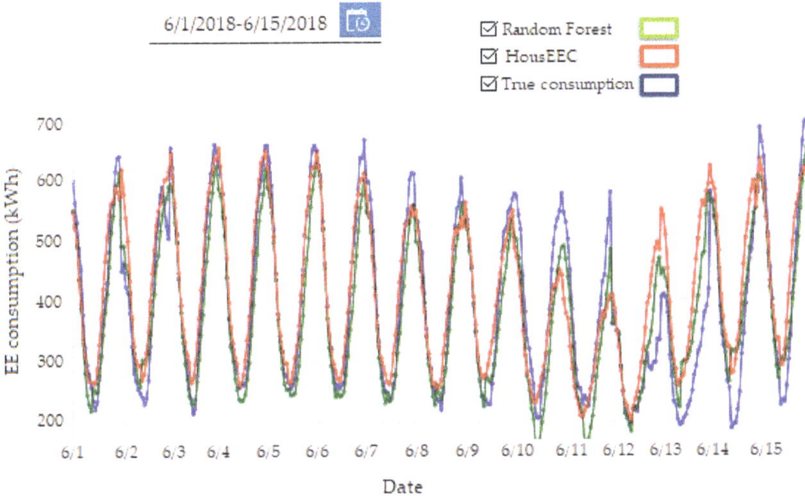

Figure 14. Historical performance comparison of predicted to true EE consumption.

When using real-time data collection devices, it is inevitable that some amount of data gets lost due to different circumstances (sensor fault, communication error, environmental disturbance, etc.). In this context, the use of techniques that deal with missing data is a crucial part of the implementation of a forecasting system. To guarantee that our forecasting system would work smoothly, we considered two cases of missing data and appropriate techniques to handle it. The first case is missing values of EE consumption for one hour for a particular household. For this case, we implemented linear interpolation, a mathematical method that adjusts a function to the existing data and uses it to extrapolate the missing data. The second case is when sensor readings are missing for two or more consecutive hours for a particular household. In this case, the missing values are replaced with the forecasted values for those hours by the HousEEC model (or the general model, if the missing values occur in the first week of the data collection process for the household; see Section 6.7).

8. Conclusions

The paper presents the HousEEC system, which provides day-ahead household EE consumption forecasting using a deep residual neural network. The experimental evaluation was performed on one of the richest datasets for household EE consumption, the Pecan Street dataset. The DL approach combines multiple sources of information by extracting features from (i) contextual data (e.g., weather, calendar), and (ii) the historical load of the particular household and all households present in the dataset. Additionally, we computed novel domain-specific time-series features that allow the system to better model the pattern of household energy consumption. Their contribution to reducing the error is shown in Table 2. Finally, we assessed performance by comparing the results achieved with our model with those of seven other ML algorithms, five DL and two benchmarks widely used in this area.

The experimental results show that in all cases, our model outperformed every other algorithm and approach, achieving RMSE of 0.44 kWh, MAE of 0.23 kWh and R^2 score of 0.90. The analysis shows the great potential of including our domain-specific historical load features in improved load forecasting. The hourly analysis showed that all customers used 263 kWh per hour on average. Our model makes 8% error on average per hour forecast (22 kWh), which is 4 percentage points better than the benchmark model results. The daily analysis showed that all households used 6140 kWh per day on average; our model makes 2% error on average per day forecast (131 kWh) and the benchmark model makes 6% (360 kWh). The comparison between end-to-end DL architectures and our proposed DL feature-based architecture showed that our method performs better, achieving significantly lower RMSE compared to the best performing end-to-end DL architecture, the temporal convolutional network. We believe that the main reason for this improvement is the domain-specific features, which give the algorithms the most relevant information derived from the raw data for future load forecasting. According to the analysis of similar studies for STLF for households, we can say that our achieved results are very promising compared to the state-of-the-art approaches. We also believe that our study shows reliable results because the method was tested on a significantly large number of households over 12 months using a 24 h forecasting horizon.

The proposed method, which predicts EE consumption on an individual household level, offers great commercial potential because it is scalable and not dependent on the current number of households in the system. In addition, predicting individual forecasts enables their aggregation, which yields better forecasting for the aggregation level compared to the conventional strategy of direct forecast of the aggregated load [25,84]. Our method also has significant value because it is not dependent on the number of households included in the system. Implementation of the system does not suffer from cold start; we addressed the cold start problem by introducing a new general model that does not use household-specific historical load features. This model is intended to provide predictions for each new household that appears in the system for the first week, until the required data for the HousEEC model is collected. Another important detail that we considered in the system implementation is the occurrence of missing data. We tackled this by using two techniques, interpolation and the use of forecasted values to fill the missing data in the EE consumption of a particular household.

We expect that the final model could perform well on other datasets which contain EE consumption data for households with similar economic status, located in places with similar weather conditions. It was trained with data from a large number of households, which make it more general, robust and able to adapt to many different EE consumption patterns. Additionally, if the HousEEC architecture is used for model re-training with new data, we expect it to show equally good performance, since it incorporates data from multiple relevant sources that affect the EE consumption of households. However, further investigation of the model's performance on other datasets is considered for future work.

Another improvement would be to introduce additional features, such as EE price, number of household members, age of users, daily schedule of users (working hours), size of household, and means of heating and cooling. We believe that these attributes would improve the accuracy; however, this requires additional private information about households, which might not be easy to obtain.

Finally, we plan to introduce the clustering of households, because there are different trends and patterns for each household in the dataset and there are large variations in the electricity consumption patterns at the household level. A clustering algorithm would group similar households into clusters and, in a way, define household profiles. This way, there will be several prediction models for different clusters of households. We believe this can increase the forecasting performance, because there are several types of users (active users who regularly go to work, older users who spend the biggest part of the day at home, etc.), and it is more difficult for the model to acquire a knowledge about the EE consumption characteristics of different users.

Author Contributions: I.K. and S.S. were the main authors who significantly contributed to the research; in particular, they were responsible for dataset analysis, implementation of methods, experimental evaluation, and writing the paper. I.I. and S.J. were responsible for defining the energy consumption prediction problem, and conceptualization of the approach and the system. M.G. was responsible for the machine learning part of the research, in particular the definitions of algorithms, the concepts, and the ideas regarding how to tackle the problem. H.G. was responsible for the research as a whole, leading and guiding the study, the problem definitions, conceptualization of the methods, definition of the experiments, and writing the paper. All authors have read and agreed to the published version of the manuscript.

Funding: This research received no external funding.

Acknowledgments: We are truly grateful to Pecan Street Inc. for giving us the opportunity to use their dataset for our research purposes. We are also thankful for the support of the NVIDIA Corporation and their generous donation of a Titan Xp GPU that we utilized for this study.

Conflicts of Interest: The authors declare no conflict of interest.

Appendix A

Table A1. Related datasets.

#	Dataset	User Type: Industrial (I)/Household (H)	Collection Period (Years)	Data Sampling Resolution	No. of Users	Weather Data	Public Access
1	Pecan Street [85]	H	4 *	1 min	1000 *	Yes	Yes **
2	REFIT [86]	H	2	8 s	20	Yes	Yes
3	PLAID [87]	H	0.5	1 s	56	No	Yes
4	UK-DALE [88]	H	2	1 s	5	No	Yes
5	GREEND [89]	H	1	1 s	8	No	Yes
6	ECO [90]	H	0.7	1 s	6	No	Yes
7	REDD [91]	H	0.3	1 s	10	No	Yes
8	IHPEC [75]	H	4	1 min	1	No	Yes
9	HES [92]	H	1	2 min	24	No	No
10	CER [93]	H	2	30 min	5000	No	No
11	DOE [94]	I	2	1 h	11	Yes	Yes
12	EnerNOC [95]	I	1	5 min	100	Yes	Yes
13	GEFCom [96]	I	4.5	1 h	1	Yes	Yes
14	Industrial Machines [97]	I	0.3	1 s	1	No	Yes
15	NREL RSF Measured Data [98]	I	1	1 h	1	No	Yes

* Ongoing collection. ** Public access for research purposes by university members.

Appendix B

Table A2. Contextual features.

Weather Features	Interaction Features	Calendar Features
T_t	$T_t \times H$	Day of week
T_t^2	$T_t^2 \times H$	Day of month
T_t^3	$T_t^3 \times H$	Month
T_{t-24}	$T_t \times M$	Hour
T_{t-25}	$T_t^2 \times M$	Holiday
T_{t-26}	$T_t^3 \times M$	Working day
T_{t-48}	$T_t \times D$	
T_{t-49}	$T_t^2 \times D$	
T_{t-50}	$T_t^3 \times D$	
T_{t-72}	$D \times H$	
T_{t-96}	$T_{davg} \times H$	
T_{t-120}	$T_{davg}^2 \times H$	
T_{t-144}	$T_{davg}^3 \times H$	
T_{t-168}	$T_{davg} \times M$	
T_{davg}*	$T_{davg}^2 \times M$	
T_{davg}^2	$T_{davg}^3 \times M$	
T_{davg}^3		
humidity		
wind speed		
precipitation		
apparent temperature		

T, temperature; Tdavg, daily average temperature; H, hour; D, day of week; M, month.

Appendix C

The Vanilla multiple regression benchmark model is a load forecasting model that uses multiple sources of data to predict future load; in particular, polynomials of temperature and their interaction with calendar features. The model can be defined as follows:

$$
\begin{aligned}
L_t = \ & \beta_0 + \beta_1 Trend_t + \beta_2 M_t + \beta_3 W_t + \beta_4 H_t + \beta_5 W_t H_t + \beta_6 T_t + \beta_7 T_t^2 + \beta_8 T_t^3 \\
& + \beta_9 M_t T_t + \beta_{10} M_t T_t^2 + \beta_{11} M_t T_t^3 + \beta_{12} H_t T_t + \beta_{13} H_t T_t^2 + \beta_{14} H_t T_t^3
\end{aligned}
\tag{A1}
$$

where L_t is the load forecast for time t; βi are the coefficients calculated using the ordinary least square method; $Trend_t$ is an increasing number which presents a linear trend at time t; M_t, W_t and H_t are the month of the year, day of the week and hour of the day for time t, respectively; and T_t is the temperature for time t.

The final benchmarking Vanilla model used in this work is defined as follows:

$$
\begin{aligned}
L_t = \ \beta_0 + \beta_1 M_t \ & + \beta_2 W_t + \beta_3 H_t + \beta_4 W_t H_t + \beta_5 T_t + \beta_6 T_t^2 + \beta_7 T_t^3 + \beta_8 M_t T_t \\
& + \beta_9 M_t T_t^2 + \beta_{10} M_t T_t^3 + \beta_{11} H_t T_t + \beta_{12} H_t T_t^2 + \beta_{13} H_t T_t^3 + \beta_{14} L_{t-26} \\
& + \beta_{15} L_{t-25} + \beta_{16} L_{t-24} + \beta_{17} T_{t-26} + \beta_{18} T_{t-25} + \beta_{19} T_{t-24} \\
& + \beta_{20} T_{davg} H + \beta_{21} T_{davg}^2 H + \beta_{22} T_{davg}^3 H + \beta_{23} T_{davg} M + \beta_{24} T_{davg}^2 M \\
& + \beta_{25} T_{davg}^3 M + \beta_{26} T_{t-26} H + \beta_{27} T_{t-26}^2 H + \beta_{28} T_{t-26}^3 H + \beta_{29} T_{t-25} H \\
& + \beta_{30} T_{t-25}^2 H + \beta_{31} T_{t-25}^3 H + \beta_{32} T_{t-24} H + \beta_{33} T_{t-24}^2 H + \beta_{34} T_{t-24}^3 H \\
& + \beta_{35} T_{t-26} M + \beta_{36} T_{t-26}^2 M + \beta_{37} T_{t-26}^3 M + \beta_{38} T_{t-25} M + \beta_{39} T_{t-25}^2 M \\
& + \beta_{40} T_{t-25}^3 M + \beta_{41} T_{t-24} M + \beta_{42} T_{t-24}^2 M + \beta_{43} T_{t-24}^3 M
\end{aligned}
\tag{A2}
$$

where T_{davg} is the average daily temperature from two days before the forecasted day. This formula represents the benchmark for obtaining the forecasts for the instances from the first interval, from midnight to 09:00. Analogously, for the instances form the second interval (from 10:00 to midnight), instead of using values referring to the time before 24, 25 and 26 h, we used values referring to the time before 48, 49 and 50 h from the forecasted hour. Additionally, we removed the trend variable, since our formulation of the forecasting problem does not meet the requirements for its calculation.

Autoregressive Integrated Moving Average (ARIMA) is one of the most commonly used methods for time-series forecasting. In general, the ARIMA model is noted as ARIMA(p,d,q), where the p parameter is an integer that confirms how many lagged series are going to be used to forecast periods ahead; d parameter tells how many differencing orders are going to be used to make the series stationary; and q is the number of lagged forecast error terms in the prediction equation. Seasonal Autoregressive Integrated Moving Average (SARIMA) is seasonal ARIMA and it is used with time series with seasonality. This model is generally termed as SARIMA(p,d,q) $\times$ (P,D,Q)^S.

Appendix D

- Linear regression attempts to model the relationship between the features and the dependent variable (in our case EE consumption) by fitting a linear equation to observed data. It learns a model by fitting a linear equation to the training data. The model optimizes a function so that the square of the errors is minimized.
- K-nearest neighbors (KNN) is an algorithm that uses feature-vector similarity to predict the value of interest. This means that for each feature vector in the test data, it finds the K-nearest neighbors in the training set and computes the average of the target class. This average value is then used as a prediction of the model. In our experiments, we used the Euclidean and Manhattan metrics for calculation of the distance between feature vectors. The empirical analysis showed that the Manhattan distance is more appropriate, and it was therefore used in the experiments.
- Decision tree regressor is a machine learning model used to predict a target by learning decision rules from features. Decision trees are constructed via an algorithmic approach that identifies

ways to split a data set based on different conditions. After training the model, as a result we have a tree with decision nodes with two or more branches representing values for the chosen feature, and leaf nodes representing a numerical prediction of EE consumption.

- Random forest consists of a large number of individual decision trees that operate as an ensemble. This method uses bagging to combine many decision trees as parallel estimators. The result is based on the majority vote of the results received from each decision tree. Random forests reduce the risk of overfitting and give higher accuracy than a single decision tree. The two concepts that make it random are bootstrapping and feature randomness.

- Support vector machines (SVMs) are characterized by the use of kernel functions, which are used to transform feature vectors into higher dimensional space, in which a separation hyperplane is learned to best fit the training data. We tested several kernels, and empirically chose the linear kernel function, which was used in the experiments.

- Gradient boosting is an algorithm which uses boosting method to combine individual decision trees. Boosting means combining a learning algorithm in series to achieve a strong learner from many sequentially connected weak learners.

- XGBoost is an implementation of gradient boosted decision trees designed for speed and performance. It implements regularization and it offers possibilities for handling missing values. It also uses parallelization of tree construction, which makes the training much faster.

References

1. Chen, J.-F.; Wang, W.-M.; Huang, C.-M. Analysis of an adaptive time-series autoregressive moving-average (ARMA) model for short-term load forecasting. *Electr. Power Syst. Res.* **1995**, *34*, 187–196. [CrossRef]
2. Huang, S.-J.; Shih, K.-R. Short-term load forecasting via ARMA model identification including non-gaussian process considerations. *IEEE Trans. Power Syst.* **2003**, *18*, 673–679. [CrossRef]
3. Pappas, S.; Ekonomou, L.; Karamousantas, D.; Chatzarakis, G.; Katsikas, S.; Liatsis, P. Electricity demand loads modeling using AutoRegressive Moving Average (ARMA) models. *Energy* **2008**, *33*, 1353–1360. [CrossRef]
4. Lee, Y.-S.; Tong, L.-I. Forecasting time series using a methodology based on autoregressive integrated moving average and genetic programming. *Knowl.-Based Syst.* **2011**, *24*, 66–72. [CrossRef]
5. Chakhchoukh, Y.; Panciatici, P.; Mili, L. Electric Load Forecasting Based on Statistical Robust Methods. *IEEE Trans. Power Syst.* **2010**, *26*, 982–991. [CrossRef]
6. Papalexopoulos, A.; Hesterberg, T. A regression-based approach to short-term system load forecasting. *IEEE Trans. Power Syst.* **1990**, *5*, 1535–1547. [CrossRef]
7. Göb, R.; Lurz, K.; Pievatolo, A. Electrical load forecasting by exponential smoothing with covariates. *Appl. Stoch. Model. Bus. Ind.* **2013**, *29*, 629–645. [CrossRef]
8. Lu, Q.C.; Grady, W.M.; Crawford, M.M.; Anderson, G.M. An adaptive nonlinear predictor with orthogonal escalator structure for short-term load forecasting. *IEEE Trans. Power Syst.* **1989**, *4*, 158–164. [CrossRef]
9. Vazquez, R.; Amaris, H.; Alonso, M.; López, G.; Moreno, J.I.; Olmeda, D.; Coca, J. Assessment of an Adaptive Load Forecasting Methodology in a Smart Grid Demonstration Project. *Energies* **2017**, *10*, 190. [CrossRef]
10. Al-Hamadi, H.; Soliman, S. Short-term electric load forecasting based on Kalman filtering algorithm with moving window weather and load model. *Electr. Power Syst. Res.* **2004**, *68*, 47–59. [CrossRef]
11. Takeda, H.; Tamura, Y.; Sato, S. Using the ensemble Kalman filter for electricity load forecasting and analysis. *Energy* **2016**, *104*, 184–198. [CrossRef]
12. Li, G.; Cheng, C.-T.; Lin, J.-Y.; Zeng, Y. Short-Term Load Forecasting Using Support Vector Machine with SCE-UA Algorithm. In Proceedings of the Third International Conference on Natural Computation (ICNC 2007), Haikou, China, 24–27 August 2007; Volume 1, pp. 290–294. [CrossRef]
13. Mohandes, M. Support vector machines for short-term electrical load forecasting. *Int. J. Energy Res.* **2002**, *26*, 335–345. [CrossRef]
14. Dudek, G. Short-Term Load Forecasting Using Random Forests. *Adv. Intell. Syst. Comput.* **2015**, *323*, 821–828. [CrossRef]

15. Cheng, Y.-Y.; Chan, P.P.; Qiu, Z.-W. Random forest based ensemble system for short term load forecasting. In Proceedings of the 2012 International Conference on Machine Learning and Cybernetics, Xian, China, 15–17 July 2012; Volume 1, pp. 52–56.
16. Mandal, P.; Senjyu, T.; Funabashi, T. Neural networks approach to forecast several hour ahead electricity prices and loads in deregulated market. *Energy Convers. Manag.* **2006**, *47*, 2128–2142. [CrossRef]
17. Hong, T.; Pinson, P.; Fan, S. Global Energy Forecasting Competition 2012. *Int. J. Forecast.* **2014**, *30*, 357–363. [CrossRef]
18. Hong, T.; Wang, P.; White, L. Weather station selection for electric load forecasting. *Int. J. Forecast.* **2015**, *31*, 286–295. [CrossRef]
19. Hong, T. Short Term Electric Load Forecasting. Ph.D. Thesis, Graduate Faculty of North Carolina State University, Raleigh, NC, USA, 2010.
20. Ben Taieb, S.; Hyndman, R.J. A gradient boosting approach to the Kaggle load forecasting competition. *Int. J. Forecast.* **2014**, *30*, 382–394. [CrossRef]
21. Nedellec, R.; Cugliari, J.; Goude, Y. GEFCom2012: Electric load forecasting and backcasting with semi-parametric models. *Int. J. Forecast.* **2014**, *30*, 375–381. [CrossRef]
22. Hosein, S.; Hosein, P. Load forecasting using deep neural networks. In Proceedings of the 2017 IEEE Power & Energy Society Innovative Smart Grid Technologies Conference (ISGT), Washington, DC, USA, 23–26 April 2017; pp. 1–5.
23. Acharya, S.K.; Wi, Y.M.; Lee, J. Short-Term Load Forecasting for a Single Household Based on Convolution Neural Networks Using Data Augmentation. *Energies* **2019**, *12*, 3560. [CrossRef]
24. Rahman, A.; Srikumar, V.; Smith, A.D. Predicting electricity consumption for commercial and residential buildings using deep recurrent neural networks. *Appl. Energy* **2018**, *212*, 372–385. [CrossRef]
25. Kong, W.; Dong, Z.Y.; Jia, Y.; Hill, D.J.; Xu, Y.; Zhang, Y. Short-Term Residential Load Forecasting Based on LSTM Recurrent Neural Network. *IEEE Trans. Smart Grid* **2017**, *10*, 841–851. [CrossRef]
26. Marino, D.L.; Amarasinghe, K.; Manic, M. Building energy load forecasting using Deep Neural Networks. In Proceedings of the IECON 2016—42nd Annual Conference of the IEEE Industrial Electronics Society, Florence, Italy, 23–26 October 2016; pp. 7046–7051.
27. Bouktif, S.; Fiaz, A.; Ouni, A.; Serhani, M.A. Single and Multi-Sequence Deep Learning Models for Short and Medium Term Electric Load Forecasting. *Energies* **2019**, *12*, 149. [CrossRef]
28. Tian, C.; Ma, J.; Zhang, C.; Zhan, P. A Deep Neural Network Model for Short-Term Load Forecast Based on Long Short-Term Memory Network and Convolutional Neural Network. *Energies* **2018**, *11*, 3493. [CrossRef]
29. Liang, Y.; Niu, D.; Hong, W.-C. Short term load forecasting based on feature extraction and improved general regression neural network model. *Energy* **2019**, *166*, 653–663. [CrossRef]
30. Zheng, H.; Yuan, J.; Chen, L. Short-Term Load Forecasting Using EMD-LSTM Neural Networks with a Xgboost Algorithm for Feature Importance Evaluation. *Energies* **2017**, *10*, 1168. [CrossRef]
31. Gund, D.A.; Eduardo, G. Shor T-Term Load Forecasting for Industrial Customers Using Fasart and Fasback Neuro-Fuzzy Systems. In Proceedings of the 14th IEEE Power Systems Computation Conference, Sevilla, Spain, 24–28 June 2002; IEE: New York, NY, USA, 2002; pp. 24–28.
32. Zhu, R.; Guo, W.; Gong, X. Short-Term Load Forecasting for CCHP Systems Considering the Correlation between Heating, Gas and Electrical Loads Based on Deep Learning. *Energies* **2019**, *12*, 3308. [CrossRef]
33. Berk, K.; Hoffmann, A.; Müller, A. Probabilistic forecasting of industrial electricity load with regime switching behavior. *Int. J. Forecast.* **2018**, *34*, 147–162. [CrossRef]
34. Bracale, A.; Carpinelli, G.; De Falco, P.; Hong, T. Short-term industrial load forecasting: A case study in an Italian factory. In Proceedings of the 2017 IEEE PES Innovative Smart Grid Technologies Conference Europe (ISGT-Europe), Torino, Italy, 26–29 September 2017; pp. 1–6.
35. Porteiro, R.; Nesmachnow, S.; Hernández-Callejo, L. Short Term Load Forecasting of Industrial Electricity Using Machine Learning. In *Proceedings of the Education and Technology in Sciences*; Springer Science and Business Media LLC: Basel, Switzerland, 2020; pp. 146–161.
36. Shahzadeh, A.; Khosravi, A.; Nahavandi, S. Improving load forecast accuracy by clustering consumers using smart meter data. In Proceedings of the 2015 International Joint Conference on Neural Networks (IJCNN), Killarney, Ireland, 12–17 July 2015; pp. 1–7.

37. Quilumba, F.; Lee, W.-J.; Huang, H.; Wang, D.Y.; Szabados, R.L. Using Smart Meter Data to Improve the Accuracy of Intraday Load Forecasting Considering Customer Behavior Similarities. *IEEE Trans. Smart Grid* **2014**, *6*, 911–918. [CrossRef]

38. Sevlian, R.; Rajagopal, R. A scaling law for short term load forecasting on varying levels of aggregation. *Int. J. Electr. Power Energy Syst.* **2018**, *98*, 350–361. [CrossRef]

39. Hedén, W. Predicting Hourly Residential Energy Consumption using Random Forest and Support Vector Regression: An Analysis of the Impact of Household Clustering on the Performance Accuracy. Master's Thesis, KTH Royal Institute of Technology, School of Engineering Sciences, Stockholm, Sweden, 2016; pp. 1–74.

40. Shi, H.; Xu, M.; Li, R. Deep Learning for Household Load Forecasting—A Novel Pooling Deep RNN. *IEEE Trans. Smart Grid* **2017**, *9*, 5271–5280. [CrossRef]

41. Ghofrani, M.; Hassanzadeh, M.; Etezadi-Amoli, M.; Fadali, M.S. Smart meter based short-term load forecasting for residential customers. In Proceedings of the 2011 North American Power Symposium, Boston, MA, USA, 4–6 August 2011; pp. 1–5. [CrossRef]

42. Hsiao, Y.-H. Household Electricity Demand Forecast Based on Context Information and User Daily Schedule Analysis From Meter Data. *IEEE Trans. Ind. Informatics* **2014**, *11*, 33–43. [CrossRef]

43. Maciejowska, K.; Nitka, W.; Weron, T. Day-Ahead vs. Intraday—Forecasting the Price Spread to Maximize Economic Benefits. *Energies* **2019**, *12*, 631. [CrossRef]

44. Wang, Y.; Wu, L. Improving economic values of day-ahead load forecasts to real-time power system operations. *IET Gener. Transm. Distrib.* **2017**, *11*, 4238–4247. [CrossRef]

45. Energy Metering Systems|eGauge. Available online: https://www.egauge.net/ (accessed on 18 March 2020).

46. Day-ahead Market|Nord Pool. Available online: https://www.nordpoolgroup.com/the-power-market/Day-ahead-market/ (accessed on 18 March 2020).

47. Fahad, M.U.; Arbab, N. Factor Affecting Short Term Load Forecasting. *J. Clean Energy Technol.* **2014**, *2*, 305–309. [CrossRef]

48. López, M.; Sans, C.; Verdú, S.V.; Senabre, C. Classification of Special Days in Short-Term Load Forecasting: The Spanish Case Study. *Energies* **2019**, *12*, 1253. [CrossRef]

49. Cox, D.R. Interaction. *Int. Stat. Rev. Rev. Int. Stat.* **1984**, *52*, 1. [CrossRef]

50. Humeau, S.; Wijaya, T.K.; Vasirani, M.; Aberer, K. Electricity load forecasting for residential customers: Exploiting aggregation and correlation between households. In Proceedings of the 2013 Sustainable Internet and ICT for Sustainability (SustainIT), Palermo, Italy, 30–31 October 2013; pp. 1–6. [CrossRef]

51. Fan, S.; Methaprayoon, K.; Lee, W.-J. Multiregion Load Forecasting for System With Large Geographical Area. *IEEE Trans. Ind. Appl.* **2009**, *45*, 1452–1459. [CrossRef]

52. Dahl, M.; Brun, A.; Kirsebom, O.S.; Andresen, G. Improving Short-Term Heat Load Forecasts with Calendar and Holiday Data. *Energies* **2018**, *11*, 1678. [CrossRef]

53. Bengio, Y. Learning Deep Architectures for AI. *Found. Trends Mach. Learn.* **2009**, *2*, 1–127. [CrossRef]

54. Xiong, W.; Droppo, J.; Huang, X.; Seide, F.; Seltzer, M.; Stolcke, A.; Yu, D.; Zweig, G. The microsoft 2016 conversational speech recognition system. In Proceedings of the 2017 IEEE International Conference on Acoustics, Speech and Signal Processing (ICASSP), New Orleans, LA, USA, 5–9 March 2017; pp. 5255–5259.

55. Wu, Y.; Schuster, M.; Chen, Z.; Le, Q.V.; Norouzi, M.; Macherey, W.; Krikun, M.; Cao, Y.; Gao, Q.; Macherey, K.; et al. Google's NMT. *arXiv* **2016**, arXiv:1609.08144.

56. He, K.; Zhang, X.; Ren, S.; Sun, J. Deep Residual Learning for Image Recognition. In Proceedings of the 2016 IEEE Conference on Computer Vision and Pattern Recognition (CVPR), New Orleans, LA, USA, 5–9 March 2016; pp. 770–778.

57. He, K.; Gkioxari, G.; Dollár, P.; Girshick, R. Mask R-CNN. In Proceedings of the 2017 IEEE International Conference on Computer Vision (ICCV), Venice, Italy, 22–29 October 2017; pp. 2980–2988.

58. Pramono, S.H.; Rohmatillah, M.; Maulana, E.; Hasanah, R.N.; Hario, F. Deep Learning-Based Short-Term Load Forecasting for Supporting Demand Response Program in Hybrid Energy System. *Energies* **2019**, *12*, 3359. [CrossRef]

59. Chen, K.; Chen, K.; Wang, Q.; He, J.; Hu, J.; He, J. Short-Term Load Forecasting With Deep Residual Networks. *IEEE Trans. Smart Grid* **2018**, *10*, 3943–3952. [CrossRef]

60. He, K.; Zhang, X.; Ren, S.; Sun, J. Identity Mappings in Deep Residual Networks. In *Proceedings of the Applications of Evolutionary Computation*; Springer Science and Business Media LLC: Berlin, Germany, 2016; Volume 9908, pp. 630–645.

61. Chen, D.; Hu, F.; Nian, G.; Yang, T. Deep Residual Learning for Nonlinear Regression. *Entropy* **2020**, *22*, 193. [CrossRef]

62. Barnston, A.G. *Correspondence among the Correlation, RMSE, and Meidke Foresast Verification Measures*; Refinement of the Neidke Score. Wea. Forecasting; American Meteorological Society (AMS): Boston, MA, USA, 1992; Volume 7, pp. 699–709.

63. Hyndman, R.J.; Koehler, A.B. Another look at measures of forecast accuracy. *Int. J. Forecast.* **2006**, *22*, 679–688. [CrossRef]

64. Barten, A.P. The coefficient of determination for regression without a constant term. *The Logic of Multiparty Systems* **1987**, *15*, 181–189. [CrossRef]

65. Geurts, M.; Box, G.E.P.; Jenkins, G.M. Time Series Analysis: Forecasting and Control. *J. Mark. Res.* **1977**, *14*, 269. [CrossRef]

66. Freedman, D.A. *Statistical Models: Theory and Practice*; Cambridge University Press (CUP): Cambridge, UK, 2005.

67. Aha, D.W.; Kibler, D.; Albert, M.K. Instance-Based Learning Algorithms. *Mach. Learn.* **1991**, *6*, 37–66. [CrossRef]

68. Wang, Y.; Witten, I.H. Induction of model trees for predicting continuous classes. In Proceedings of the 9th European Conference on Machine Learning Poster Papers, Prague, Czech Republic, 23–25 April 1997; Springer: Prague, Czech Republic, 1997.

69. Breiman, L. Random Forest. *Mach. Learn.* **2001**, *45*, 5–32. [CrossRef]

70. Shevade, S.; Keerthi, S.S.; Bhattacharyya, C.; Murthy, K. Improvements to the SMO algorithm for SVM regression. *IEEE Trans. Neural Netw.* **2000**, *11*, 1188–1193. [CrossRef]

71. Friedman, J.H. Greedy function approximation: A gradient boosting machine. *Ann. Stat.* **2001**, 1189–1232. [CrossRef]

72. Friedman, J.H. Stochastic gradient boosting. *Comput. Stat. Data Anal.* **2002**, *38*, 367–378. [CrossRef]

73. Chen, T.; Guestrin, C. XGBoost: A scalable tree boosting system. In Proceedings of the ACM SIGKDD International Conference on Knowledge Discovery and Data Mining, San Franciso, CA, USA, 22–27 August 2016.

74. Gasparin, A.; Lukovic, S.; Alippi, C. Deep Learning for Time Series Forecasting: The Electric Load Case. *arXiv* **2019**, arXiv:1907.09207.

75. IHPEC—UCI Machine Learning Repository: Individual Household Electric Power Consumption Data Set. Available online: https://archive.ics.uci.edu/ml/datasets/individual+household+electric+power+consumption (accessed on 22 March 2020).

76. Lusis, P.; Khalilpour, K.R.; Andrew, L.; Liebman, A. Short-term residential load forecasting: Impact of calendar effects and forecast granularity. *Appl. Energy* **2017**, *205*, 654–669. [CrossRef]

77. Muralitharan, K.; Sakthivel, R.; Vishnuvarthan, R. Neural network based optimization approach for energy demand prediction in smart grid. *Neurocomputing* **2018**, *273*, 199–208. [CrossRef]

78. Yildiz, B.; Bilbao, J.I.; Dore, J.; Sproul, A. Short-term forecasting of individual household electricity loads with investigating impact of data resolution and forecast horizon. *Renew. Energy Environ. Sustain.* **2018**, *3*, 3. [CrossRef]

79. Ali, S.; Mansoor, H.; Arshad, N.; Khan, I. Short Term Load Forecasting using Smart Meter Data. In Proceedings of the Tenth ACM International Conference on Future Energy Systems—e-Energy '19, Phoenix, AZ, USA, 25–28 June 2019; pp. 419–421.

80. Ganz, K.; Hinterstocker, M.; Von Roon, S. Day-ahead probabilistic load forecasting for individual electricity consumption—Assessment of point-and interval-based methods. In Proceedings of the 2019 IEEE PES Innovative Smart Grid Technologies Europe (ISGT-Europe), Bucharest, Romania, 29 September–2 October 2019; pp. 1–5.

81. Gerossier, A.; Girard, R.; Kariniotakis, G.; Michiorri, A. Probabilistic day-ahead forecasting of household electricity demand. *CIRED—Open Access Proc. J.* **2017**, *2017*, 2500–2504. [CrossRef]

82.	Voss, M.; Bender-Saebelkampf, C.; Albayrak, S. Residential Short-Term Load Forecasting Using Convolutional Neural Networks. In Proceedings of the 2018 IEEE International Conference on Communications, Control, and Computing Technologies for Smart Grids (SmartGridComm), Aalborg, Denmark, 29–31 October 2018; pp. 1–6. [CrossRef]

83.	Wijaya, T.K.; Vasirani, M.; Humeau, S.; Aberer, K. *Residential Electricity Load Forecasting: Evaluation of Individual and Aggregate Forecasts*; EPFL: Lausanne, Switzerland, 2014; pp. 1–22.

84.	Fallah, S.N.; Ganjkhani, M.; Shamshirband, S.; Chau, K. Wing Computational intelligence on short-term load forecasting: A methodological overview. *Energies* **2019**, *12*, 393. [CrossRef]

85.	Dataport|Login|Signup. Available online: https://dataport.pecanstreet.org/ (accessed on 18 March 2020).

86.	REFIT Datasets—REFIT. Available online: https://www.refitsmarthomes.org/datasets/ (accessed on 18 March 2020).

87.	Plug Load Appliance Identification Dataset (PLAID)|Energy.duke.edu. Available online: https://energy.duke.edu/content/plug-load-appliance-identification-dataset-plaid (accessed on 18 March 2020).

88.	UK Domestic Appliance-Level Electricity (UK-DALE) Dataset|Jack Kelly. Available online: https://jack-kelly.com/data/ (accessed on 18 March 2020).

89.	GREEND Download|SourceForge.net. Available online: https://sourceforge.net/projects/greend/ (accessed on 18 March 2020).

90.	EE202B—ECO Dataset. Available online: https://sites.google.com/view/activities-prediction-202b/project-homepage/eco-dataset (accessed on 18 March 2020).

91.	REDD. Available online: http://redd.csail.mit.edu/ (accessed on 18 March 2020).

92.	UKERC Energy Data Centre. Available online: https://ukerc.rl.ac.uk/DC/cgi-bin/edc_search.pl?GoButton=Detail&WantComp=26&&RELATED=1 (accessed on 18 March 2020).

93.	ISSDA|Commission for Energy Regulation (CER). Available online: http://www.ucd.ie/issda/data/commissionforenergyregulationcer/ (accessed on 18 March 2020).

94.	Long-Term Energy Consumption & Outdoor air Temperature for 11 Commercial Buildings—Datasets—OpenEI Datasets. Available online: https://openei.org/datasets/dataset/consumption-outdoor-air-temperature-11-commercial-buildings (accessed on 18 March 2020).

95.	EnerNOC Open :: Data. Available online: https://open-enernoc-data.s3.amazonaws.com/anon/index.html (accessed on 18 March 2020).

96.	Global Energy Forecasting Competition 2012—Load Forecasting Kaggle. Available online: https://www.kaggle.com/c/global-energy-forecasting-competition-2012-load-forecasting/data (accessed on 18 March 2020).

97.	Industrial Machines Dataset for Electrical Load Disaggregation|IEEE DataPort. Available online: https://ieee-dataport.org/open-access/industrial-machines-dataset-electrical-load-disaggregation (accessed on 18 March 2020).

98.	NREL RSF Measured Data 2011—Datasets—OpenEI Datasets. Available online: https://openei.org/datasets/dataset/nrel-rsf-measured-data-2011 (accessed on 18 March 2020).

Article

Short-Term Load Forecasting for a Single Household Based on Convolution Neural Networks Using Data Augmentation

Shree Krishna Acharya [1], Young-Min Wi [2] and Jaehee Lee [3,*]

[1] Department of Electronics Engineering, Mokpo National University, Muan 58554, Korea; krishna089@mokpo.ac.kr
[2] School of Electrical and Electronic Engineering, Gwangju University, Gwangju 61743, Korea; ymwi@gwangju.ac.kr
[3] Department of Information and Electronics Engineering, Mokpo National University, Muan 58554, Korea
* Correspondence: jaehee@mokpo.ac.kr; Tel.: +82-61-450-2431

Received: 16 August 2019; Accepted: 15 September 2019; Published: 17 September 2019

Abstract: Advanced metering infrastructure (AMI) is spreading to households in some countries, and could be a source for forecasting the residential electric demand. However, load forecasting of a single household is still a fairly challenging topic because of the high volatility and uncertainty of the electric demand of households. Moreover, there is a limitation in the use of historical load data because of a change in house ownership, change in lifestyle, integration of new electric devices, and so on. The paper proposes a novel method to forecast the electricity loads of single residential households. The proposed forecasting method is based on convolution neural networks (CNNs) combined with a data-augmentation technique, which can artificially enlarge the training data. This method can address issues caused by a lack of historical data and improve the accuracy of residential load forecasting. Simulation results illustrate the validation and efficacy of the proposed method.

Keywords: data augmentation; convolution neural network; residential load forecasting

1. Introduction

Short-term load forecasting (STLF) is an important part of power system planning and operation [1]. The STLF, which has a prediction range of one hour to 168 h, is used for controlling and scheduling of daily power system operations. Furthermore, forecasting customer-level energy consumption is essential for many potential applications in the future power system, such as demand response (DR) programs, home load scheduling with renewables, and optimal operation of energy storage systems (ESS) [2].

Statistical methods, including multiple linear regression [3,4], exponential smoothing [5], and the autoregressive integrated moving average (ARIMA) [6] are the most commonly used for the STLF. Recently, deep-learning-based forecasting techniques are gaining attention in the STLF. Recursive neural networks (RNNs) capable of learning long-term dependence are being applied to the assumption of load prediction in [7]. However, the vanishing gradient problem of RNNs makes it hard to improve forecasting accuracy. In [8,9], to overcome this problem in the RNN, the long short-term memory (LSTM) has been proposed. Some studies show the capability of LSTM for more improved forecasting performance at the system-level forecast when there is relatively long-term historical load data. In addition to the above methods, many deep learning algorithms, such as deep feed forward [10,11], deep belief network (DBN) [12], are being applied to load forecasting. In addition, some hybrid spectral methods, including wavelet analysis [13] and empirical mode decomposition (EMD) [14] with neural networks, have been proposed to remove the uncertainty of historical electrical load. However, all of these techniques were implemented in the substation- or system-level load forecasting with training from sufficient historical load data.

In the STLF at the system-level, a lot of historical data are available because small electrical load changes do not significantly affect the overall load pattern trend. However, at the household level, STLF may not have enough data for some households to capture long-term dependencies and properly train the learning network. For example, if a homeowner has recently changed, only a small amount of electrical load data will help its load forecasting. In addition, even if homeowners have not changed for a long time, the pattern of energy consumption can change with new electrical devices and lifestyle changes [15]. Therefore, there is a limit to using the previous data to increase the training data size. In [2,7], RNN and LSTM were applied to single household load prediction, but these studies did not take into account the past data shortage issues.

To address the lack of historical data for training, the convolutional neural networks (CNNs) can be an effective method for residential load forecasting, because CNN can capture short-term trends in load data when the local data points are strongly related to each other [16]. In addition, data augmentation can be one solution to handle the issues. Data augmentation can handle short-duration data collection by enlarging the size of the training data in a way that adds extras copies of training examples in training dataset and minimizes the overfitting in deep-learning techniques [17]. The overall training cost of a deep network will be reduced when input data are large and contain more similar information. With such enlarged training data, the accuracy of residential load forecasting can be improved. Technical paper [7] tries to enlarge the input data using another household's load series. However, this approach has the potential to obtain data that have different characteristics of the load series of the target household. Moreover, some target households have similar load profiles in their load series, but others do not [2]. Because of load profile diversity within and between households, existing forecasting benchmarks yield cons rather than pros. In fact, forecasting with only a single household's data may not often have sufficient information to fit a wide variety of the model capacity, particularly in deep-learning techniques. Therefore, proper data augmentation is required, which can artificially create new training data from the historical data of a single household.

Herein, the paper proposes a load-forecasting method for a single residential household based on convolutional neural networks (CNNs). The CNN is a type of deep neural network where its structure is formed by using convolution and pooling layers [18–20]. With the help of various filters, CNNs can learn the inherent information of an electric load series. The proposed forecasting introduces a novel data augmentation technique that concatenates the various residual load series generated from the electrical load of a single household. The original load series is converted to another residual load series containing uncertainty information of electrical load. Several residual load series are extracted through multiple k-means clustering to collect sufficient training data [21,22]. Among the extracted residual load series, the less uncertain and more homogeneous ones were fed to the CNNs as training data. The proposed augmentation technique can provide enough homogeneous training data to CNNs for more accurate forecasting. The proposed forecasting method was tested on ten single households for a year and is compared with the results of pooling-based augmentation [7].

The rest of this paper is organized as follows. Section 2 discusses the implementation of the proposed augmentation strategies and the context of residual load series. Similarly, Section 3 describes the proposed algorithm. Section 4 talks about the implementation procedure and discusses the simulation results. Section 5 concludes the paper.

2. Augmentation Implementation

2.1. CNN with Augmentation

Since an electric load series consists of many load profiles, the diversities between the load profiles are a major concern for forecasting using CNN filter networks. In fact, CNNs can facilitate more precisely given less diverse training load profiles rather than highly diverse ones [23]. Because of human actions, weather, and types of day, the differences between load profiles are uncertain, non-linear, and complex [2,24]. On the other hand, the amount of training data is crucial for fitting all the parameters of CNNs.

To address the issues of the uncertainty and necessities of huge training data, the state-of-the-art papers used other households load series for data augmentation. Moreover, the existing approaches do forecasting by hoping that load profiles will repeat and assume that the repetition of load profiles is often similar among households during the same time interval [7,23]. As a consequence, the forecasting accuracy is trivialized when the augmented data contains only correlated households' load series. However, the effects of human actions and the type of day are totally different in the electricity load of a single household and are too difficult to predict. Thus, incorporating other households' load series into training data inevitably burdens the learning ability of the CNN rather than optimizing it.

The augmentation strategy can be applied with single households and is valid when each deep-learning-based forecasting framework improved its performance [25,26]. To improve cognition, augmentation techniques need to grasp all the potentially uncertain information about the electric load series. Figure 1a,b describe the concept of enlarging data with the existing pooling and proposed augmentation approach for deep-learning-based forecasting. To enlarge data, the proposed method artificially generates several augmented load series, each of which needs to be extracted in a way that facilitates the CNN learning strategies. To obtain a rationale, each series should have less uncertain and more homogeneous information. The appropriate series are concatenated with each other to turn out a huge amount of training data with homogeneous information. The more homogeneous information there is, the more useful it is for a CNN network to address the granular-level load prediction. Any dissimilar augmented load series in concatenation could destroy a CNN's optimal cognition. The procedure of extracting a homogeneous information load series is vital and depends totally upon imagination [27]. The proposed augmentation strategies are described in the following section.

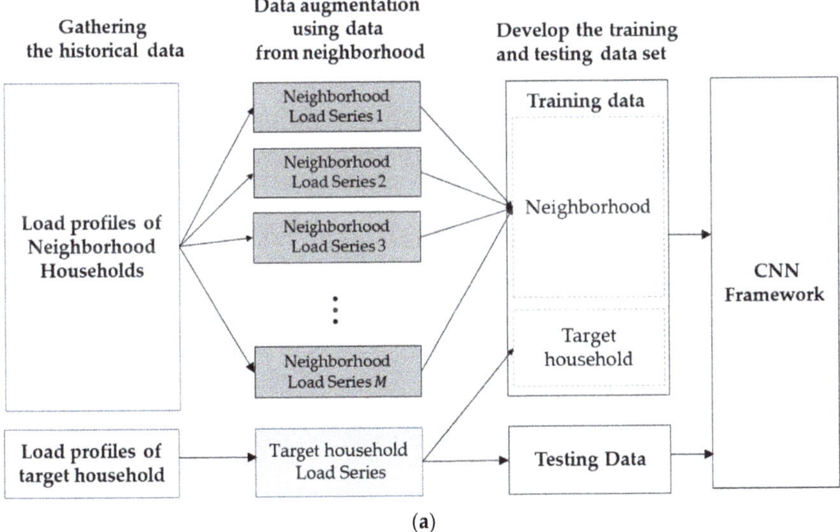

(a)

Figure 1. *Cont.*

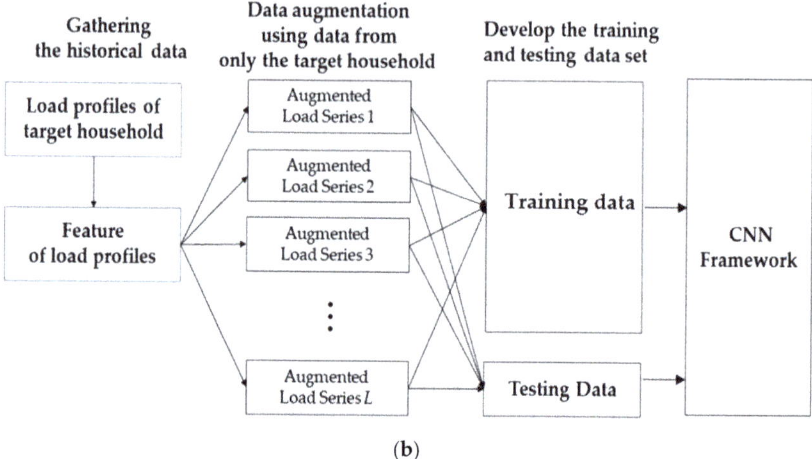

Figure 1. Comparison of data augmentation strategies: **(a)** pooling technique and **(b)** proposed augmentation technique.

2.2. Extraction of Residual Load Series

The data structure of the historical electricity load of a target household can be expressed as the following matrix:

$$P = \begin{bmatrix} p_{1,1},\, p_{2,1},\, \ldots,p_{1,t},\ldots,p_{1,T} \\ p_{2,1},\, p_{2,2},\, \ldots,p_{2,t},\ldots,\, p_{2,T} \\ \vdots \\ p_{d,1},\, p_{d,2},\, \ldots,p_{d,t},\ldots,p_{d,T} \\ \vdots \\ p_{D,1},\, p_{D,2},\, \ldots,p_{D,t},\ldots,p_{D,T} \end{bmatrix}, \tag{1}$$

where $p_{d,t}$ represents historical electric load at time t in day d; D represents the number of days of historical data, and the time period T is 24 h. This matrix of a historical load can be simplified as follows:

$$P = [p_1,\, p_2,\, p_3,\ldots,p_d,\ldots,p_D], \tag{2}$$

where vector p_d are the hourly load profiles of a target household in day d. Figure 2 shows all daily load profiles of a household for one month. In Figure 2, it can be found that the electrical load of a single household fluctuates abruptly. Generally, a smaller electricity load tends to have a higher variation, which makes it difficult to accurately forecast the residential electricity load. In CNN-based forecasting, one way to achieve high performance is to reduce the variations of training data [21].

To improve the learning ability of CNNs, one must extract new features from a residential load series, which has less volatility but still has inherent characteristics of a residential load. One way to reduce the variation is to remove the regular pattern from the load series [28] so that CNNs use only its residuals. With this approach, each load profile is decomposed into centroid and residual load profiles as follows:

$$p_d = c + r_d, \tag{3}$$

where vector c is the centroid load profile (average profile) of the historical load, and r_d is a vector of the residual load series which is the difference between a given centroid c and an actual load profile on day d. The centroid load profile can be used as the baseline of a particular group of daily load profiles, and the repetition of the centroid load profile yields an average load series for a certain duration.

Figure 3 shows an average load series, and its residual load series of a residential household for a month. In Figure 3, the average load series does not contain any uncertain information, but it shows only a regular pattern. On the other hand, the residual load series contains all the uncertain and complex information about the residential load. The residual load series is less volatile than is the original load series but still has inherent characteristics of the residential load. With training from this less volatile data, the CNNs can forecast the small residential load more accurately.

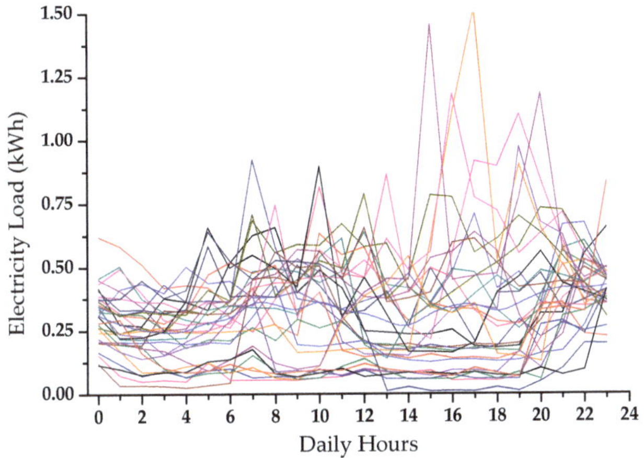

Figure 2. Daily load profile of a single household for one month.

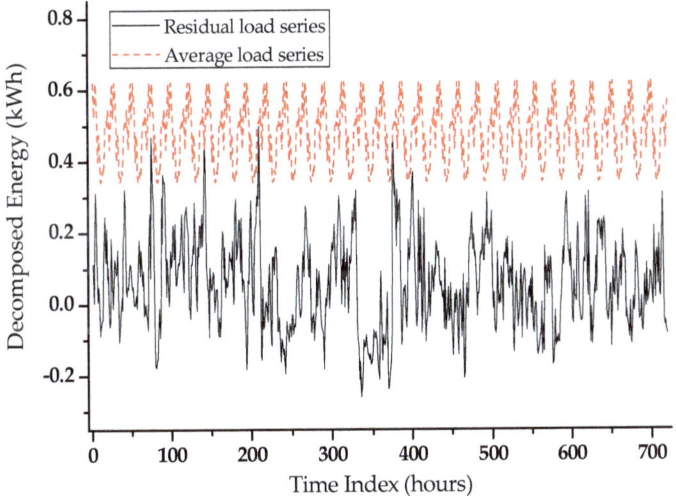

Figure 3. Average and its residual load series of a single household.

To ensure the appropriateness of the residual load series in learning networks, auto-correlation (AC) or partial auto-correlation (PAC) coefficients [29] can be used. Higher AC or PAC coefficients, out of the confidence interval, mean that the present residual load series is strongly coupled with the historical residual load series. Figure 4 shows AC and PAC coefficients of the residual load series of a selected single household with different time lagging. As shown in Figure 4, when the lags are the multiple of 24 h, the AC coefficients would peak, so that some AC coefficients are out of the confidence interval. Similarly, the PAC coefficient spikes also confirm that the residual load series has vital information about the residential electricity load. The information is in a hidden state and can be learned by using learning networks.

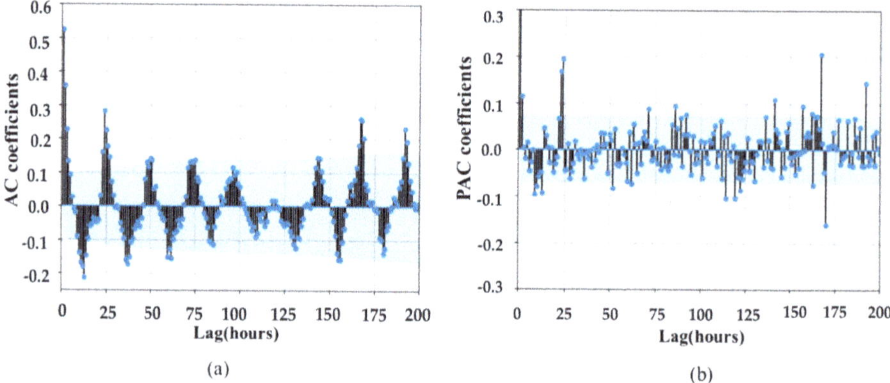

(a) (b)

Figure 4. Auto-correlation of residual load series with different time lagging: (**a**) auto-correlation (AC) coefficients, and (**b**) partial auto-correlation (PAC) coefficients.

With these residential load series, the paper proposes a CNN-based method for residential load forecasting using data augmentation. The augmentation technique and overall forecasting procedure are described in the following section.

3. Proposed Residential Load Forecasting Method

3.1. Generation of Centroid Load Profiles

To generate the different centroid load profiles from historical residential load P, multiple k-means clustering [21,22] is used in the paper. The multiple k-means clustering to generate the centroid load profiles of a residential load can be express as follows:

$$\text{Minimize} \sum_{i=1}^{k} \sum_{p_d \in S_{k,i}} \|p_d - c_{k,i}\|^2, \; k = 1, 2, \ldots, K, \tag{4}$$

where $S_{k,i}$ is the i-th partition of the load profile set P, which is generated with a clustering number of k, and $c_{k,i}$ is the centroid load profile of the corresponding partition of $S_{k,\,i}$. The clustering starts with a clustering number 1 and ends with the clustering number K. Finally, $\left(K^2 + K\right)/2$ centroid load profiles are generated. The paper used all centroids generated with clustering numbers of one to k, and the l-th generated centroid load profile is defined as c_l'. The centroid load profiles (c_l') are $\left(K^2 + K\right)/2$ daily load patterns, which are the mean values of load patterns of similar days. Figure 5 shows the centroid load profiles of a single household for one month using the multiple k-means clustering algorithm.

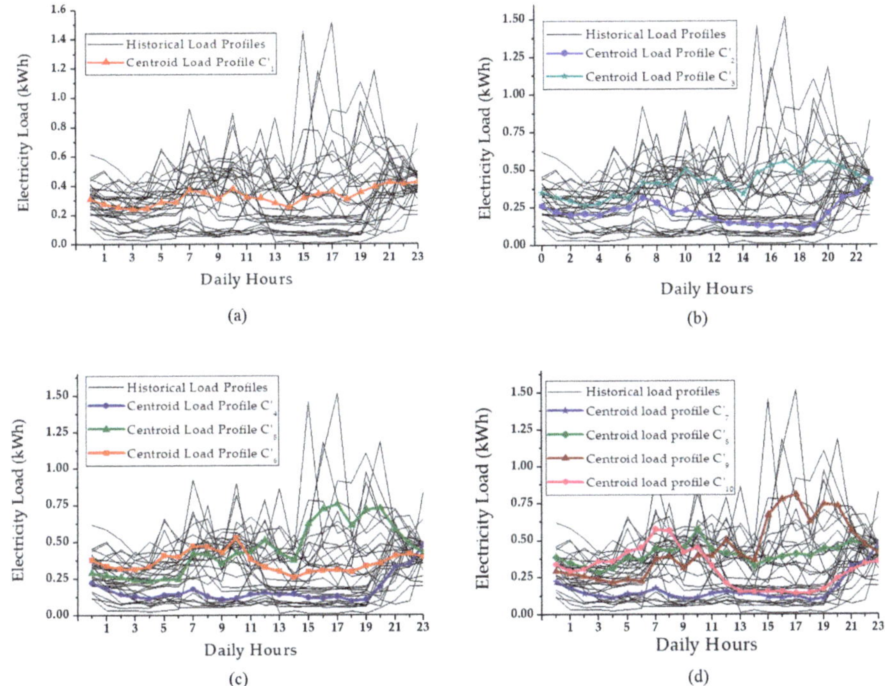

Figure 5. Centroid load profiles generated using multiple k-means clustering algorithm with different clustering number k: (**a**) $k = 1$; (**b**) $k = 2$; (**c**) $k = 3$, and (**d**) $k = 4$.

3.2. Augmentation of Homogeneous Residual Load Profile

Using multiple k-means clustering and its corresponding centroid load profile, the residential load profiles can be generated as follows:

$$r_{d,l} = p_d - c'_l, \tag{5}$$

where $r_{d,l}$ is the vector of residual load profiles of the target household on day d, which is generated with the l-th centroid load profile. This residual load profile is expected to be less volatile and less uncertain than was the original load profile. In addition, from Equation (5), the amount of training data will be increased much more by using residual load profiles as training data. Namely, one time series (p_d) of a single household load can be transformed into several time series of residual load series ($r_{d,1}$, $r_{d,2}$, ..., $r_{d,l}$).

For the CNN, the appropriate training data are concatenated with each other, and each training data should have less uncertainty and more homogeneous information. The more homogenous information there is, the more useful it will be for a CNN network to address the granular-level load prediction. Any dissimilar augmented load series could destroy the CNN's optimal cognition. Therefore, one needs to select the more homogeneous residual load profiles from among the augmented load profiles from Equation (5).

To select the more homogenous residual load profiles, the Frobenius norm [30] Φ_l of each residual load profiles are calculated as follows:

$$\Phi_l = \sqrt{\sum_{d=1}^{D} \|r_{d,l}\|^2}, \tag{6}$$

where each $r_{d,l}$ is sequentially observed with time. When residual load profiles ($r_{d,l}$) have lower Φ_l, most residuals generated with c'_l are closed to its centroid, and these augmented residual load profiles can be expected to be homogeneous. Finally, the paper used only the residual load series with lower Φ_l as training set R_{in},

$$R_{in} = \left\{ r_{d,l} \,\middle|\, \Phi_l \leq \frac{1}{L} \sum_{l=1}^{L} \Phi_l \right\}. \tag{7}$$

In addition to the training set, the paper used the residual load profiles with the lowest Φ_l as the test set for the CNN model. Figure 6 shows the structure of training and testing sets of the proposed method. In Figure 6, the residual load profile is expressed with its elements as $r_{d,l} = \left\{ r_{(1,d),l}, r_{(2,d),l}, \cdots, r_{(t,d),l}, \cdots, r_{(24,d),l} \right\}$. In the training set, several time series of residual load series ($r_{d,1}$, $r_{d,2}$, ..., $r_{d,l}$) can be used instead of one original time series (p_d) of a single household load. With this enragement of training data, forecasting accuracy for individual residential households can be improved.

3.3. CNN Model for Residential Load Forecasting

The training process is yielded by running a program with a given number of iterations. To optimize the CNN model, in each iteration, a root mean square is used for the training process as follows:

$$\text{argmin} \; \sqrt{ \frac{1}{L} \cdot \frac{1}{D} \cdot \sum_{l=1}^{L} \sum_{d=1}^{D} \left(\hat{r}_{d,l} - r_{d,l} \right)^2 }, \tag{8}$$

where $\hat{r}_{d,l}$ is the predicted vector load profile obtained from the CNN, and L represents the number of selected residual load profiles from Equation (6). A well-trained and converged CNN forecasting network is used for testing the process for predicting the future load. Since the CNN network deals with residual load profiles, the forecasting result can be generated in terms of a residual load profile. In fact, the forecast load profile $\hat{p}_{D+1}$ can be obtained by adding both the centroid load profile and the forecast residual load profile as follows:

$$\hat{p}_{D+1} = \hat{c} + \hat{r}_{D+1,p}, \tag{9}$$

where $\hat{r}_{D+1,\,p}$ represents the day-ahead forecasted residual load profile, and $\hat{c}$ represents the most appropriate centroid load profile which has the lowest Frobenius norm Φ_l.

Figure 7 explains the entire load forecasting procedure for the proposed methodology. In the first stage, centroid load profiles are generated using multiple k-means clustering, and different types of residual load profiles are extracted with corresponding centroid load profiles. In the second stage, only homogeneous residual load series are selected using the Frobenius norm for training the CNN. In the final stage, the CNN forecasting framework is used to predict day-ahead residential load profiles. The CNN implementation can be summarized into three parts: (1) initialization of the CNN parameters, (2) training the CNN model with the help of input matrix R_{in}, and (3) predicting the day-ahead load profile using the optimally trained CNN model. With the proposed forecasting method, it is expected that the CNN can cognize the characteristics of historical load profiles more accurately, so that the forecasting accuracy is improved interestingly.

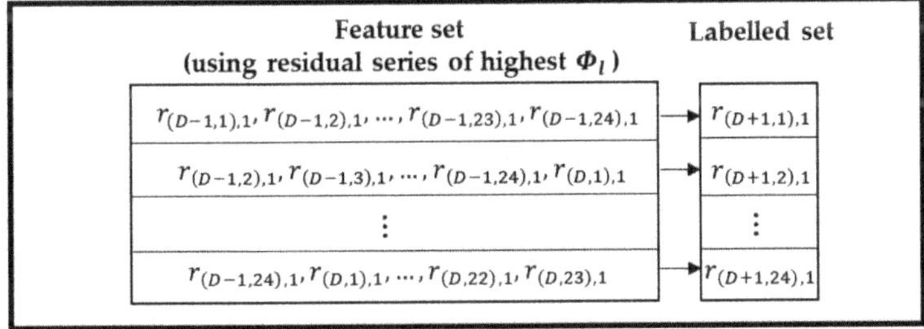

Figure 6. Structure of training and testing data for the proposed method.

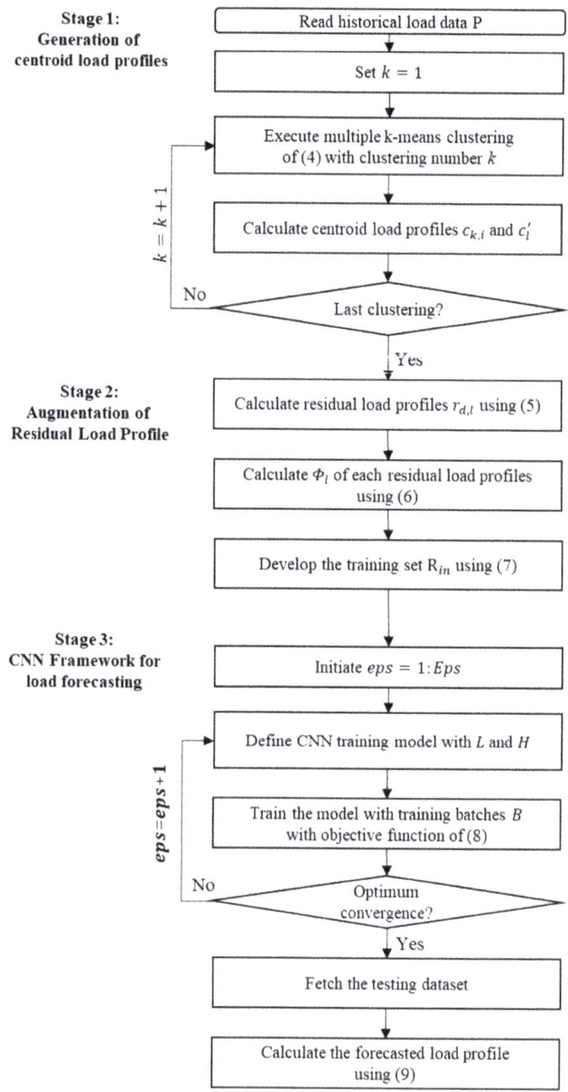

Figure 7. The overall procedure of the proposed load forecasting method for an individual household.

4. Simulation Results

4.1. Data Description and Hyper-Parameter Tuning

The proposed method was tested using hourly metering data gathered from 1181 residential households in Seoul, Korea, for one year (August 2016 to July 2017). With this dataset, the results of the proposed method are compared with the results of pooling-based augmentation [7] as well as with the results of other deep-learning models [2,7,20]. For the day-ahead load forecasting, historical load data for the last 30 days were used for the training process, and historical load data of the previous day were used for testing. To evaluate the accuracy of forecasting results, the paper employed the mean absolute percentage error (MAPE) and root mean square error (RMSE) as follows:

$$\text{MAPE} = \frac{1}{T} \sum_{t=1}^{T} \frac{\left| \hat{p}_{t,\,(D+1)} - p_{t,\,(D+1)} \right|}{p_{t,\,(D+1)}} \times 100\% , \tag{10}$$

$$\text{RMSE} = \sqrt{\frac{1}{T} \sum_{t=1}^{T} \left(\left(\hat{p}_{t,\,(D+1)} - p_{t,\,(D+1)} \right) \right)^2} . \tag{11}$$

The proposed method is developed and tested through Python with the Keras library, whose backend is Tensorflow [31,32]. To avoid over-fitting in the training process, the parameter settings in Table 1 were tested for the proposed method. The tuning process of hyper-parameters was based on [23]. The numbers of hidden layers were selected based on [33–38]. The common hyper-parameters, such as activation function, optimizer, loss function, etc., are reported in Table 1. The additional more specific parameters of CNN were settled with the size of filter 3×3, number of input filters 24, maximum pooling size 3×3, and size of strides 1. The convolution layers were followed by a fully connected layer with the rectified linear unit (ReLU) activation functions. The final fully connected layer predicted one-hour electric load at a time, which was matched with [36]. The dataset (30 days) was split into validating set (3 days), testing set (1 day), and training set (26 days).

Table 1. Hyper parameters for selected deep learning models.

Parameters	BPNN	CNN	LSTM
No. of hidden layers	2 or 3	2 or 3	2 or 3
No. of nodes per layer	32	24	20
Activation functions	ReLU	ReLU	tanh and sigmoid
No. of Epochs (iteration)	150	150	300
Optimizer	RMS-Prop	RMS-prop	RMS-prop
Loss Function	MSE	MSE	MSE
Testing samples	24-h	24-h	24-h

To tackle the overfitting in CNNs, the proposed method was preliminarily tested with different numbers of hidden layers in CNN architecture. Table 2 shows the forecasting results of ten households with six different hidden layers. The ten single households in Table 2 were randomly selected, and the results were monthly average values in July (peak season in Korea). In most households, the forecasting results with two or three hidden layers were more accurate. With these results, the paper set the number of hidden layers to be 2.

Table 2. Load Forecasting results (mean absolute percentage error (MAPE)) with a number of different hidden layers.

Household	Hidden Layer 0	Hidden Layer 1	Hidden Layer 2	Hidden Layer 3	Hidden Layer 4	Hidden Layer 5
1	23.75%	24.44%	19.26%	21.77%	23.43%	26.77%
2	11.96%	11.25%	11.17%	9.63%	11.51%	12.98%
3	32.10%	32.05%	30.08%	35.87%	38.09%	39.73%
4	9.86%	9.53%	9.65%	10.30%	11.05%	11.47%
5	13.13%	13.59%	11.72%	12.44%	15.47%	15.54%
6	12.88%	12.44%	11.77%	11.43%	12.32%	13.82%
7	11.76%	11.34%	9.77%	12.15%	11.68%	12.82%
8	14.78%	14.97%	13.53%	14.60%	15.12%	16.32%
9	22.19%	22.30%	22.06%	21.41%	22.46%	21.69%
10	37.24%	43.59%	34.72%	46.80%	47.12%	49.25%

4.2. Effects of Proposed Augmentation Method

Tables 3 and 4 show the MAPE and RMSE results of day-ahead forecasting with and without the proposed augmentation technique. The ten single households in tested Tables 3 and 4 were randomly selected, and the results were monthly average values in July (peak season in Korea).

Table 3. Load Forecasting results (MAPE) with and without proposed augmentation.

Household	Without Augmentation			With the Proposed Augmentation	
	BPNN (%)	LSTM (%)	CNN (%)	LSTM (%)	CNN (%)
1	32.17	33.49	43.40	31.36	19.26
2	20.54	21.62	24.77	16.60	9.63
3	39.52	38.15	48.48	37.41	30.08
4	15.56	14.61	18.42	15.47	9.53
5	17.36	16.50	20.85	17.15	11.72
6	17.46	16.85	20.77	16.99	11.77
7	14.61	14.85	17.01	13.36	9.77
8	20.38	20.31	24.08	18.17	13.53
9	42.40	43.11	46.03	28.89	21.41
10	53.71	57.02	66.64	51.28	34.72

Table 4. Load Forecasting results (root mean square error (RMSE)) with and without proposed augmentation.

Household	Without Augmentation			With the Proposed Augmentation	
	BPNN (kWh)	LSTM (kWh)	CNN (kWh)	LSTM (kWh)	CNN (kWh)
1	0.3601	0.3440	0.4092	0.3156	0.1666
2	0.2614	0.2864	0.3169	0.1253	0.1116
3	0.2382	0.2242	0.2570	0.2160	0.1313
4	0.0954	0.0907	0.1114	0.1397	0.0691
5	0.0790	0.0768	0.0882	0.0744	0.0506
6	0.0757	0.0728	0.0859	0.0769	0.0488
7	0.0663	0.0679	0.0743	0.0635	0.0389
8	0.1083	0.1028	0.1160	0.0960	0.0683
9	0.1423	0.1321	0.1638	0.1072	0.0661
10	0.3323	0.3074	0.3421	0.2984	0.1796

By using the proposed augmentation, the forecasting accuracy was improved by 5 percent, at least, as shown in Table 3. It can be concluded that the proposed forecasting method can significantly improve the forecasting accuracy more than can the other forecasting models without augmentation. When comparing only the cases with the proposed augmentation, the CNN provided more accurate forecasting than the LSTM. It is probably because the augmented data contain plenty of similar random information with small variations. With this information, the CNN received the opportunity for obtaining optimal convergence. On the other hand, as LSTM required more repetitive information for training to improve its performance, this augmented data was comparatively ineffective for LSTM.

Figure 8a,b show the average MAPE and RMSE of ten households, calculated using the results of daily forecasting in July. On most days, the proposed forecasting method significantly improved the forecasting accuracy more than did the other forecasting model without augmentation.

Figure 9 shows the daily average of forecasting results for four higher uncertain households. For most times, the proposed forecasting method significantly improved forecasting accuracy. This is probably because the homogeneous information from the proposed augmentation provides the opportunity for the CNN framework to obtain optimal convergence.

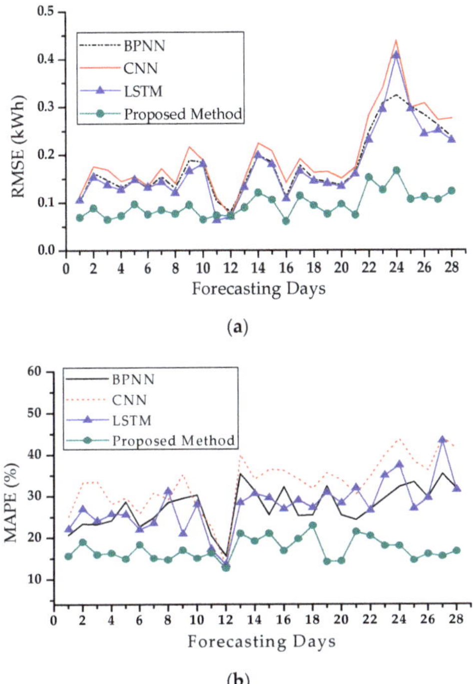

(a)

(b)

Figure 8. Average mean absolute percentage error (MAPE) and root mean square error (RMSE) of ten arbitrary selected households in July: (**a**) MAPE result, and (**b**) RMSE result.

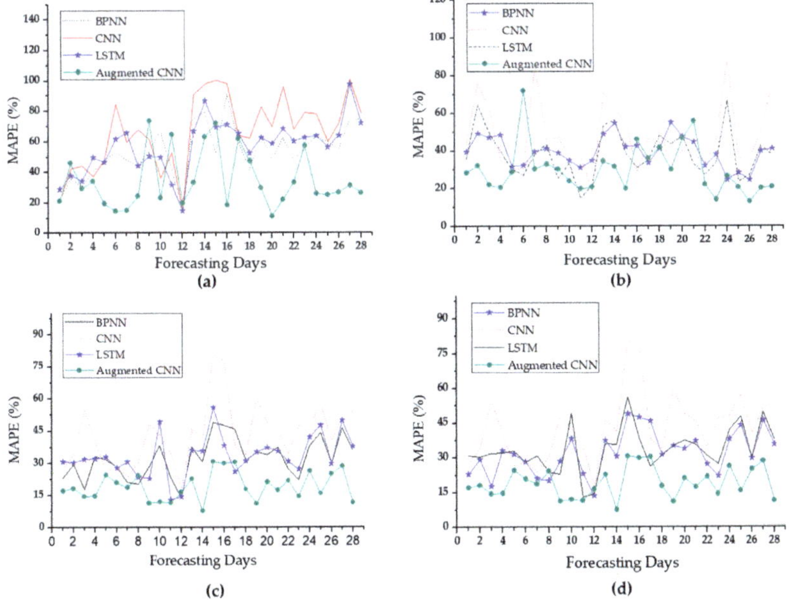

(a)

(b)

(c)

(d)

Figure 9. Daily average of MAPE of higher uncertain households from Table 3: (**a**) household 10; (**b**) household 8; (**c**) household 9 and (**d**) household 1.

4.3. Forecasting Results in Peak Day

Figure 10a,b show the forecasting results of the peak day for two selected households. One was a less uncertain household which showed the lowest monthly average MAPE in Table 2 (household 4). The other household was a more uncertain household which showed the highest monthly average MAPE in Table 2 (household 10). The peak loads of the two households in July were 1.062 kW (23:00, July 17) and 2.033 kW (18:00, July 14), respectively. During a peak day, it is expected that the residential load profiles will be highly uncertain so that the load forecasting is more challenge. To test the forecasting accuracy, the results of the proposed forecasting method were compared to the results of the forecasting models using pooling augmentation techniques [7].

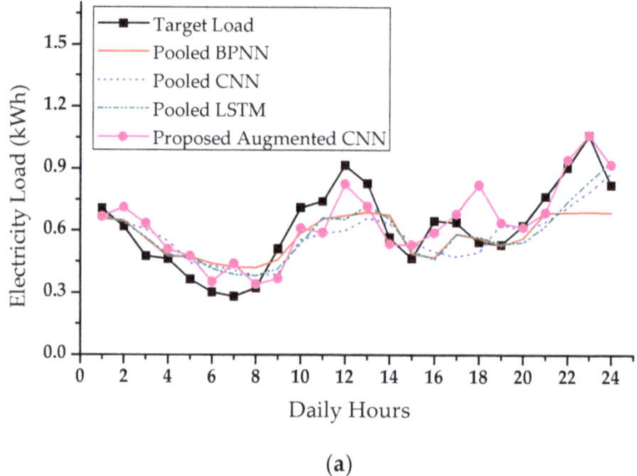

(a)

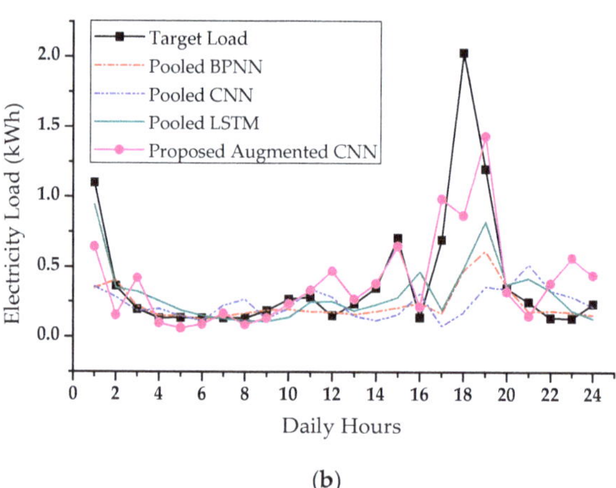

(b)

Figure 10. Day-ahead load forecasting for peak day: (**a**) lower uncertain household (household 4) and (**b**) higher uncertain household (household 10).

For pooling augmentation, the simulation used historical load data of six neighbors as an additional training set. In Figure 10a,b, the predicted load curves of the proposed method were much closer to the actual load profile at most hours of the day, for both households. Especially at the peak time of the day, the proposed forecasting model can provide significantly accurate forecasting results. On the other hand, the forecasting models using the other pooling techniques showed a worse performance for load forecasting at the peak time.

An important day for a forecasting test is the day of maximum energy. During this day, the residential load profiles would be highly uncertain at all hours of the day, so that the load forecasting is more challenging. The daily maximum energy consumption of the selected two households in July was 19.381 kWh (July 21) and 15.797 kWh (July 30), respectively. Figure 11a,b demonstrate the target load profile and predicted load profile of the less uncertain household and the highly uncertain household, respectively. For the less uncertain household, the predicted load curves of the proposed method were much closer to the actual load profile at most hours of the day. The proposed model reported 7.999% of MAPE for the less uncertain household, which was lower than the 10.6566% of MAPE from the pooled LSTM. Similarly, for the highly uncertain household, the proposed model reported 40.4058% of MAPE, which was lower than the 53.319% of MAPE from the pooled LSTM. These results strongly validate the proposed methodology for load prediction in the residential sector.

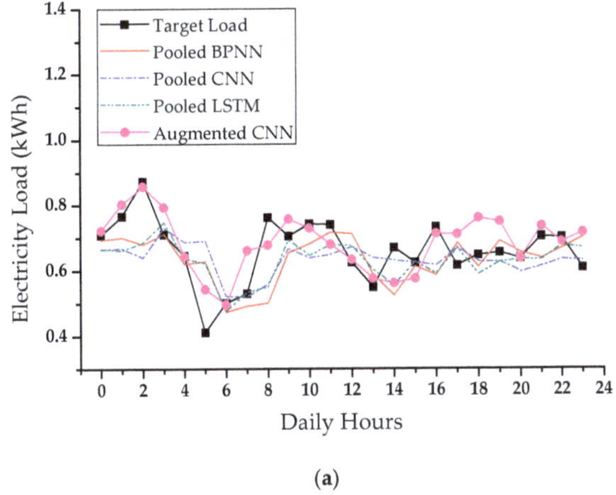

(**a**)

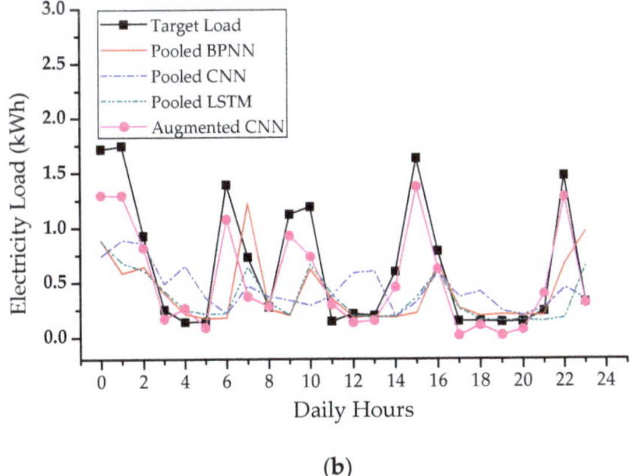

(**b**)

Figure 11. Day-ahead hourly load forecasting for the day of maximum energy consumption: (**a**) less uncertain household (household 4) and (**b**) highly uncertain household (household 10).

4.4. Monthly Results of Day-Ahead Load Forecasting

To examine the efficacy of the proposed technique for one year, this section tested the performance throughout eleven months by picking one of the best performer households and one of the worst performer households.

Tables 5 and 6 show the monthly average MAPE and RMSE results of the less uncertain household and the highly uncertain household. The test results are from September 2016 to July 2017. In Tables 5 and 6, the proposed method improved the forecasting accuracy by more than 6 percent throughout the year.

Table 5. Monthly-average of MAPE for the lower uncertain household.

Forecasting Model		Sept.	Oct.	Nov.	Dec.	Jan.	Feb.	Mar.	Apr.	May	Jun.	Jul.
Pooled BPNN	MAPE (%)	16.48	16.72	19.33	16.55	18.21	17.30	20.99	20.22	20.88	26.99	15.02
	RMSE (kWh)	0.086	0.099	0.105	0.095	0.099	0.111	0.104	0.110	0.110	0.124	0.106
Pooled CNN	MAPE (%)	15.97	20.66	21.11	18.04	19.80	18.71	22.38	22.12	22.74	28.62	15.89
	RMSE (kWh)	0.085	0.111	0.116	0.104	0.110	0.121	0.112	0.119	0.120	0.132	0.112
Pooled LSTM	MAPE (%)	14.46	16.33	18.23	15.31	17.31	16.70	23.30	18.76	19.90	26.82	14.03
	RMSE (kWh)	0.080	0.099	0.101	0.091	0.096	0.069	0.107	0.106	0.110	0.123	0.100
Proposed Method	MAPE (%)	9.662	11.65	10.53	9.50	9.91	10.34	11.25	10.95	12.07	12.25	9.65
	RMSE (kWh)	0.062	0.061	0.060	0.060	0.061	0.105	0.067	0.075	0.072	0.068	0.070

Table 6. Monthly-average of MAPE for the higher uncertain household.

Forecasting Model		Sept.	Oct.	Nov.	Dec.	Jan.	Feb.	Mar.	Apr.	May	Jun.	Jul.
Pooled BPNN	MAPE (%)	20.07	24.65	34.87	46.06	34.32	36.06	43.42	42.87	36.86	43.25	35.39
	RMSE (kWh)	0.086	0.119	0.137	0.156	0.160	0.168	0.185	0.191	0.141	0.152	0.241
Pooled CNN	MAPE (%)	18.80	28.08	34.83	55.80	39.61	37.44	48.49	53.90	40.09	50.07	41.40
	RMSE (kWh)	0.085	0.129	0.140	0.168	0.174	0.175	0.203	0.204	0.151	0.162	0.246
Pooled LSTM	MAPE (%)	16.998	24.81	32.13	46.47	34.02	33.10	43.09	47.59	33.81	45.15	35.99
	RMSE (kWh)	0.080	0.116	0.135	0.156	0.164	0.161	0.176	0.198	0.140	0.153	0.228
Proposed Method	MAPE (%)	13.83	12.79	22.26	30.47	23.51	22.09	26.21	30.89	23.05	29.53	29.12
	RMSE (kWh)	0.062	0.075	0.085	0.101	0.092	0.105	0.106	0.119	0.099	0.084	0.131

4.5. Impact of Clustering Number K

Figure 12a,b show the average MAPE and RMSE results of ten households with different clustering numbers K. In Figure 12, very accurate forecasting can be expected when the clustering number K is increased for most households. However, for some households, the MAPE is increased when the clustering number K is over 4. The higher clustering number K provides more training data to the CNNs, so that the CNNs have more chance to learn the load characteristics. However, more training data can increase the variations of the training set, which cause overfitting of the hyper-parameters of CNNs, which degrades the optimal learning of CNNs. Therefore, the optimal clustering number must be selected for each household.

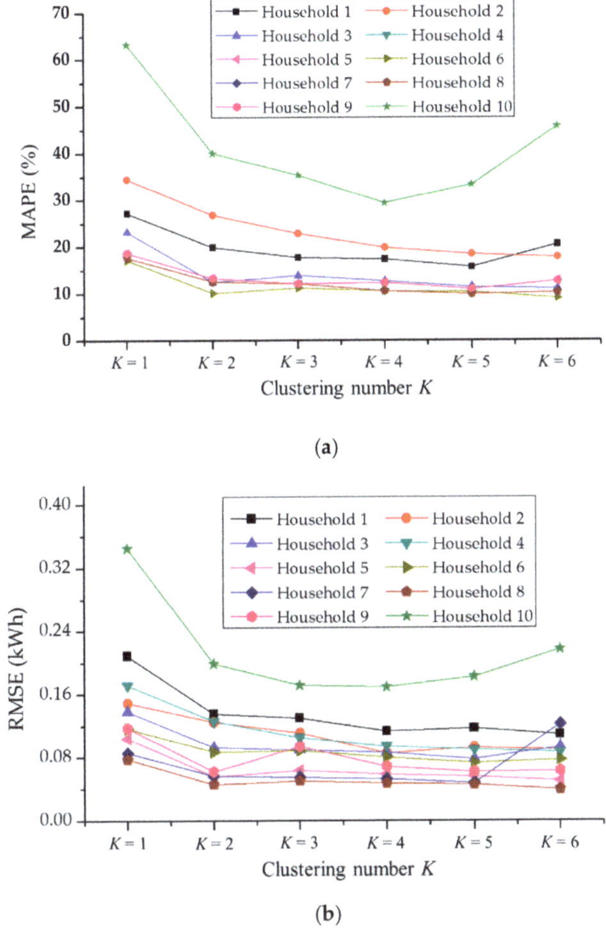

Figure 12. Effect of clustering number K on forecasting accuracy: (**a**) MAPE and (**b**) RMSE.

5. Conclusions

The paper proposed a forecasting method based on convolution neural networks (CNNs) combined with a data augmentation technique with consideration of an insufficient period of training data. The proposed data augmentation can enlarge the training data for CNNs using only a target household's own historical data, without the help of other households. The proposed forecasting method transforms a time series of a single household load data into several time series of residual loads. With this enragement of training data, forecasting accuracy for individual residential households can be improved. The test results indicated that the proposed method can deliver a notable improvement by including homogeneous information for an individual residential household's load forecasting. The proposed method can be used for energy management at the household level and evaluate the baseline of energy consumption at the household level for demand-response programs.

Author Contributions: For research, S.K.A. developed the idea of augmentation strategies for a CNN-based forecasting framework, performed the simulation, and wrote the paper. Y.-M.W. helped organize the article. J.L. provided guidance for research and revised the paper.

Funding: This work was supported by the National Research Foundation of Korea (NRF) grant funded by the Korea government (MSIT) (No. 2017R1C1B5016244). This research was supported by Korea Electric Power Corporation. (Grant number: R18XA04).

Conflicts of Interest: The authors declare no conflict of interest.

References

1. Shahidehpour, M.; Yamin, H.; Li, Z. *Market Operations in Electric Power Systems: Forecasting Scheduling and Risk Management*, 1st ed.; Wiley-IEEE Press: New York, NY, USA, 2002; pp. 21–55.
2. Kong, W.; Dong, Z.Y.; Jia, Y.; Hill, D.J.; Xu, Y.; Zhang, Y. Short-Term Residential Load Forecasting based on LSTM Recurrent Neural Network. *IEEE Trans. Smart Grid* **2017**, *10*, 841–851. [CrossRef]
3. Charytoniuk, W.; Chen, M.S.; Van Olinda, P. Nonparametric Regression Based Short-Term Load Forecasting. *IEEE Trans. Power Syst.* **1998**, *13*, 725–730. [CrossRef]
4. Song, K.-B.; Baek, Y.-S.; Hong, D.H.; Jang, G. Short-Term Load Forecasting for The Holidays Using Fuzzy Linear Regression Method. *IEEE Trans. Power Syst.* **2005**, *20*, 96–101. [CrossRef]
5. Christiaanse, W.R. Short-Term Load Forecasting Using General Exponential Smoothing. *IEEE Trans. Power Syst.* **1971**, *2*, 900–911. [CrossRef]
6. Mohamed, N.; Ahmad, M.H.; Ismail, Z. Short Term Load Forecasting Using Double Seasonal ARIMA Model. In Proceedings of the Regional Conference on Statistical Sciences 2010 (RCSS'10), Kota Bharu, Malaysia, 13 June 2010.
7. Shi, H.; Xu, M.; Li, R. Deep Learning for Household Load Forecasting-A Novel Pooling Deep RNN. *IEEE Trans. Smart Grid* **2018**, *9*, 5271–5280. [CrossRef]
8. Tian, C.; Ma, J.; Zhang, C.; Zhan, P. A Deep Neural Network Model for Short-Term Load Forecast Based on Long Short-Term Memory Network and Convolutional Neural Network. *Energies* **2018**, *11*, 3493. [CrossRef]
9. Bouktif, S.; Fiaz, A.; Ouni, A.; Serhani, M.A. Optimal Deep Learning LSTM Model for Electric Load Forecasting using Feature Selection and Genetic Algorithm: Comparison with Machine Learning Approaches. *Energies* **2018**, *11*, 1636. [CrossRef]
10. Torres, J.F.; Galicia, A.; Troncoso, A. A Scalable Approach Based on Deep Learning for Big Data Time Series forecasting. In Proceedings of the International Work-Conference on the Interplay Between Natural and Artificial Computation (IWINAC), A Coruña, Spain, 19–23 June 2017.
11. Guo, Z.; Zhou, K.; Zhang, X.; Yang, S. A Deep Learning Model for Short-Term Power Load and Probability Density Forecasting. *Energy* **2018**, *160*, 1186–1200. [CrossRef]
12. Lin, Y. Time Series Forecasting by Evolving Deep Belief Network with Negative Correlation Search. In Proceedings of the 2018 Chinese Automation Congress (CAC), Xi'an, China, 30 November–2 December 2018.
13. Vu, D.H.; Muttaqi, K.M.; Agalagaonkar, A.P. Combinatorial Approach using Wavelet Analysis and Artificial Neural Network for Short-term Load Forecasting. In Proceedings of the 2014 Australasian Universities Power Engineering Conference (AUPEC), Perth, Australia, 28 September–1 October 2014.
14. Haq, M.R.; Ni, Z. A New Hybrid Model for Short-Term Electricity Load Forecasting. *IEEE Access* **2019**, *7*, 125413–125423. [CrossRef]
15. Gram-Hanseen, G. Standby consumption in households analyzed with a practice theory approach. *J. Ind. Ecol.* **2010**, *14*, 150–165. [CrossRef]
16. Du, S.; Li, T.; Gong, X.; Yang, Y.; Horng, S.J. Traffic Flow Forecasting based on Hybrid Deep Learning Framework. In Proceedings of the 12th International Conference on Intelligent Systems and Knowledge Engineering, Nanjing, China, 24–26 November 2017.
17. Wang, Y.; Chen, Q.; Gan, D.; Yang, J.; Kirchen, D.S.; Kang, C. Deep Learning-Based Socio-demographic Information Identification from Smart Meter Data. *IEEE Trans. Smart Grid* **2018**, *10*, 2593–2602. [CrossRef]
18. LeCun, Y.; Bengio, Y. *Convolution Neural Networks for Images*; Speech and Time Series; MIT Press: Cambridge, MA, USA, 1988.

19. Pattanayek, S. *Pro Deep Learning with TensorFlow: A Mathematical Approach to Advanced Artificial Intelligence in Python*, 1st ed.; Apress: New York, NY, USA, 2017; pp. 153–222.
20. Amarasinghe, K.; Marino, D.L.; Manic, M. Deep Neural Network for Energy Load Forecasting. In Proceedings of the IEEE 26th International Symposium on Industrial Electronics (ISIE), Edinburgh, UK, 19–21 June 2017.
21. Kwac, J.; Flora, J.; Rajagopal, R. Household Energy Consumption Segmentation Using Hourly Data. *IEEE Trans. Smart Grid* **2015**, *5*, 420–430. [CrossRef]
22. Wang, X.D.; Chen, R.C.; Yan, F.; Zeng, Z.Q.; Hong, C.Q. Fast Adaptive K-means Subspace Clustering for High-Dimensional Data. *IEEE Access* **2019**, *7*, 42639–42651. [CrossRef]
23. Lai, G.; Chang, W.C.; Yang, Y.; Liu, H. Modelling Long-and Short-Term Temporal Patterns with Deep Neural Networks. In Proceedings of the 41st International ACM SIGIR Conference on Research & Development in Information Retrieval, Ann Arbor, MI, USA, 8–12 July 2018.
24. Stephen, B.; Tang, X.; Harvey, P.R.; Galloway, S.; Jennett, K.I. Incorporating Practice Theory in Sub-Profile Models for Short term aggregated Residential Load Forecasting. *IEEE Trans. Smart Grid* **2017**, *8*, 1591–1598. [CrossRef]
25. Charalambous, C.C.; Bharath, A.A. A data augmentation methodology for training machine/deep learning gait recognition algorithms. In Proceedings of the British Machine Vision Conference, York, UK, 19–22 September 2016.
26. Lemley, J.; Bazrafkan, S.; Corcoran, P. Smart Augmentation Learning an Optimal Data Augmentation Strategy. *IEEE Access* **2017**, *5*, 5858–5869. [CrossRef]
27. Goodfellow, I.; Dengio, Y.; Courville, A. *Deep Learning*, 1st ed.; The MIT Press: Cambridge, MA, USA, 2016; pp. 224–270.
28. Zhang, Y.; Chen, W.; Xu, R.; Black, J. A Cluster-Based Method for Calculating Baselines for Residential Loads. *IEEE Trans. Smart Grid* **2015**, *7*, 1–10. [CrossRef]
29. Farukh, A.; Feng, D.; Habib, S.; Rahman, U.; Rasool, A.; Yan, Z. Short Term Residential Load Forecasting: An Improved Optimal Nonlinear Auto Regressive (NARX) Method with Exponential Weight Decay Function. *Electronics* **2018**, *7*, 432.
30. Sharma, T.; Shokeen, D.; Mathur, D. Multiple K Means++ Clustering of Satellite Image Using Hadoop Map Reduce and Spark. *Int. J. Adv. Stud. Comput. Sci. Eng.* **2016**, *5*, 23–29.
31. Abadi, M.; Agarwal, A.; Barham, P.; Brevdo, E.; Chen, Z.; Citro, C. Tensor Flow: Large-Scale Machine Learning on Heterogeneous Distributed Systems. *arXiv* **2016**, arXiv:1603.04467.
32. Pedregosa, F.; Varoquax, G.; Gramfort, A. Scikit-learn: Machine Learning in Python. *J. Mach. Learn. Res.* **2011**, *12*, 2825–2830.
33. Kong, W.; Dong, Z.Y.; Luo, F.; Meng, K. Effect of Automatic Hyper-Parameter Tuning for Residential Load Forecasting via Deep Learning. In Proceedings of the 2017 Australasian Universities Power Engineering Conference (AUPEC), Melbourne, Australia, 9–22 November 2017.
34. Marino, D.L.; Amarasinghe, K.; Manic, M. Building Energy Load Forecasting Using Deep Neural Networks. In Proceedings of the 42nd Annual Conference of the IEEE Industrial Electronics Society, Florence, Italy, 24–27 October 2016.
35. Ozaki, Y.; Yano, M.; Onishi, O. Effective Hyperparameter Optimization Using Nelder–Mead Method in Deep Learning. *IPSJ Trans. Comput. Vis. Appl.* **2017**, *9*, 20. [CrossRef]
36. Torres, J.F.; Troncoso, A.; Gutierrez, D.; Martinez-Alvarez, F. Random Hyper-Parameter Search-Based Deep Neural Network for Power Consumption Forecasting. In Proceedings of the International Work-Conference on Artificial Neural Networks Neural Networks, Gran Canaria, Spain, 12–14 June 2019.
37. Neary, P. Automatic Hyper Parameter Tuning in Deep Convolutional Neural Networks Using Asynchronous Reinforcement Learning. In Proceedings of the 2018 IEEE International Conference on Cognitive Computing (ICCC), San Francisco, CA, USA, 2–7 July 2018.
38. Kim, J.Y.; Cho, S.B. Evolutionary Optimization of Hyper-Parameters in Deep Learning Models. In Proceedings of the 2019 IEEE Congress on Evolutionary Computation (CEC), Wellington, New Zealand, 10–13 June 2019.

Article

Solving the Cold-Start Problem in Short-Term Load Forecasting Using Tree-Based Methods

Jihoon Moon [1], Junhong Kim [2], Pilsung Kang [2] and Eenjun Hwang [1,*]

[1]   School of Electrical Engineering, Korea University, 145 Anam-ro, Seongbuk-gu, Seoul 02841, Korea; johnny89@korea.ac.kr
[2]   School of Industrial Management Engineering, Korea University, 145 Anam-ro, Seongbuk-gu, Seoul 02841, Korea; junhongkim@korea.ac.kr (J.K.); pilsung_kang@korea.ac.kr (P.K.)
*   Correspondence: ehwang04@korea.ac.kr; Tel.: +82-2-3290-3256

Received: 20 January 2020; Accepted: 12 February 2020; Published: 17 February 2020

Abstract: An energy-management system requires accurate prediction of the electric load for optimal energy management. However, if the amount of electric load data is insufficient, it is challenging to perform an accurate prediction. To address this issue, we propose a novel electric load forecasting scheme using the electric load data of diverse buildings. We first divide the electric energy consumption data into training and test sets. Then, we construct multivariate random forest (MRF)-based forecasting models according to each building except the target building in the training set and a random forest (RF)-based forecasting model using the limited electric load data of the target building in the test set. In the test set, we compare the electric load of the target building with that of other buildings to select the MRF model that is the most similar to the target building. Then, we predict the electric load of the target building using its input variables via the selected MRF model. We combine the MRF and RF models by considering the different electric load patterns on weekdays and holidays. Experimental results demonstrate that combining the two models can achieve satisfactory prediction performance even if the electric data of only one day are available for the target building.

Keywords: short-term load forecasting; building electric energy consumption forecasting; cold-start problem; transfer learning; multivariate random forests; random forest

---

1. Introduction

The continuing environmental problems caused by the enormous amount of carbon dioxide produced by the burning of fossil fuels, such as coal and oil, for energy production has resulted in considerable focus on smart grid technologies owing to their effective use of energy [1,2]. A smart grid is an intelligent electric power grid that combines information and communication technology with the existing electric power grid [3]. The smart grid can optimize energy use by sharing electric energy production and consumption information with consumers and suppliers in both directions and in real time [4]. The most fundamental approach for sustainable development of smart grids is electric power generation using renewable energy sources, such as photovoltaic and wind energy [5,6]. Furthermore, an energy management system (EMS) in smart grids requires an optimization algorithm for the advanced operation of an energy storage system (ESS) [7]; it also has to plan various strategies by considering consumer-side decision making [8].

Artificial intelligence (AI) technology-based applications are a highly relevant area for smart grid control and management [6–8]. In particular, short-term load forecasting (STLF) is a core technology of the EMS [9]; moreover, accurate electric load forecasting is required for stable and efficient smart grid operations [10]. From the perspective of a supplier, it is challenging to provide optimal benefits in a cost-effective analysis while storing a large amount of electric energy in the ESS; however, the smart grid can plan effectively by predicting future electric energy consumption and receiving the required energy from internal and external energy sources [11]. It is also possible to optimize the renewable energy generation process [11,12]. From the perspective of a consumer, the EMS can quickly cope with situations such as blackouts and can help to save energy costs because it confirms the electric energy consumption and peak hours during the day [12].

Electric energy consumption patterns are complicated according to the types of buildings [13]; moreover, the electric energy consumption is frequently changed owing to uncertain external factors [14]. Therefore, it is challenging to predict the exact electric energy consumption in buildings [15]. Besides, when forecasting electric energy consumption, the complex correlations associated with an electric load between the current time and the previous time should be appropriately considered [7,11]. To adequately reflect previously uncertain external factors and electric energy consumption, AI techniques can be used to predict future building electric energy consumption based on diverse information, such as historical electric loads, locations, populations, weather factors, and events [16]. Moreover, the importance of multistep-ahead electric load forecasting has increased to quickly determine new uncertainties in power systems [17].

Most AI techniques use large amounts of data to construct STLF models. However, as sufficient electric load data of buildings connected to smart grids for a short time or new/renovated buildings are not collected, it is challenging to construct STLF models using these data sets. We defined this problem of lack of data as a cold-start problem. The cold-start problem [18] can occur in computer-based information systems that require automated data modeling. In particular, this problem involves the issue where the system cannot derive inferences from insufficient information regarding users or items. In future, because of the expansion of the smart grid market, it is expected that new data sets will be collected from newly constructed or renovated buildings. Hence, EMSs require a novel building electric energy consumption forecasting model that can be applied to these buildings.

In this paper, we propose a novel STLF model that combines random forest (RF) models while considering two cases (i.e., weekdays and holidays) to solve the cold-start problem. To achieve this, we first collected sufficient electric energy consumption data sets from 15 buildings. The collected data sets were divided into training and test sets; moreover, we developed a transfer learning-based STLF model based on multivariate random forests (MRF) in the training set. We also constructed a RF-based STLF model using the building electric energy consumption data of only 24 h and then combined the two models by considering the schedule. Consequently, we assumed the building electric energy consumption for only 24 h in the test set and performed multistep-ahead hourly electric load forecasting (24 points) of the target building to prepare for uncertainty.

The rest of this paper is organized as follows: in Section 2, we review several STLF models based on AI techniques using sufficient and insufficient data sets, respectively. In Section 3, we describe the input variable configuration for STLF models. In Section 4, we describe the RF-based STLF model construction in detail. Section 5 presents and discusses the experimental results to evaluate the prediction performance of the proposed model. In Section 6, we provide a conclusion and future research directions.

2. Related Works

In this section, we introduce the research on electric energy consumption forecasting for buildings with and without sufficient data sets. Table 1 summarizes the information about the selected papers, and these studies are described in detail subsequently.

Several studies have predicted electric energy consumption for buildings with sufficient data sets based on traditional machine-learning and deep-learning (DL) methods. Candanedo et al. [19] proposed data-driven prediction models for a low-energy house using the data from home appliances, lighting, weather conditions, time factors, etc. They used multiple linear regression (MLR), support vector regression (SVR), RF, and gradient boosting machine (GBM) to construct 10 minute-interval electric energy consumption forecasting models. They confirmed that time factors were essential variables to build the prediction models; moreover, the GBM model exhibited better prediction performance than other models. Wang et al. [20] presented an hourly electric energy consumption forecasting model for two institutional buildings based on RF. They considered time factors, weather conditions, and the number of occupants as the input variables of the RF model. They compared the prediction performance of the RF model with that of the SVR model and confirmed that the RF model presented better prediction performance than the SVR model. Li et al. [21] proposed an extreme stacked autoencoder (SAE), which combined the SAE with an extreme learning machine (ELM) to improve the prediction results of building energy consumption. The electric energy consumption data were collected from one retail building in Fremont, CA. The authors predicted the building electric energy consumption at 30 min and 60 min intervals and compared the prediction performance of their proposed model with that of backward propagation neural network (BPNN), SVR, generalized radial basis function neural network, and MLR. Their proposed model demonstrated better prediction performance than other models. Almalaq et al. [22] presented a hybrid prediction model based on a genetic algorithm (GA) and long short-term memory (LSTM). GA was employed to optimize the window size and the number of hidden neurons for the LSTM model construction. Their proposed model predicted two public data sets of residential and commercial buildings and compared the prediction performance with autoregressive integrated moving average (ARIMA), decision tree (DT), k-nearest neighbor, artificial neural network (ANN), GA-ANN, and LSTM models. They confirmed that their proposed model exhibited better prediction performance than other models.

Several studies reported the construction of electric energy consumption forecasting models for buildings with insufficient data sets based on pooling, transfer learning, and data generation. Shi et al. [23] proposed a household electric load forecasting model using a pooling-based deep recurrent neural network (PDRNN). The pooling method was used to overcome the limitation of the complexity of household electric loads, such as volatility and uncertainty. Then, LSTM with five layers and 30 hidden units in each layer was employed to build an electric load-forecasting model using the pooling method. They compared the prediction performance of the PDRNN with that of ARIMA, SVR, recurrent neural network (RNN), etc. and confirmed that their proposed method outperformed ARIMA by 19.5%, SVR by 13.1%, and RNN by 6.5% in terms of the root mean square error. Ribeiro et al. [24] proposed a transfer-learning method, called Hephaestus, for cross-building electric energy consumption forecasting based on time-series multi-feature regression with seasonal and trend adjustments. Hephaestus was applied in the pre- and post-processing phases; then, standard machine learning algorithms such as ANN and SVR were used. This method adjusted the electric energy consumption data from various buildings by removing the effects of time through time-series adaptation. It also provided time-independent features through non-temporal domain adaptation. The authors confirmed that Hephaestus can improve electric energy consumption forecasting for a building by 11.2% by using additional electric energy consumption data from other buildings. Hooshmand and Sharma [25] constructed a transfer learning-based electric energy consumption forecasting model in small data set regimes. They collected publicly available electric energy consumption data and classified different types of customers. Then, normalization was utilized for training the trends and seasonality of time series efficiently. They built a convolutional neural network (CNN) architecture through the pre-training step that learns from a public data set with the same type of buildings as the target building. Subsequently, they retrained only the last fully connected layer using the data set of the target building to predict the energy consumption data of the target building. They demonstrated that their proposed model consistently demonstrated lower prediction performance error when compared to seasonal ARIMA, fresh CNN, and pre-trained CNN models. Tian et al. [26] proposed parallel building electric energy consumption forecasting models based on generative adversarial networks (GANs). They initially generated the parallel data through GAN by using a small number of the original data sets and then configured the mixed data set, which included the original data and the parallel data. Finally, they utilized the mixed data set to train several machine learning algorithms such as BPNN, ELM, and SVR. Experimental results exhibited that the parallel data consisted of similar distributions of the original data, and the prediction models trained by the mixed data set demonstrated better prediction performance than those trained using the original data, information diffusion technology, heuristic mega-trend-diffusion, and bootstrap methods.

The differences between the methods described above and our method are as follows:

The forecasting models in previous studies [19–22] presented excellent prediction performance using a sufficient data set of target buildings. However, our forecasting model can exhibit a satisfactory prediction performance even if the electric load data of the target building is insufficient.

The forecasting models in previous studies [19–24,26] could not predict the electric loads for various buildings. However, our forecasting model predicts the electric loads for 15 buildings; consequently, it can be considered as a generalized forecasting model.

The previous studies based on DL techniques [21–23,25,26] demonstrate a certain amount of computational cost to optimize the various hyperparameters of DL models for the target building. We use the RF with minimal tuning of hyperparameters to construct satisfactory forecasting models for several buildings.

The previous studies based on transfer learning [24,25] considered the types of the building to construct the forecasting models. However, we built a transfer learning-based forecasting model even without knowledge of the type of buildings. Besides, even if the electric load of only 24 h for the target building is known, our forecasting model can predict multistep electric load forecasting.

Table 1. Summary of several approaches for building electric energy consumption forecasting (MLR: multiple linear regression, SVR: support vector regression, GBM: gradient boosting machine, RF: random forest, SAE: stacked autoencoder, ELM: extreme learning machine, GA: genetic algorithm, LSTM: long short-term memory, PDRNN: pooling-based deep recurrent neural network, ANN: artificial neural network, CNN: convolutional neural network, GAN: generative adversarial network, BPNN: backward propagation neural network).

Author (Year)	Type of Target Buildings	Dataset for Target Buildings	Time Granularity	AI Techniques
Candanedo et al. [19] (2017)	Residential	Sufficient	10 min	MLR, SVR, GBM, RF
Wang et al. [20] (2018)	Educational	Sufficient	1 h	RF
Li et al. [21] (2017)	Commercial	Sufficient	30 min, 1 h	SAE, ELM
Almalaq et al. [22] (2018)	Residential, Commercial	Sufficient	Residential: 1 min, Commercial: 5 min	GA, LSTM
Shi et al. [23] (2018)	Residential	Insufficient	30 min	PDRNN
Ribeiro et al. [24] (2018)	Educational	Insufficient	Daily	ANN, SVR
Hooshmand and Sharma [25] (2019)	Commercial, Industrial, Educational	Insufficient	15 min	CNN
Tian et al. [26] (2019)	Commercial	Insufficient	30 min, 1 h	GAN, BPNN, ELM, SVR

3. Input Variable Configuration

Our goal is to predict the electric energy consumption of buildings that were recently built. To address this issue, we constructed two STLF models, one using the sufficient electric load data of other buildings and the other using the limited electric load data of the target building. In this section, we describe the construction of input variables for practical model training. Section 3.1 describes the electric energy consumption data sets of the 15 buildings that we collected. Section 3.2 shows the input variable configuration for the RF-based forecasting model using the small data set of the target building. Section 3.3 exhibits the input variable configuration for the transfer learning-based forecasting models using the sufficient data sets from different buildings.

3.1. Data Sets of Electric Energy Consumption from Buildings

We randomly received the hourly electric energy consumption data collected from smart meters connected with 15 sites from the Korea Electric Power Corporation (KEPCO). One smart meter, which is installed per site, usually represents one building or a couple of connected buildings. In this paper, we defined a smart meter as a building. The period of data collection was from 1 January 2015 to

31 July 2018. As the collected building data were provided under anonymity and because of the de-identification owing to privacy, we could not determine the types, characteristics, and locations of the buildings. The collected smart meter data had an average missing value rate of 0.6%; we imputed these missing values using linear interpolation. The statistical analysis of the collected smart meter data for each building is listed in Table 2.

Table 2. Statistics of electric energy consumption data from each building (Unit: kWh).

Building #	Minimum	1st Quartile	Median	Mean	3rd Quartile	Maximum
1	109.7	202.8	230.4	291.6	344.4	821.8
2	96.6	212.3	313.5	399.7	583.1	1148.7
3	51.6	163.0	240.9	251.5	313.0	870.0
4	27.8	429.5	547.3	572.5	694.3	1295.3
5	94.3	176.2	226.3	251.1	323.0	527.0
6	99.4	216.1	240.0	257.7	289.4	602.0
7	16.8	545.9	590.6	650.2	706.6	1453.6
8	26.8	113.2	128.8	185.0	247.7	532.4
9	187.1	423.7	492.4	518.8	593.3	1140.2
10	190.6	560.9	667.5	780.9	850.7	2478.1
11	105.8	237.8	291.7	301.7	348.2	614.2
12	570.2	973.4	1153.4	1193.7	1373.8	2464.2
13	884.6	1359.2	2222.4	2627.6	3626.5	6835.2
14	1408.0	1801.0	2161.0	2478.0	2984.0	5393.0
15	80.5	117.5	160.5	204.6	259.5	756.1

Herein, we made one assumption. Even though we have sufficient electric energy consumption data sets for other buildings, we have the electric energy consumption data set of only 24 h for the target building. Thus, as we know the electric load of the target building for only 24 h (24 points), we predicted the electric energy consumption for the subsequent 24 h.

3.2. Case 1: Time Factor-Based Forecasting Modeling

As mentioned above, we assumed that the electric energy consumption data of only 24 h for the target building was known. Hence, we explain the input variable configuration for the prediction model utilizing the electric energy consumption data of only 24 h. Generally, while constructing a STLF model, various factors, such as time factors, weather information, and historical electric loads, are reflected as input variables [27,28]. However, we cannot utilize weather information as an input variable because we do not know the location of the buildings. The historical electric load is used to reflect recent trends and patterns [28,29]; moreover, it can be applied when it is composed of sufficient data sets. However, as we assumed that the historical electric load is composed of data of only 24 h, reflecting on the historical electric load is inappropriate. This implies that we do not know any recent trends or patterns. In the case of time factors, there are various factors, namely month, day, hour, day of the week, and holiday. However, the data set was too small to effectively reflect the characteristics of months, days, days of the week, and holidays. Therefore, we considered only "hour" as an input variable to construct the STLF model.

$$Hour_x = \sin\left(\left(\frac{360}{24}\right) \times Hour\right) \tag{1}$$

$$Hour_y = \cos\left(\left(\frac{360}{24}\right) \times Hour\right) \tag{2}$$

However, in the case where 11 pm and 12 am of the subsequent day are adjacent, but the difference is 23, it is challenging to train the input variable of the STLF model effectively. Thus, we use Equations (1) and (2) to enhance the sequence data in the one-dimensional space to the continuous data in the two-dimensional space [11,30]. These variables can more adequately reflect the continuous characteristic, similar to the shape of the clock, as shown in Figure 1.

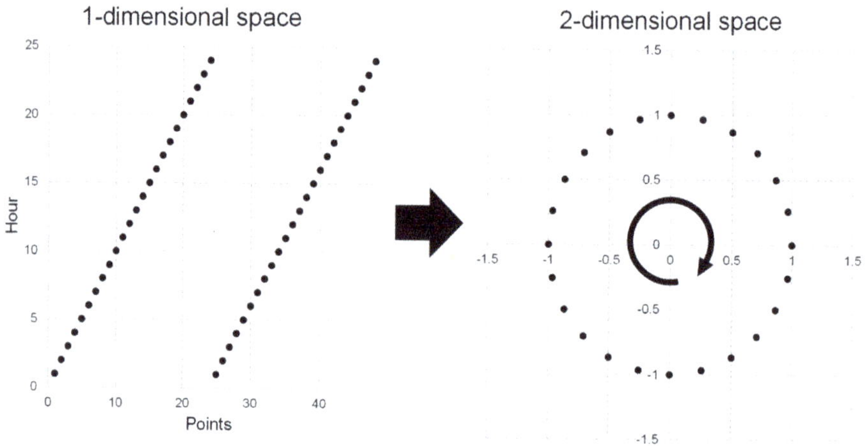

Figure 1. Example of the enhancement of one-dimensional space to two-dimensional space.

Tables 3 and 4 lists the results of the statistical analysis of the building electric energy consumption for one-dimensional and two-dimensional spaces. In Table 3, we can observe that the two-dimensional space reflects the building electric energy consumption more effectively than the one-dimensional space. In Table 4, we can see that almost of p-values of the F-statistic are $< 2.2 \times 10^{-16}$, which are highly significant. This means that at least, one of the predictor variables is significantly related to the outcome variable. We used 24-hour information to construct the STLF model and then applied the same time information to predict the building electric energy consumption over the next 24 h.

Table 3. R-squared statistics and standard error for each building.

Building #	One-Dimensional Space			Two-Dimensional Space		
	Multiple R-Squared	Adjusted R-Squared	Standard Error	Multiple R-Squared	Adjusted R-Squared	Standard Error
1	0.0062	0.0062	132.4	0.2785	0.2784	112.8
2	0.0314	0.0314	228.5	0.3395	0.3395	188.7
3	0.0378	0.0378	114.9	0.2542	0.2541	101.2
4	5.969×10^{-7}	-3.126×10^{-5}	206.7	0.0571	0.0570	200.7
5	0.0003	0.0003	88.4	0.0052	0.0052	88.2
6	0.3599	0.3599	45.0	0.2838	0.2838	47.6
7	0.0126	0.0126	179.9	0.1485	0.1485	167.1
8	0.0049	0.0048	110.6	0.2652	0.2651	95.0
9	0.3985	0.3985	91.7	0.2879	0.2879	99.8
10	0.0212	0.0211	361.5	0.2433	0.2432	317.9
11	0.0415	0.0415	81.9	0.3233	0.3232	68.8
12	0.1056	0.1055	280.3	0.2751	0.2750	252.3
13	0.0886	0.0886	1361.0	0.5962	0.5961	905.8
14	0.0899	0.0899	829.1	0.4772	0.4772	628.4
15	0.0321	0.0321	116.5	0.4514	0.4514	87.7

Table 4. F-statistics and *p*-value for each building.

Building #	One-Dimensional Space		Two-Dimensional Space	
	F-Statistics	*p*-Value	F-Statistics	*p*-Value
1	197	$< 2.2 \times 10^{-16}$	6057	$< 2.2 \times 10^{-16}$
2	1018	$< 2.2 \times 10^{-16}$	8067	$< 2.2 \times 10^{-16}$
3	1234	$< 2.2 \times 10^{-16}$	5348	$< 2.2 \times 10^{-16}$
4	0.019	0.8911	950	$< 2.2 \times 10^{-16}$
5	8.994	0.002711	82.410	$< 2.2 \times 10^{-16}$
6	1.765×10^4	$< 2.2 \times 10^{-16}$	6220	$< 2.2 \times 10^{-16}$
7	400	$< 2.2 \times 10^{-16}$	2737	$< 2.2 \times 10^{-16}$
8	153	$< 2.2 \times 10^{-16}$	5664	$< 2.2 \times 10^{-16}$
9	2.08×10^4	$< 2.2 \times 10^{-16}$	6345	$< 2.2 \times 10^{-16}$
10	678	$< 2.2 \times 10^{-16}$	5046	$< 2.2 \times 10^{-16}$
11	1359	$< 2.2 \times 10^{-16}$	7498	$< 2.2 \times 10^{-16}$
12	3705	$< 2.2 \times 10^{-16}$	5956	$< 2.2 \times 10^{-16}$
13	3052	$< 2.2 \times 10^{-16}$	2.317×10^4	$< 2.2 \times 10^{-16}$
14	3102	$< 2.2 \times 10^{-16}$	1.433×10^4	$< 2.2 \times 10^{-16}$
15	1040	$< 2.2 \times 10^{-16}$	1.291×10^4	$< 2.2 \times 10^{-16}$

3.3. Case 2: Transfer Learning-Based Forecasting Modeling

In addition to the electric energy consumption data set of the target building for only 24 h, we also have sufficient electric energy consumption data sets of other buildings. Hence, we can reflect on various characteristics to use as input variables for a transfer learning-based model construction. Consequently, we used a time factor and historical electric load as input variables. We first divided the electric energy consumption data of all buildings into training and test sets. In the training set, we used the electric energy consumption data of different buildings to train the transfer learning-based STLF models and these models were applied to the electric energy consumption data of the target building when it was used in the test set. For time factors, we used the month, week, day, hour, day of the week, and holiday information. In the case of months, weeks, days, and days of the week, we enhanced the time factors from the one-dimensional space to two-dimensional space, as shown in Equations (3) to (10) [30,31]. Here *WN* is the week number based on the ISO 8601 standard [32], and *LDM* represents the last day of the month.

$$Day_x = \sin\left(\left(\frac{360}{LDM}\right) \times Day\right) \tag{3}$$

$$Day_y = \cos\left(\left(\frac{360}{LDM}\right) \times Day\right) \tag{4}$$

$$Week_x = \sin\left(\left(\frac{360}{WN}\right) \times Week\right) \tag{5}$$

$$Week_y = \cos\left(\left(\frac{360}{WN}\right) \times Week\right) \tag{6}$$

$$Month_x = \sin\left(\left(\frac{360}{12}\right) \times Month\right) \tag{7}$$

$$Month_y = \cos\left(\left(\frac{360}{12}\right) \times Month\right) \tag{8}$$

$$Day\_of\_the\_Week_x = \sin\left(\left(\frac{360}{7}\right) \times Day\_of\_the\_Week\right) \tag{9}$$

$$Day\_of\_the\_Week_y = \cos\left(\left(\frac{360}{7}\right) \times Day\_of\_the\_Week\right) \tag{10}$$

The holidays included Saturdays, Sundays, and national holidays and exhibited predominantly low electric energy consumption during working hours, unlike working days [33]. In the case of holidays, we used one-hot encoding to reflect the property as "1" on the holiday and "0" otherwise. Hence, we applied nine time factors as input variables at the prediction time point.

In addition, we applied not only the historical electric load data but also the day of the week and holiday, which are a part of the data, to reflect both the historical electric load of the building and its characteristics. Consequently, we configured 15 input variables at the prediction point time. Table 5 lists the information on these variables; where D represents the day.

Table 5. List of input variables for the transfer learning-based forecasting model.

No.	Input Variable	Variable Type
1	$Hour_x$	Continuous $[-1, 1]$
2	$Hour_y$	Continuous $[-1, 1]$
3	Day_x	Continuous $[-1, 1]$
4	Day_y	Continuous $[-1, 1]$
5	$Week_x$	Continuous $[-1, 1]$
6	$Week_y$	Continuous $[-1, 1]$
7	$Month_x$	Continuous $[-1, 1]$
8	$Month_y$	Continuous $[-1, 1]$
9	$Day\_of\_the\_Week_x$	Continuous $[-1, 1]$
10	$Day\_of\_the\_Week_y$	Continuous $[-1, 1]$
11	$Holiday$	Binary [1: Holiday, 0: Weekday]
12	$Electric\_Load_{D-1}$	Continuous
13	$Day\_of\_the\_Week_{x,D-1}$	Continuous $[-1, 1]$
14	$Day\_of\_the\_Week_{y,D-1}$	Continuous $[-1, 1]$
15	$Holiday_{Day-1}$	Binary [1: Holiday, 0: Weekday]

As our goal is to predict all-time points 24 h later, we constructed all the input variables by utilizing each input variable for the prediction time point. Thus, we used a total of 360 input variables, i.e., (15 (number of input variables) × 24 (prediction time points)) for the STLF model construction, as shown in Figure 2.

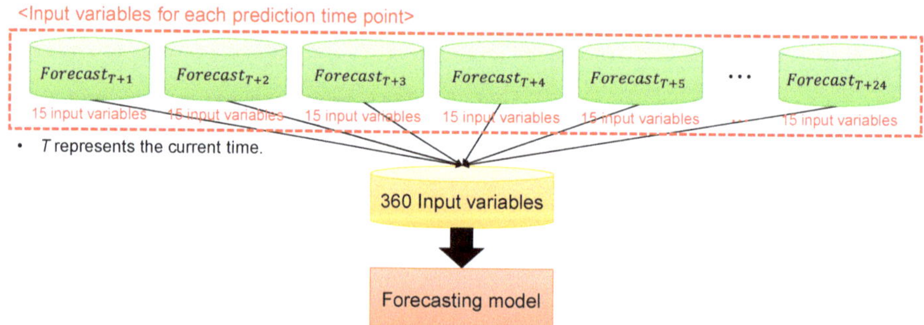

Figure 2. Input variable configuration for the transfer learning-based forecasting model.

4. Forecasting Model Construction

In this section, we describe the STLF model construction using the limited data set of the target building and the data sets of other buildings; moreover, we also present the method to select the prediction value derived from the transfer learning-based model. In addition, we combined the two STLF models and thus present a total of 15 STLF models. Figure 3 illustrates a brief system architecture of our proposed model.

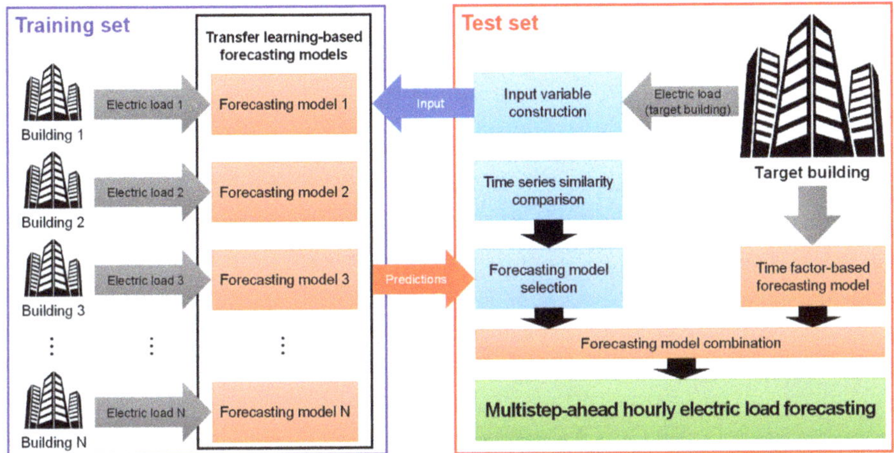

Figure 3. Overall system architecture.

4.1. Case 1: Time Factor-Based Forecasting Model

To construct an STLF model using only "hour" as an input (independent) variable, we used one statistical technique and two machine learning algorithms. Even though SVM, DL, and boosting methods exhibit excellent prediction performance in STLF [7–9,34–37], they require a significant amount of time to optimize the various hyperparameters and also require sufficient data sets. We did not consider these methods because we constructed an STLF model using a data set from the building electric energy consumption data of only 24 h. Thus, we considered MLR, DT, and RF, which allow simple model construction and exhibit satisfactory prediction performance [38,39]. We used two-time factors, namely, $Hour_x$ and $Hour_y$, as independent variables and the electric energy consumption for the target building as the dependent (output) variable. Therefore, we predicted multistep-ahead hourly electric loads using the time factor-based forecasting model via a sliding window time series analysis, as shown in Figure 4.

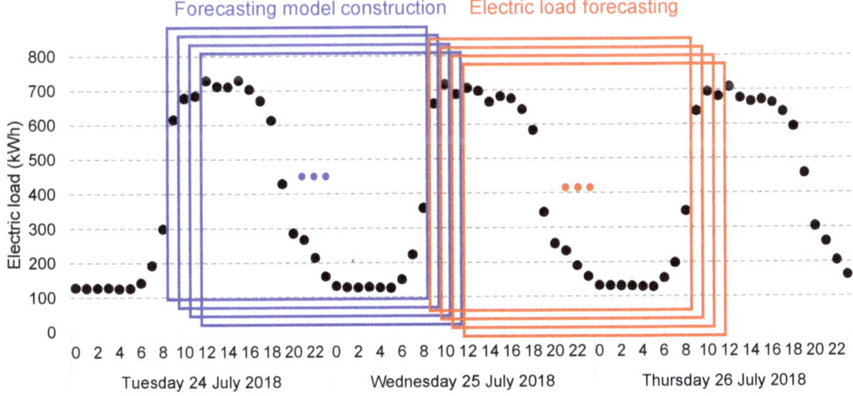

Figure 4. Example of multistep-ahead electric load forecasting via a sliding window method.

4.1.1. Multiple Linear Regression (MLR)

MLR is a common statistical technique that is widely used in many STLF models [39]. MLR [39] analyzes the relationship between a continuous dependent variable and one or more independent variables. The value of the dependent variable can be predicted using a range of values of the independent variables based on an identity function that describes, as closely as possible, the relationship between these variables. In addition, MLR determines the overall fit of the prediction model and the relative contribution of each independent variable to the total variance. Equation (11) represents the method to construct the MLR-based STLF model. Y_i is the forecast energy consumption at time i and β_0 is the intercept of population Y. β_1 and β_2 are population slope coefficients. By defining the weights of the MLR model based on β_1 and β_2 at the prediction time, we can construct a more sophisticated STLF model. Herein, β_0, β_1, and β_2 are calculated when the prediction model is built by using the electric loads at the previous one day from the prediction points. Because our model focuses on the prediction of multistep electric loads using a sliding window method, the weights of β_0, β_1, and β_2 are adjusted every hour.

$$Y_i = \beta_0 + \beta_1 \times Hour_x + \beta_2 \times Hour_y \tag{11}$$

4.1.2. Decision Tree (DT)

A DT [40] is used to construct classification or regression models in the form of a tree structure. It separates a data set into smaller subsets while an associated DT is being incrementally extended. The final result is a tree with a decision node that has two or more branches and a leaf node that denotes a classification or decision. The topmost decision node in a tree corresponds to the best predictor, called root node. DTs can present a higher explanatory power because they determine the independent variables that have a more powerful impact when predicting the values of the target variable [41]. However, continuous variables (i.e., building electric energy consumption) used in the prediction of the time series are considered as discontinuous values. Hence, prediction errors are likely to occur near the boundary of separation.

4.1.3. Random Forest (RF)

RF [41] is an ensemble learning method that combines different DTs that classify a new instance by the majority vote. Each DT node utilizes a subset of attributes randomly selected from the original set of characteristics. As RF can handle large amounts of data effectively, it exhibits a high prediction performance in the field of STLF [14]. Besides, when compared to other AI techniques, such as DL and gradient-boosting algorithms, RF requires less fine tuning of its hyperparameters [42]. The basic hyperparameters of RF include the number of trees to grow (nTree) and the number of variables randomly sampled as candidates at each split (mTry). The correct value for nTree is usually not of much concern because increasing the number of trees in the RF raises the computational cost and does not contribute significantly to prediction performance improvement [14,41]. However, it is interesting to note that picking too small a value of mTry can lead to overfitting. We consider only two input variables, mTry and nTree, which are set to 2 and 128 [43], respectively.

4.2. Case 2: Transfer Learning-Based Forecasting Model

Transfer learning [44] is a process that utilizes the knowledge gained while solving one issue to a different but related issue. To achieve this, a base machine-learning method is first trained on a base data set or task. Then, the method repurposes or transfers the learned features to the second target data set or task. This process can operate if the features are general and the intentions are suitable for both base and target tasks rather than being specific to the base task. We previously configured the input variable for transfer learning-based STLF model construction. We first divided the training and test sets. In the training set, we used electric energy consumption data of other buildings to train the transfer learning-based forecasting model. Then, the model predicted the multistep electric load for the target building using its input variables in the test set. Finally, we selected appropriate prediction results for the target building from the other 14 forecasting models. We describe the details in the subsections below.

4.2.1. Multivariate Random Forests (MRF)

Neural network techniques have been used to predict multistep electric loads [45–47]. However, they require scaling (e.g., robust normalization, standardization, and min-max normalization) before building a prediction model [11]. Thus, when the transfer learning-based STLF model is applied to the target building, the electric load of the target building can exhibit a different range than the electric load of other buildings. Consequently, if STLF models that are trained by scaling are used to predict future electric load for the target building, it is challenging to accurately apply them into the electric load range of the target building.

Consequently, we used MRF (when the number of output features > one) for transfer learning-based STLF model construction. MRF [48] can provide multivariate outcomes and can generate an ensemble of multivariate regression trees through bootstrap resampling and predictors subsampling for univariate RF. In MRF, node cost is measured as the sum of squares of the Mahalanobis distance, whereas in univariate trees (i.e., RF), the node cost is measured as the Euclidean distance (ED) [49]. MRF can provide excellent prediction performance without scaling for multiple output predictions [50,51]. Therefore, we used the input variables without scaling to train the MRF models and predicted the electric load of the target building using the input variable of the target building without scaling. To construct the MRF models, we set mTry and nTree as 120 (number of input variables/3) and 128, respectively [14]. Thus, when a total of 14 prediction results are derived from each MRF model, we did not use all of them and instead selected only one prediction result that exhibits the most similar time series through the similarity analysis between the target building and the other buildings. Then, we used the prediction result as the model that demonstrates the most similar time series to predict the electric load of the target building. Here, we consider a total of three techniques and describe the details in the subsections below.

4.2.2. Similarity Measures

We used three similarity measures, i.e., Pearson correlation coefficient (PCC), cosine similarity (CS), and ED, to analyze the similarity of the time series between the target building and other buildings. These methods are commonly used for similarity analysis [13].

PCC [52], which is a measure of the linear correlation between two variables, x and y, is the covariance of these variables divided by the product of the standard deviations in the data of equal or proportional scales. It has a value between +1 and −1, according to the Cauchy–Schwarz inequality, where +1 is a perfect positive linear correlation, 0 indicates no linear correlation, and −1 is a perfect negative linear correlation. For the paired data $\{(x_1, y_1), \cdots, (x_n, y_n)\}$ consisting of n pairs, r_{xy}, which is a substituting estimation of the covariances and variances based on a sample, is defined according to Equation (12). Here, n is the sample size, x_i and y_i are the individual sample points indexed with time i. $\bar{x}$ and $\bar{y}$ are the sample means of x and y, respectively. For our experiments, because we considered hourly electric load for only 24 h, n is 24. x_i and y_i are the hourly electric loads indexed with i of the target and different buildings, respectively. Herein, we apply the prediction model of the building whose PCC is closest to one by comparing the target building with other buildings.

$$r_{xy} = \frac{\sum_{i=1}^{n} (x_i - \bar{x}) \times (y_i - \bar{y})}{\sqrt{\sum_{i=1}^{n} (x_i - \bar{x})^2 \times \sum_{i=1}^{n} (y_i - \bar{y})^2}} \tag{12}$$

CS [53] indicates the similarity between vectors measured using the cosine of the angle between two vectors in the inner space. The cosine when the angle is 0° is one, and the cosine of all other angles is smaller than one. Therefore, this value is used to determine the similarity of the direction, not the magnitude of the vector. The value is +1 when the two vectors are in exactly the same direction. The value is 0 when the angle is 90°, and the value is −1 when the vectors are in completely opposite directions, i.e., an angle of 180°. At this time, the size of the vector does not affect the value. The CS can be applied to any number of dimensions and is often used to measure similarity in a multidimensional amniotic space. Given the vector values of the attributes A and B, CS ($\cos(\theta)$) can be expressed by the scalar product and magnitude of the vector, as shown in Equation (13). For our experiments, as mentioned earlier, n is 24. A_i and B_i are the hourly electric loads indexed with i of the target and different buildings, respectively. Herein, we apply the prediction model of the building whose CS is closest to one by comparing the target building with other buildings.

$$\text{similarity} = \cos(\theta) = \frac{A \cdot B}{\|A\|\|B\|} = \frac{\sum_{i=1}^{n} A_i \times B_i}{\sqrt{\sum_{i=1}^{n}(A_i)^2} \times \sqrt{\sum_{i=1}^{n}(B_i)^2}} \tag{13}$$

ED [54,55] is a common method for calculating the distance between two points. This distance can be used to define the Euclidean space, and the corresponding norm is called the Euclidean norm. A generalized term for the Euclidean norm is the L^2 norm or L^2 distance [55]. In a Cartesian coordinate system, where there are points $p = (p_1, p_2, \cdots, p_n)$ and $q = (q_1, q_2, \cdots, q_n)$ in Euclidean n space, the distance between two points p and q is calculated using two Euclidean norms, which is defined according to Equation (14). The distance between any two points on the real line is the absolute value of the numerical difference of their coordinates. It is common to identify the name of a point with its Cartesian coordinate. When the two points p and q are the same, the distance value is zero. For our experiments, as mentioned earlier, n is 24. p_i and q_i are the hourly electric loads indexed with i of the target and different buildings, respectively. Herein, we apply the prediction model of the building whose ED is closest to zero by comparing the target building with other buildings.

$$d(p,q) = d(q,p) = \sqrt{(q_1 - p_1)^2 + (q_2 - p_2)^2 + \cdots + (q_n - p_n)^2} = \sqrt{\sum_{i=1}^{n}(q_i - p_i)^2} \tag{14}$$

4.3. Case 3: Combining Short-Term Load-Forecasting Models

We considered the following situations to apply more suitable forecasting models for each case. As mentioned above, the electric load patterns vary significantly depending on the days of the week and holidays [32]. For instance, in chronological sequence, the difference in electric loads between Friday and Saturday is large. The difference in electric loads between Sunday and Monday is also large. Therefore, if the time factor-based forecasting model predicts the electric load on the weekend by using the electric load on a weekday, the forecast value presents multiple error rates because it exhibits the weekday pattern. In addition, when constructing a time factor-based forecasting model by using only a low electric load, such as that on Sunday or a national holiday, it can cause high error rates because of a high electric load on Monday. To address these issues, we combined the time factor- and transfer-learning-based forecasting models. We applied the prediction models by considering two cases at the 24 prediction points and illustrated examples of the electric load forecasting using the combined STLF model in Figure 5 and hence we constructed a total of 15 STLF models, as shown in Table 6.

- Case 1: For each prediction point, we applied the time vector-based forecasting model when both the prediction point and the day before the prediction point are weekdays.
- Case 2: For each prediction point, we applied the transfer learning-based forecasting model when the prediction point and/or the day before the prediction point are holidays.

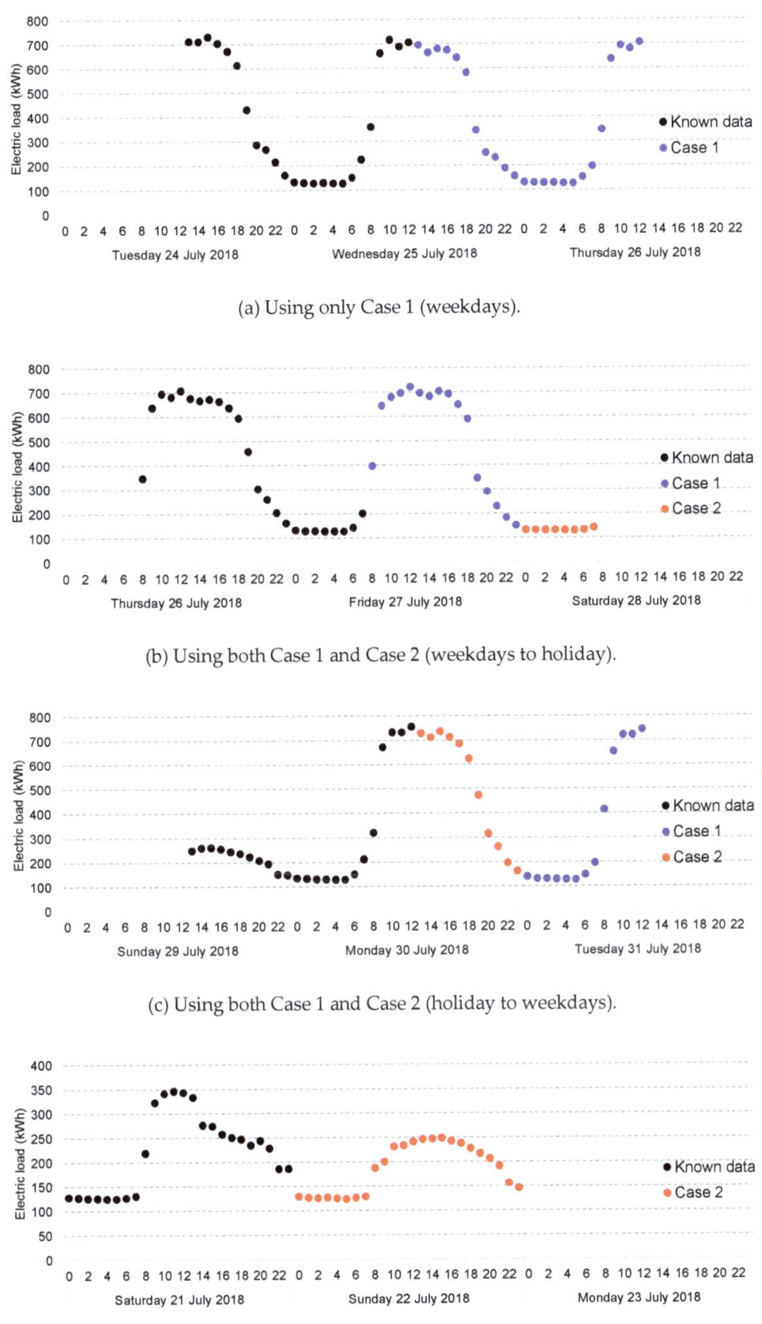

(a) Using only Case 1 (weekdays).

(b) Using both Case 1 and Case 2 (weekdays to holiday).

(c) Using both Case 1 and Case 2 (holiday to weekdays).

(d) Using only Case 2 (holidays).

Figure 5. Examples of multistep-ahead electric load forecasting using the combined short-term load forecasting (STLF) model by considering two cases.

Table 6. Construction of various forecasting models (MLR: multiple linear regression, DT: decision tree, RF: random forest, MRF: multivariate random forests, PCC: Pearson correlation coefficient, CS: cosine similarity, ED: Euclidean distance).

Model #	Methods	Description	
M01	MLR	Time factor-based forecasting model using MLR	
M02	DT	Time factor-based forecasting model using DT	
M03	RF	Time factor-based forecasting model using RF	
M04	MRF_PCC	MRF-based transfer learning model by applying PCC to analyze the time series similarity	
M05	MRF_CS	MRF-based transfer learning model by applying CS to analyze the time series similarity	
M06	MRF_ED	MRF-based transfer learning model by applying ED to analyze the time series similarity	
M07	MLR + MRF_PCC	Case 1: MLR	Case 2: MRF_PCC
M08	MLR + MRF_CS	Case 1: MLR	Case 2: MRF_CS
M09	MLR + MRF_ED	Case 1: MLR	Case 2: MRF_ED
M10	DT + MRF_PCC	Case 1: DT	Case 2: MRF_PCC
M11	DT + MRF_CS	Case 1: DT	Case 2: MRF_CS
M12	DT + MRF_ED	Case 1: DT	Case 2: MRF_ED
M13	RF + MRF_PCC	Case 1: RF	Case 2: MRF_PCC
M14	RF + MRF_CS	Case 1: RF	Case 2: MRF_CS
M15	RF + MRF_ED	Case 1: RF	Case 2: MRF_ED

5. Experimental Results

5.1. Performance Evaluation Metric

To evaluate the prediction performance of the proposed model, we used the mean absolute percentage error (MAPE), root mean square error (RMSE), and mean absolute error (MAE). The MAPE value usually presents accuracy as a percentage of the error and can be easier to comprehend than the other statistics because this number is a percentage [11,40]. The MAPE can be defined according to Equation (15). The RMSE is used to aggregate the residuals into a single measure of predictive ability, as shown in Equation (16). The RMSE is the square root of the variance, which denotes the standard error. The MAE is used to evaluate how close forecast or prediction values are to the actual observed values, as shown in Equation (17). The MAE is calculated by averaging the absolute differences between the prediction values and the actual observed values. The MAE gives the average magnitude of forecast error, while the RMSE gives more weight to the most significant errors. Lower values of MAPE, RMSE, and MAE indicate better prediction performance of the forecasting model. Here, n is the number of observations and A_t and F_t are the actual and forecast values, respectively.

$$MAPE = \frac{1}{n} \sum_{t=1}^{n} \left| \frac{A_t - F_t}{A_t} \right| \times 100 \tag{15}$$

$$RMSE = \sqrt{\frac{\sum_{t=1}^{n} (F_t - A_t)^2}{n}} \tag{16}$$

$$MAE = \frac{1}{n} \sum_{t=1}^{n} |F_t - A_t| \tag{17}$$

5.2. Prediction Performance Evaluation

To evaluate the performance of the forecasting models, we conducted the experiments with an Intel ®Core™ i7-8700k CPU with 32GB DDR4 RAM and preprocessed the datasets in RStudio version

1.1.453 with R version 3.5.1. We also carried out the construction of the forecasting models using *'tree'* (DT), *'randomForest'* (RF), and *'MultivariateRandomForest'* (MRF) packages [49,56,57].

As we collected the electric energy consumption data from a total of 15 buildings, we used 15-fold cross-validations to evaluate the prediction performance. In the collected data, the periods of the training and test sets are from 1 January 2015 to 31 July 2017 and from 1 August 2017 to 31 July, 2018, respectively.

We reported training and testing times of the MRF models for each building in Table 7. The training time represents the time to train the MRF model in each building, and the testing time indicates the time for performing all predictions from the three transfer learning-based forecasting models (i.e., M04, M05, and M06).

Table 7. Running times of the multivariate random forest (MRF) model in each building (Unit: second).

Buildings #	Training Time (s)	Testing Time (s)
1	143.38	71.75
2	144.17	72.32
3	135.72	72.49
4	140.71	71.87
5	139.45	72.53
6	153.75	72.49
7	146.83	70.14
8	139.32	69.16
9	145.62	69.09
10	138.02	68.02
11	136.02	67.79
12	137.08	68.54
13	139.33	67.77
14	136.70	67.57
15	142.65	68.94

Because MAPE is a widely used error measurement metric in the electric load-forecasting literature, we presented all the results of multistep-ahead hourly electric load forecasting accuracy using the MAPE in Tables 8–23. We also exhibited the average forecasting accuracy of multistep electric loads using RMSE and MAE results in Tables 24 and 25, respectively.

In Tables 8–25, a cooler color (blue) indicates lower MAPE, RMSE, and MAE values, while a warmer color (red) indicates higher MAPE, RMSE, and MAE values. To confirm the overall prediction performance of the forecasting models, we presented the average MAPE of different forecasting models and indicated the best accuracy in bold. In addition, a box plot for each forecasting model is shown in Figures 6–20 using MAPE values for each prediction point. This means that the box, which is located below and exhibits the smaller range, is a more stable forecasting model.

As shown in Tables 8–25 and Figures 6–20, the RF demonstrated the best prediction performance in the time factor-based forecasting models, and MRF_ED showed the best prediction performance in the transfer learning-based forecasting models. The reason why ED demonstrated the best performance is only because the distance of electric energy consumption between buildings was considered, unlike the PCC or CS; thus, it can reflect a similar range of electric energy consumption adequately when training the transfer learning-based forecasting model. Consequently, we can observe that M15 demonstrated better prediction performances than other forecasting models in most experiments. M15 is appropriate for solving the cold-start problem in STLF and used two tree-based methods; we called this model as SPROUT (solving cold start problem in short-term load forecasting using tree-based methods).

Table 8. Mean absolute percentage error (MAPE) comparison of forecasting models for Building 1 (a cooler color indicates a lower MAPE value, while a warmer color indicates a higher MAPE value).

Point	Model #														
	M01	M02	M03	M04	M05	M06	M07	M08	M09	M10	M11	M12	M13	M14	M15
1	14.2	13.4	14.3	21.9	27.6	9.8	14.5	16.6	8.4	15.8	17.9	9.7	14.6	16.7	8.5
2	15.8	15.0	16.3	20.4	25.9	9.5	13.7	15.5	8.4	14.8	16.7	9.6	14.0	15.8	8.7
3	17.0	16.5	17.0	19.4	24.7	9.4	13.1	14.7	8.5	14.2	15.8	9.6	13.1	14.7	8.5
4	18.0	18.0	17.2	18.5	23.6	9.4	12.6	13.9	8.5	13.6	15.0	9.6	12.4	13.7	8.4
5	18.6	19.2	17.4	17.9	22.8	9.4	12.1	13.2	8.5	13.2	14.3	9.7	11.9	13.0	8.3
6	18.7	20.1	17.4	17.3	22.0	9.3	11.6	12.5	8.5	12.8	13.7	9.7	11.4	12.3	8.3
7	18.8	20.8	17.6	16.8	21.2	9.3	11.2	11.9	8.5	12.5	13.2	9.7	11.0	11.7	8.3
8	18.3	21.1	17.5	16.6	20.7	9.2	11.0	11.4	8.4	12.4	12.7	9.8	10.8	11.2	8.2
9	18.2	21.1	17.8	16.4	20.4	9.2	10.9	11.0	8.4	12.2	12.4	9.7	10.8	10.9	8.3
10	17.8	20.8	17.8	16.2	20.1	9.2	10.8	10.7	8.3	12.1	12.0	9.7	10.7	10.6	8.2
11	18.1	20.5	17.9	16.1	19.9	9.2	10.7	10.5	8.3	12.0	11.8	9.6	10.6	10.4	8.2
12	18.2	20.3	17.9	16.0	19.8	9.1	10.7	10.4	8.3	12.0	11.6	9.5	10.6	10.2	8.1
13	18.2	20.2	18.0	16.1	19.9	9.1	10.9	10.5	8.4	12.0	11.6	9.6	10.7	10.3	8.3
14	18.2	20.3	17.9	16.1	20.0	9.1	10.9	10.5	8.4	12.1	11.7	9.6	10.6	10.2	8.1
15	18.1	20.5	18.1	16.3	20.2	9.1	10.9	10.5	8.4	12.3	11.9	9.8	10.7	10.3	8.2
16	18.3	20.7	17.7	16.5	20.6	9.1	11.1	10.7	8.5	12.4	12.1	9.9	10.5	10.1	8.0
17	18.3	20.7	17.8	16.6	20.8	9.1	11.1	10.8	8.6	12.5	12.1	10.0	10.5	10.2	8.0
18	18.5	20.5	18.0	16.6	21.0	9.1	11.3	11.0	8.8	12.4	12.1	9.9	10.6	10.4	8.2
19	18.5	20.2	18.2	16.6	21.2	9.1	11.3	11.1	8.9	12.2	12.0	9.8	10.8	10.6	8.4
20	18.7	20.1	18.1	16.8	21.4	9.2	11.5	11.4	9.2	12.1	12.0	9.7	10.8	10.7	8.4
21	18.4	20.5	18.2	16.8	21.6	9.3	11.3	11.3	9.0	12.1	12.1	9.8	10.9	11.0	8.6
22	18.5	21.3	18.4	17.0	22.0	9.5	11.4	11.7	9.1	12.3	12.6	10.0	11.2	11.5	8.9
23	19.0	22.4	18.3	17.3	22.5	9.8	11.7	12.2	9.4	12.8	13.4	10.5	11.1	11.7	8.9
24	20.1	23.7	19.6	17.6	23.0	10.3	12.4	13.2	10.2	13.6	14.5	11.5	12.1	12.9	9.9
Avg.	18.1	19.9	17.7	17.2	21.8	9.3	11.6	12.0	8.7	12.8	13.1	9.8	11.4	11.7	8.4

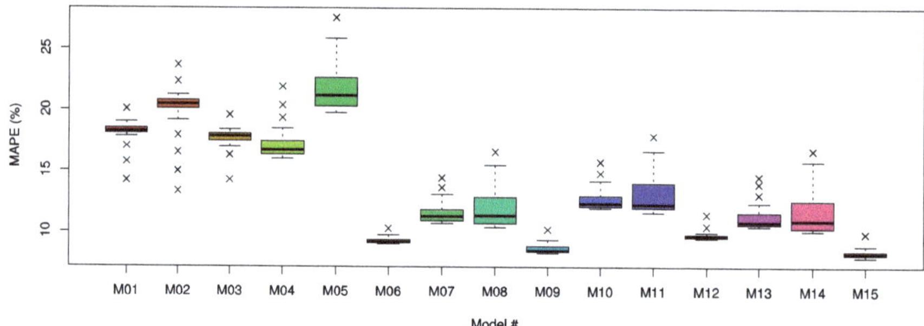

Figure 6. Box plot of the mean absolute percentage error (MAPE) of forecasting models for Building 1.

Table 9. MAPE comparison of forecasting models for Building 2 (a cooler color indicates a lower MAPE value, while a warmer color indicates a higher MAPE value).

Point	M01	M02	M03	M04	M05	M06	M07	M08	M09	M10	M11	M12	M13	M14	M15
1	30.4	24.2	27.1	45.6	46.1	19.1	40.2	40.9	18.8	39.3	40.0	17.9	37.1	37.8	15.6
2	33.3	27.8	31.2	43.2	43.9	18.6	38.8	39.7	19.1	37.8	38.6	18.0	36.0	36.8	16.2
3	36.1	31.2	32.7	41.0	41.9	18.3	37.3	38.4	19.3	36.2	37.3	18.2	34.1	35.2	16.1
4	38.1	34.3	33.4	39.0	40.0	18.0	35.9	37.1	19.5	34.8	36.0	18.4	32.5	33.8	16.2
5	39.3	37.0	33.9	37.3	38.4	17.9	34.7	36.0	19.7	33.6	34.9	18.7	31.2	32.5	16.2
6	39.2	39.1	33.9	35.8	36.9	17.7	33.5	34.9	19.8	32.5	33.9	18.9	29.9	31.3	16.2
7	40.0	40.5	34.4	34.6	35.7	17.5	32.6	34.0	19.8	31.8	33.1	19.0	29.0	30.4	16.3
8	39.3	41.2	34.5	33.6	34.7	17.5	31.8	33.2	19.8	31.2	32.6	19.2	28.4	29.8	16.4
9	39.4	41.2	35.0	32.9	34.1	17.5	31.3	32.7	19.8	30.8	32.2	19.3	28.1	29.6	16.6
10	37.9	40.6	35.0	32.3	33.6	17.7	31.0	32.5	19.9	30.3	31.9	19.3	27.8	29.3	16.8
11	38.5	39.6	35.2	32.0	33.5	17.9	30.8	32.5	20.1	30.0	31.7	19.3	27.5	29.2	16.8
12	38.6	38.3	35.4	31.9	33.5	18.2	30.8	32.6	20.3	29.7	31.5	19.2	27.5	29.2	17.0
13	38.5	37.1	35.7	32.0	33.6	18.4	30.9	32.6	20.4	29.6	31.4	19.2	27.8	29.6	17.3
14	38.7	36.2	35.3	32.2	33.9	18.6	31.2	33.0	20.6	29.8	31.6	19.2	27.8	29.6	17.3
15	38.5	35.7	35.9	32.2	34.0	18.8	31.1	32.9	20.5	30.0	31.8	19.4	28.3	30.1	17.7
16	38.5	35.6	35.2	32.0	33.8	18.8	31.2	33.1	20.8	29.9	31.8	19.6	27.7	29.6	17.4
17	38.3	35.6	35.9	31.9	33.6	19.0	30.9	32.7	20.8	29.9	31.6	19.8	28.0	29.8	17.9
18	38.0	35.7	36.0	31.9	33.6	19.3	30.8	32.6	21.0	29.8	31.5	20.0	27.9	29.7	18.1
19	38.1	36.2	36.7	31.7	33.2	19.5	30.6	32.3	21.2	29.5	31.2	20.1	28.1	29.8	18.7
20	38.2	37.2	36.5	31.4	32.7	19.6	30.3	32.0	21.3	29.1	30.8	20.1	27.6	29.3	18.6
21	38.6	38.9	37.1	30.8	32.0	19.6	29.9	31.5	21.5	28.6	30.2	20.2	27.2	28.8	18.8
22	39.6	41.3	37.4	30.3	31.4	19.8	29.6	31.2	21.9	28.5	30.1	20.8	26.8	28.4	19.1
23	41.3	44.2	37.6	30.4	31.6	20.3	29.9	31.6	22.7	29.1	30.8	21.9	26.4	28.2	19.3
24	43.9	47.3	40.8	30.5	32.0	21.0	30.4	32.5	23.9	29.8	31.9	23.3	27.4	29.5	20.8
Avg.	38.3	37.3	35.1	34.0	35.3	18.7	32.3	33.8	20.5	31.3	32.8	19.5	29.2	30.7	**17.4**

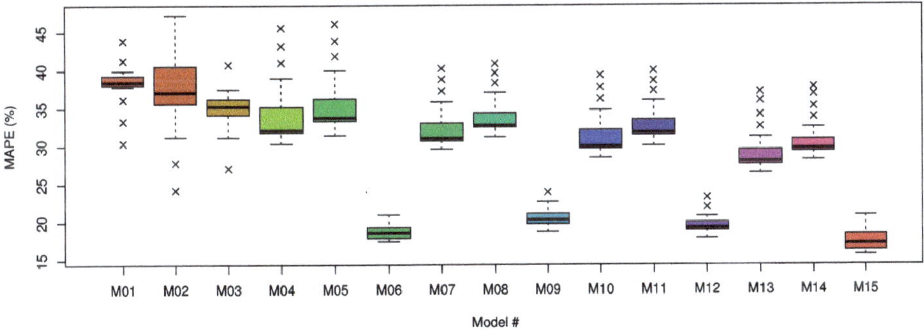

Figure 7. Box plot of the MAPE of forecasting models for Building 2.

Table 10. MAPE comparison of forecasting models for Building 3 (a cooler color indicates a lower MAPE value, while a warmer color indicates a higher MAPE value).

Point	Model #														
	M01	M02	M03	M04	M05	M06	M07	M08	M09	M10	M11	M12	M13	M14	M15
1	15.0	15.7	12.5	30.3	31.0	17.7	21.8	22.0	16.0	22.3	22.5	16.5	20.4	20.7	14.6
2	16.0	16.4	13.9	29.3	29.7	17.4	21.7	21.8	16.2	22.1	22.2	16.6	20.6	20.7	15.1
3	16.5	17.0	14.0	28.2	28.5	17.2	21.4	21.4	16.2	21.8	21.8	16.6	20.2	20.1	15.0
4	16.7	17.4	14.1	27.5	27.8	17.0	21.1	21.1	16.1	21.6	21.5	16.6	19.8	19.7	14.8
5	16.8	17.8	14.2	26.8	27.0	16.7	20.8	20.6	16.0	21.4	21.2	16.5	19.4	19.3	14.6
6	16.8	18.0	14.2	26.2	26.4	16.5	20.6	20.2	15.8	21.2	20.9	16.5	19.3	18.9	14.5
7	16.9	18.2	14.3	25.7	25.7	16.2	20.4	20.0	15.7	21.1	20.7	16.4	19.1	18.6	14.3
8	16.8	18.3	14.3	25.2	25.2	15.9	20.2	19.8	15.5	20.9	20.5	16.2	18.9	18.5	14.2
9	16.9	18.3	14.4	25.1	25.1	15.7	20.1	19.8	15.4	20.8	20.5	16.1	18.8	18.5	14.1
10	16.9	18.2	14.4	25.1	25.2	15.6	20.2	19.9	15.3	20.9	20.6	16.0	19.0	18.7	14.1
11	16.9	18.2	14.2	25.1	25.2	15.6	20.2	19.9	15.3	20.9	20.6	16.0	18.9	18.6	13.9
12	16.9	18.1	14.2	24.8	24.9	15.5	20.1	19.8	15.3	20.8	20.5	16.0	18.8	18.5	14.0
13	16.9	18.1	14.4	24.6	24.7	15.5	20.0	19.7	15.3	20.7	20.4	16.0	18.8	18.5	14.2
14	16.9	18.1	14.3	24.3	24.6	15.4	19.9	19.7	15.4	20.5	20.3	16.0	18.6	18.4	14.1
15	16.9	18.1	14.6	24.0	24.4	15.5	19.6	19.5	15.4	20.3	20.2	16.1	18.5	18.4	14.3
16	16.9	18.1	14.0	24.2	24.6	15.4	19.7	19.6	15.4	20.4	20.3	16.1	18.3	18.2	14.0
17	16.9	18.1	14.1	24.5	24.9	15.5	19.8	19.8	15.5	20.5	20.5	16.2	18.4	18.4	14.1
18	16.8	18.1	14.2	24.6	25.0	15.6	19.8	19.8	15.5	20.5	20.5	16.2	18.5	18.5	14.1
19	16.8	18.0	14.5	24.9	25.3	15.7	20.0	20.0	15.6	20.6	20.7	16.2	18.8	18.8	14.4
20	16.7	18.0	14.3	25.3	25.7	15.9	20.1	20.3	15.6	20.8	20.9	16.3	18.8	18.9	14.3
21	16.8	18.0	14.5	25.6	26.1	16.0	20.2	20.4	15.7	20.9	21.0	16.4	19.0	19.2	14.5
22	16.8	18.1	14.6	25.8	26.1	16.1	20.2	20.4	15.7	20.9	21.1	16.4	19.1	19.3	14.6
23	16.9	18.2	14.0	25.7	26.0	15.9	20.1	20.4	15.7	20.9	21.1	16.5	18.6	18.9	14.3
24	17.1	18.4	14.5	25.3	25.6	15.8	20.0	20.3	15.8	20.8	21.1	16.6	18.7	19.0	14.5
Avg.	16.7	17.9	14.2	25.8	26.0	16.1	20.3	20.3	15.6	21.0	20.9	16.3	19.1	19.0	14.4

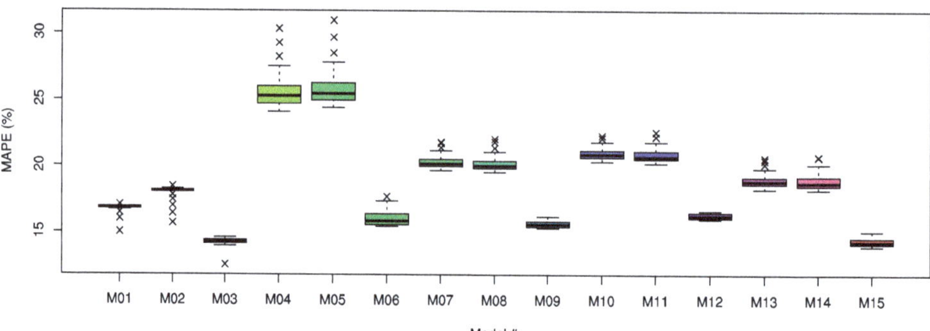

Figure 8. Box plot of the MAPE of forecasting models for Building 3.

Table 11. MAPE comparison of forecasting models for Building 4 (a cooler color indicates a lower MAPE value, while a warmer color indicates a higher MAPE value).

Point	Model #														
	M01	M02	M03	M04	M05	M06	M07	M08	M09	M10	M11	M12	M13	M14	M15
1	21.9	18.4	19.6	29.2	25.3	16.7	21.0	19.7	15.2	21.9	20.6	16.1	20.3	19.0	14.5
2	23.8	20.9	22.2	27.8	24.3	16.2	20.6	19.7	15.2	21.3	20.3	15.9	20.0	19.0	14.6
3	25.2	23.2	23.4	26.5	23.3	15.6	20.2	19.4	15.1	20.8	20.0	15.7	19.3	18.6	14.3
4	26.2	25.2	23.8	25.5	22.5	15.3	19.7	19.1	15.0	20.4	19.7	15.6	18.7	18.1	14.0
5	26.9	26.8	24.2	24.6	21.8	15.0	19.3	18.7	14.8	20.0	19.5	15.6	18.3	17.7	13.8
6	26.4	28.0	24.2	24.0	21.2	14.9	18.9	18.3	14.7	19.8	19.2	15.6	17.9	17.4	13.7
7	26.6	28.8	24.5	23.4	20.7	14.8	18.6	18.0	14.6	19.5	19.0	15.5	17.6	17.1	13.6
8	26.3	29.1	24.5	22.9	20.3	14.8	18.3	17.8	14.4	19.3	18.8	15.4	17.4	16.8	13.5
9	26.4	29.1	24.7	22.5	20.2	14.7	18.1	17.6	14.3	19.1	18.6	15.3	17.2	16.8	13.4
10	26.1	28.9	24.6	22.2	20.0	14.7	17.9	17.5	14.2	18.9	18.6	15.2	17.1	16.7	13.3
11	26.3	28.5	24.8	22.0	20.0	14.7	17.8	17.5	14.1	18.8	18.5	15.1	16.9	16.6	13.2
12	26.3	28.1	24.7	21.8	20.0	14.7	17.8	17.5	14.1	18.7	18.4	15.1	16.8	16.6	13.2
13	26.1	27.7	24.7	21.6	20.0	14.7	17.7	17.5	14.1	18.6	18.4	15.0	16.8	16.6	13.2
14	26.2	27.5	24.7	21.4	20.0	14.7	17.6	17.4	14.1	18.5	18.3	15.0	16.7	16.5	13.2
15	26.2	27.5	24.8	21.3	20.0	14.6	17.6	17.4	14.1	18.5	18.3	15.0	16.7	16.5	13.2
16	26.3	27.5	24.7	21.3	20.0	14.6	17.5	17.3	14.2	18.4	18.2	15.1	16.5	16.3	13.2
17	26.3	27.5	24.9	21.4	20.1	14.6	17.6	17.3	14.3	18.4	18.2	15.1	16.6	16.3	13.3
18	26.3	27.4	24.9	21.6	20.3	14.7	17.7	17.4	14.3	18.5	18.2	15.1	16.7	16.4	13.3
19	26.2	27.4	25.1	21.9	20.5	14.7	17.8	17.4	14.4	18.5	18.1	15.1	16.9	16.5	13.5
20	26.3	27.6	25.1	22.2	20.7	14.8	17.8	17.3	14.4	18.5	18.0	15.1	17.0	16.5	13.6
21	26.2	28.1	25.3	22.3	20.9	15.0	17.9	17.3	14.4	18.6	18.0	15.1	17.1	16.5	13.7
22	26.4	29.0	25.4	22.4	20.9	15.3	18.0	17.2	14.5	18.9	18.1	15.4	17.3	16.5	13.8
23	27.0	30.4	25.5	22.4	20.9	15.5	18.2	17.4	14.9	19.2	18.4	15.9	17.2	16.4	13.9
24	28.2	32.0	27.0	22.3	20.7	15.7	18.4	17.6	15.4	19.6	18.8	16.6	17.6	16.8	14.5
Avg.	26.1	27.3	24.5	23.1	21.0	15.0	18.4	17.9	14.5	19.3	18.8	15.4	17.5	17.0	**13.6**

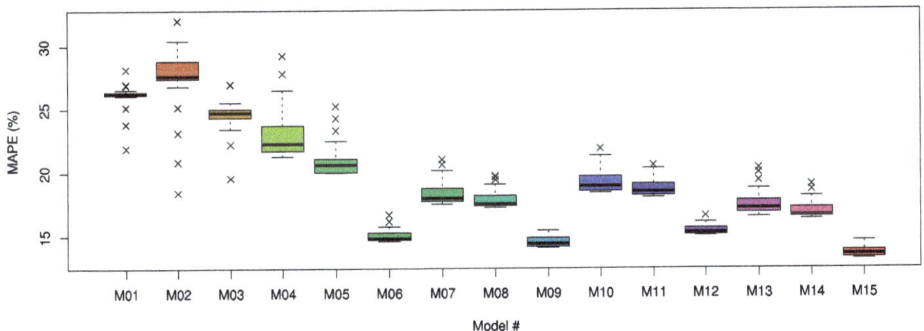

Figure 9. Box plot of the MAPE of forecasting models for Building 4.

Table 12. MAPE comparison of forecasting models for Building 5 (a cooler color indicates a lower MAPE value, while a warmer color indicates a higher MAPE value).

Point	M01	M02	M03	M04	M05	M06	M07	M08	M09	M10	M11	M12	M13	M14	M15
1	4.5	3.9	4.4	77.3	81.0	25.4	39.9	36.4	12.5	39.6	36.1	12.3	39.9	36.4	12.5
2	5.0	4.4	5.0	76.9	80.6	25.2	40.1	36.6	12.7	39.8	36.3	12.4	40.1	36.6	12.7
3	5.3	4.9	5.2	76.4	80.2	25.0	40.2	36.7	12.8	40.0	36.5	12.6	40.2	36.7	12.8
4	5.6	5.3	5.3	76.0	79.9	24.7	40.2	36.8	12.8	40.0	36.6	12.7	40.0	36.6	12.7
5	5.8	5.7	5.4	75.5	79.5	24.5	40.1	36.8	12.8	40.0	36.7	12.7	39.8	36.6	12.6
6	5.7	5.9	5.4	75.1	79.2	24.4	39.9	36.7	12.7	39.9	36.8	12.7	39.7	36.5	12.5
7	5.8	6.1	5.4	74.7	78.8	24.2	39.7	36.6	12.6	39.8	36.7	12.7	39.5	36.4	12.4
8	5.7	6.1	5.4	74.4	78.5	24.0	39.4	36.5	12.4	39.6	36.7	12.6	39.3	36.3	12.3
9	5.8	6.1	5.5	74.2	78.3	23.9	39.3	36.5	12.4	39.5	36.6	12.5	39.2	36.3	12.2
10	5.7	6.1	5.5	74.0	78.1	23.9	39.2	36.4	12.3	39.4	36.5	12.5	39.1	36.2	12.1
11	5.7	6.0	5.5	73.8	77.9	23.9	39.1	36.3	12.2	39.2	36.4	12.4	39.0	36.2	12.1
12	5.7	5.9	5.5	73.6	77.6	23.9	39.0	36.3	12.2	39.1	36.3	12.3	38.9	36.1	12.1
13	5.7	5.8	5.5	73.4	77.3	23.9	38.9	36.2	12.2	38.9	36.2	12.2	38.8	36.1	12.0
14	5.7	5.7	5.5	73.1	76.9	23.8	38.7	36.1	12.2	38.7	36.1	12.1	38.6	36.0	12.0
15	5.8	5.7	5.5	72.9	76.6	23.8	38.6	36.0	12.2	38.5	36.0	12.1	38.4	35.9	12.1
16	5.7	5.7	5.5	72.7	76.4	23.8	38.5	36.0	12.2	38.4	35.9	12.2	38.3	35.8	12.1
17	5.7	5.7	5.6	72.7	76.4	23.9	38.4	36.0	12.3	38.3	36.0	12.3	38.2	35.9	12.2
18	5.7	5.7	5.6	72.7	76.3	23.9	38.3	36.2	12.4	38.2	36.2	12.4	38.2	36.1	12.3
19	5.7	5.7	5.6	72.7	76.3	24.0	38.3	36.5	12.5	38.3	36.5	12.5	38.2	36.4	12.4
20	5.7	5.8	5.6	72.8	76.4	24.0	38.3	36.8	12.6	38.3	36.8	12.6	38.2	36.7	12.5
21	5.8	5.9	5.6	72.8	76.4	24.1	38.3	37.0	12.7	38.3	37.1	12.7	38.2	36.9	12.6
22	5.8	6.0	5.6	72.9	76.4	24.1	38.3	37.3	12.8	38.4	37.4	12.9	38.2	37.2	12.7
23	6.0	6.2	5.7	72.9	76.5	24.2	38.4	37.7	13.0	38.5	37.8	13.1	38.2	37.5	12.8
24	6.2	6.4	5.8	72.8	76.3	24.3	38.5	38.0	13.3	38.6	38.1	13.4	38.3	37.8	13.1
Avg.	5.7	5.7	5.5	74.0	77.8	24.2	39.1	36.6	12.5	39.1	36.6	12.5	38.9	36.5	12.4

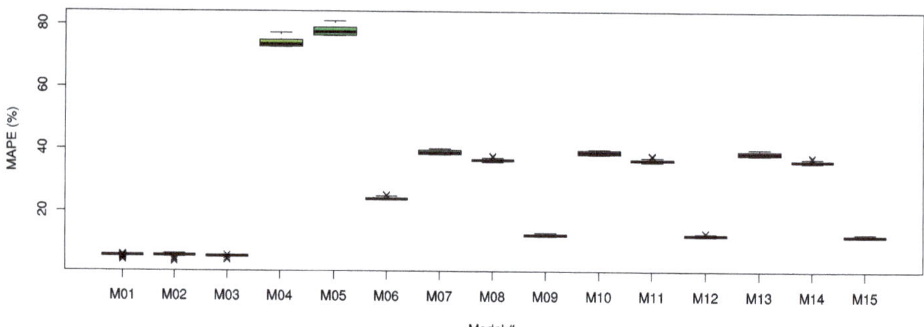

Figure 10. Box plot of the MAPE of forecasting models for Building 5.

Table 13. MAPE comparison of forecasting models for Building 6 (a cooler color indicates a lower MAPE value, while a warmer color indicates a higher MAPE value).

Point	Model #														
	M01	M02	M03	M04	M05	M06	M07	M08	M09	M10	M11	M12	M13	M14	M15
1	7.8	11.7	7.0	39.9	40.0	9.7	23.6	23.7	8.2	26.0	26.0	10.5	23.1	23.1	7.6
2	8.0	11.9	7.4	39.8	39.9	9.5	23.7	23.7	8.2	26.0	26.0	10.5	23.3	23.3	7.8
3	8.1	11.9	7.4	39.7	39.8	9.4	23.7	23.8	8.2	26.0	26.0	10.5	23.2	23.2	7.7
4	8.2	12.0	7.3	39.7	39.8	9.3	23.8	23.8	8.2	26.0	26.1	10.4	23.2	23.2	7.6
5	8.2	12.0	7.4	39.5	39.6	9.2	23.7	23.8	8.1	26.0	26.0	10.4	23.1	23.2	7.5
6	8.2	12.0	7.4	39.4	39.5	9.1	23.7	23.8	8.1	26.0	26.1	10.4	23.1	23.2	7.5
7	8.3	12.0	7.3	39.3	39.4	9.0	23.7	23.7	8.1	26.0	26.0	10.4	23.0	23.1	7.4
8	8.3	12.1	7.4	39.1	39.2	9.0	23.6	23.6	8.0	25.9	25.9	10.3	23.0	23.1	7.5
9	8.3	12.1	7.5	39.1	39.2	8.9	23.5	23.6	8.0	25.8	25.9	10.3	23.0	23.1	7.5
10	8.3	12.0	7.5	39.1	39.2	8.9	23.5	23.5	8.0	25.8	25.9	10.3	23.0	23.1	7.5
11	8.3	12.0	7.3	39.2	39.2	8.8	23.4	23.5	7.9	25.8	25.8	10.2	22.9	22.9	7.3
12	8.2	12.0	7.3	39.2	39.3	8.8	23.4	23.5	7.9	25.8	25.8	10.2	22.8	22.9	7.3
13	8.2	12.0	7.5	39.3	39.3	8.8	23.4	23.4	7.8	25.8	25.8	10.2	23.0	23.0	7.4
14	8.2	12.0	7.4	39.3	39.4	8.7	23.4	23.4	7.8	25.8	25.8	10.1	22.9	23.0	7.3
15	8.2	12.0	7.5	39.2	39.3	8.7	23.3	23.3	7.7	25.7	25.7	10.1	23.0	23.0	7.4
16	8.2	12.0	7.1	39.1	39.2	8.7	23.3	23.3	7.7	25.7	25.7	10.1	22.6	22.7	7.1
17	8.2	12.0	7.1	39.1	39.1	8.7	23.3	23.3	7.7	25.7	25.7	10.1	22.7	22.8	7.2
18	8.2	12.0	7.1	39.1	39.1	8.7	23.3	23.3	7.7	25.7	25.7	10.1	22.8	22.8	7.2
19	8.2	12.0	7.4	39.0	39.1	8.8	23.3	23.4	7.8	25.7	25.7	10.1	23.0	23.0	7.4
20	8.2	12.0	7.3	39.0	39.1	8.9	23.3	23.4	7.8	25.7	25.7	10.1	22.9	22.9	7.3
21	8.2	12.0	7.4	39.0	39.1	8.9	23.3	23.4	7.8	25.6	25.7	10.1	22.9	22.9	7.4
22	8.2	12.0	7.6	39.0	39.1	9.0	23.3	23.3	7.8	25.6	25.6	10.1	23.0	23.0	7.5
23	8.3	12.1	7.0	39.0	39.0	9.0	23.3	23.3	7.8	25.5	25.5	10.1	22.6	22.6	7.2
24	8.5	12.2	7.5	38.8	38.9	9.0	23.3	23.3	8.0	25.5	25.5	10.2	22.8	22.8	7.5
Avg.	8.2	12.0	7.3	39.3	39.3	9.0	23.5	23.5	7.9	25.8	25.8	10.2	22.9	23.0	7.4

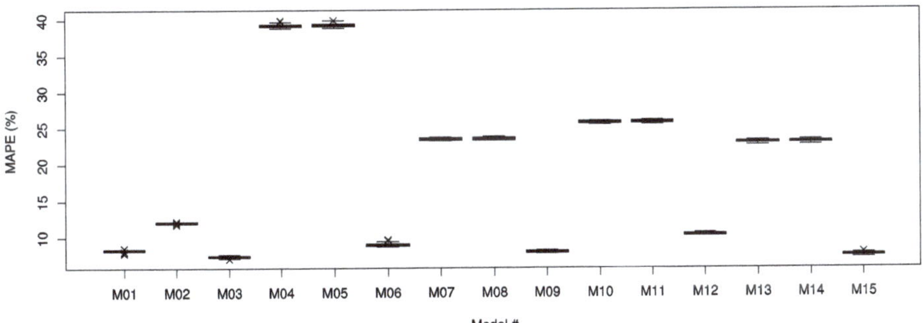

Figure 11. Box plot of the MAPE of forecasting models for Building 6.

Table 14. MAPE comparison of forecasting models for Building 7 (a cooler color indicates a lower MAPE value, while a warmer color indicates a higher MAPE value).

Point	Model #														
	M01	M02	M03	M04	M05	M06	M07	M08	M09	M10	M11	M12	M13	M14	M15
1	10.8	9.5	9.8	26.6	23.2	7.9	17.1	16.2	7.3	17.5	16.6	7.7	16.5	15.7	6.7
2	11.7	10.6	11.1	26.0	22.5	7.8	17.1	16.2	7.5	17.4	16.5	7.8	16.7	15.7	7.0
3	12.4	11.6	11.5	25.3	21.8	7.7	17.1	16.0	7.5	17.4	16.3	7.9	16.4	15.4	6.9
4	12.9	12.5	11.7	24.6	21.2	7.7	17.0	15.8	7.6	17.3	16.1	7.9	16.3	15.1	6.9
5	13.3	13.3	11.9	23.8	20.5	7.7	16.8	15.6	7.6	17.1	15.9	8.0	16.0	14.8	6.9
6	13.1	13.9	11.9	23.1	19.8	7.6	16.6	15.3	7.7	17.0	15.7	8.1	15.8	14.5	6.9
7	13.2	14.3	12.0	22.5	19.2	7.7	16.4	15.1	7.7	16.9	15.6	8.2	15.7	14.3	6.9
8	13.0	14.5	11.9	21.9	18.7	7.7	16.3	14.8	7.7	16.8	15.4	8.2	15.5	14.1	6.9
9	13.1	14.5	12.1	21.4	18.4	7.7	16.2	14.7	7.7	16.7	15.3	8.2	15.5	14.0	7.0
10	12.9	14.3	12.1	20.9	18.0	7.7	16.0	14.6	7.7	16.5	15.1	8.2	15.4	13.9	7.0
11	13.0	14.1	12.1	20.5	17.6	7.8	16.0	14.5	7.8	16.4	15.0	8.2	15.2	13.8	7.0
12	13.0	13.9	12.1	20.1	17.3	7.8	16.0	14.5	7.8	16.4	14.9	8.2	15.2	13.8	7.0
13	13.0	13.8	12.2	19.9	17.2	7.9	16.0	14.5	7.9	16.4	14.9	8.3	15.3	13.8	7.2
14	13.0	13.9	12.1	19.7	17.1	8.0	16.0	14.5	7.9	16.4	14.9	8.4	15.2	13.7	7.1
15	12.9	14.0	12.2	19.7	17.2	8.1	15.9	14.4	7.9	16.5	15.0	8.5	15.2	13.7	7.2
16	13.0	14.1	12.0	19.6	17.2	8.1	15.9	14.5	8.0	16.5	15.0	8.5	15.0	13.5	7.0
17	12.9	14.1	12.1	19.5	17.2	8.2	15.9	14.4	7.9	16.4	15.0	8.5	15.0	13.5	7.0
18	13.0	14.0	12.1	19.5	17.4	8.2	15.9	14.4	8.0	16.3	14.9	8.4	14.9	13.5	7.1
19	13.0	13.8	12.2	19.5	17.6	8.2	15.9	14.5	8.0	16.2	14.8	8.3	15.0	13.6	7.1
20	13.2	13.7	12.2	19.6	17.8	8.1	16.0	14.7	8.1	16.0	14.7	8.1	15.0	13.6	7.0
21	13.0	13.8	12.3	19.6	18.2	8.0	15.7	14.5	7.8	15.9	14.7	8.0	14.9	13.7	7.0
22	13.0	14.2	12.3	19.6	18.4	8.0	15.5	14.5	7.8	15.7	14.7	8.1	14.8	13.8	7.1
23	13.2	14.9	12.2	19.2	18.3	8.1	15.2	14.5	8.0	15.6	14.9	8.4	14.3	13.6	7.1
24	13.8	15.7	13.0	18.8	18.1	8.3	15.2	14.8	8.4	15.7	15.3	8.9	14.4	14.0	7.7
Avg.	12.9	13.6	12.0	21.3	18.7	7.9	16.1	14.9	7.8	16.6	15.3	8.2	15.4	14.1	7.0

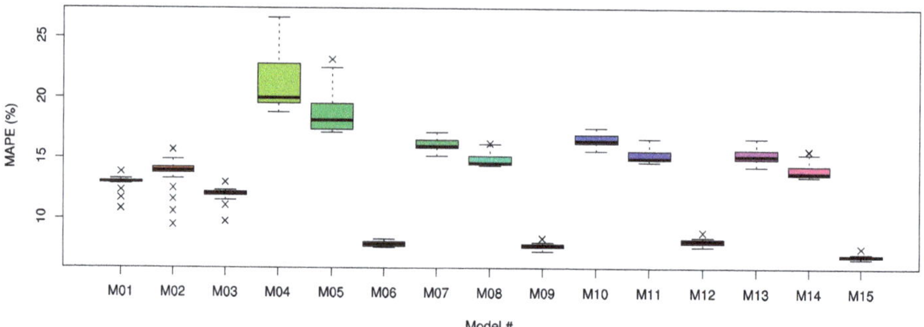

Figure 12. Box plot of the MAPE of forecasting models for Building 7.

Table 15. MAPE comparison of forecasting models for Building 8 (a cooler color indicates a lower MAPE value, while a warmer color indicates a higher MAPE value).

Point	Model #														
	M01	M02	M03	M04	M05	M06	M07	M08	M09	M10	M11	M12	M13	M14	M15
1	25.9	25.6	25.1	42.8	37.9	18.7	33.7	30.4	16.4	37.3	34.0	20.0	33.8	30.5	16.5
2	28.6	28.6	28.8	39.4	35.2	17.8	31.0	28.3	15.8	34.4	31.7	19.3	31.5	28.8	16.4
3	30.8	31.3	30.2	36.9	33.4	17.2	29.1	26.9	15.6	32.4	30.3	18.9	29.2	27.0	15.7
4	32.4	33.9	30.7	34.9	31.8	16.9	27.6	25.9	15.4	30.9	29.2	18.7	27.6	25.8	15.3
5	33.4	36.0	31.2	32.9	30.3	16.6	26.4	24.9	15.3	29.8	28.3	18.6	26.3	24.9	15.2
6	33.3	37.7	31.2	31.3	29.1	16.5	25.4	24.1	15.1	28.9	27.6	18.6	25.3	24.0	15.0
7	33.4	38.7	31.5	29.9	27.9	16.4	24.5	23.3	15.0	28.1	26.9	18.5	24.4	23.2	14.9
8	32.5	39.3	31.4	28.7	26.8	16.2	23.6	22.4	14.7	27.2	26.1	18.4	23.5	22.4	14.7
9	32.5	39.3	31.7	27.5	25.8	16.1	22.6	21.6	14.6	26.3	25.3	18.2	22.7	21.7	14.6
10	31.9	38.9	31.7	26.4	24.8	16.0	21.5	20.6	14.4	25.2	24.3	18.0	21.6	20.7	14.4
11	32.2	38.3	31.9	25.4	23.9	16.0	20.6	19.8	14.2	24.2	23.3	17.8	20.6	19.8	14.2
12	32.5	37.8	31.8	24.5	23.1	15.9	19.7	19.0	14.1	23.2	22.4	17.5	19.7	18.9	14.0
13	32.5	37.6	32.0	23.8	22.4	15.7	19.1	18.5	14.2	22.4	21.7	17.5	19.0	18.3	14.1
14	32.5	37.8	31.8	23.4	22.0	15.6	18.5	17.9	14.1	21.9	21.2	17.5	18.1	17.5	13.8
15	32.1	38.2	32.0	23.1	21.6	15.5	17.6	17.0	13.7	21.4	20.9	17.6	17.6	17.1	13.8
16	32.5	38.5	31.6	22.9	21.3	15.3	17.4	16.8	13.9	21.0	20.5	17.6	16.7	16.2	13.3
17	32.4	38.5	31.8	22.8	21.1	15.1	17.0	16.5	13.9	20.6	20.2	17.5	16.3	15.8	13.2
18	32.7	38.2	31.9	23.1	21.2	15.0	17.0	16.6	14.1	20.3	19.9	17.4	16.3	15.9	13.4
19	32.7	37.7	32.3	23.6	21.5	15.0	17.1	16.7	14.2	20.1	19.6	17.2	16.6	16.2	13.7
20	33.4	37.4	32.2	24.4	22.1	15.1	17.8	17.3	14.7	20.0	19.6	16.9	16.8	16.3	13.7
21	32.7	37.9	32.4	25.2	22.7	15.2	17.5	17.1	14.2	20.1	19.7	16.8	17.2	16.7	13.8
22	32.9	39.3	32.6	25.9	23.3	15.3	17.9	17.4	14.2	20.5	20.1	16.8	17.8	17.3	14.0
23	33.6	41.3	32.3	26.9	24.3	15.7	18.7	18.2	14.5	21.7	21.2	17.4	18.1	17.6	13.8
24	35.5	43.8	34.6	27.8	25.3	16.5	20.3	19.8	15.7	23.4	22.9	18.8	20.0	19.4	15.4
Avg.	32.3	37.1	31.4	28.1	25.8	16.1	21.7	20.7	14.7	25.1	24.0	18.0	21.5	20.5	14.5

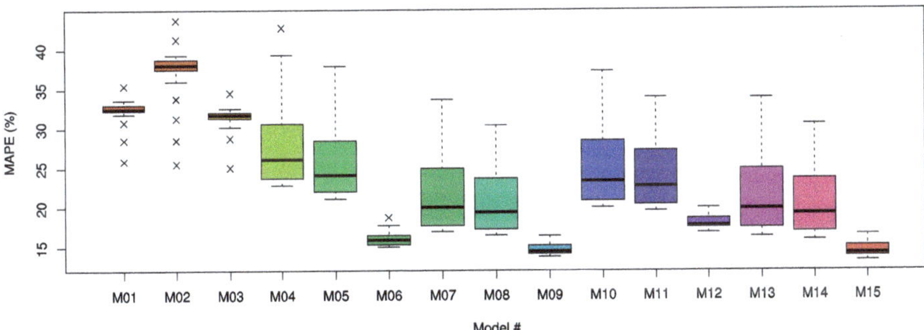

Figure 13. Box plot of the MAPE of forecasting models for Building 8.

Table 16. MAPE comparison of forecasting models for Building 9 (a cooler color indicates a lower MAPE value, while a warmer color indicates a higher MAPE value).

Point	Model #														
	M01	M02	M03	M04	M05	M06	M07	M08	M09	M10	M11	M12	M13	M14	M15
1	8.3	13.0	8.1	12.2	12.2	10.8	10.0	10.0	8.7	13.1	13.1	11.8	9.9	9.9	8.6
2	8.5	13.2	8.7	11.9	11.9	10.7	9.8	9.9	8.8	12.9	12.9	11.8	10.0	10.0	8.9
3	8.7	13.3	8.6	11.7	11.8	10.7	9.7	9.8	8.8	12.8	12.8	11.9	9.7	9.8	8.8
4	8.7	13.3	8.5	11.6	11.6	10.7	9.6	9.7	8.9	12.7	12.7	12.0	9.6	9.6	8.9
5	8.9	13.3	8.5	11.5	11.6	10.7	9.6	9.6	9.0	12.6	12.7	12.0	9.5	9.6	8.9
6	8.9	13.3	8.5	11.5	11.6	10.6	9.5	9.6	9.0	12.6	12.6	12.0	9.5	9.5	8.9
7	9.0	13.3	8.5	11.5	11.5	10.6	9.5	9.5	9.0	12.5	12.6	12.0	9.4	9.4	8.9
8	9.0	13.3	8.6	11.5	11.5	10.6	9.4	9.4	9.0	12.5	12.5	12.0	9.4	9.4	8.9
9	9.0	13.3	8.7	11.4	11.4	10.5	9.3	9.3	8.9	12.4	12.4	12.0	9.4	9.4	9.0
10	9.0	13.3	8.7	11.4	11.4	10.4	9.2	9.3	8.8	12.4	12.4	11.9	9.4	9.4	8.9
11	9.0	13.3	8.4	11.3	11.3	10.3	9.2	9.2	8.7	12.3	12.3	11.8	9.2	9.2	8.7
12	9.0	13.3	8.4	11.3	11.3	10.2	9.2	9.2	8.7	12.3	12.3	11.8	9.2	9.2	8.7
13	9.0	13.3	8.7	11.2	11.2	10.2	9.2	9.2	8.7	12.3	12.3	11.7	9.3	9.4	8.8
14	9.0	13.3	8.5	11.2	11.2	10.1	9.2	9.2	8.6	12.2	12.2	11.7	9.2	9.2	8.7
15	9.0	13.3	8.7	11.1	11.1	10.1	9.2	9.2	8.6	12.2	12.2	11.6	9.4	9.4	8.8
16	9.0	13.3	8.2	11.1	11.1	10.1	9.1	9.2	8.6	12.2	12.2	11.6	9.0	9.0	8.4
17	9.0	13.3	8.2	11.0	11.1	10.1	9.1	9.2	8.6	12.2	12.2	11.6	9.1	9.1	8.5
18	9.0	13.3	8.3	11.0	11.0	10.1	9.2	9.2	8.6	12.2	12.2	11.6	9.1	9.1	8.6
19	9.0	13.3	8.6	10.9	10.9	10.0	9.2	9.2	8.6	12.1	12.1	11.6	9.3	9.3	8.8
20	9.0	13.3	8.5	10.9	10.9	10.0	9.1	9.2	8.7	12.1	12.1	11.6	9.2	9.2	8.7
21	9.0	13.3	8.6	10.9	10.9	10.1	9.1	9.1	8.7	12.1	12.1	11.6	9.2	9.2	8.8
22	9.0	13.3	8.8	10.8	10.9	10.1	9.1	9.2	8.7	12.1	12.1	11.6	9.3	9.4	8.9
23	9.0	13.4	8.2	10.8	10.8	10.2	9.2	9.2	8.7	12.1	12.1	11.6	8.9	8.9	8.5
24	9.2	13.5	8.7	10.7	10.7	10.2	9.3	9.3	8.9	12.2	12.2	11.8	9.3	9.4	8.9
Avg.	8.9	13.3	8.5	11.3	11.3	10.3	9.3	9.4	8.8	12.4	12.4	11.8	9.4	9.4	8.8

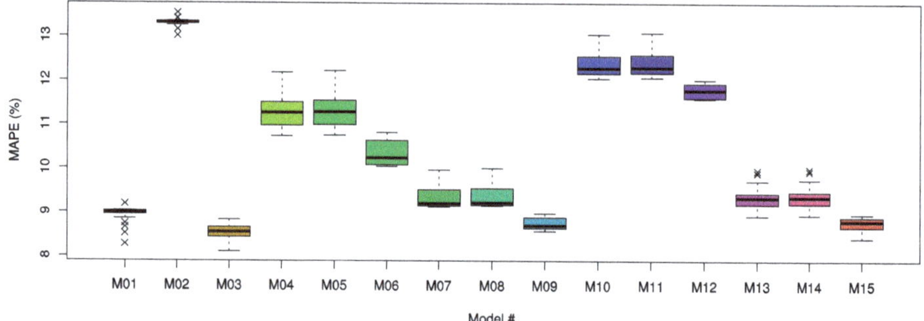

Figure 14. Box plot of the MAPE of forecasting models for Building 9.

Table 17. MAPE comparison of forecasting models for Building 10 (a cooler color indicates a lower MAPE value, while a warmer color indicates a higher MAPE value).

Point	Model #														
	M01	M02	M03	M04	M05	M06	M07	M08	M09	M10	M11	M12	M13	M14	M15
1	12.0	11.1	10.0	37.7	33.4	15.0	21.7	20.9	13.9	21.6	20.7	13.7	20.7	19.9	12.8
2	13.2	11.9	11.4	35.7	31.9	14.4	21.0	20.4	13.8	20.7	20.1	13.5	20.1	19.4	12.8
3	13.9	12.7	11.8	34.1	30.5	14.0	20.2	19.7	13.6	19.9	19.5	13.3	19.1	18.6	12.5
4	14.3	13.5	11.9	32.5	29.2	13.6	19.3	19.0	13.3	19.1	18.8	13.2	18.1	17.8	12.2
5	14.6	14.1	12.0	31.0	27.9	13.4	18.4	18.2	13.1	18.3	18.1	13.1	17.2	17.0	11.9
6	14.6	14.5	12.0	29.6	26.8	13.3	17.5	17.4	13.0	17.5	17.5	13.0	16.3	16.2	11.8
7	14.6	14.8	12.1	28.5	26.0	13.2	16.8	16.8	13.0	16.9	16.9	13.1	15.7	15.6	11.8
8	14.5	15.0	12.1	27.8	25.4	13.4	16.4	16.4	13.0	16.6	16.6	13.2	15.3	15.3	11.9
9	14.5	15.0	12.2	27.4	25.1	13.7	16.2	16.3	13.2	16.4	16.4	13.4	15.2	15.2	12.2
10	14.4	14.9	12.3	27.2	25.1	14.0	16.2	16.3	13.5	16.3	16.5	13.6	15.2	15.3	12.4
11	14.4	14.7	12.2	27.5	25.2	14.2	16.4	16.5	13.6	16.5	16.7	13.8	15.3	15.4	12.6
12	14.4	14.5	12.3	28.0	25.4	14.4	16.7	16.7	13.7	16.8	16.9	13.9	15.7	15.7	12.7
13	14.3	14.4	12.4	28.3	25.5	14.5	16.8	16.8	13.7	17.0	17.0	13.9	15.9	15.9	12.8
14	14.3	14.4	12.3	28.4	25.5	14.5	16.9	17.0	13.7	17.1	17.1	13.9	15.9	16.0	12.7
15	14.3	14.3	12.5	28.6	25.5	14.5	17.0	17.1	13.6	17.2	17.3	13.8	16.2	16.2	12.7
16	14.4	14.3	12.1	28.8	25.7	14.4	17.2	17.2	13.5	17.4	17.4	13.7	16.1	16.1	12.5
17	14.4	14.3	12.2	28.9	25.7	14.4	17.4	17.3	13.5	17.5	17.5	13.7	16.3	16.3	12.4
18	14.3	14.3	12.2	29.2	25.9	14.5	17.7	17.5	13.5	17.8	17.6	13.7	16.6	16.4	12.5
19	14.2	14.3	12.4	29.6	26.2	14.6	18.0	17.7	13.6	18.1	17.8	13.7	17.0	16.8	12.6
20	14.3	14.4	12.4	30.2	26.7	14.8	18.3	18.0	13.6	18.5	18.1	13.7	17.4	17.0	12.6
21	14.3	14.5	12.5	30.7	27.2	14.9	18.8	18.4	13.7	19.0	18.6	13.9	17.9	17.5	12.7
22	14.4	14.8	12.6	31.0	27.6	15.2	19.3	18.8	13.9	19.5	19.0	14.1	18.3	17.8	12.9
23	14.7	15.1	12.4	30.9	27.8	15.4	19.6	19.2	14.2	19.8	19.4	14.4	18.3	17.8	12.8
24	15.1	15.6	12.8	30.7	27.7	15.5	19.9	19.4	14.4	20.1	19.6	14.6	18.5	18.1	13.0
Avg.	14.3	14.2	12.1	30.1	27.0	14.3	18.1	17.9	13.6	18.2	18.0	13.7	17.0	16.8	12.5

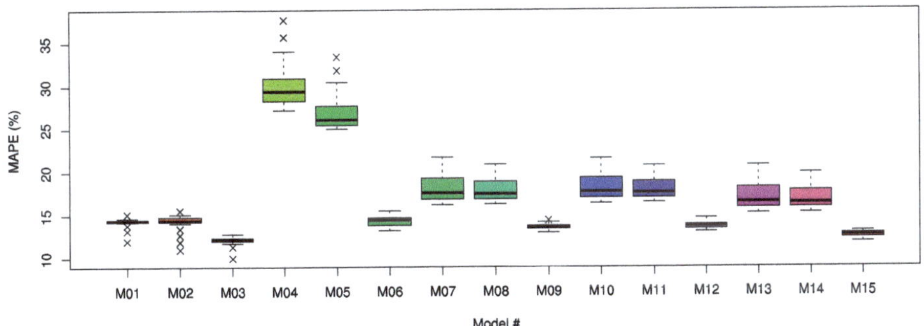

Figure 15. Box plot of the MAPE of forecasting models for Building 10.

Table 18. MAPE comparison of forecasting models for Building 11 (a cooler color indicates a lower MAPE value, while a warmer color indicates a higher MAPE value).

Point	Model #														
	M01	M02	M03	M04	M05	M06	M07	M08	M09	M10	M11	M12	M13	M14	M15
1	9.4	7.6	8.5	18.5	21.4	8.4	13.6	14.3	8.2	12.8	13.5	7.4	12.8	13.6	7.5
2	10.3	8.5	9.6	18.2	21.1	8.4	13.7	14.4	8.4	12.8	13.5	7.6	13.0	13.8	7.8
3	10.9	9.3	10.0	17.9	20.7	8.3	13.6	14.3	8.6	12.8	13.5	7.7	12.9	13.6	7.8
4	11.4	10.0	10.1	17.5	20.3	8.2	13.5	14.2	8.6	12.7	13.4	7.8	12.7	13.4	7.8
5	11.7	10.5	10.2	17.2	19.9	8.2	13.4	14.0	8.6	12.7	13.3	7.9	12.5	13.1	7.8
6	11.9	10.9	10.2	16.9	19.5	8.1	13.2	13.8	8.7	12.6	13.1	8.0	12.4	12.9	7.8
7	12.0	11.2	10.3	16.6	19.1	8.1	13.1	13.7	8.7	12.4	13.0	8.0	12.1	12.7	7.8
8	11.9	11.3	10.3	16.3	18.8	8.1	12.9	13.5	8.7	12.3	12.9	8.1	12.0	12.6	7.8
9	11.9	11.3	10.4	16.1	18.5	8.1	12.7	13.3	8.7	12.1	12.7	8.1	11.9	12.5	7.8
10	11.6	11.2	10.4	15.8	18.3	8.1	12.6	13.2	8.7	12.0	12.6	8.1	11.7	12.3	7.8
11	11.7	11.1	10.3	15.6	18.0	8.2	12.4	13.1	8.7	11.8	12.4	8.0	11.5	12.2	7.8
12	11.7	11.0	10.4	15.3	17.8	8.2	12.3	13.0	8.7	11.6	12.3	8.0	11.4	12.1	7.8
13	11.7	10.9	10.5	15.2	17.6	8.2	12.2	12.9	8.7	11.5	12.2	8.0	11.4	12.1	7.9
14	11.8	11.0	10.3	15.0	17.4	8.2	12.2	12.8	8.7	11.5	12.1	8.0	11.3	11.9	7.8
15	11.6	11.0	10.4	14.9	17.3	8.3	12.1	12.7	8.6	11.4	12.1	8.0	11.3	12.0	7.9
16	11.7	11.0	10.2	14.8	17.2	8.3	12.1	12.8	8.7	11.4	12.1	8.0	11.1	11.8	7.7
17	11.7	11.0	10.3	14.7	17.1	8.3	11.9	12.7	8.7	11.3	12.1	8.0	11.0	11.8	7.8
18	11.6	11.0	10.2	14.6	17.1	8.3	11.8	12.7	8.7	11.2	12.1	8.0	11.0	11.8	7.8
19	11.6	10.9	10.4	14.5	17.0	8.3	11.7	12.6	8.6	11.1	12.0	8.0	11.0	11.9	7.9
20	11.5	10.7	10.3	14.5	16.9	8.4	11.7	12.6	8.6	11.1	12.0	8.1	11.0	11.9	7.9
21	11.5	10.7	10.3	14.4	16.8	8.5	11.7	12.6	8.7	11.1	12.0	8.1	11.0	11.9	8.0
22	11.7	10.6	10.4	14.4	16.7	8.6	11.8	12.7	8.8	11.2	12.1	8.2	11.1	12.0	8.1
23	11.7	10.7	10.2	14.4	16.6	8.7	12.0	12.9	9.0	11.4	12.3	8.5	11.0	11.9	8.0
24	11.9	10.8	10.4	14.4	16.6	8.8	12.2	13.1	9.3	11.6	12.6	8.8	11.3	12.3	8.5
Avg.	11.5	10.6	10.2	15.7	18.2	8.3	12.5	13.2	8.7	11.9	12.6	8.0	11.7	12.4	**7.9**

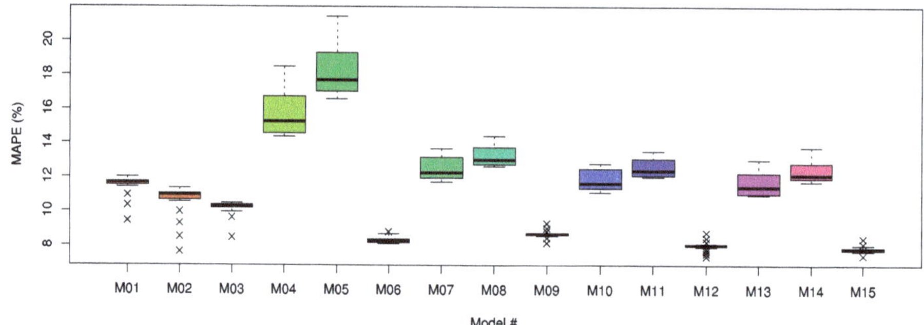

Figure 16. Box plot of the MAPE of forecasting models for Building 11.

Table 19. MAPE comparison of forecasting models for Building 12 (a cooler color indicates a lower MAPE value, while a warmer color indicates a higher MAPE value).

Point	Model #														
	M01	M02	M03	M04	M05	M06	M07	M08	M09	M10	M11	M12	M13	M14	M15
1	8.7	7.1	7.4	23.7	35.6	28.9	17.4	21.3	15.7	16.8	20.7	15.2	16.3	20.2	14.6
2	9.3	7.8	8.3	23.2	34.9	28.5	17.4	21.1	15.8	16.8	20.4	15.1	16.4	20.0	14.7
3	9.8	8.5	8.6	22.7	34.2	28.3	17.2	20.7	15.8	16.6	20.1	15.1	16.1	19.6	14.7
4	10.2	9.1	8.7	22.2	33.6	28.1	17.0	20.4	15.7	16.5	19.8	15.2	15.9	19.3	14.6
5	10.4	9.5	8.8	22.0	33.3	28.0	16.9	20.1	15.7	16.4	19.6	15.2	15.8	19.0	14.6
6	10.4	9.9	8.8	21.9	32.9	27.9	16.7	19.9	15.7	16.3	19.4	15.3	15.6	18.8	14.6
7	10.5	10.2	8.9	21.9	32.7	27.9	16.6	19.7	15.7	16.2	19.3	15.3	15.5	18.6	14.6
8	10.4	10.3	8.9	21.8	32.5	27.8	16.5	19.6	15.7	16.2	19.3	15.4	15.4	18.6	14.7
9	10.4	10.3	9.0	21.8	32.3	27.8	16.4	19.5	15.8	16.0	19.2	15.5	15.3	18.5	14.7
10	10.2	10.2	9.0	21.7	32.2	27.8	16.3	19.6	15.9	16.0	19.2	15.6	15.3	18.6	14.9
11	10.3	10.1	9.0	21.7	32.1	27.9	16.2	19.5	16.0	15.9	19.2	15.6	15.1	18.5	14.9
12	10.4	10.0	9.0	21.7	32.1	28.0	16.2	19.6	16.1	15.8	19.2	15.7	15.1	18.5	15.0
13	10.4	9.9	9.1	21.7	32.1	28.0	16.2	19.6	16.1	15.7	19.2	15.7	15.1	18.5	15.0
14	10.3	9.8	9.0	21.6	32.0	28.0	16.0	19.5	16.0	15.7	19.1	15.7	15.0	18.4	15.0
15	10.3	9.8	9.1	21.6	32.0	28.0	16.0	19.4	16.0	15.6	19.1	15.7	15.0	18.5	15.0
16	10.3	9.8	8.9	21.4	31.9	28.0	15.8	19.4	16.0	15.5	19.1	15.7	14.7	18.3	14.9
17	10.2	9.8	9.0	21.3	31.9	28.0	15.7	19.4	16.0	15.3	19.1	15.7	14.6	18.3	15.0
18	10.2	9.8	9.1	21.1	31.8	27.9	15.5	19.4	16.0	15.2	19.0	15.7	14.5	18.3	15.0
19	10.3	9.8	9.2	21.1	31.8	27.9	15.5	19.4	16.0	15.1	19.0	15.6	14.5	18.4	15.0
20	10.4	9.9	9.1	21.1	31.7	27.8	15.4	19.3	16.0	15.0	18.9	15.6	14.4	18.3	15.0
21	10.3	10.1	9.3	21.2	31.6	27.7	15.4	19.2	16.0	15.0	18.9	15.6	14.5	18.3	15.1
22	10.4	10.3	9.3	21.5	31.6	27.6	15.5	19.2	16.1	15.2	18.9	15.7	14.6	18.3	15.1
23	10.6	10.7	9.2	21.7	31.4	27.5	15.7	19.2	16.2	15.4	18.9	15.9	14.6	18.1	15.0
24	11.1	11.2	9.7	21.7	31.2	27.3	16.0	19.3	16.3	15.7	19.0	16.0	14.9	18.2	15.2
Avg.	10.2	9.7	8.9	21.8	32.5	27.9	16.2	19.7	15.9	15.8	19.3	15.5	15.2	18.7	14.9

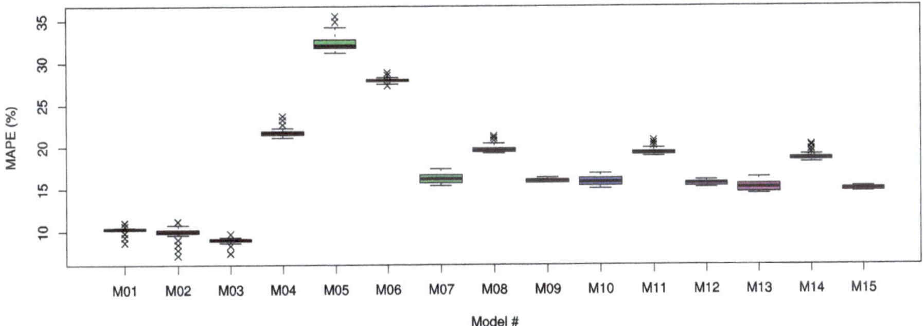

Figure 17. Box plot of the MAPE of forecasting models for Building 12.

Table 20. MAPE comparison of forecasting models for Building 13 (a cooler color indicates a lower MAPE value, while a warmer color indicates a higher MAPE value).

Point	Model #														
	M01	M02	M03	M04	M05	M06	M07	M08	M09	M10	M11	M12	M13	M14	M15
1	19.5	14.4	19.4	18.5	20.8	16.3	17.0	17.1	16.5	14.2	14.3	13.7	16.3	16.4	15.7
2	20.9	16.0	21.7	18.0	19.9	16.0	16.9	17.0	16.4	14.2	14.2	13.6	16.6	16.7	16.1
3	22.1	17.6	22.0	17.5	19.1	15.6	16.9	16.9	16.3	14.1	14.1	13.5	16.2	16.2	15.7
4	23.1	19.1	22.2	17.1	18.4	15.2	16.8	16.8	16.2	14.1	14.0	13.5	15.9	15.9	15.4
5	23.7	20.5	22.4	16.6	17.8	14.9	16.6	16.6	16.0	14.1	14.1	13.5	15.7	15.7	15.1
6	24.1	21.6	22.5	16.2	17.1	14.6	16.5	16.5	15.9	14.1	14.1	13.5	15.6	15.6	15.0
7	24.1	22.3	22.5	15.9	16.6	14.4	16.3	16.3	15.8	14.1	14.1	13.6	15.3	15.2	14.7
8	23.8	22.7	22.5	15.7	16.0	14.4	16.3	16.1	15.9	14.2	14.0	13.8	15.3	15.1	14.9
9	23.7	22.7	22.8	15.5	15.5	14.6	16.3	16.0	16.0	14.2	13.9	13.9	15.3	15.1	15.1
10	23.3	22.4	23.0	15.5	15.1	14.8	16.3	16.0	16.2	14.1	13.7	14.0	15.4	15.1	15.3
11	23.5	21.9	22.9	15.5	14.9	15.0	16.4	15.9	16.4	14.0	13.5	14.0	15.3	14.8	15.3
12	23.5	21.5	22.9	15.6	14.9	15.2	16.4	15.9	16.4	14.0	13.4	14.0	15.3	14.7	15.2
13	23.6	21.3	23.4	15.7	15.1	15.3	16.4	15.9	16.4	14.0	13.5	14.0	15.6	15.1	15.6
14	23.4	21.4	23.0	15.9	15.4	15.4	16.5	16.0	16.4	14.1	13.6	14.1	15.4	14.9	15.4
15	23.3	21.6	23.3	16.0	15.6	15.5	16.3	15.8	16.2	14.3	13.8	14.2	15.6	15.2	15.6
16	23.3	21.8	22.3	16.1	15.9	15.6	16.3	16.0	16.4	14.4	14.0	14.4	14.8	14.4	14.8
17	23.1	21.8	22.5	16.2	16.1	15.7	16.2	15.8	16.3	14.3	14.0	14.4	14.8	14.5	14.9
18	22.8	21.6	22.7	16.4	16.3	15.9	16.1	15.7	16.2	14.3	13.9	14.4	15.0	14.6	15.2
19	22.9	21.3	23.2	16.7	16.5	16.2	16.1	15.7	16.3	14.1	13.7	14.3	15.3	14.9	15.6
20	23.0	21.1	23.1	16.9	16.7	16.4	16.2	15.7	16.5	14.0	13.5	14.3	15.2	14.8	15.6
21	23.2	21.2	23.4	17.2	17.0	16.6	16.4	15.8	16.7	14.0	13.4	14.4	15.6	15.0	16.0
22	23.5	21.6	23.7	17.4	17.3	16.8	16.7	16.0	17.1	14.3	13.6	14.7	16.0	15.3	16.4
23	24.1	22.4	22.8	17.7	17.4	17.0	17.1	16.3	17.6	14.9	14.1	15.4	15.5	14.7	16.0
24	25.1	23.4	24.3	17.8	17.4	17.2	17.8	16.8	18.3	15.7	14.7	16.3	16.5	15.6	17.1
Avg.	23.2	21.0	22.7	16.6	16.8	15.6	16.5	16.2	16.4	14.2	13.9	14.1	15.6	15.2	15.5

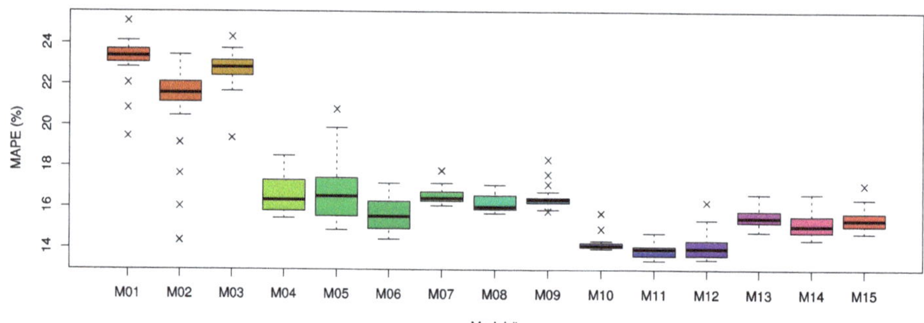

Figure 18. Box plot of the MAPE of forecasting models for Building 13.

Table 21. MAPE comparison of forecasting models for Building 14 (a cooler color indicates a lower MAPE value, while a warmer color indicates a higher MAPE value).

Point	Model #														
	M01	M02	M03	M04	M05	M06	M07	M08	M09	M10	M11	M12	M13	M14	M15
1	12.6	9.5	11.1	12.6	19.2	9.4	10.8	12.8	9.3	9.3	11.3	7.8	9.3	11.4	7.8
2	13.6	10.6	12.5	12.3	18.4	9.2	10.8	12.7	9.4	9.3	11.2	7.9	9.5	11.4	8.0
3	14.4	11.8	13.0	11.8	17.5	8.9	10.8	12.7	9.4	9.3	11.1	7.9	9.3	11.1	7.9
4	15.1	12.8	13.1	11.4	16.7	8.7	10.8	12.5	9.4	9.3	11.1	8.0	9.2	10.9	7.8
5	15.5	13.6	13.3	11.0	16.0	8.4	10.6	12.3	9.3	9.3	11.0	8.0	9.1	10.7	7.7
6	15.5	14.3	13.4	10.7	15.3	8.1	10.6	12.2	9.3	9.3	10.9	8.0	9.0	10.6	7.7
7	15.8	14.8	13.4	10.4	14.7	8.0	10.5	12.1	9.3	9.3	10.9	8.1	8.9	10.4	7.6
8	15.6	15.0	13.5	10.2	14.2	7.9	10.5	12.0	9.3	9.3	10.8	8.1	8.9	10.4	7.7
9	15.6	15.0	13.6	10.1	13.8	7.9	10.5	11.9	9.3	9.3	10.7	8.1	8.9	10.3	7.7
10	15.4	14.8	13.6	10.0	13.5	7.9	10.4	11.8	9.3	9.2	10.6	8.1	8.9	10.3	7.7
11	15.4	14.6	13.6	10.0	13.2	7.9	10.5	11.7	9.3	9.2	10.5	8.0	8.8	10.1	7.6
12	15.3	14.3	13.6	9.9	13.0	7.9	10.4	11.7	9.3	9.1	10.3	8.0	8.8	10.0	7.7
13	15.4	14.2	13.9	10.0	13.0	8.0	10.5	11.6	9.3	9.1	10.3	8.0	9.0	10.2	7.9
14	15.4	14.1	13.7	10.1	13.0	8.1	10.5	11.6	9.4	9.1	10.2	8.0	8.9	10.1	7.8
15	15.3	14.2	13.9	10.3	13.1	8.2	10.3	11.5	9.2	9.1	10.3	8.0	9.0	10.2	7.9
16	15.4	14.2	13.5	10.5	13.2	8.4	10.5	11.6	9.4	9.1	10.3	8.1	8.7	9.8	7.6
17	15.1	14.2	13.7	10.7	13.3	8.6	10.2	11.4	9.2	9.1	10.3	8.1	8.7	9.9	7.7
18	15.1	14.2	13.7	10.9	13.5	8.8	10.1	11.3	9.1	9.1	10.3	8.1	8.7	9.9	7.7
19	15.0	14.1	14.0	11.0	13.6	8.9	10.0	11.2	9.1	9.0	10.2	8.0	8.8	10.0	7.9
20	15.4	14.2	13.9	11.2	13.8	9.1	10.0	11.2	9.1	8.9	10.1	8.0	8.8	10.0	7.8
21	15.3	14.5	14.1	11.4	13.9	9.3	10.0	11.2	9.1	9.0	10.1	8.1	8.9	10.1	8.0
22	15.5	15.0	14.2	11.5	14.0	9.4	10.1	11.3	9.3	9.2	10.3	8.3	9.0	10.1	8.1
23	16.0	15.7	14.0	11.5	14.0	9.4	10.4	11.5	9.6	9.5	10.6	8.6	8.7	9.9	7.9
24	16.7	16.5	14.9	11.4	13.9	9.4	10.7	11.9	9.9	9.8	11.0	9.0	9.2	10.3	8.4
Avg.	15.2	14.0	13.6	10.9	14.5	8.6	10.4	11.8	9.3	9.2	10.6	8.1	8.9	10.3	7.8

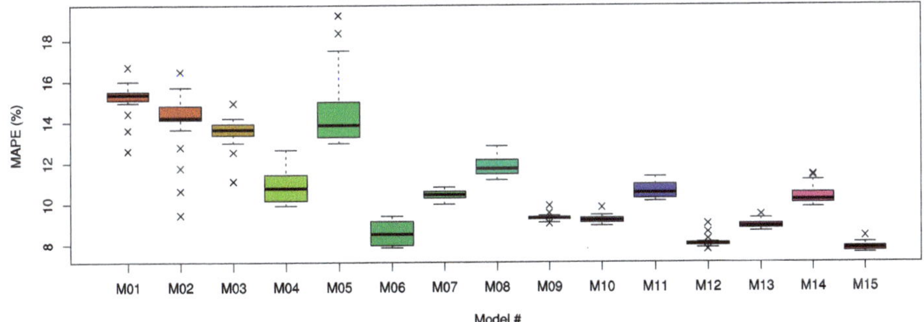

Figure 19. Box plot of the MAPE of forecasting models for Building 14.

Table 22. MAPE comparison of forecasting models for Building 15 (a cooler color indicates a lower MAPE value, while a warmer color indicates a higher MAPE value).

Point	Model #														
	M01	M02	M03	M04	M05	M06	M07	M08	M09	M10	M11	M12	M13	M14	M15
1	18.1	16.3	17.5	55.9	58.0	12.5	40.1	40.3	13.8	39.8	40.0	13.4	39.1	39.3	12.7
2	19.6	17.8	19.7	53.1	55.3	12.2	37.8	38.0	13.8	37.4	37.6	13.4	37.2	37.4	13.1
3	20.7	19.3	20.0	50.3	52.5	12.0	35.4	35.6	13.8	35.0	35.3	13.4	34.4	34.6	12.8
4	21.5	20.6	20.2	47.5	49.8	11.9	32.9	33.2	13.8	32.6	32.9	13.5	31.8	32.2	12.8
5	22.0	21.8	20.4	44.5	46.9	11.8	30.2	30.7	13.8	30.0	30.5	13.6	29.1	29.5	12.6
6	22.5	22.8	20.5	41.7	44.0	11.8	27.8	28.4	13.9	27.6	28.2	13.7	26.6	27.2	12.7
7	22.5	23.5	20.6	39.5	41.8	11.8	25.7	26.4	13.9	25.6	26.3	13.8	24.5	25.1	12.7
8	21.9	23.9	20.6	38.0	40.2	11.8	24.1	24.8	13.9	24.2	24.9	13.9	22.9	23.6	12.6
9	22.0	23.9	20.9	36.8	38.8	12.0	23.0	23.7	14.0	23.0	23.7	14.0	21.9	22.6	12.9
10	21.7	23.6	21.1	36.4	38.2	12.1	22.4	23.0	14.1	22.3	22.9	14.0	21.3	21.9	13.0
11	22.0	23.2	21.0	36.7	38.5	12.2	22.3	22.9	14.1	22.2	22.7	14.0	21.1	21.7	12.9
12	22.0	22.8	21.0	36.9	38.6	12.3	22.5	23.0	14.2	22.2	22.7	13.9	21.2	21.8	13.0
13	21.9	22.5	21.4	37.3	39.1	12.3	22.6	23.2	14.2	22.3	22.8	13.9	21.7	22.2	13.3
14	22.0	22.5	21.0	37.7	39.6	12.3	22.7	23.3	14.3	22.4	23.0	14.0	21.5	22.1	13.1
15	21.6	22.7	21.4	38.4	40.3	12.3	22.7	23.3	14.0	22.8	23.4	14.1	21.9	22.5	13.2
16	21.8	22.8	20.5	38.9	40.9	12.4	23.1	23.8	14.2	23.2	23.9	14.3	21.5	22.2	12.7
17	21.6	22.9	20.9	39.3	41.5	12.5	23.0	23.8	14.0	23.4	24.2	14.3	21.8	22.6	12.8
18	21.7	22.7	21.0	39.4	41.7	12.7	23.3	24.2	14.1	23.6	24.4	14.3	22.2	23.0	12.9
19	21.6	22.5	21.6	39.9	42.2	12.9	23.8	24.6	14.1	23.9	24.7	14.2	23.0	23.8	13.3
20	21.8	22.5	21.3	39.8	42.2	13.1	24.0	24.9	14.3	23.8	24.7	14.1	22.9	23.9	13.2
21	22.2	22.7	21.6	39.4	41.7	13.3	24.0	24.9	14.6	23.5	24.4	14.1	22.9	23.8	13.5
22	22.5	23.3	21.8	39.0	41.3	13.4	24.2	25.1	14.9	23.6	24.5	14.2	23.2	24.2	13.9
23	22.7	24.1	21.0	39.0	41.3	13.5	24.3	25.3	15.1	23.9	24.9	14.7	22.6	23.6	13.4
24	23.4	25.1	22.4	38.5	40.8	13.7	24.5	25.6	15.5	24.2	25.3	15.3	23.2	24.3	14.2
Avg.	21.7	22.3	20.8	41.0	43.1	12.5	26.1	26.8	14.2	25.9	26.6	14.0	25.0	25.6	13.1

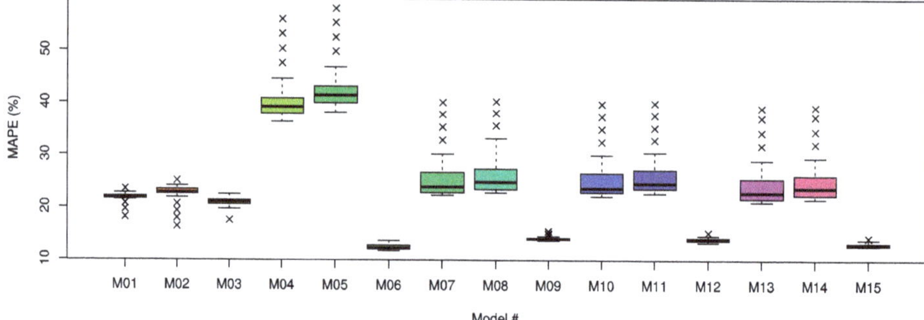

Figure 20. Box plot of the MAPE of forecasting models for Building 15.

Table 23. Average MAPE comparison of forecasting models (a cooler color indicates a lower MAPE value, while a warmer color indicates a higher MAPE value).

Building #	Model #														
	M01	M02	M03	M04	M05	M06	M07	M08	M09	M10	M11	M12	M13	M14	M15
1	18.1	19.9	17.7	17.2	21.8	9.3	11.6	12.0	8.7	12.8	13.1	9.8	11.4	11.7	8.4
2	38.3	37.3	35.1	34.0	35.3	18.7	32.3	33.8	20.5	31.3	32.8	19.5	29.2	30.7	17.4
3	16.7	17.9	14.2	25.8	26.0	16.1	20.3	20.3	15.6	21.0	20.9	16.3	19.1	19.0	14.4
4	26.1	27.3	24.5	23.1	21.0	15.0	18.4	17.9	14.5	19.3	18.8	15.4	17.5	17.0	13.6
5	5.7	5.7	5.5	74.0	77.8	24.2	39.1	36.6	12.5	39.1	36.6	12.5	38.9	36.5	12.4
6	8.2	12.0	7.3	39.3	39.3	9.0	23.5	23.5	7.9	25.8	25.8	10.2	22.9	23.0	7.4
7	12.9	13.6	12.0	21.3	18.7	7.9	16.1	14.9	7.8	16.6	15.3	8.2	15.4	14.1	7.0
8	32.3	37.1	31.4	28.1	25.8	16.1	21.7	20.7	14.7	25.1	24.0	18.0	21.5	20.5	14.5
9	8.9	13.3	8.5	11.3	11.3	10.3	9.3	9.4	8.8	12.4	12.4	11.8	9.4	9.4	8.8
10	14.3	14.2	12.1	30.1	27.0	14.3	18.1	17.9	13.6	18.2	18.0	13.7	17.0	16.8	12.5
11	11.5	10.6	10.2	15.7	18.2	8.3	12.5	13.2	8.7	11.9	12.6	8.0	11.7	12.4	7.9
12	10.2	9.7	8.9	21.8	32.5	27.9	16.2	19.7	15.9	15.8	19.3	15.5	15.2	18.7	14.9
13	23.2	21.0	22.7	16.6	16.8	15.6	16.5	16.2	16.4	14.2	13.9	14.1	15.6	15.2	15.5
14	15.2	14.0	13.6	10.9	14.5	8.6	10.4	11.8	9.3	9.2	10.6	8.1	8.9	10.3	7.8
15	21.7	22.3	20.8	41.0	43.1	12.5	26.1	26.8	14.2	25.9	26.6	14.0	25.0	25.6	13.1
Avg.	17.6	18.4	16.3	27.3	28.6	14.3	19.5	19.6	12.6	19.9	20.0	13.0	18.6	18.7	11.7

Table 24. Average root mean square error (RMSE) results of 15 forecasting models for each building (a cooler color indicates a lower RMSE value, while a warmer color indicates a higher RMSE value).

Building #	Model #														
	M01	M02	M03	M04	M05	M06	M07	M08	M09	M10	M11	M12	M13	M14	M15
1	103.9	105.2	100.3	91.3	114.2	47.0	57.1	60.6	45.1	55.6	59.2	43.2	51.9	55.7	38.3
2	185.5	192.2	179.8	127.7	130.7	91.4	114.2	117.9	90.0	121.4	124.9	99.0	106.4	110.3	79.8
3	60.7	59.1	48.7	67.1	67.6	49.8	64.2	63.9	55.2	63.6	63.3	54.5	59.0	58.6	49.0
4	187.5	185.9	177.0	147.6	137.9	92.1	121.4	118.8	97.1	115.5	112.6	89.5	110.5	107.5	83.0
5	19.4	19.3	18.6	177.5	182.8	69.7	124.4	119.9	45.1	124.4	119.9	45.1	124.3	119.8	44.9
6	36.9	28.3	24.4	111.6	111.8	33.8	82.8	82.9	34.4	80.7	80.9	29.1	79.9	80.1	26.8
7	151.7	154.7	144.2	194.9	165.5	93.4	144.7	136.5	84.1	145.1	136.9	84.8	138.4	129.8	72.7
8	93.2	93.5	88.8	67.1	60.9	44.2	52.5	50.9	44.5	49.3	47.6	40.7	45.6	43.7	36.0
9	78.5	58.5	53.6	99.6	99.8	76.4	90.1	90.3	77.6	79.1	79.3	64.4	77.9	78.1	62.9
10	148.2	169.5	130.6	287.6	247.7	148.2	209.6	195.3	144.6	216.6	202.9	154.6	204.5	189.9	137.2
11	42.7	46.8	42.1	49.5	54.3	30.1	41.5	43.0	29.2	43.7	45.2	32.2	41.0	42.6	28.5
12	183.1	191.6	174.4	321.2	484.3	421.6	260.7	327.2	265.6	264.9	330.6	269.7	257.1	324.4	262.1
13	854.8	935.6	849.9	668.6	784.1	491.5	516.5	553.3	406.3	608.5	640.0	518.1	515.3	552.2	404.7
14	580.3	632.7	569.6	484.1	611.7	358.9	368.8	426.8	306.6	423.5	474.8	370.6	357.2	416.8	292.6
15	79.2	84.3	76.9	131.6	137.7	40.3	81.2	83.4	40.8	84.5	86.6	47.0	79.7	81.9	37.7

To demonstrate the superiority of the SPROUT model, we performed several statistical tests, such as Wilcoxon signed-rank and Friedman tests [58,59]. The Wilcoxon signed-rank test [58] is used to confirm the null hypothesis to determine whether there is a significant difference between two models. If the p-value is less than the significance level, the null hypothesis is rejected, and the two models are judged to have significant differences. The Friedman test [59] is a multiple comparison test that aims to identify significant differences between the results of two or more forecasting models. To verify the results of the two tests, we used all the MAPE values (15 (number of the buildings) × 24 (prediction time points)) for each forecasting model. The results of the Wilcoxon test with the significance level set to 0.05 and the Friedman test are listed in Table 26. We can observe that the proposed SPROUT model significantly outperforms the other models because the p-value in all cases is below the significance level.

Table 25. Average mean absolute error (MAE) results of 15 forecasting models for each building (a cooler color indicates a lower MAE value, while a warmer color indicates a higher MAE value).

Building #	Model #														
	M01	**M02**	**M03**	**M04**	**M05**	**M06**	**M07**	**M08**	**M09**	**M10**	**M11**	**M12**	**M13**	**M14**	**M15**
1	58.0	53.8	51.9	54.9	70.8	29.6	37.1	38.2	29.4	34.0	35.1	26.3	32.5	33.6	24.8
2	121.1	126.8	112.9	98.9	101.6	65.6	85.1	88.0	64.9	90.0	92.9	69.8	77.0	79.9	56.9
3	43.0	41.3	34.1	52.2	52.6	37.4	47.2	47.0	40.0	46.3	46.1	39.1	42.7	42.4	35.4
4	125.7	119.4	111.0	121.3	109.9	71.3	96.1	94.1	75.4	89.7	87.6	69.0	84.0	82.0	63.3
5	13.9	13.7	13.2	151.7	160.0	52.8	81.8	77.7	29.9	81.7	77.6	29.8	81.4	77.3	29.5
6	31.1	21.4	19.2	99.1	99.2	23.3	65.8	65.8	26.7	59.8	59.9	20.8	58.5	58.6	19.4
7	89.4	85.8	78.6	143.6	119.8	55.5	104.8	96.8	55.1	102.6	94.6	53.0	96.5	88.5	46.8
8	55.1	48.5	46.6	42.3	39.0	26.5	36.0	35.0	29.1	30.7	29.7	23.8	29.4	28.4	22.6
9	66.9	45.2	43.1	65.2	65.3	54.6	66.4	66.5	60.7	51.5	51.6	45.8	51.5	51.6	45.8
10	103.1	107.3	88.2	210.9	187.9	104.6	138.8	133.3	101.8	140.3	134.8	103.2	131.4	125.8	94.3
11	30.0	33.1	29.1	39.7	44.7	22.8	30.8	32.1	22.0	33.0	34.3	24.1	30.4	31.7	21.5
12	129.2	136.2	119.0	268.9	434.4	379.0	192.6	242.7	199.6	198.0	248.2	205.0	184.1	234.3	191.1
13	536.7	602.2	549.2	455.6	515.2	368.7	346.0	358.3	304.8	411.3	423.6	370.1	359.1	371.4	317.9
14	364.3	399.2	354.2	311.5	421.8	238.8	241.3	279.6	208.7	275.2	313.6	242.7	234.4	272.7	201.8
15	47.7	48.9	44.8	87.5	92.9	26.7	48.8	50.4	28.1	50.7	52.3	30.0	46.6	48.2	25.9

Table 26. Results of the Wilcoxon and Friedman tests with SPROUT (solving cold start problem in short-term load forecasting using tree-based methods).

Compared Models	Wilcoxon Test (p-Value < 0.05)	Friedman Test
M01	$< 2.2 \times 10^{-16}$	
M02	$< 2.2 \times 10^{-16}$	
M03	1.504×10^{-8}	
M04	$< 2.2 \times 10^{-16}$	
M05	$< 2.2 \times 10^{-16}$	
M06	3.828×10^{-12}	
M07	$< 2.2 \times 10^{-16}$	Friedman chi-squared = 2547.7
M08	$< 2.2 \times 10^{-16}$	p-value $< 2.2 \times 10^{-16}$
M09	4.919×10^{-6}	
M10	$< 2.2 \times 10^{-16}$	
M11	$< 2.2 \times 10^{-16}$	
M12	1.388×10^{-7}	
M13	$< 2.2 \times 10^{-16}$	
M14	$< 2.2 \times 10^{-16}$	

5.3. Discussion

The SPROUT model demonstrated the best performance in the majority of experiments, excluding certain buildings. Thus, we analyzed these cases in detail. In Figure 21, we can observe that Buildings 3, 5, 6, 9, and 10 illustrated no significant difference in weekday and weekend electric loads. As shown in Table 2 and Figure 21f, Building 12 was unable to reflect similar electric load patterns owing to the wide range of electric energy consumption data. Hence, the transfer learning-based forecasting models demonstrated unsatisfactory prediction performance. Therefore, despite the differences in weekday and weekend patterns, the time factor-based forecasting models showed better prediction performance.

Building 13 showed that M10 to M12 presented better prediction performance than other models because MLR could predict the building electric energy consumptions better than DT and RF. In addition, as listed in Table 20, Building 13 demonstrated a wide range of electric energy consumption data and also showed the highest electric energy consumption. Therefore, even when the ED was close, it was challenging to accurately derive the results of the transfer learning-based forecasting model. The electric load patterns of Building 15 were considerably similar to those of Building 8, as shown in the *F* test in Table 27. Therefore, as M06 demonstrated accurate predictions on both weekdays and weekends, it exhibited the lowest prediction error rate.

Table 27. *F* test between Building 8 and Building 15.

Statistical List	Building 8		Building 15
Mean	185.03		204.63
Variance	12,292.48		14,017.42
F		0.88	
P (F <= f) one-tail		0	
F Critical one-tail		0.98	

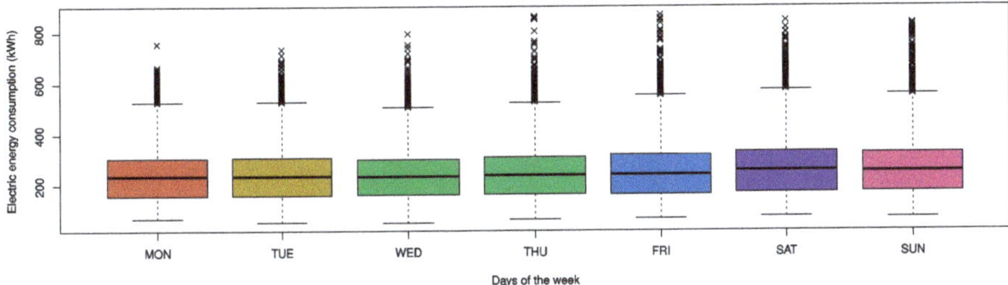

(a) Building 3

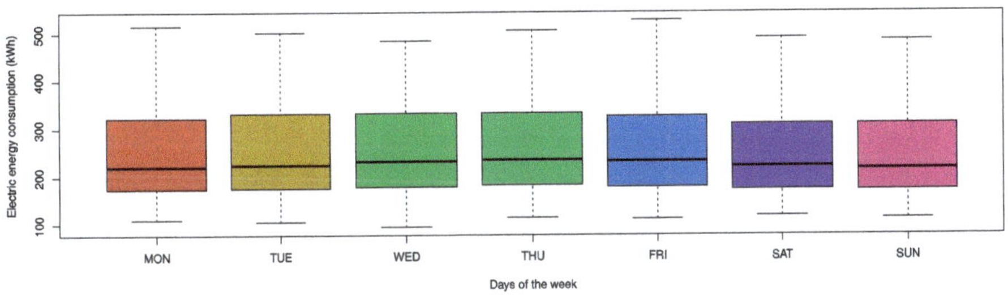

(b) Building 5

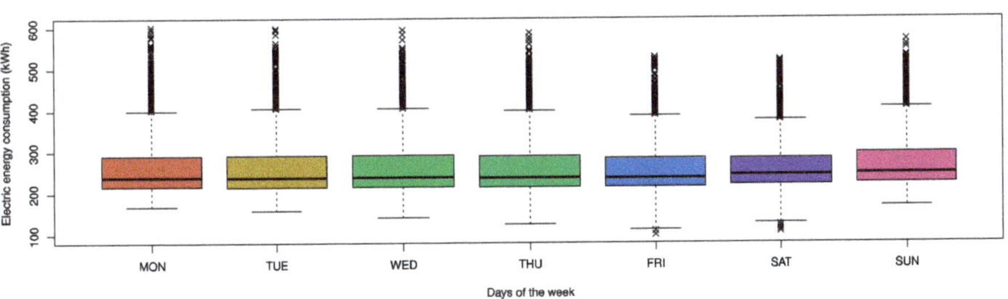

(c) Building 6

Figure 21. *Cont.*

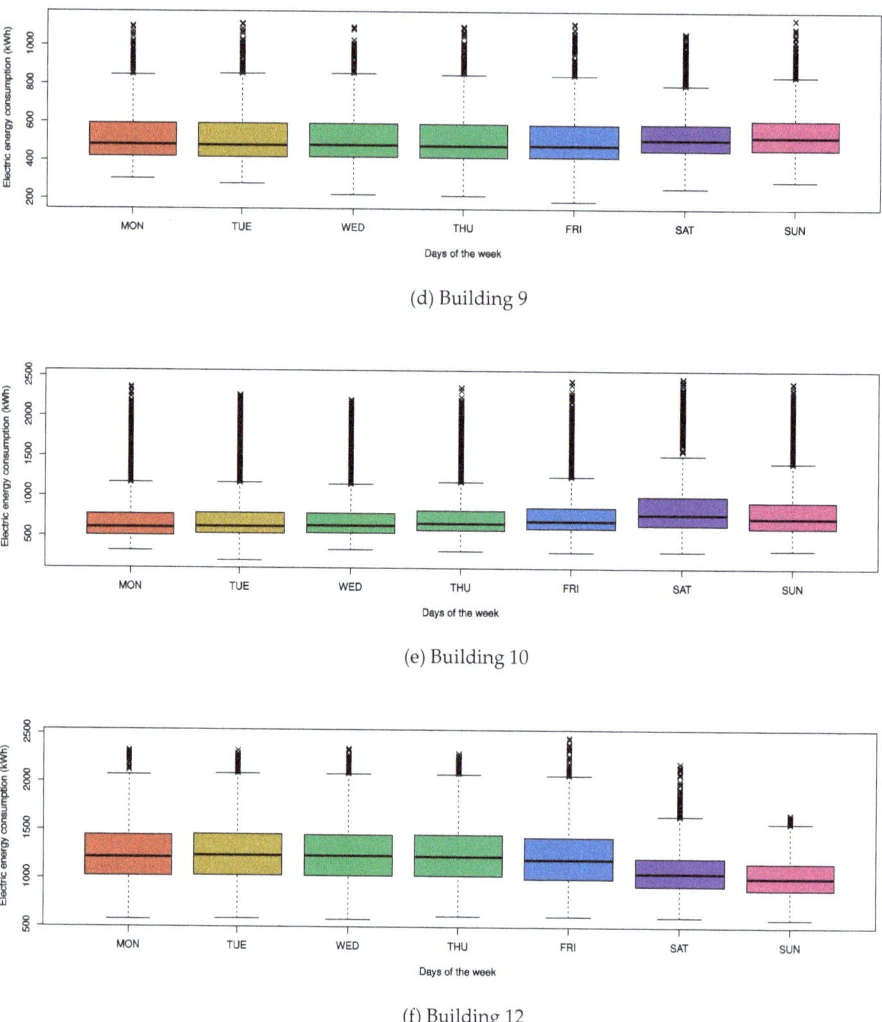

(d) Building 9

(e) Building 10

(f) Building 12

Figure 21. Box plot of electric loads according to each day of the week.

6. Conclusions

When sufficient building electric energy consumption data are not available as in newly constructed or renovated buildings, it is difficult to train and construct superior STLF models. Nevertheless, electric load-forecasting models are required for efficient power management, even by considering such limited data sets. In this paper, we proposed a novel STLF model, called SPROUT, to predict electric energy consumption for buildings with limited data sets by combining time factor- and transfer learning-based forecasting models. We used MRF to construct transfer learning-based STLF models for each building by using sufficient building electric energy data and selected the model, which exhibited the most similar time-series pattern to predict the electric load of the target building. We also constructed STLF models based on RF by using the building electric energy consumption data of only 24 h and then used the two models depending on weekdays and holidays. To verify the validity and applicability of our model, we used MLR and DT to construct time factor-based forecasting models and compared their prediction performance. The experimental results showed that the RF-based

STLF model presented a better prediction performance in the time factor-based forecasting model, and the MRF-based STLF model, by applying ED, exhibited a better prediction performance in the transfer learning-based forecasting model. By combining these models, our model (SPROUT) presented excellent prediction performances in MAPE, RMSE, and MAE results. The SPROUT showed an average MAPE value of 11.2 in the experiments and exhibited more accurate prediction performances of 5.9%p (MLR), 6.7%p (DT), and 4.6%p (RF) than time factor-based STLF models. It also showed more accurate prediction performances of 15.6%p (MRF_PCC), 16.9%p (MRF_CS), and 2.6%p (MRF_ED) than transfer learning-based STLF models. We demonstrated that the SPROUT can achieve better prediction performance than other forecasting models.

However, when electric load exhibited no significant difference in weekday and weekend electric loads in the building, the time factor-based STLF models outperformed our model. In addition, the transfer learning-based STLF models presented unsatisfactory prediction performance in the building with the highest electric energy consumption and hence our model cannot perform accurate electric load forecasting. To address these issues, we plan to consider additional electric load data over a period of time for performing weekly electric load pattern analysis and data normalization.

Author Contributions: All authors have read and agree to the published version of the manuscript. Conceptualization, J.M.; methodology, J.M. and J.K.; software, J.M.; validation, J.M., J.K., and P.K.; formal analysis, P.K. and E.H.; data curation, J.M. and J.K.; writing—original draft preparation, J.M.; writing—review and editing, E.H.; visualization, J.K.; supervision, E.H.; project administration, P.K. and E.H.; funding acquisition, P.K. and E.H.

Funding: This research was supported in part by the Korea Electric Power Corporation (grant number: R18XA05) and in part by Energy Cloud R&D Program (grant number: 2019M3F2A1073179) through the National Research Foundation of Korea (NRF) funded by the Ministry of Science and ICT.

Acknowledgments: We greatly appreciate the anonymous reviewers for their comments and suggestions.

Conflicts of Interest: The authors declare no conflict of interest.

References

1. Cao, X.; Dai, X.; Liu, J. Building energy-consumption status worldwide and the state-of-the-art technologies for zero-energy buildings during the past decade. *Energy Build.* **2016**, *128*, 198–213. [CrossRef]
2. Alani, A.Y.; Osunmakinde, I.O. Short-Term Multiple Forecasting of Electric Energy Loads for Sustainable Demand Planning in Smart Grids for Smart Homes. *Sustainability* **2017**, *9*. [CrossRef]
3. Abbasi, R.A.; Javaid, N.; Khan, S.; ur Rehman, S.; Amanullah; Asif, R.M.; Ahmad, W. Minimizing Daily Cost and Maximizing User Comfort Using a New Metaheuristic Technique. In *Workshops of the International Conference on Advanced Information Networking and Applications*; Springer: Cham, Switzerland, 2019; pp. 80–92. [CrossRef]
4. Nawaz, M.; Javaid, N.; Mangla, F.U.; Munir, M.; Ihsan, F.; Javaid, A.; Asif, M. An Approximate Forecasting of Electricity Load and Price of a Smart Home Using Nearest Neighbor. In *Conference on Complex, Intelligent, and Software Intensive Systems*; Springer: Cham, Switzerland, 2019; pp. 521–533. [CrossRef]
5. Zhou, K.; Fu, C.; Yang, S. Big data driven smart energy management: From big data to big insights. *Renew. Sustain. Energy Rev.* **2016**, *56*, 215–225. [CrossRef]
6. Bose, B.K. Artificial Intelligence Techniques in Smart Grid and Renewable Energy Systems—Some Example Applications. *Proc. IEEE* **2017**, *105*, 2262–2273. [CrossRef]
7. Kim, J.; Moon, J.; Hwang, E.; Kang, P. Recurrent inception convolution neural network for multi short-term load forecasting. *Energy Build.* **2019**, *194*, 328–341. [CrossRef]
8. Ahmad, A.S.; Hassan, M.Y.; Abdullah, M.P.; Rahman, H.A.; Hussin, F.; Abdullah, H.; Saidur, R. A review on applications of ANN and SVM for building electrical energy consumption forecasting. *Renew. Sustain. Energy Rev.* **2014**, *33*, 102–109. [CrossRef]
9. Monfet, D.; Corsi, M.; Choinière, D.; Arkhipova, E. Development of an energy prediction tool for commercial buildings using case-based reasoning. *Energy Build.* **2014**, *81*, 152–160. [CrossRef]
10. Parhizi, S.; Khodaei, A.; Shahidehpour, M. Market-Based Versus Price-Based Microgrid Optimal Scheduling. *IEEE Trans. Smart Grid* **2018**, *9*, 615–623. [CrossRef]

11. Moon, J.; Park, S.; Rho, S.; Hwang, E. A comparative analysis of artificial neural network architectures for building energy consumption forecasting. *Int. J. Distrib. Sens. Netw.* **2019**, *15*. [CrossRef]
12. Aimal, S.; Javaid, N.; Rehman, A.; Ayub, N.; Sultana, T.; Tahir, A. Data Analytics for Electricity Load and Price Forecasting in the Smart Grid. In *Workshops of the International Conference on Advanced Information Networking and Applications*; Springer: Cham, Switzerland, 2019; pp. 582–591. [CrossRef]
13. Iglesias, F.; Kastner, W. Analysis of Similarity Measures in Times Series Clustering for the Discovery of Building Energy Patterns. *Energies* **2013**, *6*, 579–597. [CrossRef]
14. Moon, J.; Kim, K.-H.; Kim, Y.; Hwang, E. A Short-Term Electric Load Forecasting Scheme Using 2-Stage Predictive Analytics. In Proceedings of the IEEE International Conference on Big Data and Smart Computing (BigComp), Shanghai, China, 15–17 January 2018; pp. 219–226. [CrossRef]
15. Ghareeb, A.; Wang, W.; Hallinan, K. Data-driven modelling for building energy prediction using regression-based analysis. In Proceedings of the Second International Conference on Data Science, E-Learning and Information Systems (DATA '19), Dubai, United Arab Emirates, 2–5 December 2019; pp. 1–5. [CrossRef]
16. Dayarathna, M.; Wen, Y.; Fan, R. Data Center Energy Consumption Modeling: A Survey. *IEEE Commun. Surv. Tutor.* **2016**, *18*, 732–794. [CrossRef]
17. Cai, M.; Pipattanasompom, M.; Rahman, S. Day-ahead building-level load forecasts using deep learning vs. traditional time-series techinques. *Appl. Energy* **2019**, *236*, 1078–1088. [CrossRef]
18. Cold Start. Available online: https://en.wikipedia.org/wiki/Cold_start (accessed on 8 January 2020).
19. Candanedo, L.M.; Feldheim, V.; Deramaix, D. Data driven prediction models of energy use of appliances in a low-energy house. *Energy Build.* **2017**, *140*, 81–97. [CrossRef]
20. Wang, Z.; Wang, Y.; Zeng, R.; Srinivasan, R.S.; Ahrentzen, S. Random Forest based hourly building energy prediction. *Energy Build.* **2018**, *171*, 11–25. [CrossRef]
21. Li, C.; Ding, Z.; Zhao, D.; Yi, J.; Zhang, G. Building Energy Consumption Prediction: An Extreme Deep Learning Approach. *Energies* **2017**, *10*. [CrossRef]
22. Almalaq, A.; Zhang, J.J. Evolutionary Deep Learning-Based Energy Consumption Prediction for Buildings. *IEEE Access* **2019**, *7*, 1520–1531. [CrossRef]
23. Shi, H.; Xu, M.; Li, R. Deep Learning for Household Load Forecasting—A Novel Pooling Deep RNN. *IEEE Trans. Smart Grid* **2018**, *9*, 5271–5280. [CrossRef]
24. Ribeiro, M.; Grolinger, K.; ElYamany, H.F.; Higashino, W.A.; Capretz, M.A.M. Transfer learning with seasonal and trend adjustment for cross-building energy forecasting. *Energy Build.* **2018**, *165*, 352–363. [CrossRef]
25. Hooshmand, A.; Sharma, R. Energy Predictive Models with Limited Data using Transfer Learning. In Proceedings of the Tenth ACM International Conference on Future Energy Systems (e-Energy '19), Phoenix, AZ, USA, 25–28 June 2019; pp. 12–16. [CrossRef]
26. Tian, C.; Li, C.; Zhang, G.; Lv, Y. Data driven parallel prediction of building energy consumption using generative adversarial nets. *Energy Build.* **2019**, *186*, 230–243. [CrossRef]
27. Dagdougui, H.; Bagheri, F.; Le, H.; Dessaint, L. Neural network model for short-term and very-short-term load forecasting in district buildings. *Energy Build.* **2019**, *203*. [CrossRef]
28. Lusis, P.; Khalilpour, K.R.; Andrew, L.; Liebman, A. Short-term residential load forecasting: Impact of calendar effects and forecast granularity. *Appl. Energy* **2017**, *205*, 654–669. [CrossRef]
29. Son, M.; Moon, J.; Jung, S.; Hwang, E. A Short-Term Load Forecasting Scheme Based on Auto-Encoder and Random Forest. In Proceedings of the 3rd International Conference on Applied Physics, System Science and Computers (APSAC), Dubrovnik, Croatia, 26–28 September 2018; pp. 138–144. [CrossRef]
30. Moon, J.; Park, J.; Hwang, E.; Jun, S. Forecasting power consumption for higher educational institutions based on machine learning. *J. Supercomput.* **2018**, *74*, 3778–3800. [CrossRef]
31. Moon, J.; Jun, S.; Park, J.; Choi, Y.-H.; Hwang, E. An Electric Load Forecasting Scheme for University Campus Buildings Using Artificial Neural Network and Support Vector Regression. *KIPS Trans. Comput. Commun. Syst.* **2016**, *5*, 293–302. [CrossRef]
32. ISO Week Date. Available online: https://en.wikipedia.org/wiki/ISO_week_date (accessed on 8 January 2020).
33. Luy, M.; Ates, V.; Barisci, N.; Polat, H.; Cam, E. Short-Term Fuzzy Load Forecasting Model Using Genetic–Fuzzy and Ant Colony–Fuzzy Knowledge Base Optimization. *Appl. Sci.* **2018**, *8*, 864. [CrossRef]
34. Oprea, S.-V.; Bara, A. Machine Learning Algorithms for Short-Term Load Forecast in Residential Buildings Using Smart Meters, Sensors and Big Data Solutions. *IEEE Access* **2019**, *7*, 177874–177889. [CrossRef]

35. Zhang, W.; Quan, H.; Srinivasan, D. Parallel and reliable probabilistic load forecasting via quantile regression forest and quantile determination. *Energy* **2018**, *160*, 810–819. [CrossRef]

36. Abbasi, R.A.; Javaid, N.; Ghuman, M.N.J.; Khan, Z.A.; Ur Rehman, S.; Amanullah. Short Term Load Forecasting Using XGBoost. In *Workshops of the International Conference on Advanced Information Networking and Applications*; Springer: Cham, Switzerland, 2019; pp. 1120–1131. [CrossRef]

37. Le, T.; Vo, M.T.; Vo, B.; Hwang, E.; Rho, S.; Baik, S.W. Improving Electric Energy Consumption Prediction Using CNN and Bi-LSTM. *Appl. Sci.* **2019**, *9*. [CrossRef]

38. Lahouar, A.; Slama, J.B.H. Day-ahead load forecast using random forest and expert input selection. *Energy Convers. Manag.* **2015**, *103*, 1040–1051. [CrossRef]

39. Yildiz, B.; Bilbao, J.I.; Sproul, A.B. A review and analysis of regression and machine learning models on commercial building electricity load forecasting. *Renew. Sust. Energ. Rev.* **2017**, *73*, 1104–1122. [CrossRef]

40. Tso, G.K.F.; Yau, K.K.W. Predicting electricity energy consumption: A comparison of regression analysis, decision tree and neural networks. *Energy* **2007**, *32*, 1761–1768. [CrossRef]

41. Moon, J.; Kim, Y.; Son, M.; Hwang, E. Hybrid Short-Term Load Forecasting Scheme Using Random Forest and Multilayer Perceptron. *Energies* **2018**, *11*. [CrossRef]

42. Ahmad, M.W.; Mourshed, M.; Rezgui, Y. Trees vs Neurons: Comparison between random forest and ANN for high-resolution prediction of building energy consumption. *Energy Build.* **2017**, *147*, 77–89. [CrossRef]

43. Oshiro, T.M.; Perez, P.S.; Baranauskas, J.A. How many trees in a random forest? In Proceedings of the International Conference on Machine Learning and Data Mining in Pattern Recognition, Berlin, Germany, 13–20 July 2012; pp. 154–168. [CrossRef]

44. Transfer Learning. Available online: https://en.wikipedia.org/wiki/Transfer_learning (accessed on 8 January 2020).

45. Qiu, X.; Ren, Y.; Suganthan, P.N.; Amaratunga, G.A. Empirical mode decomposition based ensemble deep learning for load demand time series forecasting. *Appl. Soft Comput.* **2017**, *54*, 246–255. [CrossRef]

46. Liang, Y.; Niu, D.; Hong, W.-C. Short term load forecasting based on feature extraction and improved general regression neural network model. *Energy* **2019**, *166*, 653–663. [CrossRef]

47. Baek, S.-J.; Yoon, S.-G. Short-Term Load Forecasting for Campus Building with Small-Scale Loads by Types Using Artificial Neural Network. In Proceedings of the 2019 IEEE Power & Energy Society Innovative Smart Grid Technologies Conference (ISGT), Washington, DC, USA, 18–21 February 2019; pp. 1–5. [CrossRef]

48. Segal, M.; Xiao, Y. Multivariate random forests. *WIREs Data Mining Knowl Discov.* **2011**, *1*, 80–87. [CrossRef]

49. Rahman, R.; Otridge, J.; Pal, R. IntegratedMRF: Random forest-based framework for integrating prediction from different data types. *Bioinformatics* **2017**, *33*, 1407–1410. [CrossRef]

50. Gao, X.; Pishdad-Bozorgi, P. A framework of developing machine learning models for facility life-cycle cost analysis. *Build. Res. Informat.* **2019**, 1–25. [CrossRef]

51. Bracale, A.; Caramia, P.; De Falco, P.; Hong, T. Multivariate Quantile Regression for Short-Term Probabilistic Load Forecasting. *IEEE Trans. Power Syst.* **2020**, *35*, 628–638. [CrossRef]

52. Pearson Correlation Coefficient. Available online: https://en.wikipedia.org/wiki/Pearson_correlation_coefficient (accessed on 8 January 2020).

53. Cosine Similarity. Available online: https://en.wikipedia.org/wiki/Cosine_similarity (accessed on 8 January 2020).

54. Dokmanic, I.; Parhizkar, R.; Ranieri, J.; Vetterli, M. Euclidean Distance Matrices: Essential theory, algorithms, and applications. *IEEE Signal Process. Mag.* **2015**, *32*, 12–30. [CrossRef]

55. Euclidean Distance. Available online: https://en.wikipedia.org/wiki/Euclidean_distance (accessed on 8 January 2020).

56. Ripley, B. Tree: Classification and Regression Trees R Package v.1.0-37. 2016. Available online: https://CRAN.R-project.org/package=tree (accessed on 5 February 2020).

57. Liaw, A.; Wiener, M. Classification and regression by randomForest. *R News* **2002**, *2*, 18–22.

58. Zhou, Y.; Zhou, M.; Xia, Q.; Hong, W.-C. Construction of EMD-SVR-QGA Model for Electricity Consumption: Case of University Dormitory. *Mathematics* **2019**, *7*. [CrossRef]

59. Fan, G.F.; Peng, L.L.; Hong, W.-C. Short term load forecasting based on phase space reconstruction algorithm and bi-square kernel regression model. *Appl. Energy* **2018**, *224*, 13–33. [CrossRef]

Article

Load Nowcasting: Predicting Actuals with Limited Data

Florian Ziel

House of Energy Markets and Finance, University of Duisburg-Essen, 45141 Essen, Germany;
florian.ziel@uni-due.de

Received: 20 February 2020; Accepted: 10 March 2020; Published: 20 March 2020

Abstract: We introduce the problem of load nowcasting to the energy forecasting literature. The recent load of the objective area is predicted based on limited available metering data within this area. Thus, slightly different from load forecasting, we are predicting the recent past using limited available metering data from the supply side of the system. Next, to an industry benchmark model, we introduce multiple high-dimensional models for providing more accurate predictions. They evaluate metered interconnector and generation unit data of different types like wind and solar power, storages, and nuclear and fossil power plants. Additionally, we augment the model by seasonal and autoregressive components to improve the nowcasting performance. We consider multiple estimation techniques based on the lassoand ridge and study the impact of the choice of the training/calibration period. The methodology is applied to a European TSO dataset from 2014 to 2019. The overall results show that in comparison to the industry benchmark, an accuracy improvement in terms of MAE and RMSE of about 60% is achieved. The best model is based on the ridge estimator and uses a specific non-standard shrinkage target. Due to the linear model structure, we can easily interpret the model output.

Keywords: load forecasting; electricity consumption; lasso; Tikhonov regularization; load metering; preliminary load

---

1. Introduction and Motivation

In electricity system management, there is a wide range of load forecasting literature [1]. On a high hierarchy level, usually, the transmission system operator (TSO) and sometimes the distribution system operator (DSO) are responsible for the metering and publishing of the load in the corresponding electricity system. When it comes to the details, there exists a wide range of definitions for electrical load; see, e.g., [2,3]. In many countries, there exist accounting rules for the system operator, which define the metering process for billing and management purposes. Thus, from the economic point of view, these load values are very important for the generation and consumption side such as the system operator. However, in many countries, these values are finally published with a large delay with respect to delivery. For instance, PJM published the final metered load values with a delay of up to 90 day. Similarly, in Germany, the TSO published those final metered values in accordance with the accounting rules with a similar delay of up to three months.

In practice, the system operators also publish electrical load real-time data just after delivery with a very small time lag, usually less than an hour. Those load values are often referred to as prcliminary/actual/instantaneous/estimated load, depending on the considered market. Of course, these preliminary load values should be as close as possible to the final metered load values that are computed with respect to the accounting rules for the electricity system. Still, there are usually deviations, which might deviate substantially in magnitude. For the computation of the preliminary load, the system operator usually only has limited metering data available to deduce the load values for the overall electricity system.

In this paper, we address the problem of providing more accurate preliminary load values just after delivery when there is only limited metering information in the system available. Those preliminary load values should be as close as possible to the metered load, which is derived with respect to the accounting and metering rules. The academic literature on this topic available is very limited; see [4].

We contribute to this topic and propose an efficient and robust method for nowcasting load using machine learning and data science techniques. In the data science and forecasting literature, especially in applications to economics and meteorology, the phrase nowcasting is used for predicting extremely short-term forecast or predicting the very recent past [5–9]. As mentioned, in our electricity load situation, we are exactly in the case of predicting the very recent past load values under limited data availability. Hence, we propose the phrase load nowcasting for these situations.

In this manuscript, we first introduce the nowcasting problem in detail. Then, we propose several nowcasting models that are oriented to the load forecasting literature. Afterwards, we proceed with a nowcasting study to validate the models and discuss the corresponding results, including the interpretation of the best performing model. We close with a summary and some conclusions.

2. The Nowcasting Problem

2.1. Formal Problem Description

Based on the accounting rules, the system operator has to compute the final load values of the objective region for which he/she is responsible. We denote Y_t the corresponding load values at time point t. The detailed computation depends on the regulatory details and the mentioned accounting rules of the considered electricity market. Still, independent of the market, all accounting rules that determine the load Y_t have in common that they specify the system balance, so the match of the supply and demand, the interconnection with neighboring areas, and potential grid losses.

Under the assumption of no grid losses, we could state for each time point t that:

$$Y_t = \sum_i \text{Consumption\_of\_unit}_{i,t} + \sum_i \text{Interconnector\_balance}_{i,t}$$

where $\text{Consumption\_of\_unit}_{i,t}$ is the electricity consumption of unit i and $\text{Interconnector\_balance}_{i,t}$ the imbalance of interconnector i. Obviously, both sums are taken across all consumption units and interconnectors. Of course, from the generation point of view, we can also state:

$$Y_t = \sum_i \text{Generation\_of\_unit}_{i,t} + \sum_i \text{Interconnector\_balance}_{i,t}$$

where $\text{Generation\_of\_unit}_{i,t}$ is the generated electricity of generation unit i and $\text{Interconnector\_balance}_{i,t}$. In practice, the latter is easier to compute as we have usually less production units (mainly large power plants) than consumers. Therefore, the latter is usually applied for deriving the load. Moreover, the generation units are often divided into subgroups, dependent on generation type, which could be nuclear, lignite, coal, natural gas, oil, pump storage, hydro, biomass, wind, and solar, among others. In the formulas, the generation based equation above turns into:

$$Y_t = \sum_i X_{G,i,t} + \sum_i X_{I,i,t} \tag{1}$$

where $X_{G,i,t}$ is the generation of generation unit i and $X_{I,i,t}$ the interconnector balance i. Again, the sums are taken across all units and interconnectors in the balancing area.

As introduced above, the key problem is that there is only limited metering information at the time of prediction. Therefore, some generation units or interconnectors are not metered (yet) and reduce the number of available time series. Thus, we used:

$$Y_t = L_t + \varepsilon_t \tag{2}$$

$$= \sum_{i=1}^{\mathcal{J}_G} X_{G,i,t} + \sum_{i=1}^{\mathcal{J}_I} X_{I,i,t} + \varepsilon_t \tag{3}$$

with L_t as the overall metered load across all $\mathcal{J}_G$ metered generation units and all $\mathcal{J}_I$ interconnectors balance time series datasets. The error term ε_t absorbs the missing information of Y_t, which is not covered by L_t, including potential grid losses and contaminated data. In practice, this leads usually to the fact that the sum of all available metered generation units plus the interconnector imbalance L_t is well below the targeted load Y_t. In the application example, we show below that this is about 80% of the overall load. Remember that in a perfect metering environment where $\mathcal{J}_G$ and $\mathcal{J}_I$ cover all units, it holds $L_t = Y_t$ or, equivalently, $\varepsilon_t = 0$; see (1).

Now, the prediction task is to nowcast (or forecast) Y_t by $\widehat{Y}_t$ given the available information up to time t, i.e., $X_{G,i,t}$ and $X_{I,i,t}$. A specific restriction is that recent values Y_{t-k} (e.g., Y_t, Y_{t-1}, Y_{t-2}, or Y_{t-3}) are not available for predicting Y_t. As mentioned in the Introduction, the last known values usually have a huge delay, often up to 90 days. Thus, we assume that Y_{t-K} is the last known value where K is a relatively large number. In the situation of hourly data with 90 days of publication delay, this would be $K = 24 \times 90 = 2160$.

2.2. Data and Problem Illustration

We considered a dataset ranging from 31 December 2014 to 30 April 2019 for the region of a European system operator. The data were metered in quarter-hourly resolution, and if not stated otherwise, all load values are given in MW. There were $\mathcal{J}_G = 92$ generation time series and $\mathcal{J}_I = 5$ five interconnection balance time series available. The generation time series contained seven wind power series and five solar series and a diverse collection of power plant productions of different types: nuclear, lignite, coal, natural gas (NG), oil, pump storage, hydro, and biomass. Potential missing data were replaced by the last known values. Moreover, we applied clock change adjustments to the data due to daylight savings time. Hence, for the last Sunday in March, we interpolated the missing clock change hour, and for the last Sunday in October, we averaged the doubling clock change hour.

In Figure 1, we illustrate an example of the considered dataset for the last week of April 2019. We observed that the load process Y_t exhibited the typical daily pattern with smaller values during night than during day time, and smaller values on the weekend than on working days. Additionally, we see that the process L_t (see Equation (3)) is the sum of all available meters series $X_{G,i,t}$ and $X_{I,i,t}$. Note that metering data exhibited negative values, and this held particularly for the transmission data of the interconnectors and the storages. Thus, only if all metered data were positive, the process L_t was visually that of all individual generation and interconnector data. Such a particular example period can be spotted for the last hours of Sunday in Figure 1.

Further, we observed that during the illustrated period, the generation had a substantial infeed of wind and solar power. Additionally, we see that nuclear power provided base load energy, but also some coal power plants in the last two days of April 2019. The remaining power plant contributed only little to the energy supply during this period. Finally, we want to highlight that L_t followed the same pattern as Y_t, but lied consistently below Y_t. This also motivated the first simple model for predicting Y_t.

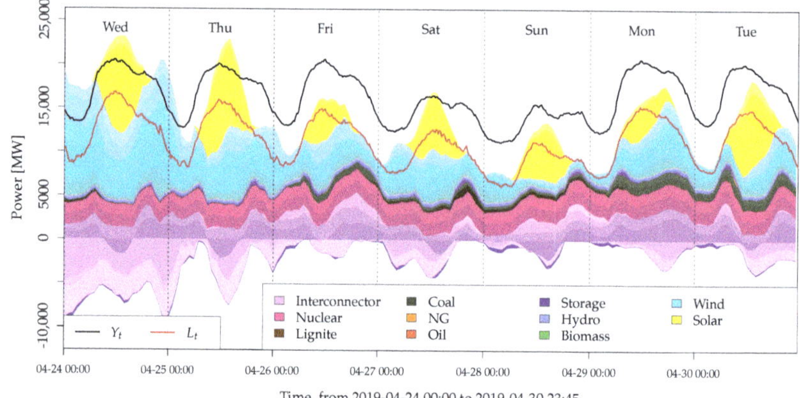

Figure 1. Time series plot of the load Y_t and the process L_t with its single components $X_{G,i,t}$ and $X_{I,i,t}$ classified by generation type in the last week of April 2019.

3. Nowcasting Models

3.1. Benchmark Model

The industry benchmark from the system operator solves the problem stated above by a linear regression on L_t motivated by Equation (2). Thus,

$$Y_t = \alpha_0 + \alpha_1 L_t + \varepsilon_t \tag{4}$$

To estimate the unknown coefficients α_0 and α_1, the industry benchmark applies ordinary least squares (OLS) to the past years data of the same month of the target time t. Thus, if we want to predict Y_t, which is in January, we take all January values for Y_t and L_t of the previous year to estimate α_0 and α_1. As we had quarter-hourly data, this was 31×96 data points. By OLS, we used $\widehat{\alpha}_0$ and $\widehat{\alpha}_1$ and computed nowcasts Y_t by $\widehat{Y}_t = \widehat{\alpha}_0 + \widehat{\alpha}_1 L_t$.

The estimation principle is visualized in Figure 2. Here, α_0 and α_1 of Model (4) were estimated using the input data from April 2018 for estimating Y_t in April 2019. Note that we will generalize this estimation method slightly and consider a broader range of training periods options in the application.

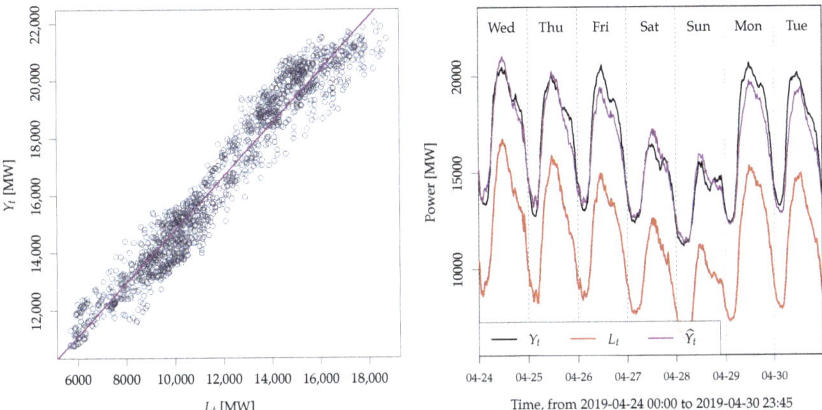

Figure 2. (Left) Scatter plot of the process L_t (see (3)) and load Y_t in April 2018 with the fitted line of Model (4). (Right) Time series plot of Y_t, L_t, and $\widehat{Y}_t = \widehat{\alpha}_0 + \widehat{\alpha}_1 L_t$ for the last week of April 2019 as in Figure 1.

3.2. Proposed Nowcasting Model

The proposed model was motivated by Equation (3). First, we imposed a linear model on the individual generation and interconnector components by:

$$Y_t = \beta_0 + \sum_{i=1}^{\mathcal{J}_G} \beta_{G,i} X_{G,i,t} + \sum_{i=1}^{\mathcal{J}_I} \beta_{I,i} X_{I,i,t} + \varepsilon_t \tag{5}$$

$$= \beta_0 + \beta_G' X_{G,t} + \beta_I' X_{I,t} + \varepsilon_t \tag{6}$$

$$= \beta_0 + \beta_F' X_{F,t} + \varepsilon_t. \tag{7}$$

with $X_{F,t} = (X_{G,t}, X_{I,t})$. We regarded this as a fundamental linear load model, as the only linear inputs were X_F, which contained all fundamental information: the generated power data $X_{G,t}$ and the interconnector imbalance $X_{I,t}$. Note that (7) can be regarded as natural extension of (4) because Model (7) turns into (4) by choosing $\beta_i = \alpha_1$ for $i > 0$.

However, we extended Model (7) by two further terms: (i) a term that contains seasonal information and (ii) a term that represents autoregressive information. In load forecasting, both terms showed high relevance; see, e.g., [10–13]. Sometimes, models with many seasonal and autoregressive components performed even very well in short-term forecasting; see, e.g., [14].

Formally, the extended model is given by:

$$Y_t = \beta_0 + \beta_F' X_{F,t} + \beta_S' X_{S,t} + \beta_A' X_{A,t} + \varepsilon_t. \tag{8}$$

$$= \beta_0 + \beta' X_t + \varepsilon_t \tag{9}$$

where $X_{S,t}$ is a vector of seasonal regressors and $X_{A,t}$ is a vector of autoregressive components of Y_t. Of course, (8) turns into (7) by choosing $\beta_S = 0$ and $\beta_A = 0$. Note that we also defined $X_t = (X_{F,t}, X_{S,t}, X_{X,t})$, which did not include the intercept. Hence, $\beta = (\beta_F, \beta_S, \beta_A)'$ did not include β_0.

It is widely known that in electricity demand, load and consumption modeling periodic features play an important role. The most important seasonalities are daily, weekly, and annual cycles. We suggested to model the three periodic components by periodic cubic by splines with periodicities S_D, S_W, and S_A, which represent a day, a week, and a (meteorologic) year, as in [15]. For out quarter-hourly data, we had $S_D = 96$, $S_W = 96 \times 7 = 672$, and $S_A = 96 \times 365.24 = 35{,}063.04$. In contrast to Fourier analysis, periodic B-splines have the advantage that the basis functions are local and allow for flexibility. When applied to positive data with positivity constraints, they also benefit from the fact that they are always positive. We chose equidistant basis functions for each period. Additionally, we specified the number basis functions B_D, B_W, and B_A for each period. For our application, we chose $B_D = 24$, $B_W = 12$, and $B_A = 24$. Thus, β_S had a length of $B_D + B_W + B_A$, which was 60 in our application.

Furthermore, we had to specify the autoregressive term in (8). We defined the autoregressive components:

$$X_{A,t} = ((Y_{t-k})_{k \in \mathcal{K}_K}, (Y_{t-k})_{k \in \mathcal{K}_A})$$

with two sets of lags $\mathcal{K}_K$ and $\mathcal{K}_A$. $\mathcal{K}_K$ contained lags around the most recent available Y_{t-K}, and $\mathcal{K}_A$ contained lags around a calendar year ago. The latter mimicked annual effects.

We specified for the most recent lags:

$$\mathcal{K}_K = \{K + (0, 1, \ldots, 8), K + S_D + (-8, -7, \ldots, 8), K + 2S_D, K + 8S_D\}$$

which contains the nine most recent known values, the values eight hours around the day before Y_{t-K}, and the lags of the past eight days at the same hour as t. Remember that $K = 90S_D$ in our application; thus, Y_{t-z} with $z = K + S_D = 91S_D = 13 \times 7S_D = 13S_W$ had the same weekday as the target Y_t. For lag around a year ago, we specified:

$$\mathcal{K}_A = \{50S_W, 52S_W + (-8, -7, \ldots, -1, 1, 2, \ldots, 8)S_D, 52S_W + (-8, -7, \ldots, 8), 54S_W\}$$

as $52S_W = 364$ is approximately one calendar year. In total, $\mathcal{K}_K$ and $\mathcal{K}_A$ contributed 54 parameters to the model.

To summarize, the overall (8) had many parameters. In our application scenario, in total, there were $5 + 12 + 80 + 60 + 54 = 211$ parameters. As this might lead to overestimation issues when applying plain OLS, we proposed the application of efficient regularization techniques to tackle the nowcasting problem adequately.

3.3. Estimation of Proposed Nowcasting Model

We will see that the estimation procedure (or training method) played an important role in an accurate nowcasting. Obviously, a natural estimation candidate for Model (8) is linear regression. However, as we had many parameters and some of them might contain useless information, this might be suboptimal. Regularization can help to address the problem. In the energy forecasting literature, the lasso (least selection and shrinkage operator) seems to be a popular choice for shrinkage and feature selection methods in linear models; see, e.g., [15–18]. An extension of the lasso is given by the elastic net, which also has been applied [19–25].

For introducing the estimation procedure, we require some further notations. Let $\{1, \ldots, T\}$ be the time points of available data for Y_t. Thus, our objective was to predict the load Y_t at time point $t = T + K$, which corresponds to the actual time point. Let $\mathbb{T} \subseteq \{1, \ldots, T\}$ be the training period of size $n_{\mathbb{T}}$. Define $\mathbb{Y}(m_0, s_0) = ((Y_t - m_0)/s_0)_{t \in \mathbb{T}}$ as the (m_0, s_0)-standardized response vector and $\mathbb{X}(m_p, s_p) = (\mathbb{X}_i(m_i, s_i))_{i \in \{1, \ldots, p\}} = ((X_{i,t} - m_i)/s_i)_{(i,t) \in \{1, \ldots, p\} \times \mathbb{T}}$ as the scaled input matrix with scaling coefficients $m_p = (m_1, \ldots, m_p)$ and $s_p = (s_1, \ldots, s_p)$ and number of input parameters p. Denote $m = (m_0, m_1, \ldots, m_p)$ and $s = (s_0, s_1, \ldots, s_p)$ the collections of all scaling coefficients.

Furthermore, denote c as a vector of the same size as β, which will be the shrinkage target. In the vast majority of applications, this is $c = 0$. The intuition behind this choice is that a specific regressor has zero impact if it contains useless information, to reduce the garbage in, garbage out problem.

With all the notations above, the elastic net estimator for β in Model (8) is given as:

$$\widehat{\beta}_{\lambda,\alpha}(m, s; c) = \underset{(\beta_0, \beta) \in \mathbb{R}^{p+1}}{\arg\min} \frac{1}{n_{\mathbb{T}}} \left\| \mathbb{Y}(m_0, s_0) - \beta_0 + \beta' \mathbb{X}(m_p, s_p) \right\|_2^2 + \lambda\alpha \|\beta - c\|_1 + \lambda \frac{1 - \alpha}{2} \|\beta - c\|_2^2 \quad (10)$$

where $\lambda, \alpha \geq 0$ are tuning parameters, p is the number of parameters (length of β), and $\| \cdot \|_1$ and $\| \cdot \|_2$ as the standard ℓ_1 and ℓ_2 norm. The tuning parameters λ and α characterize the regularization properties of the elastic net. For $\alpha = 1$, we used the popular choice of the lasso (least absolute shrinkage and selection operator), and for $\alpha = 0$, we used the ridge regression, which is also known as Tikhonov regularization. For $\lambda = 0$, we used the OLS solution, and for very large λ, we used a solution very close to the shrinkage target c. In the non-ridge case $\alpha > 0$, we even used exactly c as the solution if λ was sufficiently large. For the case of the ridge regression, we had an explicit solution available. This was:

$$\widehat{\beta}_{\lambda,0}(m, s; c) = \left(\widetilde{\mathbb{X}}(m_p, s_p)'\widetilde{\mathbb{X}}(m_p, s_p) + \mathrm{Diag}\left(\widetilde{\lambda}\right) \right)^{-1} \left(\widetilde{\mathbb{X}}(m_p, s_p)\mathbb{Y}(m_0, s_0) + \lambda\widetilde{c} \right)$$

with $\widetilde{\mathbb{X}}(m_p, s_p) = (1, \mathbb{X}(m_p, s_p))$, $\widetilde{\lambda} = (0, \lambda 1)'$, and $\widetilde{c} = (0, c)'$. In the elastic net or lasso case with $\alpha > 0$, we had efficient estimation techniques based on the coordinate descent or LARS (least angle regression) available. Both options had the drawback that they could only handle the case $c = 0$.

However, also the scaling coefficients m and s impacted the estimation substantially. Usually, the scaling coefficients m and s in (10) are standardized so that $\mathbb{Y}(m_0, s_0)$ remains unchanged by $m = 0$ and $s = 1$, and $\mathbb{X}_i(m_i, s_i)$ has mean zero and standard deviation of one, i.e., it holds that $\mathbb{X}_i(m_i, s_i)'1 = 0$ and $\|\mathbb{X}_i(m_i, s_i)\|_2 = 1$. The latter can be achieved by choosing $m_i = n_{\mathbb{T}}^{-1}\mathbb{X}_i'1$ and $s_i = \sqrt{n_{\mathbb{T}}^{-1}(\mathbb{X}_i - m_i 1)'(\mathbb{X}_i - m_i 1)}$. This scaling procedure is standard in the literature and, e.g., the default in the `glmnet` or `lars` packages in R for estimation of the elastic net and lasso estimation with $c = 0$.

Still, it turned out that for our nowcasting problem, the scaling procedure for $\mathbb{X}$ was suboptimal as we ignored historic observations. It is true that Y_T was the last known observation. However, for X_t, we knew all observations up to $T + K$, the time point when the forecast was created. Thus, we proposed to compute the scaling coefficients s_i and m_i on the larger and more recent information set $\mathbb{T}_K = \mathbb{T} \cup \{T + 1, \ldots, T + K\}$ for all $\mathbb{X}_i$. Moreover, we suggested for reasons explained in the next paragraph to scale the $\mathbb{Y}(m_0, s_0)$ by the corresponding sample mean and standard deviation $m_0 = n_{\mathbb{T}}^{-1}\mathbb{Y}'1$ and $s_0 = \sqrt{n_{\mathbb{T}}^{-1}(\mathbb{Y} - m_0 1)'(\mathbb{Y} - m_0 1)}$.

Now, we discuss the impact of the shrinkage target c in more detail. We mentioned already that the standard choice $c = 0$ was motivated by the fact that by default, a regressor has no influence. Only if a regressor contributes substantially to the explanation of the response Y_t, the estimated coefficient will deviate from zero and show a corresponding impact. If we have no further information about our regressors, this is a reasonable approach. We will apply this approach to the ridge and lasso estimator and denote them by **0-ridge** and **0-lasso**.

However, in our situation, we knew something about the fundamental relationship between our response vector Y_t and the fundamental regressors $X_{F,t}$ from Equations (1) and (7). This fundamental relationship could help to impose a suitable regularization for our model. We explain this with the following example: Suppose there is a situation where in the in-sample period or training period $\{0, \ldots, T\}$, a certain power plant or interconnector is offline; thus, all observations are zero. A reason could be that it is a new unit that just started operating somewhere after the last observation known Y_T. Then, the ridge or lasso estimators with $c = 0$ will give an estimated coefficient of zero for the corresponding unit. Hence, the power plant will have no out-of-sample contribution to the overall load even though it is operating now, at Y_{T+K}. Thus, from the fundamental point of view, it makes sense to deviate from the shrinkage target of 0 for all generation units or interconnectors. If we assume that the metered values are reasonable, eps. not contaminated by implausible data, then taking these values into account should improve the forecasting accuracy. This holds at least for the situation just explained. Hence, we proposed the choice:

$$c_C = (c_F, c_{A,S})$$

with $c_F = \mathbf{1}$ and $c_{A,S} = \mathbf{0}$ corresponding to the impact as in the perfect fundamental situation from Equation (1). Obviously, the vector c_F had a length of β_F, and $c_{A,S}$ had the aggregated length of β_A and β_F. We applied this choice for the ridge regression only and denoted it by **c-ridge**. The reason why this choice was not applied to lasso or elastic net estimators with $\alpha > 0$ was the unavailability of efficient estimation algorithms.

4. Nowcasting Study

We conducted a rolling window nowcasting study using the considered European dataset, and the design was similar to a standard rolling window forecasting study, as illustrated in Figure 3. The initial last known load value Y_T was on 29 January 2018 at 23:45. Based on historic data, we nowcast the $S_D = 96$ values $Y_{T+K+1}, \ldots, Y_{T+K+S_D}$. We considered a publication delay of $K = 90 \times 96 = 8640$ (90 days), which resulted in the first nowcast being on 30 April 2018, approximately three months later. Then, we shifted the last known load value by a day ($S_D = 96$ time points) to Y_{T+S_D} and nowcast $Y_{T+K+S_D+1}, \ldots, Y_{T+K+2S_D}$. This procedure was repeated $N = 366$ times, which gave an out-of-sample time of about a year and around $96 \times 366 = 35,136$ observations for evaluation. For the in-sample dataset, we considered for our application six choices:

(i) All available data from the past 37 months (three years plus one month):
$(365 \times 3 + 30 - 90) \times 96 = 99{,}360$ observations of Y_t, denoted as **3years**

(ii) All available data from the past 25 months (two years plus one month):
$(365 \times 2 + 30 - 90) \times 96 = 64{,}320$ observations of Y_t, denoted as **2years**

(iii) All available data from the past 13 months (one year plus one month):
$(365 + 30 - 90) \times 96 = 29{,}280$ observations of Y_t, denoted as **1year**

(iv) Data of the past year, 120 days centered around the nowcasting day of the past year:
$120 \times 96 = 11{,}520$ observations of Y_t, denoted as **4months**

(v) Data of the past year, 60 days centered around the nowcasting day of the past year:
$60 \times 96 = 5760$ observations of Y_t, denoted as **2months**

(vi) Data of the past year, 30 days centered around the nowcasting day of the past year:
$30 \times 96 = 2880$ observations of Y_t, denoted as **1month**

Option (i) used the maximum amount of data of $(365 \times 3 + 30 - 90) = 1035$ days, which was also used for illustration in Figure 3. Note that Option (vi) was very close to the industry benchmark approach, which used the data of the month of the previous year for estimating the model parameters.

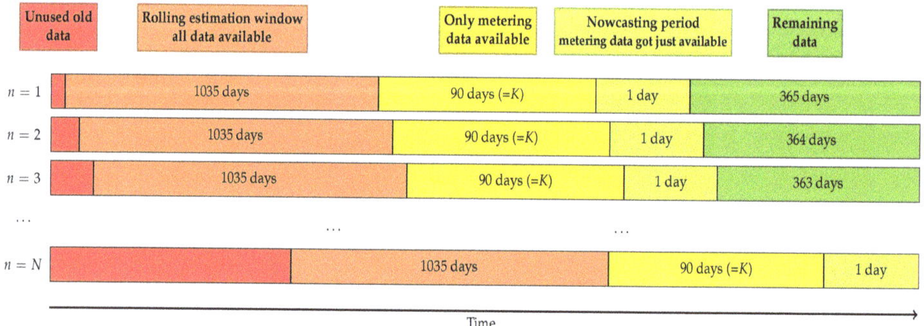

Figure 3. Illustration of the nowcasting study design.

We considered the all competing models, **benchm**, **0-lasso** (λ), **0-ridge** (λ), and **c-ridge**(λ) in the rolling window forecasting study. As emphasized, the lasso and ridge models depended on the tuning parameter λ, which we had to specify. For all models, we considered exponential grids Λ for λ; in detail: For the ridge models, we chose $\Lambda = 2^{\mathcal{L}_r}$ with $\mathcal{L}_r$ as an equidistant grid from -10 to 20 of length

100, and for the lasso models, $\Lambda = 2^{\mathcal{L}_l}$ as an equidistant grid from -30 to 3 of length 100. Of course, we did not know in advance the optimal λ. Therefore, we considered for the **0-lasso**, **0-ridge**, and **c-ridge** models a version where λ was chosen on the past performance (cumulated loss) of the the corresponding models, initializing with $\lambda = 1$ for the first prediction. We denoted the models by **0-lasso***, **0-ridge***, and **c-ridge***.

For measuring the nowcasting accuracy or measures for forecasting performance, we considered the out-of-sample MAE (mean absolute error) and the out-of-sample RMSE (root mean square error). To evaluate the forecasting accuracy also for each of the $S_D = 96$ quarter-hours separately, we defined:

$$\text{MAE} = \frac{1}{S_D} \sum_{s=1}^{S_D} \text{MAE}_s \quad \text{with} \quad \text{MAE}_s = \frac{1}{N} \sum_{n=1}^{N} |Y_{i,s} - \widehat{Y}_{i,s}| \tag{11}$$

$$\text{RMSE} = \frac{1}{S_D} \sum_{s=1}^{S_D} \text{RMSE}_s \quad \text{with} \quad \text{RMSE}_s = \sqrt{\frac{1}{N} \sum_{n=1}^{N} |Y_{i,s} - \widehat{Y}_{i,s}|^2} \tag{12}$$

Note that our models were regression based, and the forecasted value should coincide with the the expected value. Thus, the RMSE should be preferred for evaluation as it identified the true mean correctly. In contrast, the MAE was optimal for median forecasts. However, it is often used as a robust alternative to the RMSE. For more details on the evaluation of point forecasts, we refer to [26].

5. Results

5.1. Nowcasting Performance

We first discuss the overall nowcasting performance of the considered models. The out-of-sample MAE and RMSE values are given in Table 1 and 2. There, we also computed improvements in the MAE and RMSE with respect to the benchmark model **benchm** estimated on the shorted training period **1month**. Remember that **ridge*** and **lasso*** chose the tuning parameter based on the past performance, whereas ridge and lasso represented the models that gave ex-post the best prediction accuracy on the λ-grid Λ.

Table 1. Out-of-sample MAE in MW with relative improvement in % with respect to the benchmark trained on the shortest training period for all models and training periods. A heat map is used to indicate better ($\rightarrow$ green) and worse ($\rightarrow$ red) performing models.

Models →	benchm		c-ridge*		0-ridge*		0-lasso*		c-ridge		0-ridge		0-lasso	
Period ↓	MAE	Imp.	MAE	Imp.	MAE	Imp.	MAE	Imp.	MAE	Imp.	MAE	Imp.	MAE	Imp.
3years	1302.7	−18.3	453.6	58.8	483.6	56.1	509.5	53.7	452.1	58.9	481.4	56.3	507.0	53.9
2years	1328.8	−20.7	430.0	60.9	474.1	56.9	487.8	55.7	428.7	61.1	469.0	57.4	484.7	56.0
1year	1290.5	−17.2	653.9	40.6	588.7	46.5	591.0	46.3	630.5	42.7	581.7	47.2	588.8	46.5
4months	1130.2	−2.7	934.3	15.1	549.5	50.1	583.8	47.0	923.2	16.1	538.3	51.1	578.6	47.4
2months	1097.9	0.3	944.5	14.2	602.4	45.3	626.6	43.1	919.6	16.5	593.8	46.1	617.2	43.9
1month	1100.9	0.0	918.0	16.6	607.1	44.9	635.0	42.3	913.1	17.1	604.1	45.1	629.3	42.8

Table 2. Out-of-sample RMSE in MW with relative improvement in % with respect to the benchmark trained on the shortest training period for all models and training periods. A heat map is used to indicate better ($\rightarrow$ green) and worse ($\rightarrow$ red) performing models.

Models $\rightarrow$	benchm		c-ridge*		0-ridge*		0-lasso*		c-ridge		0-ridge		0-lasso	
Period $\downarrow$	RMSE	Imp.	RMSE	Imp.	RMSE	Imp.	RMSE	Imp.	RMSE	Imp.	RMSE	Imp.	RMSE	Imp.
3years	1556.0	−18.8	578.9	55.8	710.0	45.8	868.5	33.7	582.2	55.5	713.0	45.6	825.0	37.0
2years	1562.4	−19.3	**560.4**	**57.2**	705.1	46.2	759.5	42.0	**556.8**	**57.5**	699.5	46.6	721.9	44.9
1year	1460.6	−11.5	1051.3	19.7	858.9	34.4	940.9	28.2	919.9	29.8	817.2	37.6	923.3	29.5
4months	1332.9	−1.8	1185.3	9.5	776.6	40.7	960.9	26.6	1102.3	15.8	754.6	42.4	880.6	32.8
2months	1299.5	0.8	1274.3	2.7	877.1	33.0	975.9	25.5	1121.3	14.4	828.2	36.8	966.9	26.2
1month	1309.7	0.0	1147.9	12.4	850.3	35.1	917.6	29.9	1150.5	12.2	858.2	34.5	914.5	30.2

First, we observe that all ridge and lasso models showed clear improvements against the benchmark. The largest improvement of around 60% in both measures was gained by the **c-ridge** (or **c-ridge***) model calibrated on the training period of **2years**. Second, we see that the **ridge*** and **lasso*** models showed almost the same performance as **ridge** and **lasso**, which indicated that the ex-post selection of λ was not a big problem. Next, the benchmark model **benchm** with short calibration periods of **1month** and **2months** showed the best prediction accuracy against the benchmark model. In contrast, the ridge and lasso approaches showed that long training periods of **2years** and **3years** performed best. The reason was likely that the estimation of many parameters required more data to receive stable parameter estimates. Figure 4 illustrates the solution path of the ridge and lasso models for a calibration period **2years** which uses about two years of data.

Here, the $\| \cdot \|_1$-norm of $\hat{\beta}$ as a typical measure for model complexity is plotted against the MAE and RMSE score. Note that $\|\hat{\beta}\|_1$ is the sum of all absolute parameters. The solution paths for different λ values of a certain model e.g., **c-ridge** (λ) (red circle), are represented by the color intensity. The darker the color of the symbol within the solution path, the smaller λ. Thus, black symbols correspond to the OLS solution.

We observe that all three models **c-ridge** (λ), **0-ridge** (λ), and **0-lasso** (λ) converged to the the OLS solution for small λ. The OLS solution had an MAE of around 500 MW and an RMSE of slightly above 700 MW with an $\| \cdot \|_1$-norm of β of around 5.5. We clearly see that for small λ values, **0-ridge** (λ) and **0-lasso** (λ) obtained smaller β values and tended towards the **0** solution. In contrast, **c-ridge** (λ) had always a similar range of the $\| \cdot \|_1$-norm of β. The corresponding MAE and RMSE minima has a $\| \cdot \|_1$-norm around 5.2, which is a similar magnitude as the OLS solution. Thus, the parameter complexity of both solutions was comparable, but the parameters were better selected by the **c-ridge** approach due to the shrinkage towards a reasonable target, instead of **0**.

Next, we wanted to look at the intraday structure of the nowcasting errors across the 96 quarter-hours. The forecasting accuracy in term of MAE_s and $RMSE_s$ is visualized in Figure 5. There, we observe that the benchmarks exhibited a relatively clear diurnal pattern. The nowcasting error was largest during the working hours, esp. during the afternoon. For the lasso and ridge models, the daily pattern was substantially reduced. For instance, the MAE_s of **c-ridge*** varied between 383 MW and 484 MW, which was a variation of around 100 MW. The intraday MAE h variation of the MAE of the benchmark model was around 300 MW and significantly larger. However, as the overall forecasting error reduced by 60%, the relative variation of the of the MAE forecasting performance remained at a similar level.

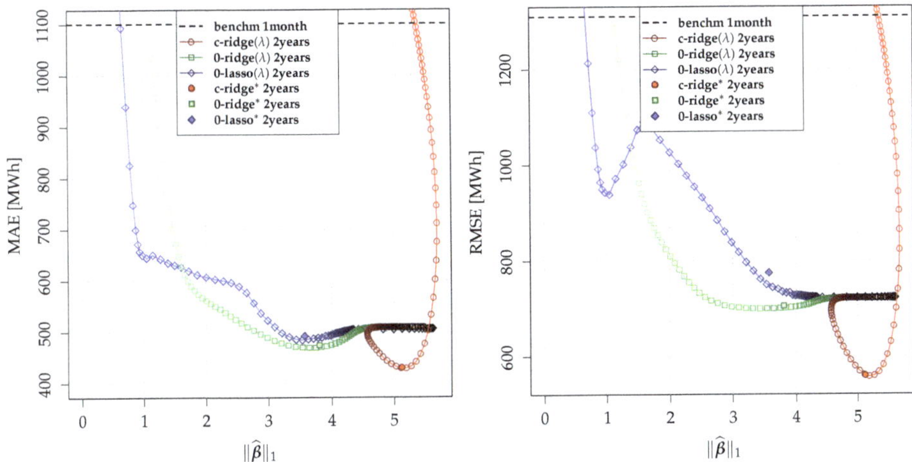

Figure 4. Graph of $\|\widehat{\beta}\|_1$ against MAE (left) and RMSE (right) of the selected lasso and ridge models, illustrating the solution paths for different λ values. The darker the color, the smaller the shrinkage (black = OLS).

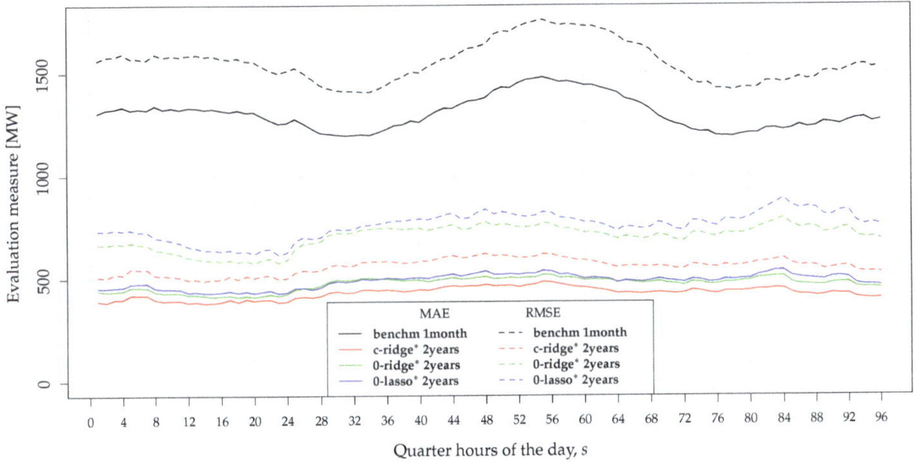

Figure 5. Intraday prediction accuracy in MAE_s and RMSE_s of selected models.

We saw that the proposed models with an in-sample sample size of about two years performed best. It was clear that the computational complexity increased with the amount of data used for training and calibration. Still, in all cases, the models allowed the implementation and application on a real-time basis due to the linear model structure. For instance, the estimation of the **c-ridge**, **0-ridge**, and **0-lasso** models on the full λ-grid with a training period of **2years** took 3.0 s, 0.5 s, and, 2.3 s, respectively. These times were measured on a standard computer using a simple CPU. The ridge models were estimated using the `solve.QP` function of the R package `quadprog`, and the lasso model was trained and calibrated using `glmnet` function of the R package `glmnet`.

5.2. Model Interpretation

As our models were linear models, it was relatively easy to interpret the parameters. The easiest way to get an understanding of the impact of each parameter in the model was to evaluate the absolute impact of parameter i with respect to the overall parameter contribution $|\widehat{\beta}_i| / \|\widehat{\beta}\|_1$. Those impacts of the **c-ridge*** model with a training period of about two years such as the benchmark model **benchm** with training period of about a month are illustrated in the bar chart in Figure 6. As the full model had many parameters, we grouped the impacts $|\widehat{\beta}_i| / \|\widehat{\beta}\|_1$ by parameter type to maintain readable results.

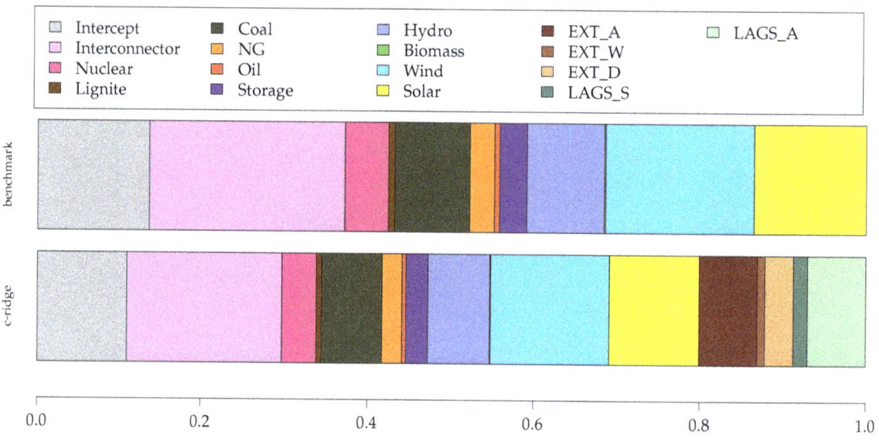

Figure 6. Bar chart of the absolute impact $|\widehat{\beta}_i| / \|\widehat{\beta}\|_1$ of Model **c-ridge*** for **2years** and **benchm** for **1month** grouped by parameter type.

Obviously, we saw that the only the c-ridgemodel had a contribution from external regressors and autoregressive impacts (EXT_A, EXT_W, and EXT_D represent the annual, weekly, and daily seasonal components; LAGS_A and LAGS_S represent the annual and short-term autoregressive lags), as the benchmark model did not take those effects into account. Here, it seemed that the annual impacts contributed substantially to the **c-ridge*** model, and this held for both types' effects from deterministic external regressors (EXT_A) and autoregressive effects (LAGS_A). Furthermore, the daily seasonal component (EXT_D) showed about a 3.5% contribution to the overall solution. For the generation units, we observed that all reduced their absolute impact in the **c-ridge*** model with respect to the benchmark model. However, all parameters remained relevant.

The interpretation by the absolute impacts $|\widehat{\beta}_i| / \|\widehat{\beta}\|_1$ was suitable for evaluation of the impact within the estimated model. However, the regressors $X_{i,t}$ lived on completely different scales. To obtain interpretable impacts with respect to the load Y_t, we had to evaluate the time series of $\widehat{\beta}_i X_{i,t}$, which represented the impact of each single component to the final model. Therefore, Figure 7 shows a time series plot of the actual load Y_t, the benchmark model **benchm** nowcasts, and the **c-ridge*** model nowcasts, along with the estimated contributions $\widehat{\beta}_i X_{i,t}$ for each regressor i.

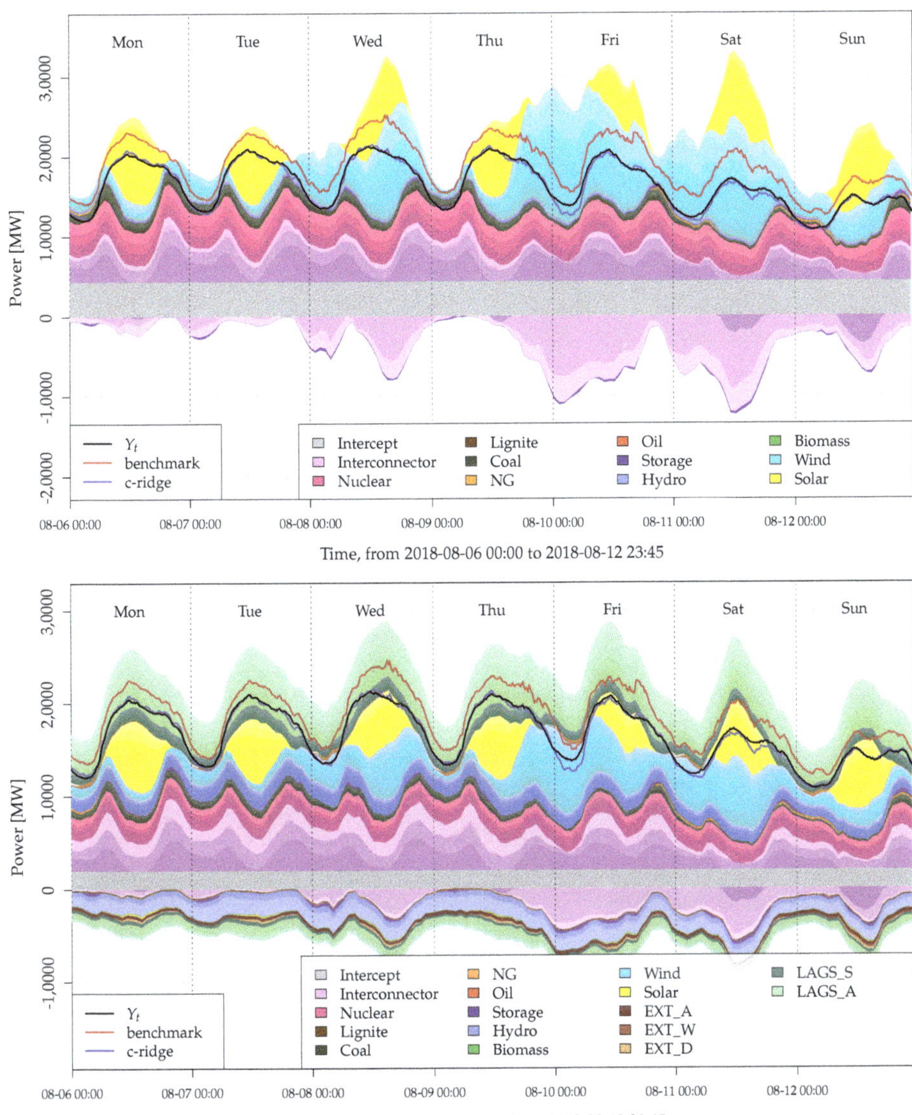

Figure 7. Time series plot of the actual load Y_t (black), with the fitted model of the benchmark model (red) and the **c-ridge*** approach (blue) on 6–12 August 2018. Additionally, the estimated impact of the single components $\widehat{\beta}_i X_{i,t}$ for the **c-ridge*** model (bottom) and benchmark model (top) classified by type with different colors is illustrated.

We observed that for both models, the interconnector, wind, and solar contributed substantially to the final solution. For the **c-ridge*** nowcast, a very important contribution to $\widehat{Y}_t$ came from the annual autoregressive impacts (LAG_A). It mainly had positive contributions, but also some negative contributions. For the **c-ridge*** nowcast, some moderate impact could be seen from the nuclear power and hydro. The latter contributed more to the negative side than to the positive, which was a bit surprising, as the fundamental model would suggest a positive impact. Furthermore, the benchmark model had no negative contribution from hydro power. All other generation types had only a minor impact for both considered models. Finally, we observed that the intercept contributed around 2000 MW to the final contribution of the **c-ridge*** model, which was about 10% of the overall load Y_t. Remember that about 80% of the load Y_t was metered (by generation units and interconnectors). Thus, from the missing 20% load, around a half (=10%) seemed to be base load.

6. Summary and Conclusions

We formally introduced the problem of load nowcasting to the energy forecasting literature. In contrast to load forecasting, the recent load of a certain balancing area was predicted based on limited available metering data within this area. Thus, we were predicting the recent past. We introduced an industry benchmark model and multiple high-dimensional linear model to tackle the nowcasting problem. The model design orientated from load forecasting problems. Next to the impacts of metered generation and interconnector units, the models had seasonal and autoregressive components to improve the prediction performance. We considered multiple estimation techniques based on lasso and ridge and studied the impact of the choice of the training/calibration period.

The overall results showed that in comparison to the industry benchmark, an accuracy improvement in terms of MAE and RMSE of about 60% was achieved. The best model was based on the ridge estimator and used a specific non-standard shrinkage target. Moreover, we highlighted that the model parameters could be interpreted. The overall results showed that the annual effects (deterministic and autoregressive) contributed significantly to the proposed ridge model.

Future research could investigate more nowcasting models, especially non-linear ones, like artificial neural networks or support vector machines. Obviously, the study could be extended to probabilistic nowcasting. The considered nowcasting models could also serve a basis for the construction of load forecasting models. Here, the generation and interconnector units $X_{i,t}$ had to be considered in a lagged manner ($X_{i,t-k}$), potentially for multiple lags. In general, many methodologies can be transferred from energy forecasting, especially from short-term load forecasting.

Finally, the model accuracy might be enriched by the use of more external information. In load forecasting, the (average) temperature of a objective area is often seen as highly relevant. Thus, the incorporation into a nowcasting model could be beneficial as well. This information can be added easily by adding the temperature (and potential non-linear transformations) as a new regressor to the model. We can also add further dummy variables that characterize known structural breaks, e.g., for changes in the regulation or reshaping of the balancing area. Furthermore, it was clear that additional metering information would improve the nowcasting accuracy. With respect to renewable energy information from wind and solar power, a finer geographical resolution might improve the forecasting accuracy, as Figure 7 shows a high importance for a few individual time series of the **c-ridge*** model with respect to the benchmark model.

Funding: This research received no external funding.

Conflicts of Interest: The author declares no conflict of interest.

References

1. Hong, T.; Fan, S. Probabilistic electric load forecasting: A tutorial review. *Int. J. Forecast.* **2016**, *32*, 914–938. [CrossRef]
2. Schumacher, M.; Hirth, L. How Much Electricity Do We Consume? A Guide to German and European Electricity Consumption and Generation Data (2015). FEEM Working Paper No. 88.2015. Available online: https://ssrn.com/abstract=2715986orhttp://dx.doi.org/10.2139/ssrn.2715986 (accessed on 20 December 2019).
3. Hirth, L.; Mühlenpfordt, J.; Bulkeley, M. The ENTSO-E Transparency Platform—A review of Europe's most ambitious electricity data platform. *Appl. Energy* **2018**, *225*, 1054–1067. [CrossRef]

4. Gerbec, D.; Gubina, F.; Toros, Z. Actual load profiles of consumers without real time metering. In Proceedings of the IEEE Power Engineering Society General Meeting, San Francisco, CA, USA, 12–16 June 2005; IEEE: Piscataway, NJ, USA, 2005; pp. 2578–2582.

5. Banbura, M.; Giannone, D.; Reichlin, L. Nowcasting. Available online: https://papers.ssrn.com/sol3/papers.cfm?abstract_id=1717887 (accessed on 20 December 2019).

6. Sun, J.; Xue, M.; Wilson, J.W.; Zawadzki, I.; Ballard, S.P.; Onvlee-Hooimeyer, J.; Joe, P.; Barker, D.M.; Li, P.W.; Golding, B.; et al. Use of NWP for nowcasting convective precipitation: Recent progress and challenges. *Bull. Am. Meteorol. Soc.* **2014**, *95*, 409–426. [CrossRef]

7. Xingjian, S.; Chen, Z.; Wang, H.; Yeung, D.Y.; Wong, W.K.; Woo, W.c. Convolutional LSTM network: A machine learning approach for precipitation nowcasting. In *Advances in Neural Information Processing Systems*; Curran Associates, Inc.: Dutchess, NY, USA, 2015; pp. 802–810.

8. Sanfilippo, A. Solar Nowcasting. In *Solar Resources Mapping*; Springer: Cham, Switzerland, 2019; pp. 353–367.

9. Sala, S.; Amendola, A.; Leva, S.; Mussetta, M.; Niccolai, A.; Ogliari, E. Comparison of Data-Driven Techniques for Nowcasting Applied to an Industrial-Scale Photovoltaic Plant. *Energies* **2019**, *12*, 4520. [CrossRef]

10. Gaillard, P.; Goude, Y.; Nedellec, R. Additive models and robust aggregation for GEFCom2014 probabilistic electric load and electricity price forecasting. *Int. J. Forecast.* **2016**, *32*, 1038–1050. [CrossRef]

11. Ziel, F. Modeling public holidays in load forecasting: A German case study. *J. Mod. Power Syst. Clean Energy* **2018**, *6*, 191–207. [CrossRef]

12. Ziel, F. Quantile regression for the qualifying match of GEFCom2017 probabilistic load forecasting. *Int. J. Forecast.* **2019**, *35*, 1400–1408. [CrossRef]

13. Kanda, I.; Veguillas, J.Q. Data preprocessing and quantile regression for probabilistic load forecasting in the GEFCom2017 final match. *Int. J. Forecast.* **2019**, *35*, 1460–1468. [CrossRef]

14. Haben, S.; Giasemidis, G.; Ziel, F.; Arora, S. Short term load forecasting and the effect of temperature at the low voltage level. *Int. J. Forecast.* **2019**, *35*, 1469–1484. [CrossRef]

15. Ziel, F.; Liu, B. Lasso estimation for GEFCom2014 probabilistic electric load forecasting. *Int. J. Forecast.* **2016**, *32*, 1029–1037. [CrossRef]

16. Dudek, G. Pattern-based local linear regression models for short-term load forecasting. *Electr. Power Syst. Res.* **2016**, *130*, 139–147. [CrossRef]

17. Takeda, H.; Tamura, Y.; Sato, S. Using the ensemble Kalman filter for electricity load forecasting and analysis. *Energy* **2016**, *104*, 184–198. [CrossRef]

18. Wang, Y.; Gan, D.; Zhang, N.; Xie, L.; Kang, C. Feature selection for probabilistic load forecasting via sparse penalized quantile regression. *J. Modern Power Syst. Clean Energy* **2019**, *7*, 1200–1209. [CrossRef]

19. Uniejewski, B.; Nowotarski, J.; Weron, R. Automated variable selection and shrinkage for day-ahead electricity price forecasting. *Energies* **2016**, *9*, 621. [CrossRef]

20. Ambach, D.; Croonenbroeck, C. Space-time short-to medium-term wind speed forecasting. *Stat. Methods Appl.* **2016**, *25*, 5–20. [CrossRef]

21. Liu, W.; Dou, Z.; Wang, W.; Liu, Y.; Zou, H.; Zhang, B.; Hou, S. Short-term load forecasting based on elastic net improved GMDH and difference degree weighting optimization. *Appl. Sci.* **2018**, *8*, 1603. [CrossRef]

22. Kath, C.; Ziel, F. The value of forecasts: Quantifying the economic gains of accurate quarter-hourly electricity price forecasts. *Energy Econ.* **2018**, *76*, 411–423. [CrossRef]

23. Narajewski, M.; Ziel, F. Econometric modelling and forecasting of intraday electricity prices. *J. Commod. Mark.* **2019**, 100107. [CrossRef]

24. Pirbazari, A.M.; Chakravorty, A.; Rong, C. Evaluating feature selection methods for short-term load forecasting. In Proceedings of the 2019 IEEE International Conference on Big Data and Smart Computing (BigComp), Kyoto, Japan, 27 February–2 March 2019; IEEE: Piscataway, NJ, USA, 2019; pp. 1–8.

25. Muniain, P.; Ziel, F. Probabilistic forecasting in day-ahead electricity markets: Simulating peak and off-peak prices. *Int. J. Forecast.* **2020**. [CrossRef]

26. Gneiting, T. Making and evaluating point forecasts. *J. Am. Stat. Assoc.* **2011**, *106*, 746–762. [CrossRef]

MDPI
St. Alban-Anlage 66
4052 Basel
Switzerland
Tel. +41 61 683 77 34
Fax +41 61 302 89 18
www.mdpi.com

Energies Editorial Office
E-mail: energies@mdpi.com
www.mdpi.com/journal/energies